Collins

CAMBRIDGE
IGCSE®
MATHS

Chris Pearce

CONTENTS

Geometry and trigonometry

NTRODUCTION

Velcome to the *Collins Cambridge IGCSE® Maths Student Book – Second Edition* which rovides in-depth coverage of every aspect of the latest Cambridge IGCSE Mathematics 0580 yllabus for examination from 2015. This page introduces you to the key features of the book.

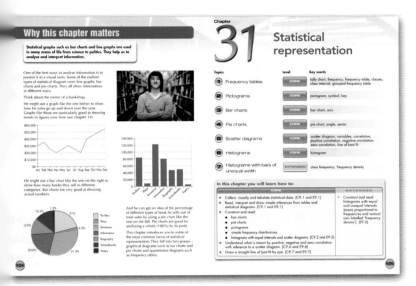

Why this chapter matters

This page is at the start of each chapter. It tells you why the mathematics in the chapter is important and how it is useful.

Chapter overviews

The overview at the start of each chapter shows what you will be studying and the key words you need to know. Syllabus references are included and if you are doing the Extended curriculum you must also cover the corresponding Core content, for example, C1.1. and E1.1.

Worked examples

Worked examples take you through questions step by step and help you understand the topic before you start the practice questions.

Practice questions and answers

Every chapter has extensive questions to help you practise the skills you need. Many of the questions require you to solve problems, which is an important part of mathematics.

Colour-coded levels

The colour coded panels at the side of the question pages show whether the questions are at core (blue) or extended level (yellow). The (E) on some topic headings shows that the content in that topic is at extended level only.

Exam preparation

Each of the four main sections in the book ends with exam questions from Cambridge past papers. These will be useful for revision. Mark schemes, written by the author, are available in the Teacher's Pack.

Why this chapter matters

A pattern is an arrangement of repeated parts. You see patterns every day in clothes, art and home furnishings. Patterns can also occur in numbers.

There are many mathematical problems that can be solved using patterns in numbers. Some numbers have fascinating features.

Here is a pattern.

$3 + 5 = 8$ (5 miles ≈ 8 km)

$5 + 8 = 13$ (8 miles ≈ 13 km)

$8 + 13 = 21$ (13 miles ≈ 21 km)

Approximately how many kilometres are there in 21 miles?

Note: ≈ means 'approximately equal to'.

In the boxes are some more patterns. Can you work out the next line of each pattern?

Now look at these numbers and see why they are special.

$4096 = (4 + 0^9)^6$

$81 = (8 + 1)^2$

$5^2 = 5 \times 5 = 25$

$50^2 = 50 \times 50 = 2500$

$500^2 = 500 \times 500 = 250\,000$

$10 \times 10 = 100$

$10 \times 10 \times 10 = 1000$

$10 \times 10 \times 10 \times 10 = 10\,000$

Some patterns have special names.
Can you pair up these patterns and the names?

4, 8, 12, 16, …	Prime numbers
1, 4, 9, 16, …	Multiples (of 4)
2, 3, 5, 7, …	Cube numbers
1, 8, 27, 64, …	Square numbers

You will look at these in more detail in this chapter.

$1 \times 1 = 1$

$11 \times 11 = 121$

$111 \times 111 = 12\,321$

Below are four sets of numbers. Think about which number links together all the other numbers in each set.
(The mathematics that you cover in 1.2 'Factors of whole numbers' will help you to work this out!)

10, 5, 2, 1

18, 9, 6, 3, 2, 1

25, 5, 1

32, 16, 8, 4, 2, 1

$1 \times 1 = 1$

$2 \times 2 = 1 + 3$

$3 \times 3 = 1 + 3 + 5$

$4 \times 4 = 1 + 3 + 5 + 7$

$1089 \times 9 = 9801$

$10\,989 \times 9 = 98\,901$

$109\,989 \times 9 = 989\,901$

Chapter

Number

Topics	Level	Key words
1 Square numbers and roots	CORE	square, square number, square root
2 Multiples of whole numbers	CORE	multiple
3 Factors of whole numbers	CORE	factor, factor pair, lowest common multiple, highest common factor
4 Prime numbers	CORE	prime number
5 Prime factorisation	CORE	product of prime factors, index (indices), prime factorisation
6 More about HCF and LCM	CORE	
7 Real numbers	CORE	natural number, integer, real number, rational number, irrational number

In this chapter you will learn how to:

CORE

- Identify and use:
  - natural numbers
  - integers (positive, negative and zero)
  - prime numbers
  - square numbers
  - common factors and common multiples
  - rational and irrational numbers (e.g. π, √2)

- real numbers. (C1.1 and E1.1)
- Express any number as a product of its prime factors. (C1.1 and E1.1)
- Find the lowest common multiple (LCM) and highest common factor (HCF) of two numbers. (C1.1 and E1.1)

What is the next number in this sequence?

    1, 4, 9, 16, 25, ...

Write each number as:

    1×1, 2×2, 3×3, 4×4, 5×5, ...

These factors can be represented by **square** patterns of dots:

From these patterns, you can see that the next pair of factors must be $6 \times 6 = 36$, therefore 36 is the next number in the sequence.

Because they form square patterns, the numbers 1, 4, 9, 16, 25, 36, ... are called **square numbers**.

When you multiply any number by itself, the answer is called the *square of the number* or the *number squared*. This is because the answer is a square number. For example:

    the square of 5 (or 5 squared) is $5 \times 5 = 25$

    the square of 6 (or 6 squared) is $6 \times 6 = 36$

There is a short way to write the square of any number. For example:

    5 squared (5×5) can be written as 5^2

    13 squared (13×13) can be written as 13^2

So, the sequence of square numbers, 1, 4, 9, 16, 25, 36, ..., can be written as:

    1^2, 2^2, 3^2, 4^2, 5^2, 6^2, ...

The **square root** of n is the number of which the square is n. This can be written as $\sqrt{n}$. For example, the square root of 16 (4) can be written as $\sqrt{16}$.

Square numbers have exact square roots, for example:

    the square root of 9 is 3: $\sqrt{9} = 3$

    the square root of 25 is 5: $\sqrt{25} = 5$

    the square root of 100 is 10: $\sqrt{100} = 10$

CORE

EXERCISE 1A

1 The square number pattern starts:

1   4   9   16   25   …

Copy and continue the pattern above until you have written down the first 20 square numbers. You may use your calculator for this.

2 Work out the answer to each of these number sentences.

$1 + 3 =$

$1 + 3 + 5 =$

$1 + 3 + 5 + 7 =$

Look carefully at the pattern of the three number sentences. Then write down the next three number sentences in the pattern and work them out.

3 Find the next three numbers in each of these number patterns. (They are all based on square numbers.) You may use your calculator.

1	4	9	16	25	36	49	64	81
a 2	5	10	17	26	37	…	…	…
b 2	8	18	32	50	72	…	…	…
c 3	6	11	18	27	38	…	…	…
d 0	3	8	15	24	35	…	…	…

ADVICE AND TIPS

Look for the connection with the square numbers on the top line.

4 a Work out the values of both expressions in each pair. You may use your calculator.

$3^2 + 4^2$ and 5^2

$5^2 + 12^2$ and 13^2

$7^2 + 24^2$ and 25^2

$9^2 + 40^2$ and 41^2

b Describe what you notice about your answers to part **a**.

5 a $13^2 = 169$. What is $\sqrt{169}$?

b Find $\sqrt{25}$

c Find $\sqrt{81}$

d Find $\sqrt{121}$

e Find $\sqrt{400}$

6 4 and 81 are square numbers with a sum of 85.

Find two different square numbers with a sum of 85.

The following exercise will give you some practice on multiples, factors, square numbers and prime numbers.

EXERCISE 1B

1 Write out the first three numbers that are multiples of both of the numbers shown.

   a 3 and 4   **b** 4 and 5   **c** 3 and 5   **d** 6 and 9   **e** 5 and 7

2 Here are four numbers.

    10   16   35   49

Copy and complete the table by putting each of the numbers in the correct box.

	Square number	Factor of 70
Even number		
Multiple of 7		

3 Arrange these four number cards to make a square number.

4 An alarm flashes every 8 seconds and another alarm flashes every 12 seconds. If both alarms flash together, how many seconds will it be before they both flash together again?

5 A bell rings every 6 seconds. Another bell rings every 5 seconds. If they both ring together, how many seconds will it be before they both ring together again?

6 From this box, choose one number that fits each of these descriptions.

   a a multiple of 3 and a multiple of 4

   b a square number and an odd number

   c a factor of 24 and a factor of 18

   d a prime number and a factor of 39

   e an odd factor of 30 and a multiple of 3

   f a number with 5 factors exactly

   g a multiple of 5 and a factor of 20

   h a prime number that is one more than a square number

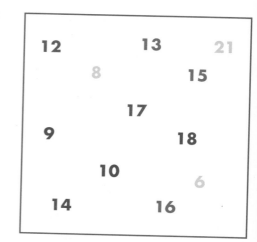

1.2 Multiples of whole numbers

When you multiply any whole number by another whole number, the answer is called a **multiple** of either of those numbers.

For example, 5 × 7 = 35, which means that 35 is a multiple of 5 and it is also a multiple of 7. Here are some other multiples of 5 and 7:

multiples of 5 are: 5   10   15   20   25   30   35   …

multiples of 7 are: 7   14   21   28   35   42   …

EXERCISE 1C

1 Write out the first five multiples of:

a 3   **b** 7   **c** 9   **d** 11   **e** 16

Remember: the first multiple is the number itself.

2 Use your calculator to see which of the numbers below are:

a multiples of 4   **b** multiples of 7   **c** multiples of 6.

72	135	102	161	197
132	78	91	216	514

> **ADVICE AND TIPS**
>
> There is no point testing odd numbers for multiples of even numbers such as 4 and 6.

3 Find the biggest number that is smaller than 100 and that is:

a a multiple of 2   **b** a multiple of 3

c a multiple of 4   **d** a multiple of 5

e a multiple of 7   **f** a multiple of 6.

4 A party of 20 people are getting into taxis. Each taxi holds the same number of passengers. If all the taxis fill up, how many people could be in each taxi? Give two possible answers.

5 Here is a list of numbers.

6   8   12   15   18   28

a From the list, write down a multiple of 9.

b From the list, write down a multiple of 7.

c From the list, write down a multiple of both 3 and 5.

6 How many numbers between 1 and 100 are multiples of both 6 and 9? List the numbers.

A **factor** of a whole number is any whole number that divides into it exactly. So:

the factors of 20 are   1   2   4   5   10   20

the factors of 12 are   1   2   3   4   6   12

Factor facts

Remember these facts.

● 1 is always a factor and so is the number itself.

● When you have found one factor, there is always another factor that goes with it – unless the factor is multiplied by itself to give the number. For example, look at the number 20:

$1 \times 20 = 20$ so 1 and 20 are both factors of 20

$2 \times 10 = 20$ so 2 and 10 are both factors of 20

$4 \times 5 = 20$ so 4 and 5 are both factors of 20.

These are called **factor pairs**.

You may need to use your calculator to find the factors of large numbers.

EXAMPLE 1

Find the factors of 36.

Look for the factor pairs of 36. These are:

$1 \times 36 = 36$ $2 \times 18 = 36$ $3 \times 12 = 36$ $4 \times 9 = 36$ $6 \times 6 = 36$

6 is a repeated factor so is counted only once.

So, the factors of 36 are 1, 2, 3, 4, 6, 9, 12, 18, 36.

EXERCISE 1D

1 What are the factors of each of these numbers?

a 10   **b** 28   **c** 18   **d** 17   **e** 25

f 40   **g** 30   **h** 45   **i** 24   **j** 16

2 What is the biggest factor that is less than 100 for each of these numbers?

a 110   **b** 201   **c** 145   **d** 117

e 130   **f** 240

CORE

3 Find the largest common factor for each pair of numbers.

a 2 and 4 **b** 6 and 10 **c** 9 and 12

d 15 and 25 **e** 9 and 15 **f** 12 and 21

g 14 and 21 **h** 25 and 30 **i** 30 and 50

j 55 and 77

4 Find the highest odd number that is a factor of 40 and a factor of 60.

Lowest common multiple

The **lowest common multiple** (LCM) of two numbers is the *smallest* number that appears in the multiplication tables of both numbers.

For example, the LCM of 3 and 5 is 15, the LCM of 2 and 7 is 14 and the LCM of 6 and 9 is 18.

EXAMPLE 2

Find the LCM of 18 and 24.

Write out the 18 times table: 18, 36, 54, (72), 90, 108, ...

Write out the 24 times table: 24, 48, (72), 96, 120, ...

You can see that 72 is the smallest (least) number in both (common) tables (multiples).

Highest common factor

The **highest common factor** (HCF) of two numbers is the *biggest* number that divides exactly into both of them.

For example, the HCF of 24 and 18 is 6, the HCF of 45 and 36 is 9 and the HCF of 15 and 22 is 1.

EXAMPLE 3

Find the HCF of 28 and 16.

Write out the factors of 28: 1, 2, (4), 7, 14, 28

Write out the factors of 16: 1, 2, (4), 8, 16

You can see that 4 is the biggest (highest) number in both (common) lists (factors).

EXERCISE 1E

1 Find the LCM of each pair of numbers.

a 24 and 56 **b** 21 and 35 **c** 12 and 28 **d** 28 and 42

e 12 and 32 **f** 18 and 27 **g** 15 and 25 **h** 16 and 36

CORE

2 Find the HCF of each pair of numbers.

a 24 and 56   **b** 21 and 35   **c** 12 and 28   **d** 28 and 42

e 12 and 32   **f** 18 and 27   **g** 15 and 25   **h** 16 and 36

i 42 and 27   **j** 48 and 64   **k** 25 and 35   **l** 36 and 54

3 The HCF of two numbers is 6.

The LCM of the same two numbers is 72.

What are the numbers?

1.4   Prime numbers

What are the factors of 2, 3, 5, 7, 11 and 13?

Notice that each of these numbers has only two factors: itself and 1. They are all examples of **prime numbers**.

So, a prime number is a whole number that has only two factors: itself and 1.

Note: 1 is *not* a prime number, since it has only one factor – itself.

The prime numbers up to 50 are:

2, 3, 5, 7, 11, 13, 17, 19, 23, 29, 31, 37, 41, 43, 47

EXERCISE 1F

CORE

1 Write down the prime numbers between 20 and 30.

2 Write down the only prime number between 90 and 100.

3 Decide which of these numbers are **not** prime numbers.

462   108   848   365   711

4 When three different prime numbers are multiplied together the answer is 105.

What are the three prime numbers?

5 A shopkeeper has 31 identical soap bars.

He is trying to arrange the bars on a shelf in rows, each with the same number of bars.

Is it possible?

Explain your answer.

Every whole number that is *not* prime can be written as the **product of prime factors**. For example:

$$35 = 5 \times 7$$
$$40 = 2 \times 2 \times 2 \times 5$$
$$465 = 3 \times 5 \times 31$$
$$5929 = 7 \times 7 \times 11 \times 11$$

5 and 7 are the prime factors of 35.

2 and 5 are the prime factors of 40.

You can use **indices** to write the product more easily. For example:

$$40 = 2^3 \times 5 \qquad \text{The small 3 is an \textbf{index}.}$$
$$5929 = 7^2 \times 11^2$$
$$96 = 2 \times 2 \times 2 \times 2 \times 2 \times 3 = 2^5 \times 3$$

Writing a number in this way is called **prime factorisation**.

You can write the numbers in a different order but you can do the factorisation in only one way. So for 96 there must be five 2s and one 3.

EXAMPLE 4

Write 630 as a product of prime factors.

Check the prime numbers (2, 3, 5, 7, …) in turn to find which of them divide into 630.

2 is a factor of 630. $630 \div 2 = 315$

Now do the same with 315.

2 is not a factor of 315 but 3 is. $315 \div 3 = 105$

3 is a factor of 105. $105 \div 3 = 35$

3 is not a factor of 35 but 5 is. $35 \div 5 = 7$

7 is a prime number, so stop there.

It is more convenient to write the divisions in a column, like this.

```
2 | 6 3 0
3 | 3 1 5
3 | 1 0 5
5 |   3 5
        7
```

Now list the prime numbers, including the final 7.

$$630 = 2 \times 3 \times 3 \times 5 \times 7 = 2 \times 3^2 \times 5 \times 7$$

EXERCISE 1G

CORE

1 Work out these numbers.

a $2^2 \times 3^2$ b $3 \times 5 \times 7$ c 2×5^3 d $2^4 \times 3 \times 17$

e $2 \times 3 \times 7 \times 17$ f 5×7^3 g $3^2 \times 11^2$ h $2^3 \times 13^2$

2 Write each of these numbers as a product of prime factors.

a 90   b 152   c 64   d 330

e 289   f 800   g 97   h 1001

3 Which of these numbers can be written as the product of exactly two prime factors?

77   124   129   245   221   103

4 $450 = 2 \times 3^2 \times 5^2$

a Use this fact to write 900 as a product of prime factors.

b Write 1800 as a product of prime factors.

c Write 1350 as a product of prime factors.

5 Match each number to the correct product of prime factors.

144               $2^2 \times 3^4$

200               $2^4 \times 3^2$

324               $2^3 \times 5^2$

500               $2^2 \times 5^3$

6 a Find the smallest number that has four different prime factors.

b Find the smallest number that has five different prime factors.

7 $1224 = 2^3 \times 3^2 \times 17$

Use this fact to write each of these numbers as a product of prime factors.

a 612   b 306   c 408

8 Which of these numbers cannot be written as a product of prime factors?

70   71   72   73   74   75   76   77   78   79

9 $539 = 7^2 \times 11$ $847 = 7 \times 11^2$ $539 \times 847 = 456\ 533$

Use these facts to write 456 533 as a product of prime factors.

You can use prime factorisation to find the LCM or the HCF of two or more numbers.

EXAMPLE 5

$810 = 2 \times 3^4 \times 5$

$252 = 2^2 \times 3^2 \times 7$

a Find the HCF of 810 and 252.

b Find the LCM of 810 and 252.

a To find the HCF, multiply all the prime numbers (including repeats) that occur in *both* lists.

There is one 2 in both numbers.

There are two 3s in both numbers.

The HCF is $2 \times 3^2 = 18$

b To find the LCM, multiply all the prime numbers (including repeats) that occur in *either* list.

You need to multiply two 2s, four 3s, one 5 and one 7.

The LCM is $2^2 \times 3^4 \times 5 \times 7 = 11\,340$

Why does this work?

You can check that 18 is a common factor: $810 \div 18 = 45 = 3^2 \times 5$

and: $252 \div 18 = 14 = 2 \times 7$

Since the answers to these divisions have no common factor, 18 is the *highest* common factor.

You can check that 11 340 is a common multiple.

$11\,340 \div 810 = 14 = 2 \times 7$

$11\,340 \div 252 = 45 = 3^2 \times 5$

Since the answers to these divisions have no common factor, 11 340 is the *lowest* common multiple.

EXAMPLE 6

Find the LCM of 21, 27 and 35.

$21 = 3 \times 7$ $27 = 3^3$ $35 = 5 \times 7$

Looking at the prime numbers that occur in *any* of these, you need three 3s, one 5 and one 7.

The LCM is $3^3 \times 5 \times 7 = 945$.

CORE

1 $72 = 2^3 \times 3^2$ $162 = 2 \times 3^4$

Use these facts to find:

a the HCF of 72 and 162

b the LCM of 72 and 162.

2 $105 = 3 \times 5 \times 7$ $245 = 5 \times 7^2$

Use these facts to find:

a the HCF of 105 and 245

b the LCM of 105 and 245.

3 **a** Write 240 as the product of prime factors.

b Write 126 as the product of prime factors.

c Find the HCF of 240 and 126.

d Find the LCM of 240 and 126.

4 **a** Write 72 and 108 as the product of prime factors.

b Find the HCF of 72 and 108.

c Find the LCM of 72 and 108.

5 **a** Find the HCF of 64 and 144.

b Find the LCM of 64 and 144.

6 **a** Find the HCF of 132 and 693.

b Find the LCM of 132 and 693.

7 $24 = 2^3 \times 3$ $60 = 2^2 \times 3 \times 5$ $36 = 2^2 \times 3^2$

Use these facts to find:

a the HCF of 24, 60 and 36

b the LCM of 24, 60 and 36.

8 **a** Find the HCF of 25, 35 and 45.

b Find the LCM of 25, 35 and 45.

9 **a** Find the HCF of 105, 135 and 375.

b Find the LCM of 105, 135 and 375.

10 **a** Find the HCF of 288, 432 and 648.

b Write the LCM of 288, 432 and 648 as a product of prime factors.

11 **a** Find the HCF of 63 and 200.

b Find the LCM of 63 and 200.

12 In question 1 you found the HCF and LCM of 72 and 162.

a Work out the product of 72 and 162.

b Work out the product of the HCF and the LCM of 72 and 162.

c Investigate whether there is a relationship between the product of the HCF and the LCM of two numbers and the product of the numbers themselves.

d Is there a similar result for the HCF and the LCM of three numbers?

Real numbers

So far you have only been looking at **natural numbers,** which are whole, positive numbers:

1, 2, 3, 4, 5....

If you also include negative whole numbers you have the **integers**:

....–3, –2, –1, 0, 1, 2, 3....

And if you include decimals too you have the **real numbers**. There are two sorts of real numbers.

Rational numbers are integers or fractions such as -17, $3\frac{5}{8}$, -8.75 or $\frac{29}{11}$.

Irrational numbers cannot be written as fractions.

Examples of irrational numbers are $\sqrt{2}$, $\sqrt{17}$ and π.

If n is a natural number that is *not* a square number, then $\sqrt{n}$ is irrational.

EXERCISE 1I

1 State whether each of these numbers is an integer or not.

    a $\sqrt{100}$ **b** $85 \div 6$ **c** 359^2

    d -7 **e** $\sqrt{20}$ **f** 6.3×10

    g 9.27×7.23 **h** $\sqrt{5} + 11$ **i** $-\sqrt{36}$

2 State whether each of these numbers is rational or irrational.

    a 12.9 **b** $\dfrac{16}{7}$ **c** $\sqrt{8}$

    d $\sqrt{9}$ **e** $\sqrt{10}$ **f** 2.83^2

    g 1.65×2.13 **h** $\pi + 2$ **i** $10 - \pi$

3 Is it possible to find two numbers, A and B, in the following cases? If so, give an example.

    a A and B are not integers, $A + B$ is an integer.

    b A and B are not integers, $A \times B$ is an integer.

    c A and B are integers, $A \times B$ is not an integer.

4 Try to find two irrational numbers that multiply to make an integer. If this is not possible, say so.

5 Try to find two irrational numbers which add up to an integer. If this is not possible, say so.

6 Try to find two irrational numbers which add up to a rational number. If this is not possible, say so.

Why this chapter matters

We use percentages and fractions in many situations in our everyday lives.

Why use fractions and percentages?

Because:

- basic percentages and simple fractions are quite easy to understand
- they are a useful way of comparing quantities
- fractions and percentages are used a lot in everyday life.

Who uses them?

Here are some examples of what you might see:

- Shops and businesses
  o Everything at half price in the sales!
  o Special offer — 10% off!

- Banks
  o Interest rates on loans 6.25%.
  o Interest rates on savings 2.5%.

- Salespeople
  o Earn 7.5% commission on sales.

- Government
  o Half of government workers are over 55.
  o Unemployment has fallen by 1%.

- Workers
  o My pay rise is 2.3%.
  o My income tax is 20%.

- Teachers
  o Your test result is 67%.
  o Three-fifths of our students gain a good mark in IGCSE mathematics.

Can you think of other examples?

You will find many everyday uses of fractions and percentages in this chapter.

2

Fractions and percentages

Topics	Level	Key words
1 Equivalent fractions	CORE	numerator, denominator, cancel, lowest terms, simplest form, proper and improper fraction, mixed number, top heavy
2 Fractions and decimals	CORE	decimal, fraction
3 Recurring decimals	EXTENDED	terminating decimal, recurring decimal
4 Percentages, fractions and decimals	CORE	percentage, decimal equivalent
5 Calculating a percentage	CORE	quantity, multiplier
6 Increasing or decreasing quantities by a percentage	CORE	multiplier
7 One quantity as a percentage of another	CORE	percentage change, percentage increase, percentage decrease, percentage profit, percentage loss
8 Simple interest and compound interest	CORE	simple interest, compound interest, principal, annual rate
9 A formula for compound interest	EXTENDED	formula
10 Reverse percentage	EXTENDED	unitary method, multiplier

In this chapter you will learn how to:

CORE	EXTENDED
• Use the language and notation of simple vulgar and decimal fractions and percentages in appropriate contexts; recognise equivalence and convert between these forms. (C1.5 and E1.5) • Calculate a given percentage of a quantity. (C1.12 and E1.12) • Express one quantity as a percentage of another. (C1.12 and E1.12) • Calculate percentage increase or decrease. (C1.12 and E1.12) • Use given data to solve problems on personal and small business finance involving earnings, simple interest and compound interest, including discount, profit and loss.(C1.16 and E1.16) • Extract data from tables and charts. (C1.16 and E1.16)	• Write recurring decimals as fractions. (E1.5) • Use a formula for compound interest. (E1.16) • Carry out calculations involving reverse percentages, for example, finding the cost price given the selling price and the percentage profit. (E1.12)

Equivalent fractions

Equivalent fractions are two or more fractions that represent the same part of a whole.

EXAMPLE 1

Complete the following.

a $\dfrac{3}{4} \longrightarrow \dfrac{\times 4}{\times 4} = \dfrac{\square}{16}$

b $\dfrac{2}{5} = \dfrac{\square}{15}$

a Multiplying the **numerator** by 4 gives 12. This means $\frac{12}{16}$ is an equivalent fraction to $\frac{3}{4}$.

b To change the **denominator** from 5 to 15, you multiply by 3. Do the same thing to the numerator, which gives $2 \times 3 = 6$. So, $\frac{2}{5} = \frac{6}{15}$.

The fraction $\frac{3}{4}$, in Example 1a, is in its **lowest terms** or **simplest form**.

This means that the only number that is a factor of both the numerator and denominator is 1.

A fraction in which the numerator (top number) is smaller than the denominator (bottom number) is called a proper fraction. An example of a proper fraction is $\frac{4}{5}$.

In an improper fraction the numerator (top number) is bigger than the denominator (bottom number). An example of an improper fraction is $\frac{9}{5}$. It is sometimes called a **top-heavy** fraction.

A mixed number is made up of a whole number and a proper fraction. An example of a mixed number is $1\frac{3}{4}$.

EXAMPLE 2

Convert $\frac{14}{5}$ into a mixed number.

$\dfrac{14}{5}$ means $14 \div 5$.

Dividing 14 by 5 gives 2 with a remainder of 4 (5 divides into 14 two times, with $\frac{4}{5}$ left over).

This means that there are 2 whole ones and $\frac{4}{5}$ left over.

So, $\dfrac{14}{5} = \dfrac{5}{5} + \dfrac{5}{5} + \dfrac{4}{5}$

$= 2\frac{4}{5}$

EXERCISE 2A

CORE

1 Copy and complete the following.

a $\dfrac{2}{5} \longrightarrow \dfrac{\times 4}{\times 4} = \dfrac{\square}{20}$

b $\dfrac{1}{4} \longrightarrow \dfrac{\times 3}{\times 3} = \dfrac{\square}{12}$

c $\dfrac{3}{8} \longrightarrow \dfrac{\times 5}{\times 5} = \dfrac{\square}{40}$

d $\dfrac{2}{3} \longrightarrow \dfrac{\times \square}{\times \square} = \dfrac{\square}{18}$

e $\dfrac{3}{4} \longrightarrow \dfrac{\times \square}{\times \square} = \dfrac{\square}{12}$

f $\dfrac{5}{8} \longrightarrow \dfrac{\times \square}{\times \square} = \dfrac{\square}{40}$

2 Copy and complete the following.

a $\dfrac{10}{15} = \dfrac{10 \div 5}{15 \div 5} = \dfrac{\square}{\square}$ **b** $\dfrac{12}{15} = \dfrac{12 \div 3}{15 \div 3} = \dfrac{\square}{\square}$ **c** $\dfrac{20}{28} = \dfrac{20 \div 4}{28 \div 4} = \dfrac{\square}{\square}$

d $\dfrac{12}{18} = \dfrac{12 \div \square}{\square \div \square} = \dfrac{\square}{\square}$ **e** $\dfrac{15}{25} = \dfrac{15 \div 5}{\square \div \square} = \dfrac{\square}{\square}$ **f** $\dfrac{21}{30} = \dfrac{21 \div \square}{\square \div \square} = \dfrac{\square}{\square}$

3 Cancel each of these fractions to its simplest form.

a $\dfrac{4}{6}$ **b** $\dfrac{5}{15}$ **c** $\dfrac{12}{18}$ **d** $\dfrac{6}{8}$ **e** $\dfrac{3}{9}$

f $\dfrac{5}{10}$ **g** $\dfrac{14}{16}$ **h** $\dfrac{28}{35}$ **i** $\dfrac{10}{20}$ **j** $\dfrac{4}{16}$

4 Put the fractions in each set in order, with the smallest first.

a $\dfrac{1}{2}, \dfrac{5}{6}, \dfrac{2}{3}$ **b** $\dfrac{3}{4}, \dfrac{1}{2}, \dfrac{5}{8}$ **c** $\dfrac{7}{10}, \dfrac{2}{5}, \dfrac{1}{2}$ **d** $\dfrac{2}{3}, \dfrac{3}{4}, \dfrac{7}{12}$ **e** $\dfrac{1}{6}, \dfrac{1}{3}, \dfrac{1}{4}$ **f** $\dfrac{9}{10}, \dfrac{3}{4}, \dfrac{4}{5}$

5 Here are four unit fractions.

$$\dfrac{1}{2} \qquad \dfrac{1}{3} \qquad \dfrac{1}{4} \qquad \dfrac{1}{5}$$

a Which two of these fractions have a sum of $\dfrac{7}{12}$?

Show clearly how you work out your answer.

b Which fraction is the biggest? Explain your answer.

6 Change each of these improper fractions into a mixed number.

a $\dfrac{7}{3}$ **b** $\dfrac{8}{3}$ **c** $\dfrac{9}{4}$ **d** $\dfrac{10}{7}$ **e** $\dfrac{12}{5}$ **f** $\dfrac{7}{5}$

7 Change each of these mixed numbers into an improper fraction.

a $3\dfrac{1}{3}$ **b** $5\dfrac{5}{6}$ **c** $1\dfrac{4}{5}$ **d** $5\dfrac{2}{7}$ **e** $4\dfrac{1}{10}$ **f** $5\dfrac{2}{3}$

g $2\dfrac{1}{2}$ **h** $3\dfrac{1}{4}$ **i** $7\dfrac{1}{6}$ **j** $3\dfrac{5}{8}$ **k** $6\dfrac{1}{3}$ **l** $9\dfrac{8}{9}$

8 Check your answers to questions **1** and **2**, using the fraction buttons on your calculator.

9 Which of these improper fractions has the largest value?

$$\dfrac{27}{4} \qquad\qquad \dfrac{31}{5} \qquad\qquad \dfrac{13}{2}$$

Show your working to justify your answer.

10 Find a mixed number that is greater than $\dfrac{85}{11}$ but smaller than $\dfrac{79}{10}$.

2.2 Fractions and decimals

Changing a decimal into a fraction

To change a **decimal** into a **fraction**, use a place-value table.

For example, $0.32 = \dfrac{32}{100}$

Units	.	Tenths	Hundredths	Thousandths
0	.	3	2	

EXAMPLE 3

Express 0.32 as a fraction.

$$0.32 = \frac{32}{100}$$

This cancels to $\frac{8}{25}$

So, $0.32 = \frac{8}{25}$

Changing a fraction into a decimal

You can change a fraction into a decimal by dividing the **numerator** by the **denominator**.

EXAMPLE 4

Express $\frac{3}{8}$ as a decimal.

$\frac{3}{8}$ means $3 \div 8$. This is a division calculation.

So, $\frac{3}{8} = 3 \div 8 = 0.375$

EXERCISE 2B

1. Change each of these decimals to a fraction, cancelling where possible.

   a 0.7   b 0.4   c 0.5   d 0.03   e 0.06

   f 0.13   g 0.25   h 0.38   i 0.55   j 0.64

2. Change each of these fractions to a decimal. Where necessary, give your answer correct to three decimal places.

   a $\frac{1}{2}$ b $\frac{3}{4}$ c $\frac{3}{5}$ d $\frac{9}{10}$ e $\frac{1}{3}$

   f $\frac{5}{8}$ g $\frac{2}{3}$ h $\frac{7}{20}$ i $\frac{7}{11}$ j $\frac{4}{9}$

3 Put each set of numbers in order, with the smallest first.

a 0.6, 0.3, $\frac{1}{2}$

b $\frac{2}{5}$, 0.8, 0.3

c 0.35, $\frac{1}{4}$, 0.15

d $\frac{7}{10}$, 0.72, 0.71

e 0.8, $\frac{3}{4}$, 0.7

f 0.08, 0.1, $\frac{1}{20}$

g 0.55, $\frac{1}{2}$, 0.4

h $1\frac{1}{4}$, 1.2, 1.23

> **ADVICE AND TIPS**
>
> Convert the fractions to decimals first.

4 Two stores sell the same T shirts at the same price of $24.

Store A has a sale and offers $\frac{1}{3}$ off the price.

Store B has a sale and offers $\frac{1}{4}$ off the price.

Which shop has the better offer in its sale?

Give a reason for your answer.

5 During April it rained on 12 days.

a For what fraction of the month did it rain? Give your answer in its simplest form.

b Give your answer as a decimal.

6 Which is bigger, $\frac{7}{8}$ or 0.87?

Show your working.

7 Which is smaller, $\frac{2}{3}$ or 0.7?

Show your working.

2.3 Recurring decimals

You can change a fraction to a decimal by dividing the numerator by the denominator.

The decimal may work out exactly. For example:

$$\frac{5}{8} = 5 \div 8 = 0.625 \qquad \frac{7}{20} = 7 \div 20 = 0.35$$

This is called a **terminating decimal**.

Alternatively, the decimal may continue. For example:

$$\frac{2}{3} = 2 \div 3 = 0.666\,66\ldots \qquad \frac{5}{11} = 5 \div 11 = 0.454\,545\ldots$$

This is called a **recurring decimal**.

There is always a repeating pattern of digits in a recurring decimal.

Here are some examples.

$\dfrac{5}{9} = 0.5555\ldots$ the 5 repeats.

$\dfrac{7}{12} = 0.583\,333\ldots$ the 3 repeats.

$\dfrac{7}{22} = 0.318\,181\,8\ldots$ the digit pair 18 repeats.

$\dfrac{6}{7} = 0.857\,142\,857\,142\,85\ldots$ the digit sequence 857 142 repeats.

You use dots over the digits to show the repeating pattern.

$\dfrac{5}{9} = 0.\dot{5}$

$\dfrac{7}{12} = 0.58\dot{3}$

$\dfrac{7}{22} = 0.3\dot{1}\dot{8}$ Note the dots over the 1 and the 8.

$\dfrac{6}{7} = 0.\dot{8}5714\dot{2}$ You just put dots over the 8 and the 2 – the first and last digits.

In the previous section, you learned how to change terminating decimals to fractions.

Changing a recurring decimal to a fraction is more difficult.

EXAMPLE 5

Write 0.777… as a fraction.

Write: $f = 0.777\ldots$

Multiply both sides by 10: $10f = 7.777\ldots$ It is easy to multiply the decimal by 10.

Now subtract: $9f = 7$ $10f - f = 9f$

                                            $7.777\ldots - 0.777\ldots = 7$

Divide by 9: $f = \frac{7}{9}$

EXAMPLE 6

Write 0.41$\dot{6}$ as a fraction.

Write: $f = 0.416\,66\ldots$

Multiply by 10: $10f = 4.1666\ldots$

Subtract: $9f = 3.75$ All the digits after 2 decimal places cancel.

Divide by 9: $f = \frac{3.75}{9} = \frac{375}{900} = \frac{5}{12}$ Simplify the fraction.

So: $0.41\dot{6} = \frac{5}{12}$

In the last two examples you multiplied by 10 because there was one recurring digit. What number will you multiply by if there are two recurring digits? Or more than two?

EXAMPLE 7

Write $0.0\dot{7}\dot{4}$ as a fraction.

Write: $\qquad\qquad\qquad\qquad f = 0.074\,074\ldots$

There are three recurring digits.

Multiply by 1000: $\qquad\qquad 1000f = 74.074\,074\ldots$

Subtract: $\qquad\qquad\qquad 999f = 74$

Divide by 999: $\qquad\qquad f = \frac{74}{999} = \frac{2}{27}$ $\qquad$ 37 is a factor of 74 and 999.

So: $\qquad\qquad\qquad 0.0\dot{7}\dot{4} = \frac{2}{27}$

EXERCISE 2C

1 Write each of these fractions as a decimal.

a $\dfrac{1}{3}$ $\qquad$ **b** $\dfrac{3}{4}$ $\qquad$ **c** $\dfrac{5}{6}$ $\qquad$ **d** $\dfrac{2}{9}$

e $\dfrac{13}{20}$ $\qquad$ **f** $\dfrac{9}{11}$ $\qquad$ **g** $\dfrac{3}{16}$ $\qquad$ **h** $\dfrac{11}{12}$

2 **a** Write $\dfrac{7}{15}$ as a recurring decimal. $\qquad$ **b** Write $\dfrac{14}{15}$ as a recurring decimal.

3 **a** Write $\dfrac{1}{9}$ as a decimal. $\qquad$ **b** Write $\dfrac{1}{6}$ as a decimal.

$\qquad$ **c** Write $\dfrac{1}{9} + \dfrac{1}{6}$ as a decimal. $\qquad$ **b** Write $\dfrac{1}{6} - \dfrac{1}{9}$ as a decimal.

4 Write 0.888… as a fraction.

5 Write $0.2\dot{4}$ as a fraction.

6 Write 0.3666… as a fraction.

7 Write $0.08\dot{3}$ as a fraction.

8 Write 2.4666… as a mixed number.

9 Write $\dfrac{3}{13}$ as a recurring decimal.

10 **a** Write $\dfrac{1}{11}$ as a recurring decimal.

$\qquad$ **b** Write $\dfrac{2}{11}$ as a recurring decimal.

$\qquad$ **c** Use your answers to **a** and **b** to predict the values of $\dfrac{3}{11}, \dfrac{4}{11}$ and $\dfrac{7}{11}$ as recurring decimals. Check whether you are correct.

EXTENDED

11 $\frac{1}{7} = 0.\dot{1}4285\dot{7}$

a Write $\frac{2}{7}$ as a recurring decimal in a similar way.

b Write $\frac{3}{7}$ as a recurring decimal.

c Without doing any more calculations, write $\frac{4}{7}$, $\frac{5}{7}$ and $\frac{6}{7}$ as recurring decimals.

12 a Here is a list of fractions.

$\frac{1}{5}$ $\quad$ $\frac{1}{6}$ $\quad$ $\frac{1}{7}$ $\quad$ $\frac{1}{8}$ $\quad$ $\frac{1}{9}$ $\quad$ $\frac{1}{10}$ $\quad$ $\frac{1}{11}$ $\quad$ $\frac{1}{12}$

Which ones are terminating decimals?

b N is a whole number. Try to find a rule for deciding whether $\frac{1}{N}$ is a terminating decimal or a recurring decimal.

2.4 Percentages, fractions and decimals

100% means the *whole* of something. You can express *part* of the whole as a **percentage**.
Per cent means 'out of 100'.

So, any percentage can be converted to a **fraction** with denominator 100.

For example:

$32\% = \frac{32}{100}$ which can be simplified by cancelling to $\frac{8}{25}$

Also, you can convert any percentage to a **decimal** by dividing the percentage number by 100. This means moving the digits two places to the right.

For example:

$65\% = 65 \div 100 = 0.65$

You can convert any decimal to a percentage by multiplying by 100%.

For example:

$0.43 = 0.43 \times 100\% = 43\%$

You can convert any fraction to a percentage by making the denominator into 100 and taking the numerator as the percentage.

For example:

$\frac{2}{5} = \frac{40}{100} = 40\%$

You can also convert fractions to percentages by dividing the numerator by the denominator and multiplying by 100%.

For example:

$$\frac{2}{5} = 2 \div 5 \times 100\% = 40\%$$

Knowing the percentage and **decimal equivalents** of the common fractions is extremely useful. Try to learn them.

$\frac{1}{2} = 0.5 = 50\%$ $\frac{1}{4} = 0.25 = 25\%$ $\frac{3}{4} = 0.75 = 75\%$ $\frac{1}{8} = 0.125 = 12.5\%$

$\frac{1}{10} = 0.1 = 10\%$ $\frac{1}{5} = 0.2 = 20\%$ $\frac{1}{3} = 0.33\ldots = 33\frac{1}{3}\%$ $\frac{2}{3} = 0.66\ldots = 66\%$

These tables show how to convert from one to the other.

Convert from percentage to:	
Decimal	**Fraction**
Divide the percentage by 100, for example: $52\% = 52 \div 100$ $= 0.52$	Make the percentage into a fraction with a denominator of 100 and simplify by cancelling if possible, for example: $52\% = \frac{52}{100} = \frac{13}{25}$

Convert from decimal to:	
Percentage	**Fraction**
Multiply the decimal by 100%, for example: $0.65 = 0.65 \times 100\%$ $= 65\%$	If the decimal has 1 decimal place put it over the denominator 10. If it has 2 decimal places put it over the denominator 100, etc. Then simplify by cancelling if possible, for example: $0.65 = \frac{65}{100}$ $= \frac{13}{20}$

Convert from fraction to:	
Percentage	**Decimal**
If the denominator is a factor of 100 multiply numerator and denominator to make the denominator 100, then the numerator is the percentage, for example: $\frac{3}{20} = \frac{15}{100}$ $= 15\%$ or convert to a decimal and change the decimal to a percentage, for example: $\frac{7}{8} = 7 \div 8$ $= 0.875$ $= 87.5\%$	Divide the numerator by the denominator, for example: $\frac{9}{40} = 9 \div 40$ $= 0.225$

EXAMPLE 8

Convert these numbers to decimals.   **a** 78%   **b** 35%   **c** $\frac{3}{25}$ **d** $\frac{7}{40}$

a   78% = 78 ÷ 100

= 0.78

c $\frac{3}{25}$ = 3 ÷ 25

= 0.12

b   35% = 35 ÷ 100

= 0.35

d $\frac{7}{40}$ = 7 ÷ 40

= 0.175

EXAMPLE 9

Convert these numbers to percentages.   **a** 0.85   **b** 0.125   **c** $\frac{7}{20}$ **d** $\frac{3}{8}$

a   0.85 = 0.85 × 100%

= 85%

c $\frac{7}{20} = \frac{35}{100}$

= 35%

b   0.125 = 0.125 × 100%

= 12.5%

d $\frac{3}{8}$ = 3 ÷ 8 × 100%

= 0.375 × 100%

= 37.5%

EXAMPLE 10

Convert these numbers to fractions.   **a** 0.45   **b** 0.4   **c** 32%   **d** 15%

a   0.45 = $\frac{45}{100}$

= $\frac{9}{20}$

c   32% = $\frac{32}{100}$

= $\frac{8}{25}$

b   0.4 = $\frac{4}{10}$

= $\frac{2}{5}$

d   15% = $\frac{15}{100}$

= $\frac{3}{20}$

EXERCISE 2D

1   Write each percentage as a fraction in its simplest form.

a 8%

b 50%

c 25%

d 35%

e 90%

f 75%

2   Write each percentage as a decimal.

a 27%

b 85%

c 13%

d 6%

e 80%

f 32%

3   Write each decimal as a fraction in its simplest form.

a 0.12

b 0.4

c 0.45

d 0.68

e 0.25

f 0.625

CORE

4 Write each decimal as a percentage.

a 0.29   b 0.55   c 0.03

d 0.16   e 0.6   f 1.25

5 Write each fraction as a percentage.

a $\frac{7}{25}$ b $\frac{3}{10}$ c $\frac{19}{20}$

d $\frac{17}{50}$ e $\frac{11}{40}$ f $\frac{7}{8}$

6 Write each fraction as a decimal.

a $\frac{9}{15}$ b $\frac{3}{40}$ c $\frac{19}{25}$

d $\frac{5}{16}$ e $\frac{1}{20}$ f $\frac{1}{8}$

7 a Convert each test score into a percentage. Give each answer to the nearest whole number.

Subject	Result	Percentage
Mathematics	38 out of 60	
English	29 out of 35	
Science	27 out of 70	
History	56 out of 90	
Technology	58 out of 75	

b If all the tests are of the same standard, which was the best result?

8 Copy and complete the table.

Percentage	Decimal	Fraction
34%		
	0.85	
		$\frac{3}{40}$
45%		
	0.3	
		$\frac{2}{3}$
84%		
	0.45	
		$\frac{3}{8}$

Calculating a percentage

To calculate a percentage of a **quantity**, you multiply the quantity by the percentage. The percentage may be expressed as either a fraction or a decimal. When finding percentages without a calculator, base the calculation on 10% (or 1%) as these are easy to calculate.

EXAMPLE 11

Calculate: **a** 10% of 54 kg **b** 15% of 54 kg.

a 10% is $\frac{1}{10}$ so $\frac{1}{10}$ of 54 kg = 54 kg ÷ 10 = 5.4 kg

b 15% is 10% + 5% = 5.4 kg + 2.7 kg = 8.1 kg

Using a percentage multiplier

You have already seen that percentages and decimals are equivalent so it is easier, particularly when using a calculator, to express a percentage as a decimal and use this to do the calculation.

For example, 13% is a **multiplier** of 0.13, 20% a multiplier of 0.2 (or 0.20) and so on.

EXAMPLE 12

a Calculate 45% of 160 cm. **b** find 52% of $460.

45% = 0.45

So 45% of 160 = 0.45 × 160 = 72 cm

52% = 0.52

So, 0.52 × 460 = 239.2

This gives $239.20.

Remember always to write a money answer with 2 decimal places.

EXERCISE 2E

1 Write down the multiplier that is equivalent to each percentage.

a 88% **b** 30% **c** 25%

d 8% **e** 115%

2 Write down the percentage that is equivalent to each multiplier.

a 0.78 **b** 0.4 **c** 0.75

d 0.05 **e** 1.1

3 Calculate each amount.

a 15% of $300

b 6% of $105

c 23% of 560 kg

d 45% of 2.5 kg

e 12% of 9 hours

f 21% of 180 cm

g 4% of $3

h 35% of 8.4 m

i 95% of $8

j 11% of 308 minutes

k 20% of 680 kg

l 45% of $360

4 An estate agent charges 2% commission on every house he sells. How much commission will he earn on a house that he sells for $120 500?

5 A store had 250 employees. During one week of a flu epidemic, 14% of the store's employees were absent.

a What percentage of the employees went into work?

b How many of the employees went into work?

6 Generally, about 20% of fans at a soccer match are women. For one match there were 42 600 fans. How many of these do you think were women?

7 At a Paris railway station, in one week 350 trains arrived. Of these trains, 5% arrived early and 13% arrived late. How many arrived on time?

8 A school estimates that 60% of the students will attend a school play. There are 1500 students in the school. The caretaker is told to put out one seat for each person expected to attend plus an extra 10% of that number in case more attend. How many seats does he need to put out?

ADVICE AND TIPS

It is not 70% of the number of students in the school.

9 A school had 850 students and the attendance record in one week was:

Monday 96%   Tuesday 98%   Wednesday 100%   Thursday 94%   Friday 88%

How many students were present each day?

10 Calculate each amount.

a 12.5% of $26

b 6.5% of 34 kg

c 26.8% of $2100

d 7.75% of $84

e 16.2% of 265 m

f 0.8% of $3000

11 Air consists of 80% nitrogen and 20% oxygen (by volume). A man's lungs have a capacity of 600 cm^3. How much of each gas will he have in his lungs when he has just taken a deep breath?

12 A factory estimates that 1.5% of all the garments it produces will have a fault in them. One week the factory produces 850 garments. How many are likely to have a fault?

13 An insurance firm sells house insurance. The annual premiums are usually 0.3% of the value of the house. What will be the annual premium for a house valued at $90 000?

14 Average prices in a shop went up by 3% last year and 3% this year. Did the actual average price of items this year rise by more, the same amount, or less than last year?

Explain how you decided.

Increasing by a percentage

There are two methods for increasing a quantity by a percentage.

Method 1

Work out the increase and add it on to the original amount.

EXAMPLE 13

Increase $6 by 5%.

Work out 5% of $6: $(5 \div 100) \times 6 = \$0.30$

Add the $0.30 to the original amount: $6 + $0.30 = $6.30

Method 2

Use a **multiplier**. An increase of 6% is equivalent to the original 100% *plus* the extra 6%. This is a total of 106% and is equivalent to the multiplier 1.06.

EXAMPLE 14

Increase $6.80 by 5%.

A 5% increase is a multiplier of 1.05

So $6.80 increased by 5% is $6.80 × 1.05 = $7.14

EXERCISE 2F

1 What multiplier is used to increase a quantity by:

   a 10%   **b** 3%   **c** 20%   **d** 7%   **e** 12%?

2 Increase each amount by the given percentage. (Use any method you like.)

   a $60 by 4%   **b** 12 kg by 8%   **c** 450 g by 5%   **d** 545 m by 10%

   e $34 by 12%   **f** $75 by 20%   **g** 340 kg by 15%   **h** 670 cm by 23%

   i 130 g by 95%   **j** $82 by 75%   **k** 640 m by 15%   **l** $28 by 8%

3 Azwan, who was on a salary of $27 500, was given a pay rise of 7%. What is his new salary?

CORE

4 In 2005 the population of a city was 1 565 000. By 2010 it had increased by 8%. What was the population of the city in 2010?

5 A small firm made the same pay increase of 5% for all its employees.

   a Calculate the new pay of each employee listed below. Each of their salaries before the increase is given.

      Caretaker, $16 500   Supervisor, $19 500

      Driver, $17 300   Manager, $25 300

   b Explain why the actual pay increases are different for each employee.

6 A bank pays 7% interest on the money that each saver keeps in the bank for a year. Allison keeps $385 in the bank for a year. How much will she have in the bank after the year?

7 In 1980 the number of cars on the roads of a town was about 102 000. Since then it has increased by 90%. Approximately how many cars are there on the roads of the town now?

8 An advertisement for a breakfast cereal states that a special-offer packet contains 15% more cereal for the same price as a normal 500 g packet. How much breakfast cereal is in a special-offer packet?

9 A headteacher was proud to point out that, since he had arrived at the school, the number of students had increased by 35%. How many students are now in the school, if there were 680 when the headteacher started at the school?

10 At a school concert there are always about 20% more girls than boys. If at one concert there were 50 boys, how many girls were there?

11 A government adds a sales tax to the price of most goods in shops. One year it is 17.5% on all electrical equipment.

Calculate the price of the following electrical equipment when sales tax of 17.5% is added.

Equipment	Pre-sales tax price
TV set	$245
Microwave oven	$72
CD player	$115
Personal stereo	$29.50

12 A television costs $400 before sales tax at 17.5% is added.

If the rate of sales tax goes up from 17.5% to 20%, how much will the cost of the television increase?

13 Bookshop BookWorms increased its prices by 5%, then increased them by 3%. Bookshop Books Galore increased its prices by 3%, then increased them by 5%.

Which shop's prices increased by the greater percentage?

a BookWorms   b Books Galore

c Both the same   d Cannot tell

Justify your choice.

14 Shop A increased its prices by 4% and then by another 4%. Shop B increased its prices by 8%.

Which shop's prices increased by the greater percentage?

a Shop A   b Shop B   c Both the same   d Cannot tell

Justify your choice.

15 A hi-fi system was priced at $420 at the start of 2008. At the start of 2009, it was 12% more expensive. At the start of 2010, it was 15% more expensive than the price at the start of 2009. What is the price of the hi-fi at the start of 2010?

Decreasing by a percentage

There are two methods for decreasing by a percentage.

Method 1

Work out the decrease and subtract it from the original amount.

EXAMPLE 15

Decrease $8 by 4%.

Work out 4% of $8: $(4 \div 100) \times 8 = \$0.32$
Subtract the $0.32 from the original amount: $\$8 - \$0.32 = \$7.68$

Method 2

Use a **multiplier**. A 7% decrease is equivalent to 7% less than the original 100%, so it represents 100% – 7% = 93% of the original. This is a multiplier of 0.93.

EXAMPLE 16

Decrease $8.60 by 5%.

A decrease of 5% is a multiplier of 0.95.

So $8.60 decreased by 5% is $\$8.60 \times 0.95 = \$8.17$

EXERCISE 2G

1 What multiplier is used to decrease a quantity by:

  a 8%   **b** 15%   **c** 25%   **d** 9%   **e** 12%?

2 Decrease each amount by the given percentage. (Use any method you like.)

  a $10 by 6%   **b** 25 kg by 8%   **c** 236 g by 10%   **d** 350 m by 3%

  e $5 by 2%   **f** 45 m by 12%   **g** 860 m by 15%   **h** 96 g by 13%

  i 480 cm by 25%   **j** 180 minutes by 35%   **k** 86 kg by 5%   **l** $65 by 42%

3 A car valued at $6500 last year is now worth 15% less. What is its value now?

4 A new diet guarantees that you will lose 12% of your mass in the first month. What mass should the following people have after one month on the diet?

  a Gracia, who started at 60 kg   **b** Pierre, who started at 75 kg

  c Greta, who started at 52 kg

5 A motor insurance firm offers no-claims discounts off the full premium, as follows.

  1 year with no claims   15% discount off the full premium
  2 years with no claims   25% discount off the full premium
  3 years with no claims   45% discount off the full premium
  4 years with no claims   60% discount off the full premium

  Mr Patel and his family are all offered motor insurance from this firm.

  Mr Patel has four years' no-claims discount and the full premium would be $440.
  Mrs Patel has one year's no-claims discount and the full premium would be $350.
  Sandeep has three years' no-claims discount and the full premium would be $620.
  Priyanka has two years' no-claims discount and the full premium would be $750.

  Calculate the actual amount each member of the family has to pay for the motor insurance.

6 A large factory employed 640 people. It had to streamline its workforce and lose 30% of the workers. How big is the workforce now?

7 On the last day of term, a school expects to have an absence rate of 6%. If the school population is 750 students, how many students will the school expect to see on the last day of term?

8 Most speedometers in cars have an error of about 5% from the true reading. When my speedometer says I am driving at 70 km/h:

  a what is the lowest speed I could be doing

  b what is the highest speed I could be doing?

9 Kerry wants to buy a sweatshirt ($19), a tracksuit ($26) and some running shoes ($56). If she joins the store's premium club which costs $25 to join she can get 20% off the cost of the goods.

Should she join or not? Give figures to support your answer.

10 A shop advertises garden ornaments at $50 but with 10% off in a sale. It then advertises an extra 10% off the sale price.

Show that this is not a decrease in price of 20%.

11 A computer system was priced at $1000 at the start of 2008. At the start of 2009, it was 10% cheaper. At the start of 2010, it was 15% cheaper than the price at the start of 2009. What is the price of the computer system at the start of 2010?

12 Show that a 10% decrease followed by a 10% increase is equivalent to a 1% decrease overall.

ADVICE AND TIPS

Choose an amount to start with.

13 A biscuit packet normally contains 300 g of biscuit and costs $1.40.

There are two special offers.

    Offer A: 20% more for the same price

    Offer B: Same amount for 20% off the normal price

Which is the better offer?

a Offer A   **b** Offer B   **c** Both the same   **d** Cannot tell

Justify your choice.

2.7 One quantity as a percentage of another

You express one quantity as a percentage of another by setting up the first quantity as a fraction of the second. You must make sure that the *units of each are the same*. Then you convert the fraction into a percentage by multiplying by 100%.

EXAMPLE 17

Express $6 as a percentage of $40.

Set up the fraction and multiply by 100%.

$$\frac{6}{40} \times 100\% = 15\%$$

EXAMPLE 18

Express 75 cm as a percentage of 2.5 m.

First, change 2.5 m to 250 cm to get a common unit.

So, the problem now becomes: Express 75 cm as a percentage of 250 cm.

Set up the fraction and multiply by 100%.

$$\frac{75}{250} \times 100\% = 30\%$$

Percentage change

A **percentage change** may be a **percentage increase** or a **percentage decrease**.

$$\text{Percentage change} = \frac{\text{change}}{\text{original amount}} \times 100$$

Use this to calculate **percentage profit** or **percentage loss** in a financial transaction.

EXAMPLE 19

Jake buys a car for $1500 and sells it for $1800. What is Jake's percentage profit?

Jake's profit is $300, so his percentage profit is:

$$\text{percentage profit} = \frac{\text{profit}}{\text{original amount}} \times 100\% = \frac{300}{1500} \times 100\% = 20\%$$

Using a multiplier (or decimal)

To use a multiplier, divide the increase by the original quantity and change the resulting decimal to a percentage.

EXAMPLE 20

Express 5 as a percentage of 40.

Set up the fraction or decimal: $5 \div 40 = 0.125$

Convert the decimal to a percentage: $0.125 = 12.5\%$

CORE

EXERCISE 2H

1 Express each amount as a percentage. Give suitably rounded figures (see page 117) where necessary.

a $5 of $20

b $4 of $6.60

c 241 kg of 520 kg

d 3 hours of 1 day

e 25 minutes of 1 hour

f 12 m of 20 m

g 125 g of 600 g

h 12 minutes of 2 hours

i 1 week of a year

j 1 month of 1 year

k 25 cm of 55 cm

l 105 g of 1 kg

2 Liam went to school with his pocket money of $2.50. He spent 80 cents at the shop. What percentage of his pocket money had he spent?

3 In Greece, there are 3 654 000 acres of agricultural land. Olives are grown on 237 000 acres of this land. What percentage of the agricultural land is used for olives?

4 During one year, it rained in Detroit on 123 days of the year. What percentage of days were wet?

5 Find the percentage profit on each transaction. Give your answers to one decimal place.

Item	Retail price (selling price)	Wholesale price (price the shop paid)
a CD player	$89.50	$60
b TV set	$345.50	$210
c Computer	$829.50	$750

6 Before Anton started to diet, his mass was 95 kg. His mass is now 78 kg. What percentage of his original mass has he lost?

7 In 2009 a city raised $14 870 000 in local tax. In 2010 it raised $15 597 000 in tax. What was the percentage increase?

8 When Ziad's team won the soccer league in 1995, they lost only four of their 42 league games. What percentage of games did they *not* lose?

9 In one year Britain's imports were as follows.

British Commonwealth	$109 530 000
USA	$138 790 000
France	$53 620 000
Other countries	$221 140 000

a What percentage of the total imports came from each source? Give your answers to 1 decimal place.

b Add up your answers to part **a**. What do you notice? Explain your answer.

10 Imran and Nadia take the same tests. Both tests are out of the same mark.

Here are their results.

	Test A	Test B
Imran	12	17
Nadia	14	20

Whose result has the greater percentage increase from test A to test B?
Show your working.

11 A supermarket advertises its cat food as shown.

A government inspector is checking the claim.

She observes that over one hour, 46 people buy cat food and 38 buy the store's own brand.

Based on these figures, is the store's claim correct?

> **8 out of 10 cat owners choose our cat food.**

12 Aya buys antiques and then sells them on the internet.

Find her percentage profit or loss on each of these items.

Item	Aya bought for:	Aya sold for:
Vase	$105	$84
Radio	$72	$90
Doll	$15	$41.25
Toy train	$50	$18

2.8 Simple interest and compound interest

Erin has a loan of $500. She agrees to pay 1.6% interest each month.

This is an example of **simple interest**.

Each month she pays 1.6% of $500 = 0.016 × 500 = $8

If she pays back the loan after six months she will pay 6 × $8 = $48 in interest.

Banks and building societies usually pay **compound interest** on savings accounts.

When compound interest is used, the interest earned each year is added to the original amount (**principal**) and the new total then earns interest at the **annual rate** in the following year. This pattern is then repeated each year while the money is in the account.

The most efficient way to calculate the total amount in the account after several years is to use a **multiplier**.

EXAMPLE 21

Elizabeth invests $400 in a savings account. The account pays compound interest at 6% each year. How much will she have in the account after three years?

The amount in the account increases by 6% each year, so the multiplier is 1.06.

After 1 year she will have	$400 × 1.06	= $424
After 2 years she will have	$424 × 1.06	= $449.44
After 3 years she will have	$449.44 × 1.06	= $476.41 (rounded)

If you calculate the differences, you can see that the amount of interest increases each year ($24, $25.44 and $26.97).

EXERCISE 2I

1 Rahul has a loan of 7000 dollars.

He pays 2% per month simple interest.

How much will he pay if he has the loan for three months?

2 Lee lends her friend 30 000 dollars.

Her friend agrees to pay simple interest of 6% per year.

How much interest will Lee earn after two years?

3 Jean has a loan of 2000 dollars.

The rate of simple interest is 8% a year.

He has paid 640 dollars interest.

How many years has he had the loan?

4 Rania puts $15 000 into a savings account where it earns 4% per annum compound interest.

  a What is her investment worth after one year?

  b What is her investment worth after two years?

5 Maria invests $1200 at 6% compound interest.

Work out the value of the investment after:

  a one year

  b two years

  c three years.

6 Amar invests $20 000 for two years at 8% simple interest.

Mona invests $20 000 for two years at 8% compound interest.

a How much does each person earn?

b Who earned more and how much more was it?

7 Luka invests $8000 at 6% compound interest for three years.

a How much is his investment worth after three years?

b How much interest has he earned after three years?

8 Mikael has a loan of $40 000.

He pays 1.6% simple interest for six months.

How much interest does he pay altogether?

9 This table shows the amount of interest paid on a loan of $12 000:

Number of years	1	2	3	4
Interest ($)	780	1560	2340	3120

a Is this simple interest or compound interest?

b What is the rate of interest per year?

10 Daniel earns 15% compound interest on an investment of $12 000.

a What will it be worth after one year?

b What will it be worth after two years?

c Show that it will be worth over $20 000 after four years.

11 This table shows the values of amounts invested at 3% compound interest for one, two and five years.

Investment ($)	Value after 1 year ($)	Value after 2 years ($)	Value after 5 years ($)
1000	1030	1060.90	1159.27
2500	2575		2898.18
5000	5150		5796.37

a Work out the two numbers missing from the table.

b Work out the interest on an amount of $5000 invested for 5 years.

Look back at Example 21.

Elizabeth invests $400 at 6% compound interest.

After three years the value, in dollars, is:

$400 \times 1.06 \times 1.06 \times 1.06 = 476.41$

You can write this calculation as 400×1.06^3.

Then you can use the power button on a calculator to work this out efficiently.

This is an example of a **formula** for finding the value of an investment. Here is the formula.

$$\text{Value of investment} = P\left(1 + \frac{r}{100}\right)^n$$

where P is the initial investment, r is the annual percentage rate and n is the number of years.

EXAMPLE 22

Jasmine invests $5500 at 2.5% per annum compound interest.

a How much is the investment worth after four years?

b How much interest has she earned after four years?

a $P = 5500$, $r = 2.5$ and $n = 4$

$$\text{Value in dollars} = 5500 \times \left(1 + \frac{2.5}{100}\right)^4$$

$$= 5500 \times 1.025^4$$

$$= 6070.97$$

b Interest in dollars $= 6070.97 - 5500 = \$570.97$

EXERCISE 2J

1 Elton invests $2000 at 4% per annum compound interest.

Work out the value of his investment after three years.

2 Carla invests $4500 at 6% compound interest.

Work out the value of her investment after four years.

3 Zak invests $4600 at 5% per annum compound interest.

Work out the value of his investment after:

a 2 years   **b** 4 years   **c** 6 years.

EXTENDED

4 Carmen invests $2500 at 7.5% per annum compound interest.

    a How much is the investment worth after 5 years?

    b How much is it worth after 8 years?

5 Marco invests $25 000 at 5.4% per annum compound interest for three years.

    How much interest does he receive?

6 Greta takes out a loan of $3500. The rate of compound interest is 2% per month.

    After 6 months she wants to pay off the loan and the interest.

    a How much must she pay altogether?

    b How much of what she pays is interest?

7 An investment is earning 10% per annum compound interest.

    How long will it take for the investment to double in value?

8 A bank offers 6% per annum compound interest on deposits.

    This table shows how much a deposit of $5000 will be worth at different times.

Initial amount ($)	Amount after 1 year ($)	Amount after 2 year ($)	Amount after 5 year ($)	Amount after 10 year ($)
5000	5300	5618	6691.13	

    Armand works out the missing number like this.

> The interest after 5 years is $6691.13 − $5000 = $1691.13
> The interest for the next five years will be the same, $1691.13
> After ten years the investment is worth $6691.13 + $1691.13 = $8382.26

    Explain why Armand is not correct.

    Then work out the correct value.

9 Luisa invests $10 000 at 5% annual interest.

    Work out how much is the investment worth after 10 years if this is:

    a simple interest

    b compound interest.

10 Credit cards usually charge monthly compound interest on any unpaid amounts.

    Greta has a loan of $1000 on her credit card.

    There is a 2% monthly charge.

    a Work out the size of Greta's loan after 12 months if she does not pay off any money.

    b Show that this is equivalent to an annual percentage of about 26.8%.

Reverse percentage questions involve working backwards from the **final amount** to find the **original amount** when you know, or can work out, the final amount as a percentage of the original amount.

Method 1: The unitary method

The **unitary method** has three steps.

Step 1: Equate the final percentage to the final value.

Step 2: Use this to calculate the value of 1%.

Step 3: Multiply by 100 to work out 100% (the original value).

EXAMPLE 23

The price of a car increased by 6% to $9116. Work out the price before the increase.

106% represents $9116.

Divide by 106.
1% represents $9116 ÷ 106

Multiply by 100.
100% represents original price: $9116 ÷ 106 × 100 = $8600

So the price before the increase was $8600.

Method 2: The multiplier method

The **multiplier** method involves fewer steps.

Step 1: Write down the multiplier.

Step 2: Divide the final value by the multiplier to give the original value.

EXAMPLE 24

In a sale the price of a freezer is reduced by 12%. The sale price is $220. What was the price before the sale?

A decrease of 12% gives a multiplier of 0.88.

Dividing the sale price by the multiplier gives $220 ÷ 0.88 = $250

So the price before the sale was $250.

EXERCISE 2K

1 Find what 100% represents in these situations.

a 40% represents 320 g

b 14% represents 35 m

c 45% represents 27 cm

d 4% represents $123

e 2.5% represents $5

f 8.5% represents $34

2 On a tiring army training session, only 28 youngsters survived the whole day. This represented 35% of the original group. How large was the original group?

3 Sales tax is added to goods and services. With sales tax at 17.5%, what is the pre-sales tax price of the following priced goods?

| T-shirt | $9.87 | Tights | $1.41 | Shorts | $6.11 |
| Sweater | $12.62 | Trainers | $29.14 | Boots | $38.07 |

4 Hugo spends $200 a month on food. This represents 24% of his monthly take-home pay. How much is his monthly take-home pay?

5 Zara's weekly pay is increased by 5% to $315. What was Zara's pay before the increase?

6 The number of workers in a factory fell by 5% to 228. How many workers were there originally?

7 In a sale the price of a TV is reduced to 500 dollars. This is a 7% reduction on the original price. What was the original price?

8 If 38% of plastic bottles in a production line are blue and the remaining 7750 plastic bottles are brown, how many plastic bottles are blue?

9 I received $385 back from the government, which represented the 17.5% purchase tax on a piece of equipment. How much did I pay for this equipment in the first place?

10 A company is in financial trouble. The workers are asked to take a 10% pay cut. Tomas works out that his pay will be $1296 per month. How much is his pay now?

11 Manza buys a car and sells it for $2940. He made a profit of 20%. What was the cost price of the car?

12 When a suit is sold in a shop the selling price is $171 and the profit is 80%. What was the cost price?

13 Tebor buys a chair. He sells it for $63 in an auction and makes a loss of 55%. What did he pay for the chair?

14 A woman's salary increased by 5% in one year. Her new salary was $19 845. How much was the increase?

15 After an 8% increase, the monthly salary of a chef was $1431. What was the original monthly salary?

16 Cassie invested some money at 4% interest per annum for two years. After two years, she had $1406.08 in the bank. How much did she invest originally?

17 A teacher asked her class to work out the original price of a cooker for which, after a 12% increase, the price was 291.20 dollars.

This is Lee's answer: 12% of 291.20 = 34.94 dollars

Original price = 291.2 − 34.94 = 256.26 ≈ 260 dollars

When the teacher read out the answer Lee ticked his work as correct. What errors has he made?

Why this chapter matters

Most jobs will require you to use some mathematics every day. Having good number skills will help you to be more successful in your job.

The mathematics used in jobs ranges from simple calculations, such as addition, subtraction, multiplication and division, to more complex calculations involving negative numbers and approximation. You will need to select the right mathematics for the job.

Jobs using mathematics

How many jobs can you think of that require some mathematics?

Here are a few ideas.

Engineer – What measurements do I need to take? How much of each type of material will be needed?

Pilot – How much fuel do I need?

Accountant – How much profit have they made?

Cashier – What coins do I need to give as change? What is the best price to sell my goods at?

Doctor – How much medicine should I prescribe?

Delivery driver – What is the best route?

Sports commentator – How many minutes are left in the game? What is his batting average?

Baker – What quantity of flour should I order?

If you already know what job you would like to do, think of what mathematics you might need for it.

The four rules

Topics

Level

Key words

 Order of operations

CORE

operation, brackets, order

2 Choosing the correct operation

CORE

3 Finding a fraction of a quantity

CORE

quantity, fraction

4 Adding and subtracting fractions

CORE

proper fraction, improper fraction, lowest terms, simplest form, denominator, mixed number, equivalent fraction

5 Multiplying and dividing fractions

CORE

numerator, reciprocal

In this chapter you will learn how to:

CORE

- Use the four rules for calculation with whole numbers, decimals, vulgar fractions and mixed numbers. (C1.8 and E1.8)
- Apply operations in the correct order, including the use of brackets. (C1.8 and E1.8)

Order of operations

Suppose you have to work out the answer to $4 + 5 \times 2$. You may say the answer is 18, but the correct answer is 14.

There is an **order** of **operations** which you must follow when working out calculations like this. The $\times$ is always done *before* the $+$.

In $4 + 5 \times 2$ this gives $4 + 10 = 14$.

Now suppose you have to work out the answer to $(3 + 2) \times (9 - 5)$. The correct answer is 20.

You have probably realised that the parts in the **brackets** have to be done *first*, giving $5 \times 4 = 20$.

So, how do you work out a problem such as $9 \div 3 + 4 \times 2$?

To answer questions like this, you *must* follow the BIDMAS (or BODMAS) rule. This tells you the order in which you *must* do the operations.

B	Brackets	**B**	Brackets
I	Indices (powers)	**O**	pOwers
D	Division	**D**	Division
M	Multiplication	**M**	Multiplication
A	Addition	**A**	Addition
S	Subtraction	**S**	Subtraction

For example, to work out $9 \div 3 + 4 \times 2$:

First divide:	$9 \div 3 = 3$	giving	$3 + 4 \times 2$
Then multiply:	$4 \times 2 = 8$	giving	$3 + 8$
Then add:	$3 + 8 = 11$		

And to work out $60 - 5 \times 3^2 + (4 \times 2)$:

First, work out the brackets:	$(4 \times 2) = 8$	giving	$60 - 5 \times 3^2 + 8$
Then the index (power):	$3^2 = 9$	giving	$60 - 5 \times 9 + 8$
Then multiply:	$5 \times 9 = 45$	giving	$60 - 45 + 8$
Then add:	$60 + 8 = 68$	giving	$68 - 45$
Finally, subtract:	$68 - 45 = 23$		

EXERCISE 3A

1 Work out each of these.

a $2 \times 3 + 5 =$ **b** $6 \div 3 + 4 =$ **c** $5 + 7 - 2 =$

d $4 \times 6 \div 2 =$ **e** $2 \times 8 - 5 =$ **f** $3 \times 4 + 1 =$

g $3 \times 4 - 1 =$ **h** $3 \times 4 \div 1 =$ **i** $12 \div 2 + 6 =$

j $12 \div 6 + 2 =$ **k** $3 + 5 \times 2 =$ **l** $12 - 3 \times 3 =$

2 Work these out. Remember: first work out the bracket.

a $2 \times (3 + 5) =$ **b** $6 \div (2 + 1) =$ **c** $(5 + 7) - 2 =$

d $5 + (7 - 2) =$ **e** $3 \times (4 \div 2) =$ **f** $3 \times (4 + 2) =$

g $2 \times (8 - 5) =$ **h** $3 \times (4 + 1) =$ **i** $3 \times (4 - 1) =$

j $3 \times (4 \div 1) =$ **k** $12 \div (2 + 2) =$ **l** $(12 \div 2) + 2 =$

3 Copy each of these and then put brackets in, where necessary, to make the answer true.

a $3 \times 4 + 1 = 15$ **b** $6 \div 2 + 1 = 4$ **c** $6 \div 2 + 1 = 2$

d $4 + 4 \div 4 = 5$ **e** $4 + 4 \div 4 = 2$ **f** $16 - 4 \div 3 = 4$

g $3 \times 4 + 1 = 13$ **h** $16 - 6 \div 3 = 14$ **i** $20 - 10 \div 2 = 5$

j $20 - 10 \div 2 = 15$ **k** $3 \times 5 + 5 = 30$ **l** $6 \times 4 + 2 = 36$

m $15 - 5 \times 2 = 20$ **n** $4 \times 7 - 2 = 20$ **o** $12 \div 3 + 3 = 2$

p $12 \div 3 + 3 = 7$ **q** $24 \div 8 - 2 = 1$ **r** $24 \div 8 - 2 = 4$

4 Ravi says that $5 + 6 \times 7$ is equal to 77.

Is he correct?

Explain your answer.

5 Three different dice give scores of 2, 3, 5. Add $\div$, $\times$, $+$ or $-$ signs to make each calculation work.

a $2 \quad 3 \quad 5 = 11$ **b** $2 \quad 3 \quad 5 = 16$ **c** $2 \quad 3 \quad 5 = 17$

d $5 \quad 3 \quad 2 = 4$ **e** $5 \quad 3 \quad 2 = 13$ **f** $5 \quad 3 \quad 2 = 30$

6 Which is smaller:

$4 + 5 \times 3$ or $(4 + 5) \times 3$?

Show your working.

7 Here is a list of numbers, some signs and one pair of brackets.

$2 \quad 5 \quad 6 \quad 18 \quad - \quad \times \quad = \quad (\quad)$

Use **all** of them to make a correct calculation.

8 Here is a list of numbers, some signs and one pair of brackets.

$3 \quad 4 \quad 5 \quad 8 \quad - \quad \div \quad = \quad (\quad)$

Use **all** of them to make a correct calculation.

When a problem is given in words you will need to decide the correct operation to use. Should you add, subtract, multiply or divide?

EXAMPLE 1

A party of 613 children and 59 adults are going on a day out to a theme park.

a How many coaches, each holding 53 people, will be needed?

b One adult gets into the theme park free for every 15 children. How many adults will have to pay to get in?

a Altogether there are 613 + 59 = 672 people.

So the number of coaches needed is 672 ÷ 53 (number of seats on each coach) = 12.67 …

13 coaches are needed (12 will not be enough).

b This is also a division, 613 ÷ 15 = 40.86 …

40 adults will get in free.

59 − 40 = 19 will have to pay.

EXERCISE 3B

1 There are 48 cans of soup in a crate. A shop had a delivery of 125 crates of soup.

a How many cans of soup were in this delivery?

b The shop is running a promotion on soup. If you buy five cans you get one free. Each can costs 39 cents. How much will it cost to get 32 cans of soup?

2 A school has 12 classes, each of which has 24 students.

a How many students are there at the school?

b The student–teacher ratio is 18 to 1. That means there is one teacher for every 18 students. How many teachers are there at the school?

3 A football club is organising travel for an away game. 1300 adults and 500 children want to go. Each coach holds 48 people and costs $320 to hire.
Tickets to the match cost $18 for adults and $10 for children.

a How many coaches will be needed?

b The club is charging adults $26 and children $14 for travel and a ticket. How much profit does the club make out of the trip?

4 A large letter costs 39 cents to post and a small letter costs 30 cents. How many dollars will it cost to send 20 large and 90 small letters?

CORE

5 Kirsty collects small models of animals. Each one costs 45 cents. She saves enough to buy 23 models but when she goes to the shop she finds that the price has gone up to 55 cents. How many can she buy now?

6 Michaela wanted to save up for a bike that costs $250. She baby-sits each week for 6 hours for $2.75 an hour, and does a Saturday job that pays $27.50. She saves three-quarters of her weekly earnings. How many weeks will it take her to save enough to buy the bike?

7 The magazine *Teen Dance* comes out every month. In a newsagent's shop the magazine costs $2.45. The annual (yearly) subscription for the magazine is $21. How much cheaper is each magazine when it is bought on subscription?

8 Paula buys a sofa. She pays a deposit of 10% of the cash price and then 36 monthly payments of $12.50. In total she pays $495. How much was the cash price of the sofa?

9 There are 125 people at a wedding. They need to get to the reception.

52 people are going by coach and the rest are travelling in cars. Each car can take up to five people.

What is the least number of cars needed to take everyone to the reception?

10 Gustav's car does 8 kilometres to each litre of fuel. He does 12 600 kilometres a year of which 4 600 is on company business.

Fuel costs 95 cents per litre.

Insurance and servicing costs $800 a year.

Gustav's company gives him an allowance of 40 cents for each kilometre he drives on company business.

How much does Gustav pay towards running his car each year?

3.3 Finding a fraction of a quantity

To do this, you simply multiply the **fraction** by the **quantity**, for example, $\frac{1}{2}$ of 30 is the same as $\frac{1}{2} \times 30$.

Remember: In mathematics 'of' is interpreted as ×.

For example, two lots of three is the same as 2×3.

EXAMPLE 2

Find $\frac{3}{4}$ of $196.

First, find $\frac{1}{4}$ by dividing by 4. Then find $\frac{3}{4}$ by multiplying your answer by 3.

$196 \div 4 = 49$ then $49 \times 3 = 147$

The answer is $147.

CORE

EXERCISE 3C

1 Calculate each amount.

  a $\frac{3}{5}$ of 30   **b** $\frac{2}{7}$ of 35   **c** $\frac{3}{8}$ of 48   **d** $\frac{7}{10}$ of 40

2 Calculate each of these quantities.

  a $\frac{3}{4}$ of $2400   **b** $\frac{2}{5}$ of 320 grams   **c** $\frac{5}{8}$ of 256 kilograms

  d $\frac{2}{3}$ of $174   **e** $\frac{5}{6}$ of 78 litres   **f** $\frac{3}{4}$ of 120 minutes

3 In each case, find out which is the larger number.

  a $\frac{2}{5}$ of 60 or $\frac{5}{8}$ of 40   **b** $\frac{3}{4}$ of 280 or $\frac{7}{10}$ of 290

  c $\frac{2}{3}$ of 78 or $\frac{4}{5}$ of 70   **d** $\frac{5}{6}$ of 72 or $\frac{11}{12}$ of 60

4 A director receives $\frac{2}{15}$ of his firm's profits. The firm made a profit of $45 600 in one year. How much did the director receive?

5 A woman left $84 000 in her will.

  She left $\frac{3}{8}$ of the money to charity.

  How much did she leave to charity?

6 Two-thirds of a person's mass is water. Paul has a mass of 78 kg. How much of his body mass is water?

7 **a** Information from the first census in Singapore showed that $\frac{2}{25}$ of the population were Indian. The total population was 10 700. How many people were Indian?

  b By 1990 the population of Singapore had grown to 3 002 800. Only $\frac{1}{16}$ of this population were Indian. How many Indians were living in Singapore in 1990?

8 Marc normally earns $500 a week. One week he is given a bonus of $\frac{1}{10}$ of his wage.

  a Find $\frac{1}{10}$ of $500.

  b How much does Marc earn altogether for this week?

9 The price of a new TV costing $360 is reduced by $\frac{1}{3}$ in a sale.

  a Find $\frac{1}{3}$ of $360.

  b How much does the TV cost in the sale?

10 A car is advertised at Lion Autos at $9000 including extras but with a special offer of one-fifth off this price.

The same car is advertised at Tiger Motors for $6000 but the extras add one-quarter to this price.

Which garage is the cheaper?

11 A jar of coffee normally contains 200 g and costs $2.

There are two special offers on a jar of coffee.

    Offer A: $\frac{1}{4}$ extra for the same price.

    Offer B: Same mass for $\frac{3}{4}$ of the original price.

Which offer is better value?

3.4   Adding and subtracting fractions

When you add two fractions with the same **denominator**, you get one of the following:

- a proper fraction that cannot be simplified, for example:

$$\frac{1}{5} + \frac{2}{5} = \frac{3}{5}$$

- a proper fraction that can be simplified to its **lowest terms** or **simplest form**, for example:

$$\frac{1}{8} + \frac{3}{8} = \frac{4}{8} = \frac{1}{2}$$

- an **improper fraction** that cannot be simplified, so it is converted to a **mixed number**, for example:

$$\frac{6}{7} + \frac{2}{7} = \frac{8}{7} = 1\frac{1}{7}$$

- an improper fraction that *can* be simplified before it is converted to a mixed number, for example:

$$\frac{5}{8} + \frac{7}{8} = \frac{12}{8} = \frac{3}{2} = 1\frac{1}{2}$$

When you subtract one fraction from another with the same denominator, you get one of the following:

- a proper fraction that cannot be simplified, for example:

$$\frac{3}{5} - \frac{1}{5} = \frac{2}{5}$$

● a proper fraction that can be simplified, for example:

$$\frac{7}{10} - \frac{1}{10} = \frac{6}{10} = \frac{3}{5}$$

Notice that $\frac{6}{10}$ and $\frac{3}{5}$ are **equivalent fractions**. They represent the same quantity.

Note: You must *always* simplify fractions by cancelling if possible.

EXAMPLE 3

Find $\frac{1}{2} + \frac{5}{8}$

These fractions *do not* have the same denominator.

However $\frac{1}{2} = \frac{4}{8}$ *so you can write:*

$$\frac{1}{2} + \frac{5}{8} = \frac{4}{8} + \frac{5}{8} = \frac{9}{8} = 1\frac{1}{8}$$

EXERCISE 3D

1 Copy and complete each of these additions.

   a $\frac{3}{7} + \frac{2}{7}$ **b** $\frac{5}{9} + \frac{2}{9}$ **c** $\frac{3}{5} + \frac{1}{5}$ **d** $\frac{3}{7} + \frac{3}{7}$

2 Copy and complete each of these subtractions.

   a $\frac{4}{7} - \frac{1}{7}$ **b** $\frac{5}{9} - \frac{4}{9}$ **c** $\frac{7}{11} - \frac{3}{11}$ **d** $\frac{9}{13} - \frac{2}{13}$

3 Copy and complete each of these additions.

   a $\frac{5}{8} + \frac{1}{8}$ **b** $\frac{3}{10} + \frac{1}{10}$ **c** $\frac{2}{9} + \frac{4}{9}$ **d** $\frac{1}{4} + \frac{1}{4}$

4 Copy and complete each of these subtractions.

   a $\frac{7}{8} - \frac{3}{8}$ **b** $\frac{7}{10} - \frac{3}{10}$ **c** $\frac{5}{6} - \frac{1}{6}$ **d** $\frac{9}{10} - \frac{1}{10}$

5 Copy and complete each of these additions. Use equivalent fractions to make the denominators the same. Show your working.

   a $\frac{1}{2} + \frac{7}{10}$ **b** $\frac{1}{2} + \frac{5}{8}$ **c** $\frac{3}{4} + \frac{3}{8}$ **d** $\frac{3}{4} + \frac{7}{8}$

   e $\frac{1}{2} + \frac{7}{8}$ **f** $\frac{1}{3} + \frac{5}{6}$ **g** $\frac{5}{6} + \frac{2}{3}$ **h** $\frac{3}{4} + \frac{1}{2}$

6 Copy and complete each of these additions. Show your working.

   a $\frac{3}{8} + \frac{7}{8}$ **b** $\frac{3}{4} + \frac{3}{4}$ **c** $\frac{2}{5} + \frac{3}{5}$ **d** $\frac{7}{10} + \frac{9}{10}$

CORE

7 Copy and complete each of these subtractions. Use equivalent fractions to make the denominators the same. Show your working.

a $\dfrac{7}{8} - \dfrac{1}{4}$ **b** $\dfrac{7}{10} - \dfrac{1}{5}$ **c** $\dfrac{3}{4} - \dfrac{1}{2}$ **d** $\dfrac{5}{8} - \dfrac{1}{4}$

e $\dfrac{1}{2} - \dfrac{1}{4}$ **f** $\dfrac{7}{8} - \dfrac{1}{2}$ **g** $\dfrac{9}{10} - \dfrac{1}{2}$ **h** $\dfrac{11}{16} - \dfrac{3}{8}$

You can only add or subtract fractions with different denominators after you have converted them to equivalent fractions with the same denominator.

EXAMPLE 4

i Find $\dfrac{2}{3} + \dfrac{1}{5}$ **ii** Find $2\dfrac{3}{4} - 1\dfrac{5}{6}$

i Note that you can change both fractions to equivalent fractions with a denominator of 15. This is the lowest common multiple of 3 and 5.

This then becomes:

$$\dfrac{2 \times 5}{3 \times 5} + \dfrac{1 \times 3}{5 \times 3} = \dfrac{10}{15} + \dfrac{3}{15}$$
$$= \dfrac{13}{15}$$

ii Split the calculation into $\left(2 + \dfrac{3}{4}\right) - \left(1 + \dfrac{5}{6}\right)$.

This then becomes:

$$= 2 - 1 + \dfrac{3}{4} - \dfrac{5}{6}$$

Note that you can change both fractions to equivalent fractions with a denominator of 12.

$$= 1 + \dfrac{9}{12} - \dfrac{10}{12}$$
$$= 1 - \dfrac{1}{12}$$
$$= \dfrac{11}{12}$$

EXERCISE 3E

1 Complete these calculations. Show your working.

a $\dfrac{1}{3} + \dfrac{1}{5}$ **b** $\dfrac{1}{3} + \dfrac{1}{4}$ **c** $\dfrac{1}{5} + \dfrac{1}{10}$

d $\dfrac{2}{3} + \dfrac{1}{4}$ **e** $\dfrac{3}{4} + \dfrac{1}{8}$ **f** $\dfrac{1}{3} + \dfrac{1}{6}$

g $\dfrac{1}{2} - \dfrac{1}{3}$ **h** $\dfrac{1}{4} - \dfrac{1}{5}$ **i** $\dfrac{1}{5} - \dfrac{1}{10}$

j $\dfrac{7}{8} - \dfrac{3}{4}$

k $\dfrac{5}{6} - \dfrac{3}{4}$

l $\dfrac{5}{6} - \dfrac{1}{2}$

m $\dfrac{5}{12} - \dfrac{1}{4}$

n $\dfrac{1}{3} + \dfrac{4}{9}$

o $\dfrac{1}{4} + \dfrac{3}{8}$

p $\dfrac{7}{8} - \dfrac{1}{2}$

q $\dfrac{3}{5} - \dfrac{8}{15}$

r $\dfrac{11}{12} + \dfrac{5}{8}$

s $\dfrac{7}{16} + \dfrac{3}{10}$

t $\dfrac{4}{9} - \dfrac{2}{21}$

u $\dfrac{5}{6} - \dfrac{4}{27}$

2 Complete these calculations. Show your working.

a $2\dfrac{1}{7} + 1\dfrac{3}{14}$

b $6\dfrac{3}{10} + 1\dfrac{4}{5} + 2\dfrac{1}{2}$

c $3\dfrac{1}{2} - 1\dfrac{1}{3}$

d $1\dfrac{7}{18} + 2\dfrac{3}{10}$

e $3\dfrac{2}{6} + 1\dfrac{9}{20}$

f $1\dfrac{1}{8} - \dfrac{5}{9}$

g $1\dfrac{3}{16} - \dfrac{7}{12}$

h $\dfrac{5}{6} + \dfrac{7}{16} + \dfrac{5}{8}$

i $\dfrac{7}{10} + \dfrac{3}{8} + \dfrac{5}{6}$

j $1\dfrac{1}{3} + \dfrac{7}{10} - \dfrac{4}{15}$

k $\dfrac{5}{14} + 1\dfrac{3}{7} - \dfrac{5}{12}$

3 In a class of children, three-quarters are Chinese, one-fifth are Malay and the rest are Indian. What fraction of the class are Indian?

4 **a** In a class election, half of the people voted for Aminah, one-third voted for Reshma and the rest voted for Peter. What fraction of the class voted for Peter?

b One of the following is the number of people in the class.

    25   28   30   32

How many people are in the class?

3.5 Multiplying and dividing fractions

What is $\dfrac{1}{2}$ of $\dfrac{1}{4}$?

The diagram shows the answer is $\dfrac{1}{8}$.

In mathematics, you always write $\dfrac{1}{2}$ of $\dfrac{1}{4}$ as $\dfrac{1}{2} \times \dfrac{1}{4}$

So you know that $\dfrac{1}{2} \times \dfrac{1}{4} = \dfrac{1}{8}$

To **multiply** fractions, you multiply the **numerators** together and you multiply the **denominators** together.

EXAMPLE 5

Work out $\frac{1}{4} \times \frac{2}{5}$.

$$\frac{1}{4} \times \frac{2}{5} = \frac{1 \times 2}{4 \times 5} = \frac{2}{20} = \frac{1}{10}$$

Sometimes you can simplify by cancelling *before* you multiply.

EXAMPLE 6

Find $\frac{3}{8} \times \frac{5}{9}$

$$\frac{3}{8} \times \frac{5}{9} = \frac{\cancel{3}^{1}}{8} \times \frac{5}{\cancel{9}_{3}} \qquad (3 \text{ is a factor of 3 and 9.})$$

$$= \frac{5}{24} \qquad \begin{matrix}(5 = 1 \times 5)\\(24 = 8 \times 3)\end{matrix}$$

To multiply mixed numbers, first change them to improper fractions.

EXAMPLE 7

Find $1\frac{3}{4} \times 2\frac{1}{2}$

$$1\frac{3}{4} \times 2\frac{1}{2} = \frac{7}{4} \times \frac{5}{2}$$

$$= \frac{35}{8}$$

$$= 4\frac{3}{8}$$

EXERCISE 3F

1 Work these out, leaving each answer in its simplest form. Show your working.

a $\frac{1}{2} \times \frac{1}{3}$ **b** $\frac{1}{4} \times \frac{2}{5}$ **c** $\frac{3}{4} \times \frac{1}{2}$ **d** $\frac{3}{7} \times \frac{1}{2}$

e $\frac{2}{3} \times \frac{4}{5}$ **f** $\frac{1}{3} \times \frac{3}{5}$ **g** $\frac{1}{3} \times \frac{6}{7}$ **h** $\frac{3}{4} \times \frac{2}{5}$

i $\frac{2}{3} \times \frac{3}{4}$ **j** $\frac{1}{2} \times \frac{4}{5}$

CORE

2 Work these out, leaving each answer in its simplest form. Show your working.

a $\dfrac{5}{16} \times \dfrac{3}{10}$ 　　b $\dfrac{9}{10} \times \dfrac{5}{12}$ 　　c $\dfrac{14}{15} \times \dfrac{3}{8}$ 　　d $\dfrac{8}{9} \times \dfrac{6}{15}$

e $\dfrac{6}{7} \times \dfrac{21}{30}$ 　　f $\dfrac{9}{14} \times \dfrac{35}{36}$

3 One-quarter of Lee's stamp collection was given to him by his sister. Unfortunately two-thirds of these were torn. What fraction of his collection was given to him by his sister and were not torn?

4 Bilal eats one-quarter of a cake, and then half of what is left. How much cake is left uneaten?

5 Work these out, giving each answer as a mixed number where possible. Show your working.

a $1\dfrac{1}{4} \times \dfrac{1}{3}$ 　　b $1\dfrac{2}{3} \times 1\dfrac{1}{4}$ 　　c $2\dfrac{1}{2} \times 2\dfrac{1}{2}$ 　　d $1\dfrac{3}{4} \times 1\dfrac{2}{3}$

e $3\dfrac{1}{4} \times 1\dfrac{1}{5}$ 　　f $1\dfrac{1}{4} \times 2\dfrac{2}{3}$ 　　g $2\dfrac{1}{2} \times 5$ 　　h $7\dfrac{1}{2} \times 4$

6 Which is larger, $\frac{3}{4}$ of $2\frac{1}{2}$ or $\frac{2}{5}$ of $6\frac{1}{2}$?

Dividing fractions

Look at the problem $3 \div \frac{3}{4}$. This is like asking, 'How many $\frac{3}{4}$s are there in 3?'

Look at the diagram.

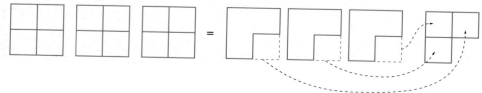

Each of the three whole shapes is divided into quarters. What is the total number of quarters divided by 3?

Can you see that you could fit the four shapes on the right-hand side of the = sign into the three shapes on the left-hand side?

i.e. $3 \div \dfrac{3}{4} = 4$

or $3 \div \dfrac{3}{4} = 3 \times \dfrac{4}{3} = \dfrac{3 \times 4}{3} = \dfrac{12}{3} = 4$

So, to divide by a fraction, you turn the fraction upside down (finding its **reciprocal**), and then multiply.

EXAMPLE 8

Work out $2\frac{1}{2} \div \frac{3}{4}$

$2\frac{1}{2} \div \frac{3}{4} = \frac{5}{2} \times \frac{4}{3}$ (change $2\frac{1}{2}$ to an improper fraction)

$\qquad\quad = \frac{5}{\cancel{2}_1} \times \frac{\cancel{4}^2}{3}$ (2 and 4 have 2 as a common factor)

$\qquad\quad = \frac{10}{3}$

$\qquad\quad = 3\frac{1}{3}$

This means that $3\frac{1}{3} \times \frac{3}{4} = 2\frac{1}{2}$

EXERCISE 3G

1   Work these out, giving your answer as a mixed number where possible. Show your working.

   a $\frac{1}{4} \div \frac{1}{3}$ **b** $\frac{2}{5} \div \frac{2}{7}$ **c** $\frac{4}{5} \div \frac{3}{4}$ **d** $\frac{3}{7} \div \frac{2}{5}$

   e $5 \div 1\frac{1}{4}$ **f** $6 \div 1\frac{1}{2}$ **g** $7\frac{1}{2} \div 1\frac{1}{2}$ **h** $3 \div 1\frac{3}{4}$

   i $1\frac{5}{12} \div 3\frac{3}{16}$ **j** $3\frac{3}{5} \div 2\frac{1}{4}$

2   A grain merchant has only thirteen and a half tonnes in stock. He has several customers who are all ordering three-quarters of a tonne. How many customers can he supply?

3   For a party, Zahar made twelve and a half litres of lemonade. His glasses could each hold $\frac{5}{16}$ of a litre. How many of the glasses could he fill from the twelve and a half litres of lemonade?

4   How many strips of ribbon, each three and a half centimetres long, can I cut from a roll of ribbon that is fifty-two and a half centimetres long?

5   Joe's stride is three-quarters of a metre long. How many strides does he take to walk the length of a bus twelve metres long?

6   Work these out, giving your answers as a mixed number where possible.

   a $2\frac{2}{9} \times 2\frac{1}{10} \times \frac{16}{35}$ **b** $3\frac{1}{5} \times 2\frac{1}{2} \times 4\frac{3}{4}$

   c $1\frac{1}{4} \times 1\frac{2}{7} \times 1\frac{1}{6}$ **d** $\frac{18}{25} \times \frac{15}{16} \div 2\frac{2}{5}$

   e $\left(\frac{2}{5} \times \frac{2}{5}\right) \times \left(\frac{5}{6} \times \frac{5}{6}\right) \times \left(\frac{3}{4} \times \frac{3}{4}\right)$ **f** $\left(\frac{4}{5} \times \frac{4}{5}\right) \div \left(1\frac{1}{4} \times 1\frac{1}{4}\right)$

CORE

Why this chapter matters

Life is full of pairs: up and down, hot and cold, left and right, light and dark, rough and smooth, to name a few. One important pairing in maths is positive and negative.

So far you have worked mostly with positive numbers, carrying out calculations with them and using them in real-life problems. But negative numbers are also important, both in maths and everyday life. Positive and negative numbers are called **directed** numbers. You can think of the + and − showing which direction they move from 0. This is important in many situations:

A negative number on a bank statement will show how much money you have spent above what you have in your bank account.

On the Celsius temperature scale zero is known as 'freezing point'. In many places temperatures fall below freezing point. You use negative numbers to represent these temperatures.

Jet pilots experience g-forces when their aircraft accelerates or decelerates quickly. Negative g-forces can be felt when an object accelerates downwards very quickly and they are represented by negative numbers.

$$5 - 9 = -4$$

When a bigger number is taken from a smaller one, the result is a negative number.

In lifts, negative numbers are used to represent floors below ground level.

Sea level can be given the value 'zero'. Mountains are described as being 'above sea level' and ocean floors as 'below sea level'. This means that depths under the sea are given using negative numbers.

As you can see, negative numbers are just as important as positive numbers and you will encounter them in your everyday life.

Directed numbers

Topics	Level	Key words

1 Introduction to directed numbers — CORE — negative, positive, directed numbers

2 Everyday use of directed numbers — CORE — profit, loss

3 The number line — CORE — number line, less than, more than, greater than

4 Adding and subtracting directed numbers — CORE

5 Multiplying and dividing directed numbers — CORE

In this chapter you will learn how to:

CORE

- Use directed numbers in practical situations. (C1.4 and E1.4)
- Add, subtract, multiply and divide directed numbers. (C1.4 and E1.4)

Negative numbers are numbers *below* zero. You meet negative numbers when the temperature falls below freezing (0 °C).

The diagram below shows a thermometer with negative temperatures. The temperature is –3 °C. This means the temperature is three degrees below zero.

The number line below shows **positive** and negative numbers.

Positive and negative numbers together are called **directed numbers**.

EXERCISE 4A

1 Write down the temperature shown by each thermometer.

a

b

c

d

e

2

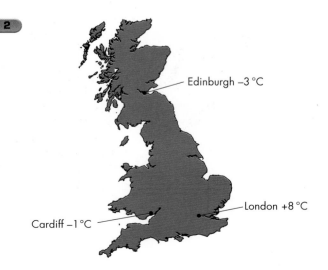

Edinburgh –3 °C

London +8 °C

Cardiff –1 °C

a How many degrees lower is the temperature in Edinburgh than in London?

b How many degrees higher is the temperature in London than in Cardiff?

CORE

3 The instructions on a bottle of de-icer say that it will stop water freezing down to –12 °C. The temperature is –4 °C.

How many more degrees does the temperature need to fall before the de-icer stops working?

4.2 Everyday use of directed numbers

There are many other situations where directed numbers are used. Here are three examples.

- When +15 m means 15 metres *above* sea level, then –15 m means 15 metres *below* sea level.

- When +2 h means 2 hours *after* midday, then –2 h means 2 hours *before* midday.

- When +$60 means a **profit** of $60, then –$60 means a **loss** of $60.

You also meet negative numbers on graphs, and you may already have plotted coordinates with negative numbers.

On bank statements and bills a negative number means you owe money. A positive number means they owe you money.

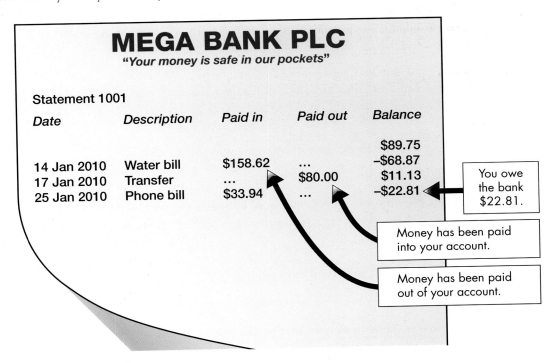

CORE

EXERCISE 4B

Copy and complete each statement.

1 If +$5 means a profit of five dollars, then …… means a loss of five dollars.

2 If +200 m means 200 metres above sea level, then …… means 200 metres below sea level.

3 If –100 m means one hundred metres below sea level, then +100 m means one hundred metres …… sea level.

4 If +5 h means 5 hours after midday, then …… means 5 hours before midday.

5 If +2 °C means two degrees above freezing point, then …… means two degrees below freezing point.

6 If +70 km means 70 kilometres north of the equator, then …… means 70 kilometres south of the equator.

7 If 10 minutes before midnight is represented by –10 minutes, then five minutes after midnight is represented by …… .

8 If a car moving forwards at 10 kilometres per hour is represented by +10 km/h, then a car moving backwards at 5 kilometres per hour is represented by …… .

9 In an office building, the third floor above ground level is represented by +3. So, the second floor below ground level is represented by …… .

10 The temperature on three days in Moscow was –7 °C, –5 °C and –11 °C.

   a Which temperature is the lowest?

   b What is the difference in temperature between the coldest and the warmest days?

11 A thermostat is set at 16 °C.

The temperature in a room at 1.00 am is –2 °C.

The temperature rises two degrees every 6 minutes.

At what time is the temperature on the thermostat reached?

Look at the **number line**.

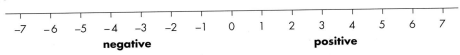

Notice that the **negative** numbers are to the left of 0 and the **positive** numbers are to the right of 0.

Numbers to the right of any number on the number line are always bigger than that number.

Numbers to the left of any number on the number line are always smaller than that number.

So, for example, you can see from a number line that:

2 is *smaller* than 5 because 2 is to the left of 5.

You can write this as 2 < 5.

–3 is *smaller* than 2 because –3 is to the *left* of 2.

You can write this as –3 < 2.

7 is *bigger* than 3 because 7 is to the *right* of 3.

You can write this as 7 > 3.

–1 is *bigger* than –4 because –1 is to the *right* of –4.

You can write this as –1 > –4.

Reminder: The **inequality** signs:

< means 'is **less than**'

> means 'is **greater than**' or 'is **more than**'

EXERCISE 4C

1 Copy each statement and put the correct symbol (< or >) in each space.

a –1 3   **b** 3 2   **c** –4 ...... –1   **d** –5 ...... –4

e 1 ...... –6   **f** –3 0   **g** –2 ...... –1   **h** 2 ...... –3

i 5 ...... –6   **j** 3 4   **k** –7 ...... –5   **l** –2 ...... –4

2

$$-1 \quad -\frac{3}{4} \quad -\frac{1}{2} \quad -\frac{1}{4} \quad 0 \quad \frac{1}{4} \quad \frac{1}{2} \quad \frac{3}{4} \quad 1$$

Copy each statement and put the correct symbol in each space.

a $\frac{1}{4}$ ...... $\frac{3}{4}$ **b** $-\frac{1}{2}$ 0   **c** $-\frac{3}{4}$ ...... $\frac{3}{4}$

d $\frac{1}{4}$ ...... $-\frac{1}{2}$ **e** -1 ...... $\frac{3}{4}$ **f** $\frac{1}{2}$ 1

CORE

CORE

3 Copy these number lines and fill in the missing numbers.

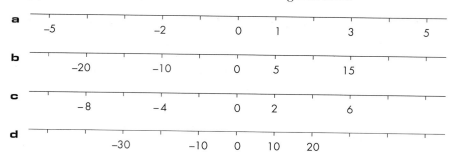

a –5   –2   0   1   3   5

b –20   –10   0   5   15

c   –8   –4   0   2   6

d   –30   –10 0 10 20

4 Here are some temperatures.

   2 °C   –2 °C   –4 °C   6 °C

Copy and complete the weather report, using these temperatures.

> The hottest place today is Barnsley with a temperature of ____,
> while in Eastbourne a ground frost has left the temperature just
> below zero at ____. In Bristol it is even colder at ____. Finally,
> in Tenby the temperature is just above freezing at ____.

4.4 Adding and subtracting directed numbers

Adding and subtracting positive numbers

These two operations can be illustrated on a thermometer scale.

- **Adding** a positive number moves the marker up the thermometer scale.

  For example,

  $-2 + 6 = 4$

- **Subtracting** a positive number moves the marker *down* the thermometer scale.

  For example,

  $3 - 5 = -2$

EXAMPLE 1

The temperature at midnight was 2 °C but then it fell by five degrees. What was the new temperature?

Falling five degrees means the calculation is $2 - 5$, which is equal to -3. So, the new temperature is -3 °C.

EXERCISE 4D

1   Find the answer to each of these.

a $2° - 4° =$	**b** $4° - 7° =$	**c** $3° - 5° =$	**d** $1° - 4° =$
e $6° - 8° =$	**f** $5° - 8° =$	**g** $-2 + 5 =$	**h** $-1 + 4 =$
i $-4 + 3 =$	**j** $-6 + 5 =$	**k** $-3 + 5 =$	**l** $-5 + 2 =$
m $-1 - 3 =$	**n** $-2 - 4 =$	**o** $-5 - 1 =$	**p** $3 - 4 =$
q $2 - 7 =$	**r** $1 - 5 =$	**s** $-3 + 7 =$	**t** $5 - 6 =$
u $-2 - 3 =$	**v** $2 - 6 =$	**w** $-8 + 3 =$	**x** $4 - 9 =$

2   At 5 am the temperature in Lisbon was -4 °C. At 11 am the temperature was 3 °C.

a   By how many degrees did the temperature rise?

b   The temperature in Madrid was two degrees lower than in Lisbon at 5 am. What was the temperature in Madrid at 5 am?

3   Here are five numbers.

4   7   8   2   5

a   Use two of the numbers to make a calculation with an answer of -6.

b   Use three of the numbers to make a calculation with an answer of -1.

c   Use four of the numbers to make a calculation with an answer of -18.

d   Use all five of the numbers to make a calculation with an answer of -12.

4   A submarine is 600 metres below sea level.

A radar system can detect submarines down to 300 metres below sea level.

To avoid detection safely, the submarine captain keeps the submarine 50 metres below the level of detection.

How many metres can the submarine climb to be safe from detection?

CORE

Adding and subtracting negative numbers

To *subtract a negative number* …

… treat the $--$ as $+$

For example: $4 - (-2) = 4 + 2 = 6$

To *add a negative number* …

… treat the $+-$ as $-$

For example: $3 + (-5) = 3 - 5 = -2$

Using your calculator

Calculations involving negative numbers can be done by using the $\boxed{(-)}$ key.

EXAMPLE 2

Work out $-6 - -2$.

Press $\boxed{(-)}$ $\boxed{6}$ $\boxed{-}$ $\boxed{(-)}$ $\boxed{2}$ $\boxed{=}$

The answer should be -4.

EXERCISE 4E

1 Write down the answer to each calculation then check your answers on a calculator.

a $-3 - 5 =$	**b** $-2 - 8 =$	**c** $-5 - 6 =$	**d** $6 - 9 =$
e $5 - 3 =$	**f** $3 - 8 =$	**g** $-4 + 5 =$	**h** $-3 + 7 =$
i $-2 + 9 =$	**j** $-6 + -2 =$	**k** $-1 + -4 =$	**l** $-8 + -3 =$
m $5 - -6 =$	**n** $3 - -3 =$	**o** $6 - -2 =$	**p** $3 - -5 =$
q $-5 - -3 =$	**r** $-2 - -1 =$	**s** $-4 - 5 =$	**t** $2 - 7 =$

2 What is the *difference* between the temperatures in each pair?

a 4 °C and −6 °C   **b** −2 °C and −9 °C   **c** −3 °C and 6 °C

3 Find what you have to *add to* 5 to get:

a 7   **b** 2   **c** 0

d −2   **e** −5   **f** −15

4 Find what you have to *subtract from* 4 to get:

a 2   **b** 0   **c** 5

d 9   **e** 15   **f** −4

CORE

CORE

5 Find what you have to *add to* –5 to get:

    a 8   **b** –3

    c 0   **d** –1

    e 6   **f** –7

6 Find what you have to *subtract from* –3 to get:

    a 7   **b** 2

    c –1   **d** –7

    e –10   **f** 1

7 You have these cards.

–9 **–8** **–4** **0** **+1** **+3** **+5**

    a Which card should you choose to make the answer to this sum as large as possible? What is the answer?

    b Which card should you choose to make the answer to part **a** as small as possible? What is the answer?

    c Which card should you choose to make the answer to this subtraction as large as possible? What is the answer?

    d Which card should you choose to make the answer to part **c** as small as possible? What is the answer?

8 The thermometer in a car is inaccurate by up to two degrees.

An ice alert warning comes on at 3 °C, according to the thermometer temperature.

If the actual temperature is 2 °C, will the alert come on?

Explain how you decide.

9 Two integers have a sum of 5.

One of the numbers is negative.

The other number is even.

What are the two numbers when the negative number is as large as possible?

The rules for multiplying and dividing two directed numbers are very easy.

- When the signs of the two numbers are the *same*, the answer is *positive*.
- When the signs of the two numbers are *different*, the answer is *negative*.

Here are some examples.

$$2 \times 4 = 8 \qquad 12 \div -3 = -4 \qquad -2 \times -3 = 6 \qquad -12 \div -3 = 4$$

A common error is to confuse, for example, -3^2 and $(-3)^2$.

$$-3^2 = -3 \times 3 = -9$$

but:

$$(-3)^2 = -3 \times -3 = +9.$$

So, this means that if a variable is introduced, for example, $a = -5$, the calculation would be:

$$a^2 = -5 \times -5 = +25$$

EXAMPLE 3

$a = -2$ and $b = -6$

Work out these numbers.

a a^2 **b** $a^2 + b^2$ **c** $b^2 - a^2$ **d** $(a - b)^2$

a $a^2 \qquad = -2 \times -2 = +4$

b $a^2 + b^2 \qquad = +4 + -6 \times -6 = 4 + 36 = 40$

c $b^2 - a^2 \qquad = 36 - 4 = 32$

d $(a - b)^2 \qquad = (-2 - -6)^2 = (-2 + 6)^2 = (4)^2 = 16$

EXERCISE 4F

1 Write down the answers.

     a -3×5 **b** -2×7 **c** -4×6

     d -2×-3 **e** -7×-2 **f** $-12 \div -6$

     g $-16 \div 8$ **h** $24 \div -3$ **i** $16 \div -4$

     j $-6 \div -2$ **k** 4×-6 **l** 5×-2

     m 6×-3 **n** -2×-8 **o** -9×-4

     p $24 \div -6$ **q** $12 \div -1$ **r** $-36 \div 9$

     s $-14 \div -2$ **t** $100 \div 4$ **u** -2×-9

CORE

2 Write down the answers.

a $-3 + -6$ **b** -2×-8 **c** $2 + -5$

d 8×-4 **e** $-36 \div -2$ **f** -3×-6

g $-3 - -9$ **h** $48 \div -12$ **i** -5×-4

j $7 - -9$ **k** $-40 \div -5$ **l** $-40 + -8$

m $4 - -9$ **n** $5 - 18$ **o** $72 \div -9$

p $-7 - -7$ **q** $8 - -8$ **r** 6×-7

3 What number do you multiply by -3 to get each number?

a 6 **b** -90 **c** -45

d 81 **e** 21

4 What number do you divide -36 by to get each number?

a -9 **b** 4 **c** 12

d -6 **e** 9

5 Evaluate these.

a $-6 + (4 - 7)$ **b** $-3 - (-9 - -3)$ **c** $8 + (2 - 9)$

6 Evaluate these.

a $4 \times (-8 \div -2)$ **b** $-8 - (3 \times -2)$ **c** $-1 \times (8 - -4)$

7 What do you get if you divide -48 by each number?

a -2 **b** -8 **c** 12 **d** 24

8 Write down six different multiplications that give the answer -12.

9 Write down six different divisions that give the answer -4.

10 Put these calculations in order from lowest to highest.

-5×4 $-20 \div 2$ $-16 \div -4$ 3×-6

11 $x = -2$, $y = -3$ and $z = -4$. Work out these numbers.

a x^2 **b** $y^2 + z^2$ **c** $z^2 - x^2$ **d** $(x - y)^2$

Why this chapter matters

> The squares and square roots of numbers are important tools in mathematics and mathematicians have helped us to use them by inventing notation.

We often need to multiply a number by itself one or more times, for example, when finding the area of a square or the volume of a cube.

You have seen in Chapter One how you can write 5^2 instead of 5×5. You can also show $5 \times 5 \times 5$ by using 5^3. This short cut is called **index notation** (see Chapter 18).

The notation for the **square root** of a number (e.g. $\sqrt{25}$) is even more convenient. Without this, you would have to write 'the number which multiplies by itself to make 25'. The sign for the **cube root** of numbers is $\sqrt[3]{}$.

The root signs are especially convenient when the roots are hard to work out and difficult to express accurately. **Square numbers** such as 4, 9, 16 and 25 have whole numbers as their square roots but most numbers have fractions. If the fractions are expressed as decimals they are sometimes recurring (that is, they never end) which means that they can never be written down accurately.

The notation we now use was only introduced in the sixteenth century (CE). One of the first people to use it in print was a German mathematician called Christoph Rudolff. The notation was simple and easy to understand and was soon widely used.

Ways of working out square roots have been developed by mathematicians over the centuries. This Babylonian tablet showing how to calculate the square root of two is 2500 years ago.

Notation can be used in calculations to represent the square roots of numbers without actually having to work them out or write them down. This makes it possible to carry out more sophisticated and accurate calculations.

Squares and cubes

opics	Level	Key words
1 Squares and square roots	**CORE**	square, square root
2 Cubes and cube roots	**CORE**	cube, cube root
3 Exponential growth and decay	**EXTENDED**	exponential growth, exponential decay

In this chapter you will learn how to:

CORE	EXTENDED
Calculate: • squares • square roots • cubes • cube roots of numbers. (C1.3 and E1.3)	• Use exponential growth and decay in relation to population and finance. (E1.17)

You know that $6 \times 6 = 36$.

This can also be written as 6^2, which you say as 'six squared'.

Another way to describe this is to say: '6 is the **square root** of 36'.
You can write the square root of 36 as $\sqrt{36}$.

In fact 36 has *two* square roots.

$$6 \times 6 = 36$$
$$\text{and} \quad -6 \times -6 = 36$$

So the square roots of 36 are 6 and –6 or $\sqrt{36}$ and $-\sqrt{36}$.

You can **square** decimals as well as whole numbers.

$$2.5^2 = 6.25 \text{ so } \sqrt{6.25} = 2.5.$$

Most calculators have a 'square' button $\boxed{x^2}$ and a 'square root' button $\boxed{\sqrt{}}$. Check that you know how to use them.

The calculator will show the positive square root, e.g. $\sqrt{36}$.

EXERCISE 5A

CORE

1 Find the value of each expression.

a 7^2

b 10^2

c 1.2^2

d 2.5^2

e 16^2

f 20^2

g 3.1^2

h 4.5^2

i $(-3)^2$

j $(-8)^2$

k 0.5^2

l $(-0.5)^2$

2 Write down the two square roots of each number.

a 9

b 100

c 121

d 1.44

e 400

f 12.25

g 1

h 10 000

3 Find these square roots.

a $\sqrt{25}$ **b** $\sqrt{36}$

c $\sqrt{100}$ **d** $\sqrt{49}$

e $\sqrt{64}$ **f** $\sqrt{2.25}$

g $\sqrt{30.25}$ **h** $\sqrt{1.44}$

i $\sqrt{400}$ **j** $\sqrt{0.25}$

4 Write down the value of each of these. You will need to use your calculator for some of them. Look for the $\boxed{\sqrt{\square}}$ key.

a 9^2 **b** $\sqrt{1600}$

c 10^2 **d** $\sqrt{196}$

e 6^2 **f** $\sqrt{225}$

g 7^2 **h** $\sqrt{144}$

i 5^2 **j** $\sqrt{441}$

5 Write down each square root. You will need to use your calculator.

a $\sqrt{576}$ **b** $\sqrt{961}$

c $\sqrt{2025}$ **d** $\sqrt{1600}$

e $\sqrt{4489}$ **f** $\sqrt{10\,201}$

g $\sqrt{12.96}$ **h** $\sqrt{42.25}$

i $\sqrt{193.21}$ **j** $\sqrt{492.84}$

6 Put these in order, starting with the smallest value.

3^2 $\sqrt{90}$ $\sqrt{50}$ 4^2

7 **a** Explain how you know that $\sqrt{40}$ is between 6 and 7.

   b Use your calculator to find $\sqrt{40}$. Write down all the decimal places on your display.

8 Between which two consecutive whole numbers does the square root of 20 lie?

9 Find two consecutive integers between which each of these square roots lies.

a $\sqrt{68}$ **b** $\sqrt{96}$

c $\sqrt{155}$ **d** $\sqrt{250}$

CORE

10 Use these number cards to make the calculation below correct.

$$\boxed{1} \quad \boxed{2} \quad \boxed{3} \quad \boxed{4} \quad \boxed{8}$$

$$\sqrt{\boxed{}\,\boxed{}\,\boxed{}} = \boxed{}\,\boxed{}$$

11 A square wall in a kitchen is being tiled.

Altogether it needs 225 square tiles.
How many tiles are there in each row?

5.2 Cubes and cube roots

You can show that $6 \times 6 \times 6 = 216$.

216 is 'six cubed'. A **cube** of a number is formed when you multiply the number by itself and then by itself again. You can write this as 6^3.

$6 \times 6 \times 6 = 6^3$ or 'six cubed'

Another way to describe this is to say: '6 is the **cube root** of 216'.

The symbol for a cube root is $\sqrt[3]{}$ so the cube root of 216 can also be written as $\sqrt[3]{216}$.

Many calculators have a button for cubes **x^3** and cube roots **$\sqrt[3]{}$**. Check that you know how to use them.

EXAMPLE 1

Find the cube roots of 64 and −64.

$4 \times 4 \times 4 = 64$ so $\sqrt[3]{64} = 4$

$-4 \times -4 \times -4 = -64$ so $\sqrt[3]{-64} = -4$

Notice that 64 and −64 have just one cube root each.

EXERCISE 5B

1 Work out these cubes.

  a 2^3 **b** 3^3 **c** 8^3

  d 10^3 **e** 1.1^3 **f** 2.5^3

  g $(-3)^3$ **h** $(-5)^3$ **i** 20^3

  j 4.1^3 **k** $(-4.1)^3$

2 Find these cube roots.

  a $\sqrt[3]{8}$ **b** $\sqrt[3]{125}$ **c** $\sqrt[3]{729}$

  d $\sqrt[3]{1}$ **e** $\sqrt[3]{27}$ **f** $\sqrt[3]{-27}$

  g $\sqrt[3]{1000}$ **h** $\sqrt[3]{3.375}$ **i** $\sqrt[3]{91.125}$

  j $\sqrt[3]{0.125}$

3 $4^3 = 64$ and $5^3 = 125$ so $\sqrt[3]{100}$ is between 4 and 5.

  Find two consecutive integers between which these cube roots lie.

  a $\sqrt[3]{200}$ **b** $\sqrt[3]{300}$

  c $\sqrt[3]{500}$ **d** $\sqrt[3]{-500}$

4 Here are four numbers.

  2^3 3^2 $\sqrt{81}$ $\sqrt[3]{729}$

  Which is the odd one out and why?

5 Find two different positive integers, A and B, such that $A^2 = B^3$.

6 Put these numbers in order, smallest first.

  2.5^3 $\sqrt{225}$ 4^2 $\sqrt[3]{2000}$

7 Copy and complete this table.

Number	Square	Cube
	100	
5		
		64
11		
	81	

8 Write these numbers in order with the smallest first.

  $\sqrt{0.8}$ $\sqrt[3]{0.8}$ 0.8^2 0.8^3

The number of visitors to a wildlife reserve in January is 600.

The number of visitors *doubles* every month.

In February the number of visitors is 600 × 2 = 1200

In March the number of visitors is 1200 × 2 = 2400

In April the number of visitors is 2400 × 2 = 4800

The number is multiplied by 2 from one month to the next.

The numbers for the four months can be written as a sequence.

600, 600 × 2, 600 × 2^2, 600 × 2^3

This is an example of **exponential growth**.

EXAMPLE 2

The cost of food increases by 10% each year.

The cost of a basket of food now is $85.

a   Work out the cost after one year.

b   Work out the cost after two years.

c   Work out the cost after three years.

    a   The multiplier to increase a number by 10% is 1.1   100% + 10% = 110% = 1.1
       The cost after one year is $85 × 1.1 = $93.50.

    b   The cost after two years is $85 × 1.1^2 or $93.50 × 1.1 = $102.85.

    c   The cost after three years is $85 × 1.1^3 or $102.85 × 1.1 = $113.14.

Raj buys a car for $56 000.

The value of his car *decreases* by 25% each year.

The multiplier for a decrease of 25% is 0.75.   100% − 25% = 75% = 0.75

The value of the car after one year is $56 000 × 0.75 = $42 000

The value of the car after two years is $42 000 × 0.75 = $31 500

The value of the car after three years is $31 500 × 0.75 = $23 625

This is an example of **exponential decay**. This is the result when the multiplier from one number to the next is less than one.

EXERCISE 5C

1 The number of bacteria on a plate is 5 million.

   a The number doubles every hour. How many bacteria are there after:

   i one hour   **ii** two hours   **iii** three hours?

   b Repeat part a, assuming that the number of bacteria now *triples* every hour.

2 There are 4000 people in a new town.

   The population increases by 50% each year.

   Work out what the population will be in:

   a one year   **b** two years   **c** three years   **d** four years.

3 There are 12 000 fish in a lake.

   Because of pollution, the number of fish halves every year.

   Work out the number of fish after:

   a one year

   b three years

   c five years.

EXTENDED

4 The pollution in the lake in question **3** has been calculated wrongly. The number of fish is actually reduced by 60% each year.

Work out the number of fish after:

a one year **b** three years **c** five years.

5 The population of a country is 20.0 million.

The population increases by 9% every 10 years.

Copy and complete this table. Round the numbers to one decimal place.

Number of years	0	10	20	30
Population (millions)	20.0			

6 Now suppose the population in question **5** reduces by 9% every year.

7 Anna has savings of $5000 in her account.

The value of her savings increases by 6% each year.

a How much will the savings be worth after four years?

b How long will it take until the savings are worth $7000?

8 Sami earns $80 000 per year.

He has been promised a pay rise of 25% each year.

a How much will he be earning after one year?

b How much will he be earning after four years?

9 The value of a flat is $150 000.

The value increases by 20% each year.

Remi says, 'The value will double in five years to $300 000 because 5 × 20% is 100%.'

Show that Remi is not correct. Then work out the correct value.

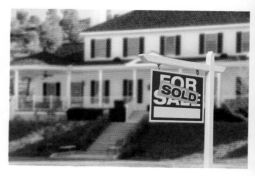

10 Carla saves $10 in January.

She says, 'Each month I shall save twice as much as I saved the month before.'

a How much will she save in February?

b How much will she save in December?

11 There are 4000 elephants in a reserve.

The numbers are reducing by a third each year.

   a Work out the number of elephants after three years.

   b Work out the number of elephants after six years.

12 In 2013 the population of Indonesia was 250 million.

The population was increasing by 1.2% per year.

Use these figures to estimate the population in 2020.

13 In Brazil in the early 1990s prices were increasing by 80% each month.

Suppose the price of a loaf of bread is $10.

   a Work out the price after one month.

   b Work out the price after two months.

   c Work out the price after six months.

   d Work out the price after a year.

EXTENDED

Sets are collections of objects. Set notation gives us a way of seeing the logical connection between sets. The mathematics of sets is very useful in designing computer circuits and electronic components.

Alice in Wonderland

You have probably heard of Alice in Wonderland. Did you know that the author, Lewis Carroll, was actually a lecturer in mathematics at the University of Oxford, England, in the nineteenth century? His real name was Charles Dodgson.

He also wrote a mathematics book called *Symbolic Logic*. Here is a problem from it.

1. All humming birds are richly coloured.
2. No large birds can live on honey.
3. Birds that do not live on honey are dull in colour.

What conclusion follows?

Symbolic Logic was about how to write sentences in symbols so that conclusions could be seen more easily. You can use set notation and Venn diagrams to do this and you will learn about them in this chapter.

Venn diagrams were invented by the logician John Venn and may be the only mathematical invention to be celebrated in a stained glass window!

A Venn diagram in glass

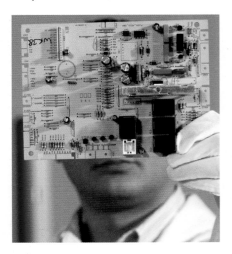

Symbolic logic has a very modern application in the design of computer circuits and the construction of electronic components. When engineers talk about NAND and NOR gates they are using ideas that were first developed in the nineteenth century for very different purposes.

Ordering and set notation

Topics	Level	Key words
1 Inequalities	**CORE**	equals, greater than, less than
2 Sets	**EXTENDED**	set, element, subset, empty set, proper subset
3 Venn diagrams	**EXTENDED**	universal set, Venn diagram, complement, intersection, union

In this chapter you will learn how to:

CORE	EXTENDED
Order quantities by magnitude and demonstrate familiarity with the symbols: $=, \neq, >, <, \geq, \leq$. (C1.6 and E1.6)	Use language, notation and Venn diagrams to describe sets and represent relationships between sets: ● Definition of sets, e.g. $A = \{x: x \text{ is a natural number}\}$ $B = \{(x, y): y = mx + c\}$ $C = \{x: a \leq x \leq b\}$ $D = \{a, b, c, \ldots\}$. (E1.2) ● Notation Number of elements in set A $n(A)$ "…is an element of…" $\in$ "…is not an element of…" $\notin$ Complement of set A A' The empty set $\varnothing$ Universal set $\mathscr{E}$ A is a subset of B $A \subseteq B$ A is a proper subset of B $A \subset B$ A is not a subset of B $A \nsubseteq B$ A is not a proper subset of B $A \not\subset B$ Union of A and B $A \cup B$ Intersection of A and B $A \cap B$. (E1.2)

You need to know the meaning of these symbols:

$=$ **equals**

$\neq$ does not equal

$>$ is **greater than**

$<$ is **less than**

$\geqslant$ is equal to or greater than

$\leqslant$ is equal to or less than.

> **ADVICE AND TIPS**
>
> The arrow tips in $<$ or $>$ signs point towards the smaller numbers

EXAMPLE 1

n is a positive integer and $n < 6$. What are the possible values of n?

Because n is positive the possible values are 1, 2, 3, 4 and 5. 6 is not included because the sign means 'less than'. If the question said $n \leqslant 6$ then 6 would be included.

Sometimes two inequalities are used together.

EXAMPLE 2

Find the possible values of x if x is an integer and $-3 < x \leqslant 3$.

This means that x is a whole number between -3 and 3.

The possible values of x are -2, -1, 0, 1, 2, 3. Note that -3 is not in the list but 3 is included because $\leqslant 3$ means x can be 'less than or equal to' 3.

EXERCISE 6A

1   Here are three symbols: $< = >$.

Put the correct symbol between the numbers in each pair.

   a   3.5 … 3.15   **b**   180 cm … 2m

   c 5×7 … 6×6 **d**   5km … 5000m

   e $\frac{1}{3}$ of 27 … $\frac{3}{4}$ of 12   **f** 4^2 … 8

   g 10×10 … 8×12 **h** $\sqrt{64}$ … 3^2

2   Here are three fractions: $\frac{1}{3}, \frac{3}{5}, \frac{1}{2}$.

Use them to fill in the gaps below to list them in order, smallest first.

… $<$ … $<$ …

3 d is the number scored when a normal six-sided dice is thrown.

List the possible values of d in each case.

a $d \geqslant 4$　　　　**b** $d < 3$　　　　**c** $d > 5$　　　　**d** $d \leqslant 5$

e $2 \leqslant d \leqslant 4$　　　**f** $3 < d < 6$　　　**g** $1 \leqslant d < 4$　　　**h** $5 < d \leqslant 6$

4 The table shows whether babies of a particular age are underweight or overweight.

Mass < 6.5 kg	Underweight
6.5 kg $\leqslant$ mass $\leqslant 8.5$ kg	Normal
Mass > 8.5 kg	Overweight

Are babies of these masses underweight, overweight or normal?

a 6.3 kg　　　　**b** 9.3 kg　　　　**c** 7.8 kg　　　　**d** 8.5 kg

5 e is an even number and $20 \leqslant e < 30$ and $e \neq 24$. List the possible values of e.

6 In a bag of balls each ball has a number n where $1 \leqslant n \leqslant 49$.

a What is the largest number on a ball?

b The number on the first ball is a multiple of 5 and $n > 40$.
What are the possible values of n?

c The number on the second ball is a multiple of 3 and $n \leqslant 10$.
What are the possible values of n?

d For the third ball $15 < n \leqslant 20$.
What are the possible values of n?

7 True or false? State which in each case.

a $3 \neq -3$　　　　**b** $-3 < -5$　　　　**c** $1.99 < 2$

d $2 < \sqrt{5} < 3$　　**e** $20^2 \leqslant 300$　　**f** 200 minutes $\geqslant$ 3 hours

8 List all the possible values for an integer x in the following cases.

a $5 < x < 9$　　　　**b** $26 \leqslant x \leqslant 28$　　　**c** $-8 < x \leqslant -4$

d $-2 \leqslant x < 2$　　　**e** $17 < x < 18$　　　**f** $32.5 \leqslant x < 33.5$

A **set** is any collection of items. They could be numbers, objects or symbols.

The items in a set are called **elements**. The elements in a set can be listed or described inside brackets like this { }.

The elements are separated by commas, e.g. {vowels in English} = {a, e, i, o, u}.

You will usually use capital letters to stand for sets, e.g. N = {1, 3, 5, 7, 9}.

You should use dots to show when you cannot list all the elements, e.g. {even numbers} = {2, 4, 6,...}.

You can describe sets using inequality symbols e.g. {integer x : $5 \leqslant x \leqslant 10$} = {5, 6, 7, 8, 9, 10} can be read as 'the set of integers between 5 and 10 inclusive'.

Describing elements

Suppose A = { 2, 4, 6, 8, 10} and B = {1, 2, 3, 4, 5, 6}.

8 is an element of A but not of B. This can be expressed as:

$8 \in A$ (8 is an element of A)

$8 \notin B$ (8 is *not* an element of B)

The number of elements in a set can be shown as n(set), e.g. the number of elements in set X is $n(X)$. So for sets A and B above, $n(A)$ = 5 and $n(B)$ = 6.

Suppose C = {6, 8, 10}. All the elements of C are also elements of A. Then you can say that C is a **subset** of A and write this as:

$C \subseteq A$.

> **ADVICE AND TIPS**
>
> $\subseteq$ is a bit like the $\leqslant$ sign we use with numbers.

Suppose P = {red, blue, yellow}. P has *eight* subsets:

{red, blue}, {red, yellow} and {blue, yellow} have two elements.

{red}, {blue}, {yellow} have one element.

You also include the **empty set**, $\varnothing$, which has no elements and P itself is the final subset.

All the subsets *except P* are called **proper subsets** and the symbol to show a proper subset is $\subset$.

So {red} $\subset P$ ({red} is a proper subset of P)

but $P \not\subset P$ (P is not a proper subset of P)

> **ADVICE AND TIPS**
>
> $\subset$ is a bit like the $<$ sign we use with numbers.

EXERCISE 6B

1 Describe these sets in words.

a {Monday, Tuesday, Wednesday...}

b {1, 3, 5, 7...}

c {Mercury, Venus, Earth, Mars...}

d {North, South, East, West}

e {1, 2, 3, 4, 5, 6}

2 List the elements of each of these sets.

    a $A = \{\text{integer } x: 5 < x < 10\}$ **b** $B = \{\text{factors of } 12\}$

    c $C = \{\text{prime numbers between 20 and 30}\}$ **d** $D = \{x: x^2 = 9\}$

3 Using the sets in question **2**, find $n(A)$, $n(B)$, $n(C)$ and $n(D)$.

4 $M = \{\text{months of the year}\}$

    a   Write down $n(M)$.   **b**   List a subset of M with three elements.

5 $X = \{6, 7, 8, 9\}$

    a   List all the subsets of X with three elements.

    b   List all the subsets of X with one element.

6 $Y = \{\text{left, right}\}$

    a   List the four subsets of Y.   **b**   Which subset is not a proper subset?

7 A set has six elements.

    a   How many subsets does it have with one element?

    b   How many subsets does it have with five elements?

8 Here are five sets.

$A = \{a, b, d, g, l, m\}$
$B = \{d, l, m, p, r\}$
$C = \{a, m, p\}$
$D = \{a, d, l, m\}$
$E = \{d, l, p\}$

    a   From the list identify:

        i   a proper subset of A **ii**   a proper subset of B.

    b X is one of the sets shown and $n(X) = 5$. Which set is X?

    c θ stands for an element with these properties:

         $\theta \notin A$ $\theta \in B$ $\theta \in C$ $\theta \notin D$ $\theta \in E$

    What letter does θ stand for?

The set that contains all the elements of a number of subsets is called a **universal set**, $\mathcal{E}$.

Take the universal set of positive single digit integers:

$\mathcal{E} = \{1, 2, 3, 4, 5, 6, 7, 8, 9\}$

This includes several subsets including:

$A = \{x: x \text{ is even}\}$

$B = \{x: x \leqslant 5)$

You can show these sets in a **Venn diagram**.

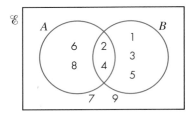

Each element of the universal set is in A, in B, in both, or in neither.

A' is the **complement** of set A and has the elements in $\mathcal{E}$ that are *not* in A:

$A' = \{1, 3, 5, 7, 9\}$

$A \cap B$ is the **intersection** of A and B and contains elements that are in A *and* B:

$A \cap B = \{2, 4\}$

$A \cup B$ is the **union** of A and B and contains elements that are in A *or* B or both:

$A \cup B = \{1, 2, 3, 4, 5, 6, 8\}$

$(A \cup B)'$ is the complement of $A \cup B$ and contains elements that are *not* in A and B together:

$(A \cup B)' = \{7, 9\}$

ADVICE AND TIPS

$\cap$ means i**n**tersection
$\cup$ means **U**nion

You can also use Venn diagrams with three sets.

Suppose you add $C = \{x: 3 \leqslant x \leqslant 7\}$ to the example.

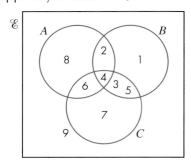

$C = \{3, 4, 5, 6, 7\}$

$A \cup B = \{1, 2, 3, 4, 5, 6, 8\}$

$(A \cup B) \cap C = \{3, 4, 5, 6\}$

You can illustrate this by shading different regions of the diagram.

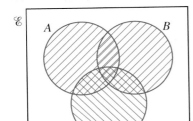

$A \cup B$ ////

C \\\\

$(A \cup B) \cap C$ ⨯⨯⨯

EXERCISE 6C

1

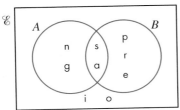

a List the elements of:

i A **ii** B **iii** $A \cap B$ **iv** $A \cup B$ **v** A'

vi $(A \cup B)'$ **vii** $A' \cap B$ **viii** $A \cap B'$.

b Write down:

i $n(A)$ **ii** $n(A \cup B)$ **iii** $n(B')$.

c $\mathscr{E}$ = {letters in the name of a city state}. What is the city state?

2

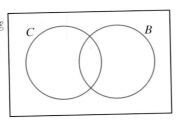

$\mathscr{E}$ = {f, r, a, c, t, i, o, n, s}

C = {r, a, t, i, o}

D = {f, i, r, s, t}

a Copy the diagram and put the elements in the correct places.

b List the elements of:

i $(C \cup D)'$ **ii** C' **iii** $C' \cup D$.

c Find:

i $n(C \cap D)$ **ii** $n((C \cap D)')$.

EXTENDED

3 $\mathscr{E}$ = {integer x: $1 \leqslant x \leqslant 12$}

A = {multiples of 2}

B = {multiplies of 3}

a Show these sets on a Venn diagram.

b Describe set $A \cap B$.

4

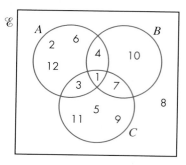

a List the elements of each of these sets.

i A

ii B

iii $A \cap B$

iv $A \cap C$

v $B \cup C$

vi C'

vii $A \cap B \cap C$

viii $(A \cup B) \cap C$

ix $(B \cup C)'$

x $(B \cup C)' \cap A$

b Complete these descriptions.

i $\mathscr{E}$ = {integer x: ...$\leqslant x \leqslant$...}

ii A = {factors of ...}

iii C = {...numbers}

5 On copies of this Venn diagram, shade each region.

a $A \cap B$

b $A' \cap B$

c $(A \cap B)'$

d $(A \cup B)'$

e $A' \cup B'$

f $A' \cap B'$

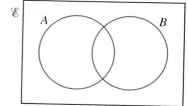

6 Which of your answers in question 5 are identical?

7 On copies of this Venn diagram, shade these regions.

a $(A \cup B) \cap C$

b $A' \cap (B \cup C)$

c $(A \cup B) \cap (A \cup C)$

d $(A \cup B \cup C)'$

e $(B \cup C)' \cap A$

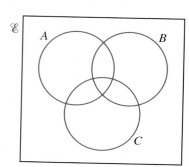

8 In this Venn diagram:

$\mathcal{E}$ = {positive integers}

P = {prime numbers}

E = {even numbers}

S = {square numbers}.

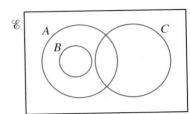

a Explain why $n(P \cap E) = 1$.

b Explain why $P \cap S = \varnothing$.

c Describe the elements of $E' \cap S$.

d Write down a number x such that $x \notin P$ and $x \notin E$ and $x \notin S$.

9 Look at this Venn diagram.

Say whether each of these statements is true or false.

a $A \subseteq B$

b $B \subseteq A$

c $A \cap B = B$

d $A \cup B = B$

e $B \cap C = \varnothing$

f $B \cup C = \mathcal{E}$

10

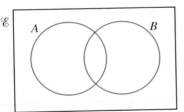

$n(A) = 50$ $n(B) = 40$ $n(A \cap B) = 27$

What is $n(A \cup B)$?

11 A and B are two sets. Write, as simply as possible:

a $A \cup A'$ **b** $B \cap B'$ **c** $((A \cup B)')'$ **d** $(A \cap B) \cap B$ **e** $(A \cup B) \cap A$.

12 $\mathcal{E}$ = {x: x is a positive integer}

E = {x: x is a multiple of 2}

T = {x: x is a multiple of 3}

F = {x: x is a multiple of 5}

a Draw a Venn diagram and place these numbers in it: 10, 11, 12, 13, 14, 15.

b Complete this description.

$E \cap T$ = {x: x is a multiple of .....}

c Write a description of $T \cap F$.

d Find an element of $E \cap T \cap F$.

e Write a description of $E \cap T \cap F$.

Why this chapter matters

We use ratio, proportion and speed in our everyday lives to help us compare two or more pieces of information.

Ratio and proportions are often used to compare sizes; speed is used to compare distances with the time taken to travel them.

A 100-m sprinter

Speed

When is a speed fast?

On 16 August 2009 Usain Bolt set a new world record for the 100-m sprint of 9.58 seconds. This is an average speed of 37.6 km/h.

The sailfish is the fastest fish and can swim at 110 km/h.

The cheetah is the fastest land animal and can travel at 121 km/h.

The fastest bird is the swift which can travel at 170 km/h.

Sailfish

Cheetah

Swift

Ratio and proportion facts

- Russia is the largest country.

- Vatican City is the smallest country.

- The area of Russia is nearly 39 million times the area of Vatican City.

- Monaco has the most people per square mile.

- Mongolia has the least people per square mile.

- The number of people per square mile in Monaco to the number of people in Mongolia is in the ratio 10 800 : 1.

- Japan has the highest life expectancy.

- Sierra Leone has the lowest life expectancy.

- On average people in Japan live over twice as long as people in Sierra Leone.

- Taiwan has the most mobile phones per 100 people (106.5).

- This is approximately four times that of Thailand (26.04).

Chapter

Ratio, proportion and rate

Topics	Level	Key words
Ratio		ratio, cancel, simplest form, map scale
Increases and decreases using ratios		increase, decrease
Speed		average, speed, distance, time
Rates		rate
Direct proportion		unitary method, direct proportion, single unit value
Inverse proportion		inverse proportion

In this chapter you will learn how to:

CORE

- Demonstrate an understanding of:
  - ratio
  - direct and inverse proportion. (C1.11 and E1.11)
- Use common measures of rate. (C1.11 and E1.11)
- Divide a quantity in a given ratio. (C1.11 and E1.11)
- Use scales in practical situations. (C1.11 and E1.11)
- Calculate average speed. (C1.11 and E1.11)

EXTENDED

- Increase and decrease a quantity by a given ratio. (E1.11)

A **ratio** is a way of comparing the sizes of two or more quantities.

A ratio can be expressed in a number of ways. For example, if Tasnim is five years old and Ziad is 20 years old, the ratio of their ages is:

Tasnim's age : Ziad's age

which is: 5 : 20

which simplifies to: 1 : 4   (dividing both sides by 5)

A ratio is usually given in one of these three ways.

Tasnim's age : Ziad's age	or	5 : 20	or	1 : 4	
Tasnim's age to Ziad's age	or	5 to 20	or	1 to 4	
$\dfrac{\text{Tasnim's age}}{\text{Ziad's age}}$	or	$\dfrac{5}{20}$	or	$\dfrac{1}{4}$	

Common units

When working with a ratio involving different units, *always convert them to the same units*. A ratio can be simplified only when the units of each quantity are the *same*, because the ratio itself has no units. Once the units are the same, the ratio can be simplified or **cancelled** like a fraction.

For example, you must convert the ratio of 125 g to 2 kg to the ratio of 125 g to 2000 g, so that you can simplify it.

125 : 2000

Divide both sides by 25:   5 : 80

Divide both sides by 5:   1 : 16   The ratio 125 : 2000 can be simplified to 1 : 16.

EXAMPLE 1

Express 25 minutes : 1 hour as a ratio in its simplest form.

The units must be the same, so change 1 hour into 60 minutes.

25 minutes : 1 hour = 25 minutes : 60 minutes

= 25 : 60   Cancel the units (minutes)

= 5 : 12   Divide both sides by 5

So, 25 minutes : 1 hour simplifies to 5 : 12

Ratios as fractions

You can express ratio in its **simplest form** as portions of a quantity by expressing the whole numbers in the ratio as fractions with the same denominator (bottom number).

EXAMPLE 2

A garden is divided into lawn and shrubs in the ratio $3 : 2$.

What fraction of the garden is covered by: **a** lawn **b** shrubs?

You find the denominator (bottom number) of the fraction by adding the numbers in the ratio (that is, $2 + 3 = 5$).

a The lawn covers $\frac{3}{5}$ of the garden.

b The shrubs cover $\frac{2}{5}$ of the garden.

EXERCISE 7A

1 Express each of these ratios in its simplest form.

    a $6 : 18$ **b** $5 : 20$

    c $16 : 24$ **d** $24 : 12$

    e $20 : 50$ **f** $12 : 30$

    g $25 : 40$ **h** $150 : 30$

2 Write each of these ratios of quantities in its simplest form. (Remember that you must always express both parts in a common unit before you simplify.)

    a 40 minutes : 5 minutes   **b** 3 kg : 250 g

    c 50 minutes to 1 hour   **d** 1 hour to 1 day

    e 12 cm to 2.5 mm   **f** 1.25 kg : 500 g

    g 75 cents : \$2   **h** 400 m: 2 km

3 A length of wood is cut into two pieces in the ratio $3 : 7$. What fraction of the original length is the longer piece?

4 Tareq and Hassan find a bag of marbles that they share between them in the ratio of their ages. Tareq is 10 years old and Hassan is 15 years old. What fraction of the marbles did Tareq get?

5 Mona and Petra share a pizza in the ratio $2 : 3$. They eat it all.

    a What fraction of the pizza did Mona eat?

    b What fraction of the pizza did Petra eat?

CORE

CORE

6 A camp site allocates space to caravans and tents in the ratio 7 : 3. What fraction of the total space is given to:

a the caravans

b the tents?

7 In a safari park at feeding time, the elephants, the lions and the chimpanzees are given food in the ratio 10 to 7 to 3. What fraction of the total food is given to:

a the elephants

b the lions

c the chimpanzees?

8 Paula wins three-quarters of her tennis matches. She loses the rest.

What is the ratio of wins to losses?

9 Three brothers share some cash.

The ratio of Marco's and Dani's share is 1 : 2.

The ratio of Dani's and Paulo's share is 1 : 2.

What is the ratio of Marco's share to Paulo's share?

Dividing amounts in a given ratio

To divide an amount in a given ratio, you first look at the ratio to see how many parts there are altogether.

For example, the ratio 4 : 3 has 4 parts and 3 parts giving 7 parts altogether.

7 parts is the whole amount.

1 part can then be found by dividing the whole amount by 7.

3 parts and 4 parts can then be calculated from 1 part.

EXAMPLE 3

Divide $28 in the ratio 4 : 3

4 + 3 = 7 parts altogether.

So 7 parts = $28.

Dividing by 7:

1 part = $4

4 parts = 4 × $4 = $16 and 3 parts = 3 × $4 = $12

So $28 divided in the ratio 4 : 3 is $16 : $12

Map scales

Map scales are often given as ratios in the form $1 : n$.

EXAMPLE 4

A map of New Zealand has a scale of 1 : 900 000.

The distance on the map from Auckland to Hamilton is 11.5 centimetres.

What is the actual distance?

1 cm on the map = 900 000 centimetres on the ground.

$\qquad\qquad$ = 9000 metres (100 centimetres = 1 metre)

$\qquad\qquad$ = 9 kilometres (1000 metres = 1 kilometre).

The distance is 11.5 × 9 kilometres

$\qquad\qquad$ = 103.5 kilometres.

EXERCISE 7B

1 Divide each according to the given ratio.

a 400 g in the ratio 2 : 3

b 280 kg in the ratio 2 : 5

c 500 in the ratio 3 : 7

d 1 km in the ratio 19 : 1

e 5 hours in the ratio 7 : 5

f $100 in the ratio 2 : 3 : 5

g $240 in the ratio 3 : 5 : 12

h 600 g in the ratio 1 : 5 : 6

i $5 in the ratio 7 : 10 : 8

j 200 kg in the ratio 15 : 9 : 1

2 The ratio of female to male members of a sports club is 7 : 3. The total number of members of the group is 250.

a How many members are female?

b What percentage of members are male?

3 A store sells small and large TV sets.

The ratio of small : large is 2 : 3.

The total stock is 70 sets.

a How many small sets are in stock?

b How many large sets are in stock?

4 When a supermarket checked a total of 357 confectionery products for sugar content, they found the ratio of products without sugar to those with sugar was 1 : 16.

How many of those products contained no sugar?

CORE

CORE

5 Joshua, Aicha and Mariam invest $10 000 in a company.

The ratio of the amounts they invest is:

Joshua : Aicha : Mariam = 5 : 7 : 8

How much does each of them invest?

6 Rewrite each of these scales as a ratio in the form 1 : n.

a 1 cm to 4 km   **b** 4 cm to 5 km   **c** 2 cm to 5 km

d 4 cm to 1 km   **e** 5 cm to 1 km   **f** 2.5 cm to 1 km

g 8 cm to 5 km   **h** 10 cm to 1 km   **i** 5 cm to 3 km

7 A map has a scale of 1 cm to 10 km.

a Rewrite the scale as a ratio in its simplest form.

b What is the actual length of a lake that is 4.7 cm long on the map?

c How long will a road be on the map if its actual length is 8 km?

> **ADVICE AND TIPS**
>
> 1 km = 1000 m
>   = 100 000 cm

8 A map has a scale of 2 cm to 5 km.

a Rewrite the scale as a ratio in its simplest form.

b How long is a path that measures 0.8 cm on the map?

c How long should a 12 km road be on the map?

9 The scale of a map is 5 cm to 1 km.

a Rewrite the scale as a ratio in the form 1 : n.

b How long is a wall that is shown as 2.7 cm on the map?

c The distance between two points is 8 km; how far will this be on the map?

10 You can simplify a ratio by changing it into the form 1 : n. For example, 5 : 7 can be rewritten as:

$$\frac{5}{5} : \frac{7}{5} = 1 : 1.4$$

Rewrite each of these ratios in the form 1 : n.

a 5 : 8   **b** 4 : 13   **c** 8 : 9

d 25 : 36   **e** 5 : 27   **f** 12 : 18

g 5 hours : 1 day   **h** 4 hours : 1 week   **i** £4 : £5

Calculating with ratios when only part of the information is known

EXAMPLE 5

A fruit drink is made by mixing orange squash with water in the ratio 2 : 3.
How much water needs to be added to 5 litres of orange squash to make the drink?

   2 parts is 5 litres.

   Dividing by 2:
      1 part is 2.5 litres
      3 parts = 2.5 litres × 3 = 7.5 litres

   So 7.5 litres of water is needed to make the drink.

EXAMPLE 6

Two business partners, Lubna and Adama, divided their total profit in the ratio 3 : 5.
Lubna received $2100. How much did Adama get?

   Lubna's $2100 was $\frac{3}{8}$ of the total profit. (Check that you know why.)
   $\frac{1}{8}$ of the total profit = $2100 ÷ 3 = $700

   So Adama's share, which was $\frac{5}{8}$, amounted to $700 × 5 = $3500.

EXERCISE 7C

1 Sean, aged 15, and Ricki, aged 10, shared some sweets in the same ratio as their ages. Sean had 48 sweets.

   a Simplify the ratio of their ages.

   b How many sweets did Ricki have?

   c How many sweets did they share altogether?

2 A blend of tea is made by mixing Lapsang with Assam in the ratio 3 : 5. I have a lot of Assam tea but only 600 g of Lapsang. How much Assam do I need to make the blend using all the Lapsang?

3 The ratio of male to female spectators at a hockey game is 4 : 5. 4500 men watched the match. What was the total attendance at the game?

4 A teacher always arranged the content of each of his lessons as 'teaching' and 'practising learnt skills' in the ratio 2 : 3.

   a If a lesson lasted 35 minutes, how much teaching would he do?

   b If he decided to teach for 30 minutes, how long would the lesson need to be?

CORE

CORE

5 A 'good' children's book has pictures and text in the ratio 17 : 8. In a book I have just looked at, the pictures occupy 23 pages.

   a Approximately how many pages of text should this book have to be a 'good' children's book?

   b What percentage of a 'good' children's book will be text?

6 Three business partners, Ren, Shota and Fatima, put money into a business in the ratio 3 : 4 : 5. They shared any profits in the same ratio. Last year, Fatima received $3400 from the profits. How much did Ren and Shota receive last year?

7 **a** Iqra is making a drink from lemonade, orange and ginger ale in the ratio 40 : 9 : 1. If Iqra has only 4.5 litres of orange, how much of the other two ingredients does she need to make the drink?

   b Another drink made from lemonade, orange and ginger ale uses the ratio 10 : 2 : 1.

   Which drink has a larger proportion of ginger ale, Iqra's or this one? Show how you work out your answer.

8 There is a group of boys and girls waiting for school buses. 25 girls get on the first bus. The ratio of boys to girls at the stop is now 3 : 2. 15 boys get on the second bus. There are now the same number of boys and girls at the bus stop. How many students altogether were originally at the bus stop?

9 A jar contains 100 cm³ of a mixture of oil and water in the ratio 1 : 4. Enough oil is added to make the ratio of oil to water 1 : 2. How much water must be added to make the ratio of oil to water 1 : 3?

7.2 Increases and decreases using ratios

Sometimes increases and decreases can be expressed in terms of ratios.

Suppose a recipe for six people requires 450 g of flour.
How much is needed for 10 people?

You need to **increase** 450 g in the ratio 10 : 6 = 5 : 3.
Think of 450 as 3 parts and you need to find 5 parts.

You need to find $\frac{5}{3}$ of 450 g.

   $450 \times \frac{5}{3} = 750$ g

If the recipe was to be changed to feed four people you would need to **decrease** the amount of flour in the ratio 4 : 6 = 2 : 3.

You need to find $\frac{2}{3}$ of 450 g.

   $450 \times \frac{2}{3} = 300$ g

EXERCISE 7D

1 Increase 200 in each ratio.

    a 3 : 1   **b** 3 : 2   **c** 10 : 1   **d** 7 : 4   **e** 6 : 5   **f** 11 : 10

2 Decrease 80 in each ratio

    a 1 : 4   **b** 3 : 4   **c** 1 : 10   **d** 7 : 10   **e** 1 : 5   **f** 4 : 5

3 A projector enlarges an image in the ratio 20 : 1.

    a What will be the size of an enlargement of a picture that is 8 cm by 6 cm?

    b What will be the size of the image if the ratio is changed to 15 : 2?

4 A photocopier enlarges a photograph in the ratio 5 : 4.

    a What is the size of the enlargement of a photograph that is 10 cm by 12 cm?

    b When the setting of the photocopier is changed, the size of the enlargement is 15 cm by 18 cm. Write the setting as a ratio, as simply as possible.

5 A photograph measures 12 cm by 20 cm.

It is made smaller in the ratio 3 : 4.

    a What are the dimensions of the new photograph?

    b The new photograph is again made smaller in the ratio 3 : 4. What are the dimensions now?

6 $5000 is invested and after a year the value has increased in the ratio 3 : 2.

    a What is the value of the investment now?

    b What is the percentage increase over one year?

The value of the investment continues to grow and in the second year it increases in the ratio 5 : 4.

    c What is the value after two years?

    d What is the percentage increase in the second year?

    e What is the overall percentage increase over two years?

    f Show that the overall increase could be written as a ratio as 15 : 8.

7 Prices are going up in the ratio 6 : 5.

    a Show that this is a 20% increase.

    b Write a 10% increase as a ratio.

    c Write a 10% decrease as a ratio.

The relationship between **speed, time** and **distance** can be expressed in three ways:

$$\text{speed} = \frac{\text{distance}}{\text{time}} \qquad \text{distance} = \text{speed} \times \text{time} \qquad \text{time} = \frac{\text{distance}}{\text{speed}}$$

In problems relating to speed, you usually mean **average** speed, as it would be unusual to maintain one exact speed for the whole of a journey.

This diagram will help you remember the relationships between distance (D), time (T) and speed (S).

$$D = S \times T \qquad S = \frac{D}{T} \qquad T = \frac{D}{S}$$

Units for speed include km/h (kilometres per hour, or 'the number of kilometres travelled in an hour') and m/s (metres per second).

EXAMPLE 7

Paula drove a distance of 270 kilometres in 5 hours. What was her average speed?

$$\text{Paula's average speed} = \frac{\text{distance she drove}}{\text{time she took}} = \frac{270}{5} = 54 \text{ kilometres per hour (km/h)}$$

EXAMPLE 8

Renata drove from her home to Frankfurt in $3\frac{1}{2}$ hours at an average speed of 60 km/h. How far is it from Renata's home to Frankfurt?

Since:

distance = speed × time

the distance from Renata's home to Frankfurt is given by:

60 × 3.5 = 210 kilometres

Note: You need to change the time to a decimal number and use 3.5 (*not* 3.30).

EXAMPLE 9

Maria is going to drive to Rome, a distance of 190 kilometres. She estimates that she will drive at an average speed of 50 km/h. How long will it take her?

$$\text{Maria's time} = \frac{\text{distance she covers}}{\text{her average speed}} = \frac{190}{50} = 3.8 \text{ hours}$$

Change the 0.8 hour to minutes by multiplying by 60, to give 48 minutes.

So, the time taken for Maria's journey will be 3 hours 48 minutes.

Remember: When you calculate a time and get a decimal answer, as in Example 9, *do not mistake* the decimal part for minutes. You must either:

● leave the time as a decimal number and give the unit as hours, or

● change the decimal part to minutes by multiplying it by 60 (1 hour = 60 minutes) and give the answer in hours and minutes.

EXERCISE 7E

1 A cyclist travels a distance of 90 kilometres in 5 hours.

What was her average speed?

2 How far along a road would you travel if you drove at 110 km/h for 4 hours?

3 I can drive from my home to see my aunt in about 6 hours. The distance is 315 kilometres.

What is my average speed?

4 The distance from Leeds to London is 350 kilometres.

The train travels at an average speed of 150 km/h.

If I catch the 9.30 am train in London, at what time should I expect to arrive in Leeds?

5 How long will an athlete take to run 2000 metres at an average speed of 4 metres per second?

6 Copy and complete this table.

	Distance travelled	Time taken	Average speed
a	150 km	2 hours	
b	260 km		40 km/h
c		5 hours	35 km/h
d		3 hours	80 km/h
e	544 km	8 hours 30 minutes	
f		3 hours 15 minutes	100 km/h
g	215 km		50 km/h

7 Eliot drove a distance of 660 kilometres, in 7 hours 45 minutes.

a Change the time 7 hours 45 minutes to a decimal.

b What was the average speed of the journey? Round your answer to 1 decimal place.

ADVICE AND TIPS

Remember to convert time to a decimal if you are using a calculator, for example, 8 hours 30 minutes is 8.5 hours.

ADVICE AND TIPS

km/h means kilometres per hour.
m/s means metres per second.

CORE

CORE

8 Johan drives home from his son's house in 2 hours 15 minutes. He says that he drives at an average speed of 70 km/h.

a Change the 2 hours 15 minutes to a decimal.

b How far is it from Johan 's home to his son's house?

9 The distance between Paris and Le Mans is 200 km. The express train between Paris and Le Mans travels at an average speed of 160 km/h.

a Calculate the time taken for the journey from Paris to Le Mans, giving your answer as a decimal number of hours.

b Change your answer to part **a** to hours and minutes.

10 The distance between two cities is 420 kilometres.

a What is the average speed of a journey from one to the other if it takes 8 hours 45 minutes?

b If Sam covered the distance at an average speed of 63 km/h, how long would it take him?

11 A train travels at 50 km/h for 2 hours, then slows down to do the last 30 minutes of its journey at 40 km/h.

a What is the total distance of this journey?

b What is the average speed of the train over the whole journey?

12 Suni runs and walks the 6 kilometres from home to work each day. She runs the first 4 kilometres at a speed of 16 km/h, then walks the next 2 kilometres at a steady 8 km/h.

a How long does it take Suni to get to work?

b What is her average speed?

> **ADVICE AND TIPS**
>
> **Remember** that there are 3600 seconds in an hour and 1000 metres in a kilometre. So to change from km/h to m/s multiply by 1000 and divide by 3600.

13 Change these speeds to metres per second.

a 36 km/h **b** 12 km/h **c** 60 km/h

d 150 km/h **e** 75 km/h

14 Change these speeds to kilometres per hour.

a 25 m/s **b** 12 m/s **c** 4 m/s

d 30 m/s **e** 0.5 m/s

> **ADVICE AND TIPS**
>
> To change from m/s to km/h multiply by 3600 and divide by 1000.

15 A train travels at an average speed of 18 m/s.

a Express its average speed in km/h.

b Find the approximate time the train would take to travel 500 m.

c The train set off at 7.30 on a 40 km journey. At approximately what time will it reach its destination?

> **ADVICE AND TIPS**
>
> To convert a decimal fraction of an hour to minutes, just multiply by 60.

16 A cyclist is travelling at an average speed of 24 km/h.

    a What is this speed in metres per second?

    b What distance does he travel in 2 hours 45 minutes?

    c How long does it take him to travel 2 km?

    d How far does he travel in 20 seconds?

17 How much longer does it take to travel 100 kilometres at 65 km/h than at 70 km/h?

7.4   Rates

The idea of speed can be applied more generally to show how quantities change.

Suppose the temperature of some hot water goes down by five degrees (5°) every minute.

Then you can say that the temperature changes at a **rate** of 5 degrees per minute.

It would go down by 20 degrees in $\frac{20}{5} = 4$ minutes.

It would go down 1 degree in $\frac{1}{5} = 0.2$ minutes.

EXERCISE 7F

1 A plant is growing at a rate of 0.5 cm per day.

    a How much will it grow in a week?

    b How long will it take to grow 10 cm?

2 Water is flowing out of a tap at a rate of 3 litres every minute.

    a How much will flow out in 5 minutes?

    b How many seconds will it take for one litre to flow out?

3 A tank contains 500 litres of water.

    The water is leaking at a rate of 2 litres per day.

    a How long would it take for half the water to leak out?

    b How much would leak out in 6 hours?

4 Rainforest in one area is being cleared at a rate of 50 m^2 a day.

    How long will it take to clear 80 000 m^2?

5 Scientists estimate that in a particular country sea levels have risen by 50 cm in 1000 years. What is the average rate of increase in mm per year?

6 Arturo says he read a book with 320 pages in 8 hours.

a What is his rate of reading in pages per hour?

b How long would he take to read 200 pages?

c How many minutes does he take to read one page?

7 The cost of travelling in a taxi is 5 dollars per kilometre for the first 10 km and 4 dollars per kilometre after that.

a What will a 10 km journey cost?

b What will a 20 km journey cost?

c What is the average cost in dollars per kilometre for a 20 kilometre journey?

d What is the average cost in dollars per kilometre for a 50 kilometre journey?

8 A car maker claims that the fuel consumption rate for a particular car is 10.6 litres/100 km.

a How much fuel will it use to travel 300 km?

b How much fuel will it use to travel 50 km?

c What is the fuel consumption in litres per 10 km?

d What is the fuel consumption in litres per kilometre?

e How far can it travel on 100 litres of fuel?

9 A car's average CO_2 (carbon dioxide) emission rate is 145 g per km.

a How many *kilograms* of CO_2 will be emitted if the car travels 100 km?

b A driver estimates that he will drive the car 20 000 km in a year. How much CO_2 will be emitted in that time?

c Estimate the driver's average rate of CO_2 emission in kilograms per day.

7.5 Direct proportion

Suppose you buy 12 items that each cost the *same*. The total amount you spend is 12 times the cost of one item.

So the total cost is in **direct proportion** to the number of items bought. The cost of a single item (the unit cost) is the constant factor that links the two quantities.

Direct proportion is not only concerned with costs. Any two related quantities can be in direct proportion to each other.

The best way to solve all problems involving direct proportion is to start by finding the **single unit value**. This method is called the **unitary method**, because it refers to a single unit value. Work through Examples 10 and 11 to see how it is done.

Remember: Before solving a direct proportion problem, think about it carefully to make sure that you know how to find the required single unit value.

EXAMPLE 10

If eight pens cost $2.64, what is the cost of five pens?

First, find the cost of one pen. This is $2.64 ÷ 8 = $0.33

So, the cost of five pens is $0.33 × 5 = $1.65

EXAMPLE 11

Eight loaves of bread will make packed lunches for 18 people. How many packed lunches can be made from 20 loaves?

First, find how many lunches one loaf will make.

One loaf will make 18 ÷ 8 = 2.25 lunches.

So, 20 loaves will make 2.25 × 20 = 45 lunches.

EXERCISE 7G

1 If 30 matches have a total mass of 45 g, what would be the total mass of 40 matches?

2 Five bars of chocolate together cost $2.90. Find the total cost of nine bars.

3 Eight men can chop down 18 trees in a day. How many trees can 20 men chop down in a day?

4 Find the cost of 48 eggs when 15 eggs can be bought for $2.10.

5 Seventy maths textbooks cost a total of $875.

    a How much will 25 of the same maths textbooks cost?

    b How many of these maths textbooks can you buy for $100?

ADVICE AND TIPS

Remember to work out the value of one unit each time. Always check that answers are sensible.

6 A lorry uses 80 litres of fuel on a trip of 280 kilometres.

    a How much fuel would the same lorry use on a trip of 196 kilometres?

    b How far would the lorry travel on a full tank of 100 litres of fuel?

CORE

CORE

7 During the winter, I find that 200 kg of coal keeps my fire burning for 12 weeks.

a If I want a fire all through the winter (18 weeks), how much coal will I need to buy?

b Last year I bought 150 kg of coal. For how many weeks did I have a fire?

8 It takes a photocopier 16 seconds to produce 12 copies. How long will it take to produce 30 copies?

9 A recipe for 12 biscuits uses:

200 g margarine   400 g sugar

500 g flour   300 g ground rice

a What quantities are needed for:

i 6 biscuits   **ii** 9 biscuits   **iii** 15 biscuits?

b What is the maximum number of biscuits I could make if I had just 1 kg of each ingredient?

10 Pieter the baker sells bread rolls in packs of 6 for $2.30.

Paulo the baker sells bread rolls in packs of 10 for $3.50.

I have $10 to spend on bread rolls.

If I want to buy as many bread rolls as possible from one shop, which shop should I use? Show your working.

7.6 Inverse proportion

As your speed *increases*, the time you take to travel a fixed distance *decreases*. This is an example of **inverse proportion**.

EXAMPLE 12

At an average speed of 6 km/h a walk takes 5 hours.

How long would it take at an average speed of 8 km/h?

The total distance is 6 × 5 = 30 kilometres.

At 8 km/h the time taken to travel 30 kilometres is:

$$\frac{30}{8} = 3\frac{3}{4}$$

= 3 hours 45 minutes.

The time taken for the journey falls as the speed rises so it is in inverse proportion to the speed.

EXAMPLE 13

When a teacher shares out some sweets among 12 students, each student gets 10 sweets.

How many will each get if there are 20 students?

The total number of sweets is $10 \times 12 = 120$.

If 20 students share 120 sweets, each student gets $120 \div 20 = 6$ sweets.

Because there are more students, each one gets fewer sweets.

The number of sweets each student gets is in inverse proportion to the number of students.

EXERCISE 7H

1 A car travelling at 30 km/h takes 40 minutes to complete a journey.

How long would it take if it travelled at 60 km/h?

2 Leo has some books that are 2 cm wide and he can fit 40 on a shelf.

How many books 5 cm wide could fit on the same shelf?

3 Chairs are available at three prices: $30, $40 and $60.

Lejla can afford to buy 24 at $60 each.

a How many can she afford to buy at $40 each?

b How many can she afford to buy at $30 each?

4 Some friends are sharing the cost of a taxi.

If there are three friends it will cost them $18 each.

How much will it cost if there are five friends?

5 Eight people can pick all the grapes in a vineyard in ten days.

a How long will it take five people?

b How many people are needed to pick them all in four days?

6 Copy and complete this table, showing how long it will take a cyclist to complete a journey at different average speeds.

Average speed (km/h)	10	20	30	40
Time in hours and minutes		2 hours 30 minutes		

7 The number of people that can be safely carried in a lift depends on their average mass. A lift will safely carry 8 people with an average mass of 75 kg.

a How many people with an average mass of 100 kg can safely travel in the lift?

b How many children with an average mass of 40 kg can safely travel in the lift?

CORE

Why this chapter matters

How accurate are we?

In real life it is not always sensible to use exact values. Sometimes it would be impossible to have exact measurements. People often round values without realising it. Rounding is done so that values are sensible.

Is it exactly 23 km to Utrecht and exactly 54 km to Amsterdam?

Does this box contain exactly 750 g of rice when full?

Was her time exactly 13.4 seconds?

Does the school have exactly 1500 students?

Imagine if people tried to use exact values all the time. Would life seem strange?

Chapter

Estimation and limits of accuracy

Topics	Level	Key words
1 Rounding whole numbers	CORE	approximation, rounded up, rounded down
2 Rounding decimals	CORE	round, digit, decimal place
3 Rounding to significant figures	CORE	significant figure
4 Upper and lower bounds	CORE	upper bound, lower bound, limits of accuracy
5 Upper and lower bounds for calculations	EXTENDED	upper bound, lower bound

In this chapter you will learn how to:

CORE

- Make estimates of numbers, quantities and lengths. (C1.9 and E1.9)
- Give approximations to specified numbers of significant figures and decimal places and round off answers to reasonable accuracy in the context of a given problem. (C1.9 and E1.9)
- Give appropriate upper and lower bounds for data given to a specified accuracy (e.g. measured lengths). (C1.10 and E1.10)

EXTENDED

- Obtain appropriate upper and lower bounds to solutions of simple problems (e.g. the calculation of the perimeter or the area of a rectangle) given data to a specified accuracy. (E1.10)

You use rounded information, or **approximations** all the time. Look at the examples on the right. Each actual figure is either above or below the approximation shown here.

How do you round numbers up or down?

If you want to round a number to the nearest multiple of ten, you round it up if it ends in 5, 6, 7, 8 or 9, and round it down if it ends in 1, 2, 3 or 4. For example:

- 25, 26, 27, 28 and 29 are **rounded up** to 30.

- 24, 23, 22 and 21 are **rounded down** to 20.

So a box with approximately 30 mints could contain any number from 25 to 34.

You can round numbers to the nearest multiple of 10, 100, 1000 and so on. The number of runners (23 000) in the report on the marathon is rounded to the nearest 1000:

- the smallest number of people actually running would be 22 500 (22 500–22 999 are rounded up to 23 000)

- the largest number of people running would be 23 499 (23 500 would be rounded up to 24 000).

So, there could actually be from 22 500 to 23 499 people in the marathon.

EXERCISE 8A

1 Round each of these numbers to the nearest 10.

| **a** 24 | **b** 57 | **c** 78 | **d** 54 | **e** 96 |
| **f** 21 | **g** 88 | **h** 66 | **i** 14 | **j** 26 |

2 Round each of these numbers to the nearest 100.

| **a** 240 | **b** 570 | **c** 780 | **d** 504 | **e** 967 |
| **f** 112 | **g** 645 | **h** 358 | **i** 998 | **j** 1050 |

3 Round each of these numbers to the nearest 1000.

| **a** 2400 | **b** 5700 | **c** 7806 | **d** 5040 | **e** 9670 |
| **f** 1120 | **g** 6450 | **h** 3499 | **i** 9098 | **j** 1500 |

CORE

CORE

4

> **Welcome to Elsecar**
> **Population 800**
> (to the nearest 100)

> **Welcome to Hoyland**
> **Population 1200**
> (to the nearest 100)

> **Welcome to Jump**
> **Population 600**
> (to the nearest 100)

Which of these sentences could be true and which must be false?

a There are 789 people living in Elsecar. **b** There are 1278 people living in Hoyland.

c There are 550 people living in Jump. **d** There are 843 people living in Elsecar.

e There are 1205 people living in Hoyland. **f** There are 650 people living in Jump.

5 These are the average attendance figures in four football leagues in 2008–9:

England Premier League	35 600
Germany Bundesliga	42 565
Italy Serie A	25 303
Spain La Liga	29 124

a Which were the highest and lowest?

b Round each number to the nearest thousand.

c The figure for Ligue 1 in France was 25 000 to the nearest thousand. What were the largest and smallest actual values for Ligue 1?

6 Matthew and Vincenza are playing a game with whole numbers.

a What is the smallest number Matthew could be thinking of?

> I am thinking of a number. Rounded to the nearest 10 it is 380.

> I am thinking of a different, smaller number. Rounded to the nearest 100 it is 400.

b Matthew's number is the smallest possible. How many possible values are there for Vincenza's number?

7 The number of adults attending a comedy show is 80 to the nearest 10.

The number of children attending is 50 to the nearest 10.

Katya says that 130 adults and children attended the comedy show.

Give an example to show that she may **not** be correct.

Rounding decimals

Decimal places

When a number is written in decimal form, the **digits** to the right of the decimal point are called **decimal places**. For example:

79.4 is written 'with one decimal place'

6.83 is written 'with two decimal places'

0.526 is written 'with three decimal places'.

To **round** a decimal number to a particular number of decimal places, take these steps.

● Count along the decimal places from the decimal point and look at the first digit to be removed.

● When the value of this digit is less than five, just remove the unwanted places.

● When the value of this digit is five or more, add 1 onto the digit in the last decimal place then remove the unwanted places.

Here are some examples.

5.852 rounds to 5.85 to two decimal places

7.156 rounds to 7.16 to two decimal places

0.274 rounds to 0.3 to one decimal place

15.3518 rounds to 15.4 to one decimal place

EXERCISE 8B

1 Round each number to one decimal place.

a 4.83	**b** 3.79	**c** 2.16	**d** 8.25
e 3.673	**f** 46.935	**g** 23.883	**h** 9.549
i 11.08	**j** 33.509		

ADVICE AND TIPS

Just look at the value of the digit in the second decimal place.

2 Round each number to two decimal places.

a 5.783	**b** 2.358	**c** 0.977	**d** 33.085	**e** 6.007
f 23.5652	**g** 91.7895	**h** 7.995	**i** 2.3076	**j** 23.9158

3 Round each of these to the number of decimal places (dp) indicated.

a 4.568 (1 dp)	**b** 0.0832 (2 dp)	**c** 45.715 93 (3 dp)	**d** 94.8531 (2 dp)
e 602.099 (1 dp)	**f** 671.7629 (2 dp)	**g** 7.1124 (1 dp)	**h** 6.903 54 (3 dp)
i 13.7809 (2 dp)	**j** 0.075 11 (1 dp)		

4 Round each of these to the nearest whole number.

a 7.82	**b** 3.19	**c** 7.55	**d** 6.172	**e** 3.961
f 7.388	**g** 1.514	**h** 46.78	**i** 23.19	**j** 96.45

CORE

5 Anna puts the following items in her shopping basket: bread $3.20, meat $8.95, cheese $6.16 and butter $3.90.

By rounding each price to the nearest dollar, work out an estimate of the total cost of the items.

6 Which of these are correct roundings of the number 3.456?

　　3　　3.0　　3.4　　3.40　　3.45　　3.46　　3.47　　3.5　　3.50

7 When a number is rounded to three decimal places the answer is 4.728.

Which of these could be the number?

　　4.71　　4.7275　　4.7282　　4.73

8.3 Rounding to significant figures

You will often use **significant figures** (sf) when you want to *approximate* a number with a lot of digits in it. You frequently use this technique with calculator answers.

A calculator gives $\sqrt{800}$ as 28.284 271 25…

Rounded to　four sf this is　28.28
　　　　　　three sf it is　28.3
　　　　　　two sf it is　28
　　　　　　one sf it is　30

The steps taken to round a number to a given number of significant figures are very similar to those used for rounding to a given number of decimal places.

● Count the digits from the left. If you are rounding to 2 sf, count two digits, for 3 sf count three digits, and so on. When the original number is less than 1, start counting from the first non-zero digit.

● Look at the next digit to the right. When the value of this next digit is less than 5, leave the digit you counted to the same. However, if the value of this next digit is equal to or greater than 5, add 1 to the digit you counted to.

● Ignore all the other digits, but put in enough zeros to keep the number the right size (value).

For example, look at this table, which shows some numbers rounded to one, two and three significant figures, respectively.

Number	Rounded to 1 sf	Rounded to 2 sf	Rounded to 3 sf
45 281	50 000	45 000	45 300
568.54	600	570	569
7.3782	7	7.4	7.38
8054	8000	8100	8050
99.8721	100	100	99.9
0.7002	0.7	0.70	0.700

EXERCISE 8C

CORE

1 Round each number to 1 significant figure.

a 46 313   **b** 57 123   **c** 30 569   **d** 94 558   **e** 85 299

f 0.5388   **g** 0.2823   **h** 0.005 84   **i** 0.047 85   **j** 0.000 876

k 9.9   **l** 89.5   **m** 90.78   **n** 199   **o** 999.99

2 Round each number to 2 significant figures.

a 56 147   **b** 26 813   **c** 79 611   **d** 30 578   **e** 14 009

f 1.689   **g** 4.0854   **h** 2.658   **i** 8.0089   **j** 41.564

k 0.8006   **l** 0.458   **m** 0.0658   **n** 0.9996   **o** 0.009 82

3 Round each of these to the number of significant figures (sf) indicated.

a 57 402 (1 sf)   **b** 5288 (2 sf)   **c** 89.67 (3 sf)

d 105.6 (2 sf)   **e** 8.69 (1 sf)   **f** 1.087 (2 sf)

g 0.261 (1 sf)   **h** 0.732 (1 sf)   **i** 0.42 (1 sf)

j 0.758 (1 sf)   **k** 0.185 (1 sf)   **l** 0.682 (1 sf)

4 What are the lowest and the highest numbers of sweets that may be in these jars?

a   **b**   **c**

5 What are the least and the greatest numbers of people that live in these towns?

Satora   population 800 (to 1 significant figure)

Nimral   population 1200 (to 2 significant figures)

Korput   population 165 000 (to 3 significant figures)

6 There are 500 fish in a pond, correct to 1 sf. What is the least possible number of fish that could be taken from the pond so that there are 400 fish in the pond, correct to 1 sf?

7 Rani says that the population of Bikran is 132 000 to the nearest thousand. Vashti says that the population of Bikran is 130 000. Explain why Vashti could also be correct.

Any recorded measurements have usually been rounded.

The true value will be somewhere between the **lower bound** and the **upper bound**.

The lower and upper bounds are sometimes known as the **limits of accuracy**.

A journey of D kilometres that is measured as 26 kilometres to the nearest kilometre could be anything between 25.5 and 26.5 kilometres:

    25.5 would round up to 26.

    26.5 would round up to 27 but anything less would round down to 26.

This means that 26.5 is the upper bound and 25.5 is the lower bound of D.

You can write $25.5 \leqslant D < 26.5$ which means that D is greater than or equal to 25.5 but less than 26.5.

EXAMPLE 1

A stick of wood measures 32 cm, to the nearest centimetre.

What are the lower and upper bounds of the actual length of the stick?

The lower bound is 31.5 cm as this is the lowest value that rounds to 32 cm to the nearest centimetre.

The upper bound is 32.5 cm as anything lower rounds to 32 cm to the nearest centimetre. 32.5 cm would round to 33 cm.

You can write:

    $31.5 \leqslant$ length of stick in cm < 32.5

Note the use of $<$ for the upper bound.

EXAMPLE 2

A time of 53.7 seconds is accurate to 1 decimal place.

What are the upper and lower bounds for the time?

The lower bound is 53.65 seconds and the upper bound is 53.75 seconds.

So $53.65 \leqslant$ time in seconds < 53.75

CORE

EXERCISE 8D

1 Write down the upper and lower bounds of each measurement.

a A length measured as 7 cm to the nearest cm.

b A mass measured as 120 g to the nearest 10 g.

c A length measured as 3400 km to the nearest 100 km.

d A speed measured as 50 km/h to the nearest km/h.

e An amount given as $6 to the nearest dollar.

f A length given as 16.8 cm to the nearest tenth of a centimetre.

g A mass measured as 16 kg to the nearest kg.

h A football crowd of 14 500 to the nearest 100.

i A distance given as 55 km to the nearest km.

j A distance given as 55 km to the nearest 5 km.

2 Write down the upper and lower bounds for each of these values, which are rounded to the given degree of accuracy. Use inequalities to show your answer.
For example, part a should be 5.5 ⩽ length in cm < 6.5.

a 6 cm (1 significant figure) **b** 17 kg (2 significant figures)

c 32 min (2 significant figures) **d** 238 km (3 significant figures)

e 7.3 m (1 decimal place) **f** 25.8 kg (1 decimal place)

g 3.4 h (1 decimal place) **h** 87 g (2 significant figures)

i 4.23 mm (2 decimal places) **j** 2.19 kg (2 decimal places)

k 12.67 minutes (2 decimal places) **l** 25 m (2 significant figures)

m 40 cm (1 significant figure) **n** 600 g (2 significant figures)

o 30 minutes (1 significant figure) **p** 1000 m (2 significant figures)

q 4.0 m (1 decimal place) **r** 7.04 kg (2 decimal places)

s 12.0 s (1 decimal place) **t** 7.00 m (2 decimal places)

3 Write down the lower and upper bounds of each of these values, which are rounded to the accuracy stated.

a 8 m (1 significant figure) **b** 26 kg (2 significant figures)

c 25 minutes (2 significant figures) **d** 85 g (2 significant figures)

e 2.40 m (2 decimal places) **f** 0.2 kg (1 decimal place)

g 0.06 s (2 decimal places) **h** 300 g (1 significant figure)

i 0.7 m (1 decimal place) **j** 366 d (3 significant figures)

k 170 weeks (2 significant figures) **l** 210 g (2 significant figures)

4 A chain is 30 m long, to the nearest metre.

A chain is needed to fasten a boat to a harbour wall. The distance to the wall is also 30 m, to the nearest metre.

Which statement is definitely true? Explain your decision.

A: The chain will be long enough.

B: The chain will not be long enough.

C: It is impossible to tell whether or not the chain is long enough.

5 A bag contains 2.5 kg of soil, to the nearest 100 g.

What is the least amount of soil in the bag?

Give your answer in kilograms and grams.

6 Chang has 40 identical marbles. Each marble has a mass of 65 g (to the nearest gram).

a What is the greatest possible mass of one marble?

b What is the least possible mass of one marble?

c What is the greatest possible mass of all the marbles?

d What is the least possible mass of all the marbles?

8.5 Upper and lower bounds for calculations (E)

When rounded values are used for a calculation, you can find upper and lower bounds for the result of the calculation.

EXAMPLE 3

The dimensions of this rectangle are given to the nearest centimetre.

Find the lower bound for the perimeter and the upper bound for the area.

27 cm
21 cm

The upper and lower bounds for the sides are:

26.5 ≤ length in cm < 27.5

20.5 ≤ width in cm < 21.5

Perimeter = 2 × (length + width)
The lower bound will be found from the lower bounds of the length and width.
Lower bound of perimeter = 2 × (26.5 + 20.5) = 94 cm.

Area = length × width
The upper bound will be found from the upper bounds of the length and width.
Upper bound of area = 27.5 × 21.5 = 591.25 cm²

EXAMPLE 4

A car travels 125 km (to the nearest km) and uses 16.1 litres of fuel (correct to one decimal place).

Find the upper and lower bounds of the fuel consumption in km/litre.

$124.5 \leqslant$ distance in kilometres < 125.5

$16.05 \leqslant$ fuel in litres < 16.15

The fuel consumption is distance ÷ fuel used.

To find the upper bound of this, use:

upper bound of the distance ÷ lower bound of the fuel used

upper bound is $125.5 \div 16.05 = 7.8193\ldots\ldots$

To find the lower bound of fuel consumption, use:

lower bound of distance ÷ upper bound of fuel used:

$= 124.5 \div 16.15 = 7.7089\ldots\ldots$

So $7.709 \leqslant$ fuel consumption in km/litre < 7.819 when the answers are rounded to three decimal places.

When solving a problem, write down the upper and lower bounds for the values given and then decide which to use to find the solution.

EXERCISE 8E

1 Boxes have a mass of 7 kg, to the nearest kilogram.

What are the upper and lower bounds for the total mass of 10 of these boxes?

2 Books each have a mass of 1200 g, to the nearest 100 g.

a What is the greatest possible mass of 10 books?
Give your answer in kilograms.

b A trolley can safely hold up to 25 kg of books.
How many books can safely be put on the trolley?

3 Jasmine says, 'My mass is 45 kilos.' Yolander says, 'My mass is 53 kilos.' Both are measured to the nearest kilogram.

What is the greatest possible difference between their masses?

Show how you worked out your answer.

4 For each of these rectangles, find the upper and lower bounds for the perimeter. The measurements are shown to the level of accuracy indicated in brackets.

a 5 cm × 9 cm (nearest cm)   **b** 4.5 cm × 8.4 cm (1 decimal place)

c 7.8 cm × 18 cm (2 significant figures)

5 Calculate the upper and lower bounds for the areas of each rectangle in question **4**.

EXTENDED

6 A cinema screen is measured as 6 m by 15 m, to the nearest metre. Calculate the upper and lower bounds for the area of the screen.

7 The measurements, to the nearest centimetre, of a box are given as 10 cm × 7 cm × 4 cm. Calculate the upper and lower bounds for the volume of the box.

8 Mr Sparks is an electrician. He has a 50-m roll of cable, correct to the nearest metre. He uses 10 m on each job, to the nearest metre.

If he does four jobs, what is the maximum amount of cable he could have left?

9 Jon and Matt are exactly 7 kilometres apart. They are walking towards each other. Jon is walking at 4 km/h and Matt is walking at 2 km/h.

Both speeds are given to the nearest kilometre per hour.

Without doing any time calculations, decide whether it is possible for them to meet in 1 hour. Justify your answer.

10 The area of a rectangular field is given as 350 m^2, to the nearest 10 m^2. One length is given as 16 m, to the nearest metre. Find the upper and lower bounds for the other length of the field.

11 A stopwatch records the time for the winner of a 100-metre race as 14.7 seconds, measured to the nearest one-tenth of a second.

a What are the upper and lower bounds of the winner's time?

b The length of the 100-metre track is correct to the nearest 1 m. What are the upper and lower bounds of the lengths of the track?

c What is the greatest possible average speed of the winner, with a time of 14.7 seconds in the 100-metre race?

12 A model car travels 40 m, measured to one significant figure, at a speed of 2 m/s, measured to one significant figure. Find the upper and lower bounds of the time taken.

13 The population of Japan is 127 000 000 to the nearest million.

The area of Japan is 378 000 km^2 to the nearest 1000 km^2.

The population density for any country is the total population divided by the area.

Find the upper and lower bounds for the population density of Japan in people/km^2. Round your answers to two decimal places.

14 An engineer testing a car's CO_2 (carbon dioxide) emissions measures 26 kg of CO_2 when it is driven 150 km.

The mass is given to the nearest kilogram.

The distance is given to the nearest kilometre.

Find the upper and lower bounds for the CO_2 emissions in grams/kilometre. Round your answers to one decimal place.

Why this chapter matters

Very large and very small numbers can often be difficult to read. Scientists use standard form as a shorthand way of representing numbers.

The planets

Mercury is the closest planet to the Sun (and is very hot). It orbits 60 million km (6×10^7 km) away from the Sun.

Venus rotates the opposite way to the other planets and has a diameter of 12 100 km (1.21×10^4 km).

Earth takes 365 days to orbit the Sun and 24 hours to complete a rotation.

Mars has the largest volcano in the solar system. It is almost 600 km across and rises 24 km above the surface. This is five times bigger than the biggest volcano on Earth.

Jupiter is made of gas. It has no solid land so visiting it is not recommended! It has a huge storm which rages across its surface. This is about 8 km high, 40 000 km long and 14 000 km wide. It looks like a red spot and is called 'the Great Red Spot'.

Saturn is the largest planet in the solar system. It is about 120 000 km across (1.2×10^5 km) and 1400 million km from the Sun (1.4×10^9 km).

Uranus takes 84 days to orbit the Sun.

Neptune is similar to Jupiter in that it is a gas planet and has violent storms. Winds can blow at up to 2000 km per hour, so a cloud can circle Neptune in about 16 hours.

Pluto is the furthest planet from the Sun. Some astronomers dispute whether it can be classed as a planet. The average surface temperature on Pluto is about −230 °C.

The mass of an electron is about 0.000 000 000 000 000 000 000 000 000 000 91 kg.
This is written 9.1×10^{-31} kg.

The mass of the Earth is about 5 970 000 000 000 000 000 000 000 kg.
This is written 5.97×10^{24} kg.

Standard form

Topics

Level

Key words

 Standard form

CORE

standard form, index

 Calculating with standard form

CORE

In this chapter you will learn how to:

CORE

- Use the standard form $A \times 10^{n}$ when n is a positive or negative integer and $1 \leqslant A < 10$. (C1.7 and E1.7)

Standard form

Powers of ten:

$$100 = 10 \times 10 = 10^2$$
$$1000 = 10 \times 10 \times 10 = 10^3$$

Extending this idea:

$$10\,000 = 10 \times 10 \times 10 \times 10 = 10^4$$
$$100\,000 = 10^5$$
$$1\,000\,000 = 10^6$$

and so on.

The power of 10 is called the **index**.

Standard form is a way of writing very large and very small numbers using powers of 10. In this form, a number is given a value between 1 and 10 multiplied by a power of 10. That is,

$A \times 10^n$ where $1 \leqslant A < 10$, and n is a whole number.

Follow through these examples to see how numbers are written in standard form.

$$52 = \qquad 5.2 \times 10 = \mathbf{5.2 \times 10^1}$$
$$73 = \qquad 7.3 \times 10 = \mathbf{7.3 \times 10^1}$$
$$625 = \quad 6.25 \times 100 = \mathbf{6.25 \times 10^2} \qquad \text{The numbers in bold are in standard form.}$$
$$389 = \quad 3.89 \times 100 = \mathbf{3.89 \times 10^2}$$
$$3147 = 3.147 \times 1000 = \mathbf{3.147 \times 10^3}$$

When writing a number in this way, you must always follow two rules.

- The first part must be a number between 1 and 10 (1 is allowed but 10 isn't).
- The second part must be a whole-number (negative or positive) power of 10. Note that you would *not normally* write the power 1.

Standard form on a calculator

A number such as 123 000 000 000 is obviously difficult to key into a calculator. Instead, you enter it in standard form (assuming you are using a scientific calculator):

$$123\,000\,000\,000 = 1.23 \times 10^{11}$$

The key strokes to enter this into your calculator could be something like this:

1 **·** **2** **3** **×10ˣ** **1** **1**

Your calculator display will display the number either as an ordinary number, if there is enough space, or in standard form. Make sure you know how to use standard form on your calculator.

Some calculators may use a different symbol for **×10ˣ**, for example, **EXP**

Standard form of numbers less than 1

You will need to use a negative index for numbers between 0 and 1:

$$0.1 = 10^{-1}$$
$$0.01 = 10^{-2}$$
$$0.001 = 10^{-3}$$
$$0.0001 = 10^{-4}$$

and so on.

For example:

$$0.000\,729 = 7.29 \times 0.0001$$
$$= 7.29 \times 10^{-4} \text{ in standard form}$$

These numbers are written in standard form. Make sure that you understand how they are formed.

a $0.4 = 4 \times 10^{-1}$ **b** $0.05 = 5 \times 10^{-2}$ **c** $0.007 = 7 \times 10^{-3}$

d $0.123 = 1.23 \times 10^{-1}$ **e** $0.007\,65 = 7.65 \times 10^{-3}$ **f** $0.9804 = 9.804 \times 10^{-1}$

g $0.0098 = 9.8 \times 10^{-3}$ **h** $0.000\,0078 = 7.8 \times 10^{-6}$

On a calculator you would enter 1.23×10^{-6}, for example, as:

$$\boxed{1}\ \boxed{\bullet}\ \boxed{2}\ \boxed{3}\ \boxed{\times 10^x}\ \boxed{(-)}\ \boxed{6}$$

Try entering some of the numbers **a** to **h** (above) into your calculator for practice.

EXERCISE 9A

1 These numbers are in standard form. Write them out in full.

a 2.5×10^2 **b** 3.45×10 **c** 4.67×10^{-3} **d** 3.46×10

e 2.0789×10^{-2} **f** 5.678×10^3 **g** 2.46×10^2 **h** 7.6×10^3

i 8.97×10^5 **j** 8.65×10^{-3} **k** 6×10^7 **l** 5.67×10^{-4}

2 Write these numbers in standard form.

a 250	**b** 0.345	**c** 46 700
d 3 400 000 000	**e** 20 780 000 000	**f** 0.000 567 8
g 2460	**h** 0.076	**i** 0.000 76
j 0.999	**k** 234.56	**l** 98.7654
m 0.0006	**n** 0.005 67	**o** 56.0045

In questions **3** to **5**, write the numbers given in each statement in standard form.

3 One year, 27 797 runners completed the New York marathon.

CORE

4 The largest number of dominoes ever toppled by one person is 321 197, although a team set up and toppled 4 491 863.

5 The asteroid *Phaethon* comes within 12 980 000 miles of the Sun. The asteroid *Pholus*, at its furthest point, is a distance of 2997 million miles from the Earth. The closest an asteroid ever came to Earth was 93 000 miles from the planet.

6 How many times bigger is 3.2×10^6 than 3.2×10^4?

7 The speed of sound (Mach 1) is 1236 kilometres per hour.

A plane travelling at Mach 2 would be travelling at twice the speed of sound.

How many kilometres would a plane travelling at Mach 3 cover in 1 minute?

8 Here are the distances of some planets from the Sun:

Jupiter	778 million kilometres
Mercury	58 million kilometres
Pluto	5920 million kilometres

Write these distances in standard form.

9.2 Calculating with standard form

Calculations involving very large or very small numbers can be done more easily if you use standard form.

You can enter numbers in a scientific calculator in standard form. The way you do this may be different for different models of calculator. Make sure you know how to do this with your calculator.

EXAMPLE 1

A pixel on a computer screen is 2×10^{-2} cm long by 7×10^{-3} cm wide.

What is the area of the pixel?

The area is calculated as length times width.

$$\text{Area in cm}^2 = 2 \times 10^{-2} \times 7 \times 10^{-3}$$
$$= 14 \times 10^{-5} \text{ (Use a calculator to do this.)}$$

The answer is not in standard form as the first part is not between 1 and 10, so you need to change it to standard form.

$$14 \times 10^{-5} = 0.00014 = 1.4 \times 10^{-4}$$

So area $= 1.4 \times 10^{-4}$ cm^2

If you use a calculator you can enter the numbers directly without any rearranging.
Your calculator may give you the answer in standard form.

EXAMPLE 2

The distance from the Sun to the Earth is 150 million kilometres.

Light travels at 3.00×10^5 km/second.

How long does light from the Sun take to reach the Earth?

Distance $= 150\,000\,000 = 1.5 \times 10^8$ km

Time $=$ distance $\div$ speed

$\quad = (1.5 \times 10^8) \div (3.00 \times 10^5)$

$\quad = 500$ seconds (Use a calculator to work this out.)

$\quad = 8$ minutes 20 seconds

EXERCISE 9B

1 These numbers are not in standard form. Write them in standard form.

a 56.7×10^2

b 0.06×10^4

c 34.6×10^{-2}

d 0.07×10^{-2}

e 56×10

f $2 \times 3 \times 10^5$

g $2 \times 10^2 \times 35$

h 160×10^{-2}

i 23 million

2 Work these out. Give your answers in standard form.

a $2 \times 10^4 \times 5.4 \times 10^3$

b $1.6 \times 10^2 \times 3 \times 10^4$

c $2 \times 10^4 \times 6 \times 10^4$

d $2 \times 10^{-4} \times 5.4 \times 10^3$

e $1.6 \times 10^{-2} \times 4 \times 10^4$

f $2 \times 10^4 \times 6 \times 10^{-4}$

g $7.2 \times 10^{-3} \times 4 \times 10^2$

h $(5 \times 10^3)^2$

i $(2 \times 10^{-2})^3$

3 Work these. Give your answers in standard form.

a $(5.4 \times 10^4) \div (2 \times 10^3)$

b $(4.8 \times 10^2) \div (3 \times 10^4)$

c $(1.2 \times 10^4) \div (6 \times 10^4)$

d $(2 \times 10^{-4}) \div (5 \times 10^3)$

e $(1.8 \times 10^4) \div (9 \times 10^{-2})$

f $\sqrt{(36 \times 10^{-4})}$

4 A typical adult has about 20 000 000 000 000 red blood cells. Each red blood cell has a mass of about 0.000 000 000 1 g. Write both of these numbers in standard form and work out the total mass of red blood cells in a typical adult.

5 A man puts one grain of rice on the first square of a chess board, two on the second square, four on the third, eight on the fourth and so on.

a How many grains of rice will he put on the 64th square of the board?

b How many grains of rice will there be altogether?

Give your answers in standard form.

ADVICE AND TIPS

Compare powers of 2 with the running totals. By the fourth square you have 15 grains altogether, and $2^4 = 16$.

6 The surface area of the Earth is approximately 3.2×10^8 square kilometres. The area of the Earth's surface that is covered by water is approximately 2.2×10^8 square kilometres.

a Calculate the area of the Earth's surface *not* covered by water. Give your answer in standard form.

b What percentage of the Earth's surface is not covered by water?

7 The Moon is a sphere with a radius of 1.74×10^3 kilometres. The formula for working out the surface area of a sphere is:

surface area = $4\pi r^2$

Calculate the surface area of the Moon.

8 Evaluate $\dfrac{E}{M}$ when $E = 1.5 \times 10^3$ and $M = 3 \times 10^{-2}$, giving your answer in standard form.

9 Work out the value of $\dfrac{3.2 \times 10^7}{1.4 \times 10^2}$ giving your answer in standard form, correct to 2 significant figures.

10 In one year, British Airways carried 33 million passengers. Of these, 70% passed through Heathrow Airport. On average, each passenger carried 19.7 kg of luggage. Calculate the total mass of the luggage carried by these passengers.

11 In 2013 the world population was approximately 7.14×10^9. In 2014 the world population was approximately 7.24×10^9.

a By how much did the population rise? Give your answer as an ordinary number.

b What was the percentage increase?

12 Here are four numbers written in standard form.

1.6×10^4 4.8×10^6 3.2×10^2 6.4×10^3

a Work out the smallest answer when two of these numbers are multiplied together.

b Work out the largest answer when two of these numbers are added together.
Give your answers in standard form.

13 The mass of Saturn is 5.686×10^{26} tonnes. The mass of the Earth is 6.04×10^{21} tonnes. How many times heavier is Saturn than the Earth? Give your answer in standard form to a suitable degree of accuracy.

14 A number is greater than 100 million and less than 1000 million.

Write down a possible value of the number, in standard form.

15 Here are some population figures for some countries.

Country	Population
Tunisia	1.10×10^7
Denmark	5.60×10^6
Senegal	1.39×10^7
Jamaica	2.71×10^6
Mexico	1.19×10^8
India	1.29×10^9

a Which country has the largest population?

b Which two countries have the largest difference in population?

c Find the total population of Senegal, Denmark and Jamaica. Give your answer in standard form to two significant figures.

d Complete this sentence:

The population of Mexico is approximately … times larger than the population of Denmark.

e Complete this sentence:

The population of India is approximately … times larger than the population of Jamaica.

16 This table shows the populations and the areas of five different countries.

Country	Population	Area
Russian Federation	1.44×10^8	1.71×10^7
Sri Lanka	2.07×10^7	6.56×10^4
Thailand	6.64×10^7	5.13×10^5
Togo	6.98×10^6	5.68×10^5
Iran	7.74×10^7	1.65×10^6

a Which country has the smallest population?

b Which country has the smallest area?

The population density is the population divided by the area.

c Which country has the largest population density?

d Which country has the smallest population density?

e What fraction of the area of the Russian Federation is the area of Sri Lanka? Give your answer in the form $\frac{1}{N}$.

Why this chapter matters

Technology is increasingly important in our lives. It helps us do many things more efficiently than we could without it.

Hundreds of years ago people in different countries had different systems of measurement. They were often based on the human body, for example, the length of people's hands, arms or feet, but they all varied and all had different names.

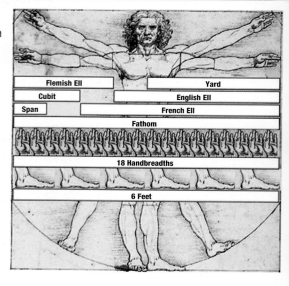

Now the world has an official standard system of measurement – the metric system. This is especially important for scientists so they can work together worldwide. It is also helpful for everyone who needs to compare lengths, masses, volumes and so on between different countries.

We also have more help now in calculating complicated measurements such as volume. Calculating aids have been used for thousands of years. In about 2000 BCE the abacus was being used in Egypt and China.

Abacuses are still widely used in China today and they were used everywhere for almost 3500 years, until John Napier devised a calculating aid called **Napier's bones**.

Napier's bones

These led to the invention of the **slide rule** by William Oughtred in 1622. This was in use until the mid-1960s. Engineers working on the first ever moon landings used slide rules to do some of their calculations.

The first **electronic computers** were produced in the mid-20th century. When the transistor was invented, the power increased and the cost and size decreased until the point where the average scientific calculator that students use in schools has more computing power than the first craft that went into space.

Applying number and using calculators

Topics	Level	Key words
1 Units of measurement	CORE	metric system, length, mass, volume, capacity
2 Converting between metric units	CORE	centimetre, millimetre, kilometre, gram, kilogram, tonne, litre, millilitre, centilitre
3 Time	CORE	24-hour clock, 12-hour clock, timetable
4 Currency conversions	CORE	exchange rate
5 Using a calculator efficiently	CORE	

In this chapter you will learn how to:

CORE

- Use current units of mass, length, area, volume and capacity in practical situations and express quantities in terms of larger or smaller units. (C4.1 and E4.1)
- Calculate times in terms of the 24-hour and 12-hour clock. (C1.14 and E1.14)
- Read clocks, dials and timetables. (C1.14 and E1.14)
- Calculate using money and convert from one currency to another. (C1.15 and E1.15)
- Use a calculator efficiently. (C1.13 and E1.13)
- Apply appropriate checks of accuracy (C1.13 and E1.13)

The **metric system** of measurement is now used in most countries of the world, except for the United States. Here is a list of the most common metric units.

Unit	How to estimate it
Length	
1 metre	A long stride for an average person
1 kilometre	Two and a half times round a school track
1 centimetre	The distance across a fingernail
Mass	
1 gram	A small coin weighs a few grams
1 kilogram	A bag of sugar
1 tonne	A saloon car
Volume/capacity	
1 litre	A full carton of orange juice
1 centilitre	A small glass is about 10 centilitres
1 millilitre	A full teaspoon is about 5 millilitres

Volume and capacity

The term 'capacity' is normally used to refer to the volume of a liquid or a gas.

For example, when referring to the volume of petrol that a car's fuel tank will hold, people may say its **capacity** is 60 litres.

EXERCISE 10A

1 Decide the metric unit you would be most likely to use to measure each of these.

   a The height of your classroom   **b** The distance from Athens to Vienna

   c The thickness of your little finger   **d** The mass of this book

   e The amount of water in a fish tank   **f** The mass of an aircraft

   g A spoonful of medicine   **h** The length of a football pitch

   i The mass of your head teacher   **j** The thickness of a piece of wire

2 Estimate the approximate metric length, mass or capacity of each of these.

   a This book (both length and mass)   **b** The length of your school hall

   c The capacity of a bucket   **d** The diameter of a coin, and its mass

   e The mass of a cat   **f** The amount of water in one raindrop

   g The dimensions of the room you are in   **h** Your own height and mass

3 Milo was asked to put up some decorative bunting from the top of each lamp post in his street. He had three sets of ladders he could use: a 2 metre, a 3.5 metre and a 5 metre ladder.

He looked at the lamp posts and estimated that they were about three times his height. He is slightly below average height for an adult male.

Which of the ladders should he use? Give a reason for your choice.

10.2 Converting between metric units

You should already know the relationships between these metric units.

Length		Mass	
10 **millimetres** = 1 **centimetre**		1000 **grams** = 1 **kilogram**	
1000 millimetres = 100 centimetres = 1 **metre**		1000 kilograms = 1 **tonne**	
1000 metres = 1 **kilometre**			
Capacity		**Volume**	
10 **millilitres** = 1 **centilitre**		1000 litres = 1 metre^3	
1000 millilitres = 100 centilitres = 1 **litre**		1 millilitre = 1 centimetre^3	

Note the equivalence between the units of capacity and volume:

1 litre = 1000 cm^3 which means   1 ml = 1 cm^3

You need to be able to convert from one metric unit to another.

Since the metric system is based on powers of 10, it should be easy for you to multiply or divide to change units. Work through the following examples.

EXAMPLE 1

Change:

a 732 cm to metres

b 840 mm to metres.

To change *small* units to *larger* units, always *divide*.

a 732 ÷ 100 = 7.32 so 732 cm = 7.32 m

b 840 ÷ 1000 = 0.84 so 840 mm = 0.84 m

EXAMPLE 2

Change:

a 1.2 m to centimetres

b 0.62 cm to millimetres.

To change large units to smaller units, always multiply.

a $1.2 \times 100 = 120$ so 1.2 m = 120 cm

b $0.62 \times 10 = 6.2$ so 0.62 cm = 6.2 mm

EXERCISE 10B

1 Fill in the gaps, using the information in this section.

a 125 cm = ... m
b 82 mm = ... cm
c 550 mm = ... m

d 4200 g = ... kg
e 5750 kg = ... t
f 85 ml = ... cl

g 755 g = ... kg
h 800 ml = ... l
i 200 cl = ... l

j 1035 l = ... m^3
k 530 l = ... m^3
l 34 km = ... m

2 Fill in the gaps, using the information in this section.

a 3.4 m = ... mm
b 13.5 cm = ... mm
c 0.67 m = ... cm

d 0.64 km = ... m
e 2.4 l = ... ml
f 5.9 l = ... cl

g 3.75 t = ... kg
h 0.94 cm^3 = ... l
i 21.6 l = ... cl

j 15.2 kg = ... g
k 14 m^3 = ... l
l 0.19 cm^3 = ... ml

3 Sarif wanted to buy two lengths of wood, each 2 m long, and 1.5 cm by 2 cm. He went to the local store where the types of wood were described as:

2000 mm × 15 mm × 20 mm

200 mm × 15 mm × 20 mm

200 mm × 150 mm × 2000 mm

200 mm × 15 mm × 2000 mm

Should he choose any of these? If so, which one?

4 How many square metres are there in a square kilometre?

> **ADVICE AND TIPS**
>
> The answer is not 1000

Times can be given using the **12-hour** or **24-hour clock**.

The 12-hour clock starts at midnight and runs to 12:00 at midday. After 12:59 it goes back to 1:00 and runs through to 12 again. So 7:45 could be 7:45 in the morning (am) or 7:45 in the evening (pm).

In everyday life you will usually use the 12-hour clock and add 'am' or 'pm' to indicate whether you mean before or after midday.

The 24-hour clock indicates the number of hours and minutes after midnight and has four digits. The first two digits are hours and the last two digits are minutes. So 1:45 pm in the 12-hour clock is 1345 in the 24-hour clock, meaning 13 hours and 45 minutes after midnight. 7:30am is 0730 but 7:30 pm is 1930 and so on.

Fifteen minutes after midnight is 0015 in the 24-hour clock and 12:15 am in the 12-hour clock.

Timetables usually use the 24-hour clock to avoid confusion.

7: 35 am = 0735   1: 42 pm = 1342   9:30 pm = 2130

EXAMPLE 3

A train left at 1135 and arrived at 1415.

How long did the journey take?

Do not use a calculator for this sort of question. Calculating 1415 − 1135 will not give the correct answer!

Break the journey into sections.

Here is one way:

1135 $\xrightarrow{\text{25 minutes}}$ 1200 $\xrightarrow{\text{2 hours}}$ 1400 $\xrightarrow{\text{15 minutes}}$ 1415

Total time = 2 hours 40 minutes.

EXERCISE 10C

1. Here is the timetable for four trains from Rome to Naples.

Rome (depart)	0900	0927	1027	1045
Naples (arrive)	1010	1130	1236	1230

   a How long does each of the journeys take in hours and minutes?

   b Which train was the high-speed express?

CORE

2 Here are the times of two trains from Rome to Venice.

Rome (depart)	0945	1036
Venice (arrive)	1333	1649

  a Write the four times as 12-hour clock times.

  b Find the length of each journey in hours and minutes.

3 A man arrived at his office at 0835 and left at 1520.

  a His journey home took 45 minutes. What time did he arrive home?

  b His journey to work in the morning took one hour and 20 minutes. What time did he leave home?

  c How long was he in the office?

4 Sunetra went on a coach trip to a forest park.

The coach left at 0830 and returned at 1855.

The journey took 2 hours and 20 minutes each way.

  a What time did Sunetra arrive at the forest park?

  b What time did she leave the forest park?

  c How long did she spend there?

5 **a** A car left at 0845 and arrived at its destination 3 hours and 25 minutes later.
    What time did it arrive?

  b On the return journey the car left at 1835 and arrived at 2125.
    How long did this return journey take?

6 Here is the timetable for a bus journey.

Lympstone	1729
Exton	1741
Topsham	1757
Digby	1809
Sowton	1823

How long did the journey take from Lympstone to:

  a Exton   **b** Digby   **c** Sowton?

7 The express train left at 1050 and arrived at 1324.

The slow train left at 1242 and arrived at 1629.

How much faster was the express train?

8 Pierre flew from Paris to Doha on a day when the clocks in Doha were one hour ahead of the clocks in Paris.

He left at 0740 and the flight took 5 hours and 35 minutes.

What was the local time when he arrived in Doha?

9 Boston is 5 hours behind London.

A flight from Boston to London left at 1935 and took 6 hours and 40 minutes.

What time did it arrive in London?

10.4 Currency conversions

Exchange rates are used to convert between one currency and another. They vary all the time, depending on what happens in the world's stock exchanges.

EXAMPLE 4

Assume the exchange rate is 1 US dollar = 0.7775 euros.

a How many euros is 210 US dollars?

b How many dollars is 850 euros?

a 210 US dollars = 210 × 0.7775 euros
= 163.275 euros

b 850 euros = 850 ÷ 0.7775 US dollars
= 1093.25 US dollars, rounding the answer to 2 decimal places.

EXERCISE 10D

1 The exchange rate is 1 euro = 9.9919 Hong Kong dollars.

Change 320 euros into Hong Kong dollars, giving your answer to 2 decimal places.

2 The exchange rate is 1 Russian rouble = 0.0328 US dollars.

How many US dollars could you get for 5000 Russian roubles?

3 Copy and complete this guide for changing US dollars into euros.

$5.55	$10	$50	$100	$250	$500	$1000
			€77.55			

4 1 US dollar = 85.7 Pakistani rupees.

If a gift costs 3686 Pakistani rupees, how many US dollars is that?

5 The exchange rate is 1 British pound to 1.2128 euros.

a A flight from London to Paris costs £185.45.

How many euros is that?

b A flight from Paris to London costs €209.50.

How many British pounds is that?

6 This table shows the conversion rates on one day between three currencies.

Currency	euro	US dollar	Japanese yen
1 euro =	1	1.2863	109.7406
1 US dollar =	0.7774	1	85.315
1 Japanese yen =	0.0091	0.0117	1

a Use the table to convert:

i 450 US dollars to euros

ii 225 euros to Japanese yen (Give your answer to the nearest yen.)

iii 37 000 Japanese yen to US dollars.

b Here are three amounts: 500 euros, 650 dollars, 54 000 Japanese yen.

Use the values in the table to put them in order, from smallest to largest, and complete this statement.

.... < .... < ....

7 On one day the exchange rate is 1 US dollar = 31.885 Taiwan dollars = 46.53 Indian rupees.

a What is 75 US dollars in Taiwan dollars?

b What is 75 US dollars in Indian rupees?

c Which is worth more, a Taiwan dollar or an Indian rupee?

d Complete this exchange rate: 1 Taiwan dollar = ... Indian rupees.

8 On 1 July 2005 the exchange rate was 1 US dollar = 8.2765 Chinese yuan.

On 1 July 2010 the exchange rate was 1 US dollar = 6.78099 Chinese yuan.

a How many *fewer* Chinese yuan could you buy for $50 on 1 July 2010 compared to 1 July 2005?

b Write the exchange rate on 1 July 2010 in the form:

1 Chinese yuan = ... US dollars.

Using a calculator efficiently

The aim of this topic is to make you aware of some of the keys on your calculator and how to use them to make calculations as efficiently as possible.

Consider the calculation $\dfrac{3.7 + 9.5}{0.38 + 0.16}$

If you just key in to your calculator 3.7 + 9.5 ÷ 0.38 + 0.16 you will not get the correct answer.

One method is to calculate the numerator (that is, 3.7 + 9.5) first, then the denominator and finally divide one by the other. Using the bracket keys, you can do it all in one operation:

(3.7 + 9.5) ÷ (0.38 +0.16) gives 24.44 (to 2 decimal places)

It is also useful to check whether the answer is reasonable.

$\dfrac{3.7 + 9.5}{0.38 + 0.16} \approx \dfrac{4 + 10}{0.5} = \dfrac{14}{0.5} = 28$ so the answer seems reasonable.

The check should be a calculation you can easily do in your head.

EXERCISE 10E

Use your calculator for this exercise. Try to key in each calculation as one continuous set, without writing down any intermediate values.

1 Work these out.

    a (10 − 2) × 180 ÷ 10   **b** 180 − (360 ÷ 5)

2 Work these out.

    a $\dfrac{1}{2}$ × (4.6 + 6.8) × 2.2   **b** $\dfrac{1}{2}$ × (2.3 + 9.9) × 4.5

3 Work out the value of each of these, when $a = 3.4$, $b = 5.6$, and $c = 8.8$.

    a $2(ab + ac + bc)$ **b** $\dfrac{a + b}{c}$ **c** $\dfrac{a}{b + c}$ **d** $\sqrt{a} + b + c$

4 Work these out, giving your answers to 2 decimal places.

    a $\sqrt{3.2^2 + 1.6^2}$ **b** $\sqrt{4.8^2 + 3.6^2}$

5 Work these out.

    a $7.8^3 + 3 \times 7.8$ **b** $5.45^3 - 2 \times 5.45 - 40$

6 Do these calculations as efficiently as you can. Check that your answers are sensible.

    a $\dfrac{15.89}{3.24 + 1.86}$ **b** $\dfrac{27}{18.1 + 17.95}$ **c** $\dfrac{383 + 936}{1.47 + 13.11}$

    d $\dfrac{0.34^2}{0.025^2}$ **e** $\sqrt{3.8 + 9.7 \times 2.8}$ **f** $\sqrt{32.4^2 - 17.1^2}$

Examination questions

Note: drawings in questions are NOT TO SCALE unless otherwise specified.

- If the degree of accuracy is not specified in the question, and if the answer is not exact, give the answer to three significant figures. Give answers in degrees to one decimal place.
- For π, use either your calculator value or 3.142.

PAPER 1

1 Write down the prime numbers between 20 and 30. [1]

Paper 11 Q2 May/June 2014

2 On a mountain, the temperature decreases by 6.5°C for every 1000 metres increase in height.

At 2000 metres the temperature is 10°C.

Find the temperature at 6000 metres. [2]

Paper 11 Q6 Oct/Nov 2012

3 A city has a population of five hundred and six thousand.

Write the size of the population:

a in figures [1]

b in standard form. [1]

Paper 11 Q4 May/June 2014

4 The width, w cm, of a carpet is 455 cm, correct to the nearest centimetre.

Complete this statement about the value of w.

...... $\leqslant w <$ ...... [2]

Paper 11 Q8 May/June 2014

5 Work out 72 cents as a percentage of 83 cents. [1]

Paper 11 Q3 Oct/Nov 2013

6 Work out 17.5% of 44 kg. [2]

Paper 11 Q10 Oct/Nov 2013

7 Pedro and Eva do their homework.

Pedro takes 84 minutes to do his homework.

The ratio Pedro's time : Eva's time = 7 : 6.

Work out the number of minutes Eva takes to do her homework. [2]

Paper 11 Q4 May/June 2013

8 Show that $1\dfrac{1}{2} \div \dfrac{3}{16} = 8$.

Do not use a calculator and show all the steps of your working. [2]

Paper 11 Q7 May/June 2013

9 Bruce invests $800 at a rate of 3% per year simple interest.

Calculate the **total** amount he has after 6 years. [3]

Paper 11 Q17 Oct/Nov 2013

10 Martina exchanged 200 Swiss francs (CHF) into euros (€).

The exchange rate was €1 = 1.14 CHF.

Calculate how much Martina received.

Give your answer correct to the nearest euro. [3]

Paper 11 Q16 May/June 2013

11 Use your calculator to find the value of $\dfrac{8.1^2 + 6.2^2 - 4.3^2}{2 \times 8.1 \times 6.2}$. [2]

Paper 11 Q9 Oct/Nov 2012

12 A football ground seats 28 750 people when it is full.

   a Write 28 750 correct to the nearest thousand. [1]

   b One day 17 250 people attend a football match.

      Work out 17 250 as a percentage of 28 750. [1]

Paper 11 Q4 Oct/Nov 2012

13 $y = \dfrac{2}{x^2} + \dfrac{x^2}{2}$

Find the value of y when $x = 6$.

Give your answer as a mixed number in its simplest form. [2]

Paper 11 Q9 May/June 2014

14 x and y are integers.

   h Find the value of x when $-7 < x < -5$. [1]

   i Find the value of y when $\dfrac{3}{4} < \dfrac{y}{16} < \dfrac{7}{8}$. [2]

Paper 11 Q11 May/June 2013

CORE

PAPER 3

1 **a** **i** 1 and 120 are factors of 120. [1]
Write down another factor of 120.

ii Find the highest common factor of 120 and 900. [2]

b 2   5   15   24   49   60   258   512
From the list, write down:

i a multiple of 30   **ii**   a square number   **iii**   the cube root of 8. [3]

c Give an example to show that the following are **not** true.

i An odd number multiplied by an even number gives an odd number. [1]

ii The cube of a negative number is positive. [1]

d Use <, > or = to complete the following statements.
Each symbol may be used more than once.

i $0.5 \ldots \ldots \dfrac{3}{8}$ **ii** $1.5 \ldots \ldots 105\%$ **iii** $0.78 \ldots \ldots \dfrac{11}{14}$ [3]

Paper 31 Q2 Oct/Nov 2013

2 **a** Luka earns $475 each week.

i He works for 38 hours each week.
How much does he earn for each hour he works? [1]

ii Luka pays $175 in rent each week.
Write the amount he pays in rent as a fraction of his weekly earnings.
Give your answer in its lowest terms. [2]

iii He spends $\dfrac{7}{20}$ of his weekly earnings on bills.
How much money does he have left after paying rent and bills? [2]

b Luka's weekly earnings of $475 are increased by 6%.
Calculate his new weekly earnings. [2]

c Luka has saved $350.
He invests this for 2 years at a rate of 4% per year compound interest.
How much interest does he receive after 2 years? [3]

Paper 31 Q2 Oct/Nov 2012

3 **a** On a map, the height of Hillibar Station is 1047 m and the height of Sular Junction is 297 m.

i Calculate the difference in these heights. [1

ii The temperature falls by 1°C for every 100 m increase in height.
One day the temperature in Sular Junction is 19°C.
Work out the temperature at Hillibar Station. [1

iii Write 297 correct to the nearest ten. [1

iv Write 1047 correct to the nearest hundred. [1

b **i** Kim arrives at Hillibar Station at 1235.
The taxi to her hotel takes 27 minutes.
Work out the time Kim arrives at her hotel. [1

ii Henry takes 17 minutes to walk from his home to Sular Junction.

He must arrive there by 1043.

Work out the latest time he can leave home. [1]

c Here is part of a train timetable.

Each journey from Sular Junction to Hillibar Station takes the same time.

Sular Junction *departs*	1059	1232	1448
Hillibar Station *arrives*	1235	1408	

i Complete the timetable. [2]

ii The distance between Sular Junction and Hillibar Station is 64 km.

Calculate the average speed, in kilometres per hour, of a train between these two stations. [2]

iii Joel arrives at Sular Junction at 1148.

At what time is the next train to Hillibar Station due to depart? [1]

Paper 31 Q1 May/Jun 2013

4 **a** **i** Write down all the factors of 22. [2]

ii Write down a multiple of 13 between 30 and 50. [1]

b 1   2   6   9   15   17   19   21   27

i Write down all the prime numbers in this list. [2]

ii Write down a cube number from this list. [1]

c **i** Write 0.0035 in standard form. [1]

ii Calculate $6 \times 10^6 \div 1.5 \times 10^2$.

Write your answer in standard form. [2]

Paper 31 Q6 May/Jun 2013

EXTENDED

1 Write the following numbers correct to one significant figure.

    a 7682 [1]

    b 0.076 82 [1]

Paper 22 Q1 Oct/Nov 2012

2 Use a calculator to work out the following.

    a $3(-4 \times 6^2 - 5)$ [1]

    b $\sqrt{3} \times \tan 30° + \sqrt{2} \times \sin 45°$ [1]

Paper 22 Q2 Oct/Nov 2013

3 The mass of 1 cm³ of copper is 8.5 grams, correct to 1 decimal place.

Complete this statement about the total mass, T grams, of 12 cm³ of copper.

…… $< T <$ …… [2]

Paper 22 Q6 May/Jun 2014

4 Shade the required region in each Venn diagram.

$A' \cup B$

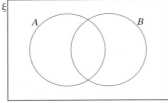

$A' \cap B'$ [2]

Paper 22 Q1 May/Jun 2013

5 Calculate $(4 \times 10^8) + (2.5 \times 10^7)$.

Give your answer in standard form. [2]

Paper 22 Q4 May/Jun 2013

6 Write the following in order of size, smallest first.

    19% $\dfrac{1}{5}$ $\sqrt{0.038}$ $\sin 11.4°$ 0.719^5 [2]

Paper 22 Q1 Oct/Nov 2013

7 Without using your calculator, work out $\dfrac{5}{6} \Big/ \dfrac{1}{2} \times 1\dfrac{1}{2} \bigcirc$

Write down all the steps in your working. [3]

Paper 22 Q8 May/Jun 2014

8 George and his friend Jane buy copies of the same book on the internet.

George pays $16.95 and Jane pays £11.99 on a day when the exchange rate is $1 = £0.626.

Calculate, in dollars, how much more Jane pays. [2]

Paper 22 Q6 May/Jun 2013

9 Joe measures the side of a square correct to 1 decimal place.

He calculates the **upper** bound for the area of the square as 37.8225 cm^2.

Work out Joe's measurement for the side of the square. [2]

Paper 22 Q8 May/Jun 2013

10 List all the **prime numbers** that satisfy this inequality.

$16 < 2 - x < 48$ [3]

Paper 22 Q11 Oct/Nov 2012

11 At the beginning of July, Kim had a mass of 63 kg.

At the end of July, his mass was 61 kg.

Calculate the percentage loss in Kim's mass. [3]

Paper 22 Q9 May/Jun 2014

12 The number of spectators at the 2010 World Cup match between Argentina and Mexico was 82 000 correct to the nearest thousand.

If each spectator paid 2600 Rand to attend the game, what is the lower bound for the total amount paid?

Write your answer in standard form. [3]

Paper 22 Q7 Oct/Nov 2012

13 Anita buys a computer for $391 in a sale.

The sale price is 15% less than the original price.

Calculate the original price of the computer. [3]

Paper 22 Q11 May/Jun 2014

PAPER 4

1 Jane and Kate share $240 in the ratio 5 : 7.

   a Show that Kate received $140. [2]

   b Jane and Kate each spent $20.

     Find the new ratio of Jane's remaining money : Kate's remaining money.

     Give your answer in its simplest form. [2]

   c Kate invests $120 for 5 years at 4% per year simple interest.

     Calculate the total amount Kate has after 5 years. [3]

   d Jane invests $80 for 3 years at 4% per year compound interest.

     Calculate the total amount Jane has after 3 years.

     Give your answer correct to the nearest cent. [3]

   e An investment of $200 for 2 years at 4% compound interest is the same as an investment of $200 for 2 years at r% per year simple interest.

     Find the value of r. [3]

   Paper 42 Q1 May/Jun 2014

2 Last year Mukthar earned $18 900.

   He did not pay tax on $5500 of his earnings.

   He paid 24% tax on his remaining earnings.

   a **i** Calculate how much tax Mukthar paid last year. [2]

      ii Calculate how much Mukthar earned each month after tax had been paid. [2]

   b This year Mukthar now earns $19 750.50.

     Calculate the percentage increase from $18 900. [2]

   c Mukthar has $1500 to invest in one of the following ways.

     • **Account A** paying simple interest at a rate of 4.1% per year

     • **Account B** paying compound interest at a rate of 3.3% per year

     Which account will be worth more after **3 years** and by how much? [5]

   Paper 42 Q1 Oct/Nov 2013

3 A tennis club has 560 members.

   a The ratio of men : women : children = 5 : 6 : 3.

     i Show that the club has 240 women members. [2]

     ii How many members are children? [1]

   b $\frac{5}{8}$ of the 240 women members play in a tournament.

     How many women do **not** play in the tournament? [2]

   c The annual membership fee in 2013 is $198 for each adult and $75 for each child.

     i Calculate the total amount the 560 members pay in 2013. [2]

     ii The adult fee of $198 in 2013 is 5.6% more than the fee in 2012.

       Calculate the adult fee in 2012. [3]

   Paper 42 Q1 May/Jun 2013

4 Noma flies from Johannesburg to Hong Kong.

Her plane leaves Johannesburg at 1845 and arrives in Hong Kong 13 hours and 25 minutes later.

The local time in Hong Kong is 6 hours ahead of the time in Johannesburg.

a At what time does Noma arrive in Hong Kong? [2]

b Noma sleeps for part of the journey.

The time she spends sleeping is given by the ratio

sleeping : awake = 3 : 4

Calculate how long Noma sleeps during the journey.

Give your answer in hours and minutes. [2]

c i The distance from Hong Kong to Johannesburg is 10 712 km.

The time taken for the journey is 13 hours and 25 minutes.

Calculate the average speed of the plane for this journey. [2]

ii The plane uses fuel at the rate of 1 litre for every 59 metres travelled.

Calculate the number of litres of fuel used for the journey from Johannesburg to Hong Kong.

Give your answer in standard form. [4]

d The cost of Noma's journey is 10 148 South African Rand (R).

This is an increase of 18% on the cost of the journey one year ago.

Calculate the cost of the same journey one year ago. [3]

Paper 42 Q7 Oct/Nov 2013

Why this chapter matters

Art

Mathematicians think that famous works of art are often based on the 'golden ratio'. This is the ratio of one part of the art to another. It seems that human brains find the 'golden ratio' very attractive.

Science

Science needs mathematics. In 1962 a space probe went off course because someone had got a mathematical formula wrong in its programming.

Sport

Is mathematics a sport? There are national and international competitions each year that use mathematics. University students compete in the annual 'Mathematics Olympiad' and there is a world Sudoku championship each year. Lots of sporting activities require maths too, such as throwing a javelin (angles).

Language

But the best description of mathematics is that it is a language.

It is the only language which people in all countries understand. Everyone understands the numbers on this stamp even if they do not speak the language of the country.

Algebra is an important part of the language of mathematics. It comes from the Arabic *al-jabr*. It was first used in a book written in 820 CE by a Persian mathematician called al-Khwarizmi.

The use of symbols grew until the 17th century when a French mathematician called Descartes developed them into the sort of algebra we use today.

Chapter

Algebraic representation and formulae

Topics	Level	Key words
The language of algebra	**CORE**	expression, symbol, variable, formula, formulae, equation, term, solving
Substitution into formulae	**CORE**	substitute
Rearranging formulae	**CORE**	rearrange, subject
More complicated formulae	**EXTENDED**	

In this chapter you will learn how to:

CORE	EXTENDED
• Use letters to express generalised numbers and express basic arithmetic processes algebraically. (C2.1 and E2.1) • Substitute numbers for words and letters in formulae. (C2.1 and E2.1) • Transform simple formulae. (C2.1 and E2.1) • Construct simple expressions and set up simple equations. (C2.1)	• Construct and transform complicated formulae and equations. (E2.1)

Algebra is based on the idea that if something works with numbers, it will work with letters. The main difference is that when you work only with numbers, the answer is also a number. When you work with letters, you get an **expression** as the answer.

Algebra follows the same rules as arithmetic, and uses the same **symbols** ($+$, $-$, $\times$ and $\div$). Below are seven important algebraic rules.

- Write '4 more than x' as $4 + x$ or $x + 4$.
- Write '6 less than p' or 'p minus 6' as $p - 6$.
- Write '4 times y' as $4 \times y$ or $y \times 4$ or $4y$. The last one of these is the neatest way to write it.
- Write 'b divided by 2' as $b \div 2$ or $\dfrac{b}{2}$.
- When a number and a letter or a letter and a letter appear together, there is a hidden multiplication sign between them. So, $7x$ means $7 \times x$ and ab means $a \times b$.
- Always write '$1 \times x$' as x.
- Write 't times t' as $t \times t$ or t^2.

Here are some algebraic words that you need to know.

Variable: This is what the letters used to represent numbers are called. These letters can take on any value, so they 'vary'.

Expression: This is any combination of letters and numbers. For example, $2x + 4y$ and $\dfrac{p - 6}{5}$ are expressions.

Equation: An equation contains an equals sign and at least one variable. The important fact is that a value can be found for the variable. This is called **solving** the equation. You will learn more about equations in another chapter.

Formula: These are like equations in that they contain an equals sign, but there is more than one variable and they are rules for working out amounts such as area or the cost of taxi fares. You refer to more than one formula as formulae.

For example, $V = x^3$, $A = \frac{1}{2}bh$ and $C = 3 + 4m$ are all **formulae**.

Term: These are the separate parts of expressions, equations or formula.

For example, in $3x + 2y - 7$, there are three terms: $3x$, $+2y$ and -7.

EXAMPLE 1

One side of this rectangle is three centimetres longer than the other.

Find a formula for the area (A) in square centimetres, and the perimeter (P) in centimetres.

$(x + 3)$ cm

x cm

Area = width × length

$A = x(x + 3)$ (Leave out the multiplication sign.)

Perimeter = distance around the outside

$P = x + (x + 3) + x + (x + 3)$

You can simplify this to:

$P = 4x + 6$

EXERCISE 11A

1 Write down the algebraic expression for:

a 2 more than x

b 6 less than x

c k more than x

d x minus t

e x added to 3

f d added to m

g y taken away from b

h p added to t added to w

i 8 multiplied by x

j h multiplied by j

k x divided by 4

l 2 divided by x

m y divided by t

n w multiplied by t

o a multiplied by a

p g multiplied by itself.

2 Asha, Bernice and Charu are three sisters. Bernice is x years old. Asha is three years older than Bernice. Charu is four years younger than Bernice.

a How old is Asha?

b How old is Charu?

3 An approximation method of converting from degrees Celsius to degrees Fahrenheit is given by this rule:

Multiply by 2 and add 30.

Using C to stand for degrees Celsius and F to stand for degrees Fahrenheit, complete this formula.

$F = \ldots\ldots$

4 Cows have four legs. Which of these formulae connects the number of legs (L) and the number of cows (C)?

a $C = 4L$

b $L = C + 4$

c $L = 4C$

d $L + C = 4$

5 a Lakmini has three bags of marbles. Each bag contains n marbles. How many marbles does she have altogether?

b Rushani gives her another three marbles. How many marbles does Lakmini have now?

c Lakmini puts one of her new marbles in each bag. How many marbles are there now in each bag?

d Lakmini takes two marbles out of each bag. How many marbles are there now in each bag?

CORE

6 Lee has *n* cubes.

- Anil has twice as many cubes as Lee.
- Reza has two more than Lee.
- Dale has three fewer than Lee.
- Chen has three more than Anil.

How many cubes does each person have?

7 **a** I go shopping with $10 and spend $6. How much do I have left?

b I go shopping with $10 and spend $*x*. How much do I have left?

c I go shopping with $*y* and spend $*x*. How much do I have left?

d I go shopping with $3*x* and spend $*x*. How much do I have left?

8 Give the total cost of:

a five books at $15 each

b *x* books at $15 each

c four books at $*A* each

d *y* books at $*A* each.

9 A boy went shopping with *A* dollars. He spent *B* dollars. How much did he have left?

10 Five ties together cost $*A*. What is the cost of one tie?

11 **a** My dad is 72 years old and I am *T* years old. How old shall we each be in *x* years' time?

b My mum is 64 years old. In two years' time she will be twice as old as I am. What age am I now?

12 I am twice as old as my son. I am *T* years old.

a How old is my son?

b How old will my son be in four years' time?

c How old was I *x* years ago?

13 Write down the perimeter of each of these figures.

a

2*x*

Square

b

4*m*

Equilateral triangle

c

3*t*

Regular hexagon

14 Write down the number of marbles each student ends up with.

Student	Action	Marbles
Andrea	Start with three bags each containing n marbles and give away one marble from each bag	
Barak	Start with three bags each containing n marbles and give away one marble from one bag	
Ahmed	Start with three bags each containing n marbles and give away two marbles from each bag	
Dina	Start with three bags each containing n marbles and give away n marbles from each bag	
Emma	Start with three bags each containing n marbles and give away n marbles from one bag	
Hana	Start with three bags each containing n marbles and give away m marbles from each bag	

15 The answer to $3 \times 4m$ is $12m$.

Write down two *different* expressions for which the answer is $12m$.

11.2 Substitution into formulae

A formula expresses the value of one variable as the other variables in the formula change.

EXAMPLE 2

The formula for the area of a trapezium is:

$$A = \frac{(a + b)h}{2}$$

Find the area of the trapezium when $a = 5$, $b = 9$ and $h = 3$.

$$A = \frac{(5 + 9) \times 3}{2} = \frac{14 \times 3}{2} = 21$$

Always **substitute** the numbers for the letters before trying to work out the value of the expression. You are less likely to make a mistake this way.

CORE

EXERCISE 11B

1 Find the value of $3x + 2$ when:

   a $x = 2$ **b** $x = 5$ **c** $x = 10$.

> **ADVICE AND TIPS**
>
> It helps to put the numbers in brackets.
> $3(2) + 2 = 6 + 2 = 8$
> $3(5) + 2 = 15 + 2 = 17$
> etc …

2 Find the value of $4k - 1$ when:

   a $k = 1$ **b** $k = 3$ **c** $k = 11$.

3 Find the value of $5 + 2t$ when:

   a $t = 2$ **b** $t = 5$ **c** $t = 12$.

4 Evaluate $15 - 2f$ when: **a** $f = 3$ **b** $f = 5$ **c** $f = 8$.

5 Evaluate $5m + 3$ when: **a** $m = 2$ **b** $m = 6$ **c** $m = 15$.

6 Evaluate $3d - 2$ when: **a** $d = 4$ **b** $d = 5$ **c** $d = 20$.

7 A taxi company uses this rule to calculate their fares.

      Fare = \$2.50 plus \$0.50 per kilometre.

   a How much is the fare for a journey of 3 km?

   b Farook pays \$9.00 for a taxi ride. How far was the journey?

   c Mayra knows that her house is 5 kilometres from town. She has \$5.50 left in her purse after a night out. Has she got enough for a taxi ride home?

8 Kaz knows that x, y and z have the values 2, 8 and 11, but she does not know which variable has which value.

   a What is the maximum value that the expression $2x + 6y - 3z$ could have?

   b What is the minimum value that the expression $5x - 2y + 3z$ could have?

> **ADVICE AND TIPS**
>
> You could just try all combinations, but if you think for a moment you will find that the $6y$ term must give the largest number. This will give you a clue to the other terms.

9 The formula for the area, A, of a rectangle with length l and width w is $A = lw$.

   The formula for the area, T, of a triangle with base b and height h is $T = \frac{1}{2}bh$.

   Find values of l, w, b and h so that $A = T$.

10 Find the value of $\dfrac{8 \times 4h}{5}$ when: **a** $h = 5$ **b** $h = 10$ **c** $h = 25$.

11 Find the value of $\dfrac{25 - 3p}{2}$ when: **a** $p = 4$ **b** $p = 8$ **c** $p = 10$.

12 Evaluate $\dfrac{x}{3}$ when: **a** $x = 6$ **b** $x = 24$ **c** $x = -30$.

13 Evaluate $\dfrac{A}{4}$ when: **a** $A = 12$ **b** $A = 10$ **c** $A = -20$.

14 Evaluate $\dfrac{12}{y}$ when: **a** $y = 2$ **b** $y = 4$ **c** $y = 6$.

15 Evaluate $\dfrac{24}{x}$ when: **a** $x = 2$ **b** $x = 3$ **c** $x = 16$.

16 A holiday cottage costs 150 dollars per day to rent.

A group of friends decide to rent the cottage for seven days.

a Which formula represents the cost of the rental for each person if there are n people in the group? Assume that they share the cost equally.

$$\dfrac{150}{n} \qquad \dfrac{150}{7n} \qquad \dfrac{1050}{n} \qquad \dfrac{150n}{n}$$

> **ADVICE AND TIPS**
>
> To check your choice in part **a**, make up some numbers and try them in the formula. For example, take $n = 5$.

b Eventually 10 people go on the holiday. When they get the bill, they find that there is a discount for a seven-day rental.

After the discount, they each find it cost them 12.50 dollars less than they expected.

How much does a 7-day rental cost?

17 **a** p is an odd number and q is an even number.

Say if each of these expressions is odd or even.

i $p + q$ **ii** $p^2 + q$ **iii** $2p + q$ **iv** $p^2 + q^2$

> **ADVICE AND TIPS**
>
> There are many answers for **b**, and **a** should give you a clue.

b x, y and z are all odd numbers.

Write an expression, using x, y and z, so that the value of the expression is always even.

18 A formula for the cost of delivery, in dollars, of orders from a warehouse is:

$$D = 2M - \dfrac{C}{5}$$

where D is the cost of the delivery, M is the distance in kilometres from the store and C is the cost of the goods to be delivered.

> **ADVICE AND TIPS**
>
> Note: a rebate is a refund of some of the money that someone has already paid for goods or services.

a How much does the delivery cost when $M = 30$ and $C = 200$?

b Rahim buys goods worth $300 and lives 10 kilometres from the store.

i The formula gives the cost of delivery as a negative value. What is this value?

ii Explain why Rahim will not get a rebate from the store.

c Maya buys goods worth $400. She calculates that her cost of delivery will be zero. What is the greatest distance that Maya could live from the store?

The **subject** of a formula is the **variable** (letter) in the formula that stands on its own, usually on the left-hand side of the equals sign. For example, x is the subject of each of these equations.

$$x = 5t + 4 \qquad\qquad x = 4(2y - 7) \qquad\qquad x = \frac{1}{t}$$

To change the existing subject to a different variable, you have to **rearrange** the formula to get that variable on the left-hand side.

EXAMPLE 3

Make m the subject of this formula. $\qquad T = m - 3$

Add 3 to both sides. $\qquad\qquad T + 3 = m$

Reverse the formula. $\qquad\qquad m = T + 3$

EXAMPLE 4

From the formula $P = 4t$, express t in terms of P.

(This is another common way of asking you to make t the subject.)

Divide both sides by 4: $\quad \dfrac{P}{4} = \dfrac{4t}{4}$

$\dfrac{P}{4} = t$

Reverse the formula: $\qquad t = \dfrac{P}{4}$

EXAMPLE 5

From the formula $C = 2m^2 + 3$, make m the subject.

Subtract 3 from both sides so that the $2m^2$ is on its own. $\quad C - 3 = 2m^2$

Divide both sides by 2: $\qquad\qquad\qquad\qquad \dfrac{C - 3}{2} = \dfrac{2m^2}{2}$

Reverse the formula: $\qquad\qquad\qquad\qquad m^2 = \dfrac{C - 3}{2}$

Take the square root on both sides: $\qquad\qquad m = \sqrt{\dfrac{C - 3}{2}}$

EXERCISE 11C

1 $T = 3k$ Make k the subject.

2 $X = y - 1$ Express y in terms of X.

3 $Q = \dfrac{p}{3}$ Express p in terms of Q.

4 $A = 4r + 9$ Make r the subject.

5 $W = 3n - 1$ Make n the subject.

6 $p = m + t$ **a** Make m the subject.   **b** Make t the subject.

7 $g = \dfrac{m}{v}$ Make m the subject.

8 $t = m^2$ Make m the subject.

9 $C = 2\pi r$ Make r the subject.

10 $A = bh$ Make b the subject.

11 $P = 2l + 2w$ Make l the subject.

12 $m = p^2 + 2$ Make p the subject.

> **ADVICE AND TIPS**
>
> **Remember** about inverse operations, and the rule 'change sides, change signs'.

CORE

13 The formula for converting temperatures in degrees Fahrenheit (F) to temperatures in degrees Celsius (C) is $C = \frac{5}{9}(F - 32)$.

     a Show that when $F = -40$, C is also equal to -40.

     b Find the value of C when $F = 68$.

     c Show that the formula can be rearranged as $F = \dfrac{9C}{5} + 32$

14 $v = u + at$ **a** Make a the subject.   **b** Make t the subject.

15 $A = \dfrac{1}{4}\pi d^2$ Make d the subject.

16 $W = 3n + t$ **a** Make n the subject.   **b** Express t in terms of n and W.

17 $x = 5y - w$ **a** Make y the subject.   **b** Express w in terms of x and y.

18 $k = 2p^2$ Make p the subject.

19 $v = u^2 - t$ **a** Make t the subject.   **b** Make u the subject.

20 $k = m + n^2$ **a** Make m the subject.   **b** Make n the subject.

21 $T = 5r^2$ Make r the subject.

22 $K = 5n^2 + w$ **a** Make w the subject.   **b** Make n the subject.

To find the value of a letter you need to make it the subject of the formula.

You often need to rearrange the formula to do this.

Some formulae will need a number of separate steps to rearrange them.

EXAMPLE 6

$$a = c + \frac{d}{y^2}$$

Make y the subject.

Subtract c from both sides: $\quad a - c = \dfrac{d}{y^2}$

Multiply both sides by y^2: $\quad y^2(a - c) = d$

Divide both sides by $a - c$: $\quad y^2 = \dfrac{d}{a - c}$

Take the square root of y^2: $\quad y = \sqrt{\dfrac{d}{a - c}}$

EXERCISE 11D

1 $a^2 + b^2 = c^2$

   a Find the value of a when $b = 6.0$ and $c = 6.5$.

   b Make a the subject.

2 $s = ut + \frac{1}{2}at^2$ is a formula used in mechanics.

   a Find the value of s when $u = -5$, $t = 4$ and $a = 10$.

   b Make a the subject.

3 $a = \dfrac{b + 2}{c}$

   a Make b the subject.

   b Make c the subject.

4 $p = \dfrac{r}{t - 3}$

Make t the subject.

5 $d = \dfrac{12}{1 + \sqrt{e}}$

Make e the subject.

EXTENDED

6 $v^2 = u^2 + 2as$

   a Find the value of v when $u = 3$, $a = 2$ and $s = 4$.

   b Make u the subject.

   c Make s the subject.

7 $T = 2\pi\sqrt{\dfrac{L}{G}}$

   a Make L the subject.

   b Show that $G = L\left(\dfrac{2\pi}{T}\right)^2$

8 $D = \pi R^2 - \pi r^2$

   a Make R the subject.

   b Make r the subject.

   c Make π the subject.

9 $3x^2 - 4y^2 = 11$

   a Find the value of x when $y = 4$.

   b Make x the subject.

   c Make y the subject.

10 $T = 2\sqrt{\dfrac{a}{c + 3}}$

   a Make a the subject.

   b Make c the subject.

11 $a^2 = b^2 + c^2 - 2bcT$

   Make T the subject.

12 $uv = fu + fv$ is a formula used in optics.

   a Find the value of f when $u = 20$ and $v = 30$.

   b Make f the subject.

   c Make u the subject.

   d Make v the subject.

Why this chapter matters

How can algebra save your life? Read on...

When you drive a car you must leave a safe distance between you and the car in front. What is a safe distance? How long will it take you to stop?

The **stopping distance** has two parts:

- **thinking distance** is the distance travelled before your brain reacts and you apply the brakes

- **braking distance** is the time it takes the car to come to a complete stop once the brakes have been applied.

Stopping distances vary according to the weather conditions, the road conditions and vehicle. This table shows stopping distances for a car with good brakes on a dry, level road with good visibility.

Typical stopping distances

32 kph	6 m	6 m	**12 metres** 3 car lengths
48 kph	9 m	14 m	**23 metres** 6 car lengths
64 kph	12 m	24 m	**36 metres** 9 car lengths
80 kph	15 m	38 m	**53 metres** 13 car lengths
96 kph	18 m	55 m	**73 metres** 18 car lengths
112 kph	21 m	75 m	**96 metres** 24 car lengths

thinking distance   braking distance   Average car length = 4 metres

Data from the UK Highway Code

You will notice that the thinking distance increases steadily with speed. It increases by 3 metres for every 16 km/h increase in speed. If you double the speed you double the thinking distance. You can show this in a simple algebraic expression:

thinking distance in metres = $0.1875x$
(where x is the speed in kilometres per hour).

The relationship between speed and braking distance is more complicated. It can be shown as:

braking distance = $0.006x^2$

So:

total stopping distance = $0.1875x + 0.006x^2$

This is a quadratic expression and you can use it to work out stopping distance at any speed (value of x). You will learn about both simple and quadratic expressions in this chapter.

Chapter

12 Algebraic manipulation

Topics	Level	Key words
1 Simplifying expressions	CORE	simplifying, coefficient, like terms
2 Expanding brackets	CORE	expand, simplification
3 Factorisation	CORE	common factors, factorise, factorisation
4 Multiplying two brackets	EXTENDED	quadratic expression, linear, quadratic expansion
5 Quadratic factorisation	EXTENDED	difference of two squares, brackets
6 Algebraic fractions	EXTENDED	cancel, single fractions

In this chapter you will learn how to:

CORE	EXTENDED
• Manipulate directed numbers. (C2.2 and E2.2) • Use brackets and extract common factors. (C2.2 and E2.2)	• Expand products of algebraic expressions. (E2.2) • Factorise where possible expressions of the form: $ax + bx + kay + kby$, $a^2x^2 - b^2y^2$; $a^2 + 2ab + b^2$; $ax^2 + bx + c$. (E2.2) • Manipulate algebraic fractions such as: $\dfrac{x}{3} + \dfrac{x-4}{2}$, $\dfrac{2x}{3} - \dfrac{3(x-5)}{2}$, $\dfrac{3a}{4} \times \dfrac{5ab}{3}$, $\dfrac{3a}{4} - \dfrac{9a}{10}$, $\dfrac{1}{x-2} + \dfrac{2}{x-3}$. (E2.3) • Factorise and simplify rational expressions such as: $\dfrac{x^2 - 2x}{x^2 - 5x + 6}$. (E2.3)

Simplifying an algebraic expression means making it neater and, usually, shorter by combining its terms where possible.

Multiplying expressions

When you multiply algebraic expressions, first you multiply the numbers, then the letters.

EXAMPLE 1

Simplify each expression.

a $2 \times t$ **b** $m \times t$ **c** $2t \times 5$ **d** $3y \times 2m$

The convention is to write the number first then the letters. The number in front of the letter is called the **coefficient**.

a $2 \times t = 2t$ **b** $m \times t = mt$ **c** $2t \times 5 = 10t$ **d** $3y \times 2m = 6my$

You would not normally be penalised for writing $2ba$ instead of $2ab$, but you should never write $ab2$ as this can be confused with powers, so *always* write the number first.

EXAMPLE 2

Simplify each expression.

a $t \times t$ **b** $3t \times 2t$ **c** $3t^2 \times 4t$ **d** $2t^3 \times 4t^2$

Multiply the same variables, using powers. The indices are added together.

a $t \times t = t^2$ (Remember: $t = t^1$)   **b** $3t \times 2t = 6t^2$

c $3t^2 \times 4t = 12t^3$ **d** $2t^3 \times 4t^2 = 8t^5$

EXERCISE 12A

1 Simplify these expressions.

a $2 \times 3t$ **b** $5y \times 3$ **c** $2w \times 4$

d $5b \times b$ **e** $2w \times w$ **f** $4p \times 2p$

g $3t \times 2t$ **h** $5t \times 3t$ **i** $m \times 2t$

j $5t \times q$ **k** $n \times 6m$ **l** $3t \times 2q$

m $5h \times 2k$ **n** $3p \times 7r$

> **ADVICE AND TIPS**
>
> Remember to multiply numbers and add indices.

2 **a** Which of these expressions are equivalent?

$2m \times 6n$ $4m \times 3n$ $2m \times 6m$ $3m \times 4n$

b The expressions $2x$ and x^2 have the same value for only two values of x. What are these values?

CORE

3 A square and a rectangle have the same area.

The rectangle has sides of length $2x$ cm and $8x$ cm.

What is the length of a side of the square?

4 Simplify these expressions.

a $y^2 \times y$	**b** $3m \times m^2$	**c** $4t^2 \times t$
d $3n \times 2n^2$	**e** $t^2 \times t^2$	**f** $h^3 \times h^2$
g $3n^2 \times 4n^3$	**h** $3a^4 \times 2a^3$	**i** $k^5 \times 4k^2$
j $-t^2 \times -t$	**k** $-4d^2 \times -3d$	**l** $-3p^4 \times -5p^2$
m $3mp \times p$	**n** $3mn \times 2m$	**o** $4mp \times 2mp$

Collecting like terms

Like terms are those that are multiples of the same variable or of the same combination of variables. For example, a, $3a$, $9a$, $\frac{1}{4}a$ and $-5a$ are all like terms.

So are $2xy$, $7xy$ and $-5xy$, and so are $6x^2$, x^2 and $-3x^2$.

Collecting like terms generally involves two steps.

● Collect like terms into groups.
● Then combine the like terms in each group.

Only like terms can be added or subtracted to simplify an expression. For example:

$a + 3a + 9a - 5a$	simplifies to	$8a$
$2xy + 7xy - 5xy$	simplifies to	$4xy$

Note that the variable does not change. You just have to combine the coefficients.

For example:

$6x^2 + x^2 - 3x^2 = (6 + 1 - 3)x^2 = 4x^2$

However, an expression such as $4p + 8t + 5x - 9$ cannot be simplified, because $4p$, $8t$, $5x$ and 9 are *not like terms*, which *cannot* be combined.

EXAMPLE 3

Simplify this expression.

$7x^2 + 3y - 6z + 2x^2 + 3z - y + w + 9$

Write out the expression: $7x^2 + 3y - 6z + 2x^2 + 3z - y + w + 9$

Then collect like terms: $\boxed{7x^2 + 2x^2}\ \boxed{+3y - y}\ \boxed{-6z + 3z}\ \boxed{+ w}\ \boxed{+ 9}$

Then combine them: $9x^2\quad +\quad 2y\quad -\quad 3z\quad + w + 9$

So, the expression in its simplest form is:

$9x^2 + 2y - 3z + w + 9$

CORE

EXERCISE 12B

1. Jared is given t, Jatan has $3 more than Jared, and Jasmin has $2t$.

   a How much more money has Jasmin than Jared?

   b How much do the three of them have altogether?

2. Write down an expression for the perimeter of each of these shapes.

   a

   b

   c

3. Write each of these expressions in a shorter form.

   a $a + a + a + a + a$

   b $c + c + c + c + c + c$

   c $4e + 5e$

   d $f + 2f + 3f$

   e $5j + j - 2j$

   f $9q - 3q - 3q$

   g $3r - 3r$

   h $2w + 4w - 7w$

   i $5x^2 + 6x^2 - 7x^2 + 2x^2$

   j $8y^2 + 5y^2 - 7y^2 - y^2$

   k $2z^2 - 2z^2 + 3z^2 - 3z^2$

> **ADVICE AND TIPS**
>
> The term a has a coefficient of 1, i.e. $a = 1a$, but you do not need to write the 1.

4. Simplify each of these expressions.

   a $3x + 4x$

   b $5t - 2t$

   c $-2x - 3x$

   d $-k - 4k$

   e $m^2 + 2m^2 - m^2$

   f $2y^2 + 3y^2 - 5y^2$

> **ADVICE AND TIPS**
>
> Remember that only **like terms** can be added or subtracted.
> If all the terms cancel out, just write 0 rather than $0x^2$, for example.

5. Simplify each of these expressions.

   a $5x + 8 + 2x - 3$

   b $7 - 2x - 1 + 7x$

   c $4p + 2t + p - 2t$

   d $8 + x + 4x - 2$

   e $3 + 2t + p - t + 2 + 4p$

   f $5w - 2k - 2w - 3k + 5w$

   g $a + b + c + d - a - b - d$

   h $9k - y - 5y - k + 10$

6. Simplify these expressions. (Be careful – two of them will not simplify.)

   a $c + d + d + d + c$

   b $2d + 2e + 3d$

   c $f + 3g + 4h$

   d $5u - 4v + u + v$

   e $4m - 5n + 3m - 2n$

   f $3k + 2m + 5p$

   g $2v - 5w + 5w$

   h $2w + 4y - 7y$

   i $5x^2 + 6x^2 - 7y + 2y$

   j $8y^2 + 5z - 7z - 9y^2$

   k $2z^2 - 2x^2 + 3x^2 - 3z^2$

7 Find the perimeter of each of these shapes, giving it in its simplest form.

a

b

c

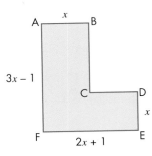

8 $3x + 5y + 2x - y = 5x + 4y$

Write down two other *different* expressions that are equal to $5x + 4y$.

9 Find the missing terms to make these equations true.

   a $4x + 5y + \ldots\ldots\ldots - \ldots\ldots\ldots = 6x + 3y$

   b $3a - 6b - \ldots\ldots\ldots + \ldots\ldots\ldots = 2a + b$

10 ABCDEF is an L-shape.

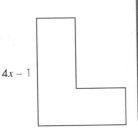

$AB = DE = x$

$AF = 3x - 1$ and $EF = 2x + 1$

> **ADVICE AND TIPS**
>
> Make sure your explanation uses expressions. Do not try to explain in words alone.

   a Explain why the length $BC = 2x - 1$.

   b Find the perimeter of the shape in terms of x.

   c If $x = 2.5$ cm, what is the perimeter of the shape?

11 A teacher asks her class to work out the perimeter of this L-shape.

Tia says: 'There is information missing so you cannot work out the perimeter.'

Maria says: 'The perimeter is $4x - 1 + 4x - 1 + 3x + 2 + 3x + 2$.'

Who is correct?

Explain your answer.

$4x - 1$

$3x + 2$

Expanding brackets

In mathematics, to '**expand**' usually means 'multiply out'. For example, expressions such as $3(y + 2)$ and $4y^2(2y + 3)$ can be expanded by multiplying them out.

Remember that there is an invisible multiplication sign between the outside number and the opening bracket. So $3(y + 2)$ is really $3 \times (y + 2)$ and $4y^2(2y + 3)$ is really $4y^2 \times (2y + 3)$.

You expand by multiplying *everything inside* the brackets by what is outside the brackets.

So in the case of the two examples above:

$$3(y + 2) = 3 \times (y + 2) = 3 \times y + 3 \times 2 = 3y + 6$$
$$4y^2(2y + 3) = 4y^2 \times (2y + 3) = 4y^2 \times 2y + 4y^2 \times 3 = 8y^3 + 12y^2$$

Look at these next examples of expansion, which show clearly how the term outside the brackets has been multiplied with the terms inside them.

$2(m + 3) = 2m + 6$ $y(y^2 - 4x) = y^3 - 4xy$

$3(2t + 5) = 6t + 15$ $3x^2(4x + 5) = 12x^3 + 15x^2$

$m(p + 7) = mp + 7m$ $-3(2 + 3x) = -6 - 9x$

$x(x - 6) = x^2 - 6x$ $-2x(3 - 4x) = -6x + 8x^2$

$4t(t^3 + 2) = 4t^4 + 8t$ $3t(2 + 5t - p) = 6t + 15t^2 - 3pt$

Note: The signs change when the quantity outside the brackets is negative. For example:

$a(b + c) = ab + ac$ $a(b - c) = ab - ac$

$-a(b + c) = -ab - ac$ $-a(b - c) = -ab + ac$

$-(a - b) = -a + b$ $-(a + b - c) = -a - b + c$

Note: A minus sign on its own in front of the brackets actually means -1, so:

$$-(x + 2y - 3) = -1 \times (x + 2y - 3) = -1 \times x + -1 \times 2y + -1 \times -3 = -x - 2y + 3$$

The effect of a minus sign outside the brackets is to change the sign of everything inside the brackets.

EXERCISE 12C

1 Expand these expressions.

a $2(3 + m)$ **b** $5(2 + l)$ **c** $3(4 - y)$ **d** $4(5 + 2k)$

e $3(2 - 4f)$ **f** $2(5 - 3w)$ **g** $5(2k + 3m)$ **h** $4(3d - 2n)$

i $t(t + 3)$ **j** $k(k - 3)$ **k** $4t(t - 1)$ **l** $2k(4 - k)$

m $4g(2g + 5)$ **n** $5h(3h - 2)$ **o** $y(y^2 + 5)$ **p** $h(h^3 + 7)$

q $k(k^2 - 5)$ **r** $3t(t^2 + 4)$ **s** $3d(5d^2 - d^3)$ **t** $3w(2w^2 + t)$

u $5a(3a^2 - 2b)$ **v** $3p(4p^3 - 5m)$ **w** $4h^2(3h + 2g)$ **x** $2m^2(4m + m^2)$

2 The local shop is offering \$1 off the price of a large tin of biscuits. Maddox wants five tins.

   a If the price of one tin is \$$t$, which of the expressions below represents how much it will cost Maddox to buy five tins?

$$5(t - 1) \qquad 5t - 1 \qquad t - 5 \qquad 5t - 5$$

   b Maddox has \$20 to spend. Will he have enough money for five tins?
Let $t = \$4.50$. Show working to justify your answer.

3 Dylan wrote the following.

$$3(5x - 4) = 8x - 4$$

Dylan has made two mistakes.

Explain the mistakes that Dylan has made.

ADVICE AND TIPS

It is not enough to give the right answer. You must try to explain why Dylan wrote 8 for 3 × 5 instead of 15.

4 The expansion $2(x + 3) = 2x + 6$ can be shown by this diagram.

   a What expansion is shown in this diagram?

   b Write down an expansion that is shown on this diagram.

Simplification

Simplification means writing down an expression as simply as possible, combining any like terms. Like terms are terms that have the same letter(s) raised to the same power and can differ only in their numerical coefficients (numbers in front). For example:

   m, $3m$, $4m$, $-m$ and $76m$ are all like terms in m

   t^2, $4t^2$, $7t^2$, $-t^2$, $-3t^2$ and $98t^2$ are all like terms in t^2

   pt, $5tp$, $-2pt$, $7pt$, $-3tp$ and $103pt$ are all like terms in pt.

Note: All the terms in tp are also like terms to all the terms in pt.

When simplifying an expression, you can only add or subtract like terms. For example:

$4m + 3m = 7m$	$3y + 4y + 3 = 7y + 3$	$4h - h = 3h$
$2t^2 + 5t^2 = 7t^2$	$2m + 6 + 3m = 5m + 6$	$7t + 8 - 2t = 5t + 8$
$3ab + 2ba = 5ab$	$5k - 2k = 3k$	$10g - 4 - 3g = 7g - 4$

Expand and simplify

When you expand the product of two brackets there are often like terms that you can collect together. You should always simplify algebraic expressions as much as possible.

EXAMPLE 4

Simplify the expression $3(4 + m) + 2(5 + 2m)$.

$3(4 + m) + 2(5 + 2m) = 12 + 3m + 10 + 4m = 22 + 7m$

EXAMPLE 5

Simplify the expression $3t(5t + 4) - 2t(3t - 5)$.

$3t(5t + 4) - 2t(3t - 5) = 15t^2 + 12t - 6t^2 + 10t = 9t^2 + 22t$

EXERCISE 12D

1 Simplify these expressions.

    a $4t + 3t$ **b** $3d + 2d + 4d$

    c $5e - 2e$ **d** $3t - t$

    e $2t^2 + 3t^2$ **f** $6y^2 - 2y^2$

    g $3ab + 2ab$ **h** $7a^2d - 4a^2d$

2 Find the missing terms to make these equations true.

    a $4x + 5y + \ldots - \ldots = 6(x - y)$

    b $3a - 6b - \ldots + \ldots = 2(a + b)$

3 The length of AB is 3 cm less than twice the length of AD.

    a Write an expression for the length of AB, in centimetres.

    EF is twice as long as AB.

    EH is 2 cm shorter than AD.

    b Find an expression for the perimeter of EFGH. Give your answer in cm.

4 Expand and simplify each expression.

    a $3(4 + t) + 2(5 + t)$ **b** $5(3 + 2k) + 3(2 + 3k)$

    c $4(3 + 2f) + 2(5 - 3f)$ **d** $5(1 + 3g) + 3(3 - 4g)$

5 Expand and simplify each expression.

   a $4(3 + 2h) - 2(5 + 3h)$

   b $5(3g + 4) - 3(2g + 5)$

   c $5(5k + 2) - 2(4k - 3)$

   d $4(4e + 3) - 2(5e - 4)$

ADVICE AND TIPS

Be careful with minus signs. For example, $-2(5e - 4) = -10e + 8$

6 Expand and simplify each expression.

   a $m(4 + p) + p(3 + m)$

   b $k(3 + 2h) + h(4 + 3k)$

   c $4r(3 + 4p) + 3p(8 - r)$

   d $5k(3m + 4) - 2m(3 - 2k)$

7 Expand and simplify each expression.

   a $t(3t + 4) + 3t(3 + 2t)$

   b $2y(3 + 4y) + y(5y - 1)$

   c $4e(3e - 5) - 2e(e - 7)$

   d $3k(2k + p) - 2k(3p - 4k)$

8 Expand and simplify each expression.

   a $4a(2b + 3c) + 3b(3a + 2c)$

   b $3y(4w + 2t) + 2w(3y - 4t)$

   c $5m(2n - 3p) - 2n(3p - 2m)$

   d $2r(3r + r^2) - 3r^2(4 - 2r)$

9 Fill in whole-number values so that this expansion is true.

$$3(\ldots x + \ldots y) + 2(\ldots x + \ldots y) = 11x + 17y$$

ADVICE AND TIPS

There is more than one answer. You don't have to give them all.

10 A rectangle with sides 5 and $3x + 2$ has a smaller rectangle with sides 3 and $2x - 1$ cut from it.

Work out the remaining area.

ADVICE AND TIPS

Write out the expression for the difference between the areas of the two rectangles and then work it out.

12.3 Factorisation

Factorisation is the opposite of expansion. When you factorise an expression you put it back into the brackets it may have come from.

In factorisation, you have to look for the **common factors** in *every* term of the expression.

EXAMPLE 6

Factorise each expression.

a $6t + 9m$ **b** $6my + 4py$ **c** $5k^2 - 25k$ **d** $10a^2b - 15ab^2$

a First look at the numerical coefficients 6 and 9.

These have a common factor of 3.

Then look at the letters, t and m.

Neither of these is a common factor as they do not appear in both terms.

You can think of the expression as $3 \times 2t + 3 \times 3m$, which gives the factorisation:

$6t + 9m = 3(2t + 3m)$

Note: You can always check a factorisation by expanding the answer.

b First look at the numbers.

These have a common factor of 2. m and p do not occur in both terms but y does, and is a common factor, so the factorisation is:

$6my + 4py = 2y(3m + 2p)$

c 5 is a common factor of 5 and 25 and k is a common factor of k^2 and k.

$5k^2 - 25k = 5k(k - 5)$

d 5 is a common factor of 10 and 15, a is a common factor of a^2 and a, b is a common factor of b and b^2.

$10a^2b - 15ab^2 = 5ab(2a - 3b)$

Note: If you multiply out each answer, you will get the expressions you started with.

EXERCISE 12E

1 Factorise these expressions.

a $6m + 12t$ **b** $9t + 3p$

c $8m + 12k$ **d** $4r + 8t$

e $mn + 3m$ **f** $5g^2 + 3g$

g $4w - 6t$

h $3y^2 + 2y$

i $4t^2 - 3t$

j $3m^2 - 3mp$

k $6p^2 + 9pt$

l $8pt + 6mp$

m $8ab - 4bc$

n $5b^2c - 10bc$

o $8abc + 6bde$

p $4a^2 + 6a + 8$

q $6ab + 9bc + 3bd$

r $5t^2 + 4t + at$

s $6mt^2 - 3mt + 9m^2t$

t $8ab^2 + 2ab - 4a^2b$

u $10pt^2 + 15pt + 5p^2t$

2 Three friends have a meal together. They each have a main meal costing $6.75 and a dessert costing $3.25.

Carla says that the bill will be $3 \times 6.75 + 3 \times 3.25$.

Suni says that she has an easier way to work out the bill as $3 \times (6.75 + 3.25)$.

a Explain why Carla's and Suni's methods both give the correct answer.

b Explain why Suni's method is better. **c** What is the total bill?

3 Factorise these expressions where possible. List those that do not factorise.

a $7m - 6t$

b $5m + 2mp$

c $t^2 - 7t$

d $8pt + 5ab$

e $4m^2 - 6mp$

f $a^2 + b$

g $4a^2 - 5ab$

h $3ab + 4cd$

i $5ab - 3b^2c$

4 Three students are asked to factorise the expression $12m - 8$. These are their answers.

Ahmed	Bernice	Craig
$2(6m - 4)$	$4(3m - 2)$	$4m(3 - \dfrac{2}{m})$

All the answers are accurately factorised, but only one is the normally accepted answer.

a Which student gave the correct answer?

b Explain why the other two students' answers are not acceptable as correct answers.

5 Explain why $5m + 6p$ cannot be factorised.

6 Alvin has correctly factorised the top and bottom of an algebraic fraction and cancelled out the terms to give a final answer of $2x$. Unfortunately some of his work has had coffee spilt on it. What was the original fraction?

$$\frac{4x}{2} = \frac{4}{2(x - 3)} = 2x$$

Multiplying two brackets

Quadratic expansion

A **quadratic expression** is one in which the highest power of the variables is 2.
For example:

$$y^2 \qquad 3t^2 + 5t \qquad 5m^2 + 3m + 8$$

An expression such as $(3y + 2)(4y - 5)$ can be expanded to give a quadratic expression.

Multiplying out such pairs of brackets is usually called **quadratic expansion**.

The rule for expanding expressions such as $(t + 5)(3t - 4)$ is similar to that for expanding single brackets: multiply everything in one set of brackets by everything in the other set of brackets.

There are several methods for doing this. Examples 7 to 9 show the three main methods: expansion, FOIL and the box method.

In the expansion method, split the terms in the first set of brackets, make each of them multiply both terms in the second set of brackets, then simplify the outcome.

EXAMPLE 7

Expand $(x + 3)(x + 4)$.

$$(x + 3)(x + 4) = x(x + 4) + 3(x + 4)$$
$$= x^2 + 4x + 3x + 12$$
$$= x^2 + 7x + 12$$

FOIL stands for First, Outer, Inner and Last. This is the order of multiplying the terms from each set of brackets.

EXAMPLE 8

Expand $(t + 5)(t - 2)$.

First terms give: $t \times t = t^2$

Outer terms give: $t \times -2 = -2t$

Inner terms give: $5 \times t = 5t$

Last terms give: $+5 \times -2 = -10$

$$(t + 5)(t - 2) = t^2 - 2t + 5t - 10$$
$$= t^2 + 3t - 10$$

The 'box method' can be used to lay out the multiplication.

EXAMPLE 9

Expand $(k - 3)(k - 2)$.

$(k - 3)(k - 2) = k^2 - 2k - 3k + 6$

$\qquad\qquad\quad = k^2 - 5k + 6$

	k	-3
k	k^2	$-3k$
-2	$-2k$	$+6$

Warning: Be careful with the signs. This is the main place where mistakes are made in questions involving the expansion of brackets.

EXERCISE 12F

Expand the expressions in questions **1–17**.

1 $(x + 3)(x + 2)$

2 $(t + 4)(t + 3)$

3 $(w + 1)(w + 3)$

4 $(m + 5)(m + 1)$

5 $(k + 3)(k + 5)$

6 $(a + 4)(a + 1)$

7 $(x + 4)(x - 2)$

8 $(t + 5)(t - 3)$

9 $(w + 3)(w - 1)$

10 $(f + 2)(f - 3)$

11 $(g + 1)(g - 4)$

12 $(y + 4)(y - 3)$

13 $(x - 3)(x + 4)$

14 $(p - 2)(p + 1)$

15 $(k - 4)(k + 2)$

16 $(y - 2)(y + 5)$

17 $(a - 1)(a + 3)$

> **ADVICE AND TIPS**
>
> A common error is to get minus signs wrong.
> $-2x - 3x = -5x$ but
> $-2x \times -3 = +6x$

EXTENDED

The expansions of the expressions in questions **18–26** follow a pattern. Work out the first few and try to spot the pattern that will allow you immediately to write down the answers to the rest.

18 $(x + 3)(x - 3)$

19 $(t + 5)(t - 5)$

20 $(m + 4)(m - 4)$

21 $(t + 2)(t - 2)$

22 $(y + 8)(y - 8)$

23 $(p + 1)(p - 1)$

24 $(5 + x)(5 - x)$

25 $(7 + g)(7 - g)$

26 $(x - 6)(x + 6)$

EXTENDED

27 This rectangle is made up of four parts with areas of x^2, $2x$, $3x$ and 6 square units.

Work out expressions for the sides of the rectangle, in terms of x.

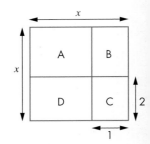

28 This square has an area of x^2 square units.
It is split into four rectangles.

a Copy and complete the table below to show the dimensions and area of each rectangle.

Rectangle	Length	Width	Area
A	$x - 1$	$x - 2$	$(x - 1)(x - 2)$
B			
C			
D			

b Add together the areas of rectangles B, C and D.

Expand any brackets and collect terms together.

c Use the results to explain why $(x - 1)(x - 2) = x^2 - 3x + 2$.

29 a Expand $(x - 3)(x + 3)$.

b Use the result in **a** to write down the answers to these. (Do not use a calculator or do a long multiplication.)

   i 97×103 **ii** 197×203

Quadratic expansion with non-unit coefficients

All the algebraic 'squared' terms (such as x^2) in Exercise 12F have a **coefficient** of 1 or –1. The next two examples show what to do if you have to expand brackets containing terms with coefficients that are not 1 or –1.

EXAMPLE 10

Expand $(2t + 3)(3t + 1)$.

$$(2t + 3)(3t + 1) = 6t^2 + 2t + 9t + 3$$
$$= 6t^2 + 11t + 3$$

	$2t$	$+3$
$3t$	$6t^2$	$+9t$
$+1$	$+2t$	$+3$

EXAMPLE 11

Expand $(4x - 1)(3x - 5)$.

$$(4x - 1)(3x - 5) = 4x(3x - 5) - (3x - 5) \quad \textbf{[Note:} (3x - 5) \text{ is the same as } 1(3x - 5).\textbf{]}$$
$$= 12x^2 - 20x - 3x + 5$$
$$= 12x^2 - 23x + 5$$

EXERCISE 12G

Expand the expressions in questions **1–21**.

1 $(2x + 3)(3x + 1)$

2 $(3y + 2)(4y + 3)$

3 $(3t + 1)(2t + 5)$

4 $(4t + 3)(2t - 1)$

5 $(5m + 2)(2m - 3)$

6 $(4k + 3)(3k - 5)$

7 $(3p - 2)(2p + 5)$

8 $(5w + 2)(2w + 3)$

9 $(2a - 3)(3a + 1)$

10 $(4r - 3)(2r - 1)$

11 $(3g - 2)(5g - 2)$

12 $(4d - 1)(3d + 2)$

13 $(5 + 2p)(3 + 4p)$

14 $(2 + 3t)(1 + 2t)$

15 $(4 + 3p)(2p + 1)$

16 $(6 + 5t)(1 - 2t)$

17 $(4 + 3n)(3 - 2n)$

18 $(2 + 3f)(2f - 3)$

19 $(3 - 2q)(4 + 5q)$

20 $(1 - 3p)(3 + 2p)$

21 $(4 - 2t)(3t + 1)$

> **ADVICE AND TIPS**
>
> Always give answers in the form $\pm ax^2 \pm bx \pm c$ even if the quadratic coefficient is negative.

22 Expand these expressions.

a $(x + 1)(x - 1)$

b $(2x + 1)(2x - 1)$

c $(2x + 3)(2x - 3)$

d Use the results in parts **a**, **b** and **c** to write down the expansion of $(3x + 5)(3x - 5)$.

23 **a** Without expanding the brackets, match each expression on the left with an expression on the right. One is done for you.

$(3x - 2)(2x + 1)$ $4x^2 - 4x + 1$

$(2x - 1)(2x - 1)$ $6x^2 - x - 2$

$(6x - 3)(x + 1)$ $6x^2 + 7x + 2$

$(4x + 1)(x - 1)$ $6x^2 + 3x - 3$

$(3x + 2)(2x + 1)$ $4x^2 - 3x - 1$

b Taking any expression on the left, explain how you can match it with an expression on the right without expanding the brackets.

EXTENDED

EXERCISE 12H

Try to spot the pattern in each of the expressions in questions **1–15** so that you can immediately write down the expansion.

1 $(2x + 1)(2x - 1)$

2 $(3t + 2)(3t - 2)$

3 $(5y + 3)(5y - 3)$

4 $(4m + 3)(4m - 3)$

5 $(2k - 3)(2k + 3)$

6 $(4h - 1)(4h + 1)$

7 $(2 + 3x)(2 - 3x)$

8 $(5 + 2t)(5 - 2t)$

9 $(6 - 5y)(6 + 5y)$

10 $(a + b)(a - b)$

11 $(3t + k)(3t - k)$

12 $(2m - 3p)(2m + 3p)$

13 $(5k + g)(5k - g)$

14 $(ab + cd)(ab - cd)$

15 $(a^2 + b^2)(a^2 - b^2)$

16 Imagine a square of side a units with a square of side b units cut from one corner.

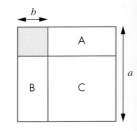

   a What is the area remaining after the small square is cut away?

   b The remaining area is cut into rectangles, A, B and C, and rearranged as shown.

     Write down the dimensions and area of the rectangle formed by A, B and C.

   c Explain why $a^2 - b^2 = (a + b)(a - b)$.

17 Explain why the areas of the shaded regions are the same.

Expanding squares

Whenever you see a **linear** bracketed term squared you must write the brackets down twice and then use whichever method you prefer to expand.

EXAMPLE 12

Expand $(x + 3)^2$.

$$(x + 3)^2 = (x + 3)(x + 3)$$
$$= x(x + 3) + 3(x + 3)$$
$$= x^2 + 3x + 3x + 9$$
$$= x^2 + 6x + 9$$

EXAMPLE 13

Expand $(3x - 2)^2$.

$$(3x - 2)^2 = (3x - 2)(3x - 2)$$
$$= 9x^2 - 6x - 6x + 4$$
$$= 9x^2 - 12x + 4$$

EXERCISE 12I

Expand the squares in questions **1–24** and simplify.

ADVICE AND TIPS

Remember always write down the brackets twice. Do not try to take any short cuts.

1 $(x + 5)^2$

2 $(m + 4)^2$

3 $(6 + t)^2$

4 $(3 + p)^2$

5 $(m - 3)^2$

6 $(t - 5)^2$

7 $(4 - m)^2$

8 $(7 - k)^2$

9 $(3x + 1)^2$

10 $(4t + 3)^2$

11 $(2 + 5y)^2$

12 $(3 + 2m)^2$

13 $(4t - 3)^2$

14 $(3x - 2)^2$

15 $(2 - 5t)^2$

16 $(6 - 5r)^2$

17 $(x + y)^2$

18 $(m - n)^2$

19 $(2t + y)^2$

20 $(m - 3n)^2$

21 $(x + 2)^2 - 4$

22 $(x - 5)^2 - 25$

23 $(x + 6)^2 - 36$

24 $(x - 2)^2 - 4$

25 A teacher asks her class to expand $(3x + 1)^2$.

Marcela's answer is $9x^2 + 1$.

Paulo's answer is $3x^2 + 6x + 1$.

a Explain the mistakes that Marcela has made.

b Explain the mistakes that Paulo has made.

c Work out the correct answer.

26 Use the diagram to show algebraically and diagrammatically that:

$$(2x - 1)^2 = 4x^2 - 4x + 1$$

EXTENDED

Factorisation involves putting a quadratic expression back into its **brackets** (if possible). Start with the factorisation of quadratic expressions of the type:

$$x^2 + ax + b$$

where a and b are integers.

Sometimes it is easy to put a quadratic expression back into its brackets, at other times it seems hard. However, there are some simple rules that will help you to factorise.

- The expression inside each set of brackets will start with an algebraic unknown such as x, and the signs in the quadratic expression show which signs to put after the unknowns.

- When the second sign in the expression is a plus, the signs in both sets of brackets are the same as the first sign.

    $x^2 + ax + b = (x + ?)(x + ?)$ Since everything is positive.
    $x^2 - ax + b = (x - ?)(x - ?)$ Since negative × negative = positive

- When the *second* sign is a *minus*, the signs in the brackets are *different*.

    $x^2 + ax - b = (x + ?)(x - ?)$ Since positive × negative = negative
    $x^2 - ax - b = (x + ?)(x - ?)$

- Next, look at the *last* number, b, in the expression. When multiplied together, the two numbers in the brackets must give b.

- Finally, look at the coefficient *of x*, which is a. The *sum* of the two *numbers* in the brackets will give a.

EXAMPLE 14

Factorise $x^2 - x - 6$.

Because of the signs you know the brackets must be $(x + ?)(x - ?)$.

Two numbers that have a product of -6 and a sum of -1 are -3 and $+2$.

So, $x^2 - x - 6 = (x + 2)(x - 3)$

EXAMPLE 15

Factorise $x^2 - 9x + 20$.

Because of the signs you know the brackets must be $(x - ?)(x - ?)$.

Two numbers that have a product of $+20$ and a sum of -9 are -4 and -5.

So, $x^2 - 9x + 20 = (x - 4)(x - 5)$

EXERCISE 12J

Factorise the expressions in questions **1–40**.

1 $x^2 + 5x + 6$

2 $t^2 + 5t + 4$

3 $m^2 + 7m + 10$

4 $k^2 + 10k + 24$

5 $p^2 + 14p + 24$

6 $r^2 + 9r + 18$

7 $w^2 + 11w + 18$

8 $x^2 + 7x + 12$

9 $a^2 + 8a + 12$

10 $k^2 + 10k + 21$

11 $f^2 + 22f + 21$

12 $b^2 + 20b + 96$

13 $t^2 - 5t + 6$

14 $d^2 - 5d + 4$

15 $g^2 - 7g + 10$

16 $x^2 - 15x + 36$

17 $c^2 - 18c + 32$

18 $t^2 - 13t + 36$

19 $y^2 - 16y + 48$

20 $j^2 - 14j + 48$

21 $p^2 - 8p + 15$

22 $y^2 + 5y - 6$

23 $t^2 + 2t - 8$

24 $x^2 + 3x - 10$

25 $m^2 - 4m - 12$

26 $r^2 - 6r - 7$

27 $n^2 - 3n - 18$

28 $m^2 - 7m - 44$

29 $w^2 - 2w - 24$

30 $t^2 - t - 90$

31 $h^2 - h - 72$

32 $t^2 - 2t - 63$

33 $d^2 + 2d + 1$

34 $y^2 + 20y + 100$

35 $t^2 - 8t + 16$

36 $m^2 - 18m + 81$

37 $x^2 - 24x + 144$

38 $d^2 - d - 12$

39 $t^2 - t - 20$

40 $q^2 - q - 56$

> **ADVICE AND TIPS**
>
> First decide on the signs in the brackets, then look at the numbers.

41 This rectangle is made up of four parts. Two of the parts have areas of x^2 and 6 square units.

The sides of the rectangle are of the form $x + a$ and $x + b$.

There are two possible answers for a and b.

Work out both answers and copy and complete the areas in the other parts of the rectangle.

Difference of two squares

In Exercise 12H, you multiplied out, for example, $(a + b)(a - b)$ and obtained $a^2 - b^2$. This type of quadratic expression has only two terms, both of which are perfect squares separated by a minus sign. This type of expression is called the **difference of two squares**. You should have found that all the expansions in Exercise 12H involved the differences of two squares.

The exercise illustrates a system of factorisation that will *always* work for the difference of two squares such as these.

$$x^2 - 9 \qquad x^2 - 25 \qquad x^2 - 4 \qquad x^2 - 100$$

There are three conditions that must be met if the difference of two squares works.

● There must be two terms.

● They must separated by a negative sign.

● Each term must be a perfect square, such as x^2 and n^2.

When these three conditions are met, the factorisation is:

$$x^2 - n^2 = (x + n)(x - n)$$

EXAMPLE 16

Factorise $x^2 - 36$.

● Recognise the difference of two squares x^2 and 6^2.

● So it factorises to $(x + 6)(x - 6)$.

Expanding the brackets shows that they do come from the original expression.

EXAMPLE 17

Factorise $9x^2 - 169$.

● Recognise the difference of two squares $(3x)^2$ and 13^2.

● So it factorises to $(3x + 13)(3x - 13)$.

EXERCISE 12K

Each of the expressions in questions **1–9** is the difference of two squares. Factorise them.

1 $x^2 - 9$

2 $t^2 - 25$

3 $m^2 - 16$

4 $9 - x^2$

5 $49 - t^2$

6 $k^2 - 100$

7 $4 - y^2$

8 $x^2 - 64$

9 $t^2 - 81$

10 a A square has a side of x units.

What is the area of the square?

b A rectangle, A, 2 units wide, is cut from the square and placed at the side of the remaining rectangle, B.

A square, C, is then cut from the bottom of rectangle A to leave a final rectangle, D.

i What is the height of rectangle B? **ii** What is the width of rectangle D?

iii What is the area of rectangle B plus rectangle A? **iv** What is the area of square C?

c By working out the area of rectangle D, explain why $x^2 - 4 = (x + 2)(x - 2)$.

11 a Expand and simplify: $(x + 2)^2 - (x + 1)^2$.

b Factorise: $a^2 - b^2$.

c In your answer for part **b**, replace a with $(x + 2)$ and b with $(x + 1)$. Expand and simplify the answer.

d What can you say about the answers to parts **a** and **c**?

e Simplify: $(x + 1)^2 - (x - 1)^2$.

Each of the expressions in questions **12–20** is the difference of two squares. Factorise them.

12 $x^2 - y^2$ **13** $x^2 - 4y^2$ **14** $x^2 - 9y^2$ **15** $9x^2 - 1$ **16** $16x^2 - 9$

17 $25x^2 - 64$ **18** $4x^2 - 9y^2$ **19** $9t^2 - 4w^2$ **20** $16y^2 - 25x^2$

Factorising $ax^2 + bx + c$

We can adapt the method for factorising $x^2 + ax + b$ to take into account the factors of the coefficient of x^2.

EXAMPLE 18

Factorise $3x^2 + 8x + 4$.

- First, note that both signs are positive. So the signs in the brackets must be $(?x + ?)(?x + ?)$.
- As 3 has only 3 and 1 as factors, the brackets must be $(3x + ?)(x + ?)$.
- Next, note that 4 can be factorised as 4×1 and 2×2.
- Now find which pair of factors of 4 combine with 3 and 1 to give 8.

$$\begin{array}{c|cc} ③ & 4 & ② \\ ① & 1 & ② \end{array}$$

You can see that the combination 3×2 and 1×2 adds up to 8.

- So, the complete factorisation becomes $(3x + 2)(x + 2)$.

EXAMPLE 19

Factorise $6x^2 - 7x - 10$.

- First, note that both signs are negative. So the signs in the brackets must be $(?x + ?)(?x - ?)$.

- As 6 can be factorised as 6×1 and 3×2, the brackets could be $(6x \pm ?)(x \pm ?)$ or $(3x \pm ?)(2x \pm ?)$.

- Next, note that 10 can be factorised as 5×2 and 1×10.

- Now find which pair of factors of 10 combine with the factors of 6 to give -7.

$$\begin{array}{cc|cc} 3 & ⑥ & \pm 1 & �±2 \\ 2 & ① & \pm 10 & �±5 \end{array}$$

You can see that the combination 6×-2 and 1×5 adds up to -7.

- So, the complete factorisation becomes $(6x + 5)(x - 2)$.

Although this seems to be very complicated, it becomes quite easy with practice and experience.

EXERCISE 12L

Factorise the expressions in questions **1–12**.

1 $2x^2 + 5x + 2$

2 $7x^2 + 8x + 1$

3 $4x^2 + 3x - 7$

4 $24t^2 + 19t + 2$

5 $15t^2 + 2t - 1$

6 $16x^2 - 8x + 1$

7 $6y^2 + 33y - 63$

8 $4y^2 + 8y - 96$

9 $8x^2 + 10x - 3$

10 $6t^2 + 13t + 5$

11 $3x^2 - 16x - 12$

12 $7x^2 - 37x + 10$

13 This rectangle is made up of four parts, with areas of $12x^2$, $3x$, $8x$ and 2 square units.

Work out expressions for the sides of the rectangle, in terms of x.

$12x^2$	$3x$
$8x$	2

14 Three students are asked to factorise the expression $6x^2 + 30x + 36$.
These are their answers.

Adam	**Bella**	**Cara**
$(6x + 12)(x + 3)$	$(3x + 6)(2x + 6)$	$(2x + 4)(3x + 9)$

All the answers are correctly factorised.

a Explain why one quadratic expression can have three different factorisations.

b Which of these is the most complete factorisation?

$2(3x + 6)(x + 3)$ $6(x + 2)(x + 3)$ $3(x + 2)(2x + 6)$

Explain your choice.

c What is the geometrical significance of the answers to parts **a** and **b**?

EXTENDED

The following four rules are used to work out the value of fractions.

Addition: $\dfrac{a}{b} + \dfrac{c}{d} = \dfrac{ad + bc}{bd}$

Subtraction: $\dfrac{a}{b} - \dfrac{c}{d} = \dfrac{ad - bc}{bd}$

Multiplication: $\dfrac{a}{b} \times \dfrac{c}{d} = \dfrac{ac}{bd}$

Note that a, b, c and d can be numbers, other letters or algebraic expressions. Remember:

● use brackets, if necessary
● factorise if you can
● **cancel** if you can.

EXAMPLE 20

Simplify each expression. **a** $\dfrac{1}{x} + \dfrac{x}{2y}$ **b** $\dfrac{2}{b} - \dfrac{a}{2b}$

a Using the addition rule: $\dfrac{1}{x} + \dfrac{x}{2y} = \dfrac{(1)(2y) + (x)(x)}{(x)(2y)} = \dfrac{2y + x^2}{2xy}$

b Using the subtraction rule: $\dfrac{2}{b} - \dfrac{a}{2b} = \dfrac{(2)(2b) - (a)(b)}{(b)(2b)} = \dfrac{4b - ab}{2b^2}$

$$= \dfrac{b(4 - a)}{2b^2} = \dfrac{4 - a}{2b}$$

Note: There are different ways of working out fraction calculations. Part **b** could have been done by making the denominator of each fraction the same.

$$\dfrac{(2)2}{(2)b} - \dfrac{a}{2b} = \dfrac{4 - a}{2b}$$

EXAMPLE 21

Simplify each expression. **a** $\dfrac{x}{3} \times \dfrac{x + 2}{x - 2}$ **b** $\dfrac{x}{3} \div \dfrac{2x}{7}$

a Using the multiplication rule: $\dfrac{x}{3} \times \dfrac{x + 2}{x - 2} = \dfrac{(x)(x + 2)}{(3)(x - 2)} = \dfrac{x^2 + 2x}{3x - 6}$

Remember that the line that separates the top from the bottom of an algebraic fraction acts as brackets as well as a division sign. Note that it is sometimes preferable to leave an algebraic fraction in a factorised form.

b Using the division rule: $\dfrac{x}{3} \div \dfrac{2x}{7} = \dfrac{(x)(7)}{(3)(2x)} = \dfrac{7}{6}$

EXAMPLE 22

Write $\dfrac{3}{x-1} - \dfrac{2}{x+1}$ as a **single fraction** as simply as possible.

Using the subtraction rule:

$$\frac{3}{x-1} - \frac{2}{x+1} = \frac{3(x+1) - 2(x-1)}{(x-1)(x+1)}$$

$$= \frac{3x + 3 - 2x + 2}{(x-1)(x+1)}$$

$$= \frac{x+5}{(x-1)(x+1)}$$

EXAMPLE 23

Simplify this expression. $\dfrac{2x^2 + x - 3}{4x^2 - 9}$

Factorise the numerator and denominator: $\dfrac{(2x+3)(x-1)}{(2x+3)(2x-3)}$

Denominator is the difference of two squares.

Cancel any common factors: $\dfrac{\cancel{(2x+3)}(x-1)}{\cancel{(2x+3)}(2x-3)}$

If at this stage there isn't a common factor on top and bottom, you should check your factorisations.

The remaining fraction is the answer: $\dfrac{(x-1)}{(2x-3)}$

EXERCISE 12M

1 Simplify each of these.

a $\dfrac{x}{2} + \dfrac{x}{3}$ **b** $\dfrac{3x}{4} + \dfrac{x}{5}$ **c** $\dfrac{3x}{4} + \dfrac{2x}{5}$ **d** $\dfrac{x}{2} + \dfrac{y}{3}$

e $\dfrac{xy}{4} + \dfrac{2}{x}$ **f** $\dfrac{x+1}{2} + \dfrac{x+2}{3}$ **g** $\dfrac{2x+1}{2} + \dfrac{3x+1}{4}$ **h** $\dfrac{x}{5} + \dfrac{2x+1}{3}$

i $\dfrac{x-2}{2} + \dfrac{x-3}{4}$ **j** $\dfrac{x-4}{4} + \dfrac{2x-3}{2}$

2 Simplify each of these.

a $\dfrac{x}{2} - \dfrac{x}{3}$ **b** $\dfrac{3x}{4} - \dfrac{x}{5}$ **c** $\dfrac{3x}{4} - \dfrac{2x}{5}$ **d** $\dfrac{x}{2} - \dfrac{y}{3}$

e $\dfrac{xy}{4} - \dfrac{2}{y}$ **f** $\dfrac{x+1}{2} - \dfrac{x+2}{3}$ **g** $\dfrac{2x+1}{2} - \dfrac{3x+1}{4}$ **h** $\dfrac{x}{5} - \dfrac{2x+1}{3}$

i $\dfrac{x-2}{2} - \dfrac{x-3}{4}$ **j** $\dfrac{x-4}{4} - \dfrac{2x-3}{2}$

3 Simplify each of these.

a $\dfrac{x}{2} \times \dfrac{x}{3}$ **b** $\dfrac{2x}{7} \times \dfrac{3y}{4}$ **c** $\dfrac{4x}{3y} \times \dfrac{2y}{x}$ **d** $\dfrac{4y^2}{9x} \times \dfrac{3x^2}{2y}$

e $\dfrac{x}{2} \times \dfrac{x-2}{5}$ **f** $\dfrac{x-3}{15} \times \dfrac{5}{2x-6}$ **g** $\dfrac{2x+1}{2} \times \dfrac{3x+1}{4}$ **h** $\dfrac{x}{5} \times \dfrac{2x+1}{3}$

i $\dfrac{x-2}{2} \times \dfrac{4}{x-3}$ **j** $\dfrac{x-5}{10} \times \dfrac{5}{x^2-5x}$

4 Simplify each of these. Factorise and cancel where appropriate.

a $\dfrac{3x}{4} + \dfrac{x}{4}$ **b** $\dfrac{3x}{4} - \dfrac{x}{4}$ **c** $\dfrac{3x}{4} \times \dfrac{x}{4}$

d $\dfrac{3x}{4} \div \dfrac{x}{4}$ **e** $\dfrac{3x+1}{2} + \dfrac{x-2}{5}$ **f** $\dfrac{3x+1}{2} - \dfrac{x-2}{5}$

g $\dfrac{3x+1}{2} \times \dfrac{x-2}{5}$ **h** $\dfrac{x^2-9}{10} \times \dfrac{5}{x-3}$ **i** $\dfrac{2x^2}{9} - \dfrac{2y^2}{3}$

5 Write these expressions as single fractions as simply as possible.

a $\dfrac{2}{x+1} + \dfrac{5}{x+2}$ **b** $\dfrac{4}{x-2} + \dfrac{7}{x+1}$ **c** $\dfrac{3}{4x+1} - \dfrac{4}{x+2}$

d $\dfrac{2}{2x-1} - \dfrac{6}{x+1}$ **e** $\dfrac{3}{2x-1} - \dfrac{4}{3x-1}$

6 For homework a teacher asks her class to simplify the expression $\dfrac{x^2-x-2}{x^2+x-6}$.

This is Tom's answer:

$$\dfrac{\cancel{x^2}-x-\cancel{2}^{\,-1}}{\cancel{x^2}+x-\cancel{6}_{\,+3}}$$

$$= \dfrac{-x-1}{x+3} = \dfrac{x+1}{x+3}$$

When she marked the homework, the teacher was in a hurry and only checked the answer, which was correct.

Tom made several mistakes. What are they?

7 An expression of the form $\dfrac{ax^2+bx-c}{dx^2-e}$ simplifies to $\dfrac{x-1}{2x-3}$.

What was the original expression?

8 Write these expressions as single fractions.

a $\dfrac{4}{x+1} + \dfrac{5}{x+2}$ **b** $\dfrac{18}{4x-1} - \dfrac{1}{x+1}$ **c** $\dfrac{2x-1}{2} - \dfrac{6}{x+1}$ **d** $\dfrac{3}{2x-1} - \dfrac{4}{3x-1}$

9 Simplify these expressions.

a $\dfrac{x^2+2x-3}{2x^2+7x+3}$ **b** $\dfrac{4x^2-1}{2x^2+5x-3}$ **c** $\dfrac{6x^2+x-2}{9x^2-4}$ **d** $\dfrac{4x^2+x-3}{4x^2-7x+3}$ **e** $\dfrac{4x^2-25}{8x^2-22x+5}$

Why this chapter matters

We use equations to explain some of the most important things in the world.

Three of the most important are shown on this page.

Why does the Moon keep orbiting the Earth and not fly off into space?

This is explained by Newton's law of universal gravitation, which describes the gravitational attraction between two bodies:

$$F = G \times \frac{m_1 \times m_2}{r^2}$$

where F is the force between the bodies, G is the gravitational constant, m_1 and m_2 are the masses of the two bodies and r is the distance between them.

Why don't planes fall out of the sky?

This is explained by Bernoulli's principle, which states that as the speed of a fluid increases, its pressure decreases. This is what causes the difference in air pressure between the top and bottom of an aircraft wing, as shown in the diagram on the left.

In its simplest form, the equation can be written as:

$$p + q = p_0$$

where p = static pressure, q = dynamic pressure and p_0 is the total pressure.

How can a small amount of plutonium have enough energy to wipe out a city?

This is explained by Einstein's theory of special relativity, which connects mass and energy in the equation:

$$E = mc^2$$

where E is the energy, m is the mass and c is the speed of light. As the speed of light is nearly 300 000 kilometres per second, the amount of energy in a small mass is huge. If this can be released, it can be used for good (as in nuclear power stations) or harm (as in nuclear bombs).

Chapter

13 Solutions of equations and inequalities

Topics	Level	Key words
Solving linear equations	CORE	equation, variable, solution, brackets, solve
Setting up equations	CORE	
Solving quadratic equations by factorisation	EXTENDED	linear equations, quadratic expressions, quadratic equation, factors, difference of two squares
Solving quadratic equations by the quadratic formula	EXTENDED	quadratic formula, coefficients, constant, soluble
Solving quadratic equations by completing the square	EXTENDED	completing the square
Simultaneous equations	CORE	simultaneous linear equations, eliminate, substitute
Solving inequalities	EXTENDED	inequality

In this chapter you will learn how to:

CORE	EXTENDED
• Solve simple linear equations in one unknown. (C2.5 and E2.5) • Solve simultaneous linear equations in two unknowns. (C2.5 and E2.5)	• Solve quadratic equations by factorisation, completing the square or by use of the formula. (E2.5) • Solve simple linear inequalities. (E2.5)

A teacher gave these instructions to her class.

What algebraic expression represents the teacher's statement?

- Think of a number.
- Double it.
- Add 3.

This is what two of her students said.

Can you work out Kim's answer and the number that Freda started with?

I chose the number 5.

Kim's answer will be:

$2 \times 5 + 3 = 13.$

My final answer was 10.

Freda's answer can be set up as an **equation**.

An equation is formed when an expression is put equal to a number or another expression. You should be able to deal with equations that have only one **variable** or letter.

The **solution** to an equation is the value of the variable that makes the equation true.
For example, the equation for Freda's answer is

$2x + 3 = 10$

where x represents Freda's number.

The value of x that makes this true is $x = 3\frac{1}{2}$.

Kim

Freda

To **solve** an equation such as $2x + 3 = 10$, do the same thing to each side of the equation until you have x on its own.

$$2x + 3 = 10$$

Subtract 3 from both sides: $\quad 2x + 3 - 3 = 10 - 3$

$$2x = 7$$

Divide both sides by 2: $\qquad \dfrac{2x}{2} = \dfrac{7}{2}$

$$x = 3\frac{1}{2}$$

Here is another example.

Mary had two bags, each of which contained the same number of marbles. She also had five spare marbles.

She put the two bags and the five spare marbles on scales and balanced them with 17 single marbles.

How many marbles were there in each bag?

If x is the number of marbles in each bag, then the equation representing Mary's balanced scales is:

$2x + 5 = 17$

Take five marbles from each pan:

$2x + 5 - 5 = 17 - 5$

$2x = 12$

Now halve the number of marbles on each pan.

That is, divide both sides by 2:

$$\frac{2x}{2} = \frac{12}{2}$$

$x = 6$

Checking the answer gives $2 \times 6 + 5 = 17$, which is correct.

EXAMPLE 1

Solve each of these equations by 'doing the same to both sides'.

a $3x - 5 = 16$ **b** $\frac{x}{2} + 2 = 10$

a Add 5 to both sides

$3x - 5 + 5 = 16 + 5$

$3x = 21$

Divide both sides by 3.

$$\frac{3x}{3} = \frac{21}{3}$$

$x = 7$

Checking the answer gives:

$3 \times 7 - 5 = 16$

which is correct.

b Subtract 2 from both sides

$\frac{x}{2} + 2 - 2 = 10 - 2$

$\frac{x}{2} = 8$

Multiply both sides by 2.

$\frac{x}{2} \times 2 = 8 \times 2$

$x = 16$

Checking the answer gives:

$16 \div 2 + 2 = 10$

which is correct.

EXERCISE 13A

CORE

1 Solve each of these equations by 'doing the same to both sides'. Remember to check that each answer works for its original equation.

a $x + 4 = 60$

b $3y - 2 = 4$

c $3x - 7 = 11$

d $5y + 3 = 18$

e $7 + 3t = 19$

f $5 + 4f = 15$

g $3 + 6k = 24$

h $4x + 7 = 17$

i $5m - 3 = 17$

j $\dfrac{w}{3} - 5 = 2$

k $\dfrac{x}{8} + 3 = 12$

l $\dfrac{m}{7} - 3 = 5$

m $\dfrac{x}{5} + 3 = 3$

n $\dfrac{h}{7} + 2 = 1$

o $\dfrac{w}{3} + 10 = 4$

p $\dfrac{x}{3} - 5 = 7$

q $\dfrac{y}{2} - 13 = 5$

r $\dfrac{f}{6} - 2 = 8$

s $\dfrac{x + 3}{2} = 5$

t $\dfrac{t - 5}{2} = 3$

u $\dfrac{3x + 10}{2} = 8$

v $\dfrac{2x + 1}{3} = 5$

w $\dfrac{5y - 2}{4} = 3$

x $\dfrac{6y + 3}{9} = 1$

y $\dfrac{2x - 3}{5} = 4$

z $\dfrac{5t + 3}{4} = 1$

> **ADVICE AND TIPS**
>
> Be careful with negative numbers.

2

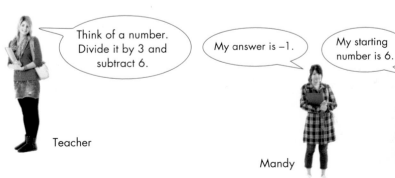

Teacher: Think of a number. Divide it by 3 and subtract 6.

Mandy: My answer is −1.

Andy: My starting number is 6.

a What answer did Andy get?

b What number did Mandy start with?

3 A teacher asked her class to solve the equation $2x - 1 = 7$.

Mustafa wrote:

$$2x - 1 = 7$$
$$2x - 1 - 1 = 7 - 1$$
$$2x = 6$$
$$2x - 2 = 6 - 2$$
$$x = 4$$

Elif wrote:

$$2x - 1 = 7$$
$$2x - 1 + 1 = 7 + 1$$
$$2x = 8$$
$$2x \div 2 = 8 \div 2$$
$$x = 4$$

When the teacher read out the correct answer of 4, both students ticked their work as correct.

a Which student used the correct method?

b Explain the mistakes the other student made.

Brackets

When you have an equation that contains **brackets**, you first must multiply out the brackets and then solve the resulting equation.

EXAMPLE 2

Solve the equation $5(x + 3) = 25$.

First multiply out the brackets to get:
$$5x + 15 = 25$$

Subtract 15: $5x = 25 - 15 = 10$

Divide by 5: $\dfrac{5x}{5} = \dfrac{10}{5}$

$$x = 2$$

An alternative method is to divide by the number outside the brackets.

EXAMPLE 3

Solve the equation $3(2x - 7) = 15$.

Divide both sides by 3: $2x - 7 = 5$

Add 7: $2x = 12$

Divide by 2: $x = 6$

Make sure you can use both methods.

CORE

1 Solve each of these equations. Some of the answers may be decimals or negative numbers. Remember to check that each answer works for its original equation. Use your calculator if necessary.

a $2(x + 5) = 16$

b $5(x - 3) = 20$

c $3(t + 1) = 18$

d $4(2x + 5) = 44$

e $2(3y - 5) = 14$

f $5(4x + 3) = 135$

g $4(3t - 2) = 88$

h $6(2t + 5) = 42$

i $2(3x + 1) = 11$

j $4(5y - 2) = 42$

k $6(3k + 5) = 39$

l $5(2x + 3) = 27$

m $9(3x - 5) = 9$

n $2(x + 5) = 6$

o $5(x - 4) = -25$

p $3(t + 7) = 15$

q $2(3x + 11) = 10$

r $4(5t + 8) = 12$

> **ADVICE AND TIPS**
>
> Once the brackets have been expanded the equations become straightforward. Remember to multiply *everything* inside the brackets with what is outside.

2 Fill in values for a, b and c so that the answer to this equation is $x = 4$.

$$a(bx + 3) = c$$

3 My son is x years old. In five years' time, I will be twice his age and both our ages will be multiples of 10. The sum of our ages will be between 50 and 100. How old am I now?

> **ADVICE AND TIPS**
>
> Set up an equation and put it equal to 60, 70, 80, etc. Solve the equation and see if the answer fits the conditions.

Equations with the variable on both sides

When a letter (or variable) appears on both sides of an equation, collect all the terms containing the letter on one side. This is usually the left-hand side of the equation. When there are more of the letters on the right-hand side, it may be easier to turn the equation round. When an equation contains brackets, you must multiply them out first.

EXAMPLE 4

Solve this equation. $\qquad 5x + 4 = 3x + 10$

There are more xs on the left-hand side, so leave the equation as it is.

Subtract $3x$ from both sides: $\qquad 2x + 4 = 10$

Subtract 4 from both sides: $\qquad 2x = 6$

Divide both sides by 2: $\qquad x = 3$

EXAMPLE 5

Solve this equation. $2x + 3 = 6x - 5$

There are more xs on the right-hand side, so turn the equation round.

$$6x - 5 = 2x + 3$$

Subtract $2x$ from both sides: $4x - 5 = 3$

Add 5 to both sides: $4x = 8$

Divide both sides by 4: $x = 2$

EXAMPLE 6

Solve this equation. $3(2x + 5) + x = 2(2 - x) + 2$

Multiply out both brackets: $6x + 15 + x = 4 - 2x + 2$

Simplify both sides: $7x + 15 = 6 - 2x$

There are more xs on the left-hand side, so leave the equation as it is.

Add $2x$ to both sides: $9x + 15 = 6$

Subtract 15 from both sides: $9x = -9$

Divide both sides by 9: $x = -1$

EXERCISE 13C

1 Solve each of these equations.

a $2x + 3 = x + 5$ **b** $5y + 4 = 3y + 6$

c $4a - 3 = 3a + 4$ **d** $5t + 3 = 2t + 15$

e $7p - 5 = 3p + 3$ **f** $6k + 5 = 2k + 1$

g $4m + 1 = m + 10$ **h** $8s - 1 = 6s - 5$

> **ADVICE AND TIPS**
>
> **Remember:** 'Change sides, change signs'. Show all your working. Rearrange *before* you simplify. If you try to do these at the same time you could get it wrong.

2 Hasan says: Miriam says:

I am thinking of a number. I multiply it by 3 and subtract 2.

I am thinking of a number. I multiply it by 2 and add 5.

Hasan and Miriam find that they both thought of the same number and both got the same final answer.

What number did they think of?

> **ADVICE AND TIPS**
>
> Set up expressions; make them equal and solve.

CORE

CORE

3 Solve each of these equations.

a $2(d + 3) = d + 12$ b $5(x - 2) = 3(x + 4)$

c $3(2y + 3) = 5(2y + 1)$ d $3(h - 6) = 2(5 - 2h)$

e $4(3b - 1) + 6 = 5(2b + 4)$ f $2(5c + 2) - 2c = 3(2c + 3) + 7$

4 Explain why the equation $3(2x + 1) = 2(3x + 5)$ cannot be solved.

> **ADVICE AND TIPS**
>
> Expand the brackets and collect terms on one side as usual. What happens?

5 Explain why these are an infinite number of solutions to this equation.

$$2(6x + 9) = 3(4x + 6)$$

13.2 Setting up equations

You can use equations to represent situations, so that you can solve real-life problems. You can solve many real-life problems by setting them up as linear equations and then solving the equation.

EXAMPLE 7

The rectangle shown has a perimeter of 40 cm.

Find the value of x.

The perimeter of the rectangle is:
$$3x + 1 + x + 3 + 3x + 1 + x + 3 = 40$$
This simplifies to: $8x + 8 = 40$
Subtract 8 from both sides: $8x = 32$
Divide both sides by 8: $x = 4$

EXERCISE 13D

Set up an equation to represent each situation described below. Then solve the equation. Remember to check each answer.

1 Every day, from Monday to Saturday a man buys a daily paper for d cents. He buys a Sunday paper for 1.80 dollars. His weekly paper bill is 7.20 dollars.

What is the price of his daily paper?

2 The diagram shows a rectangle.

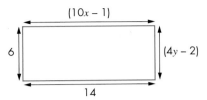

a What is the value of x? **b** What is the value of y?

3 In this rectangle, the length is 3 cm more than the width. The perimeter is 12 cm.

a What is the value of x?

b What is the area of the rectangle?

4 Masha has two bags, each of which contains the same number of sweets. She eats four sweets. She then finds that she has 30 sweets left. How many sweets were there in each bag to start with?

5 Flooring costs $12.75 per square metre.

The shop charges $35 for fitting. The final bill was $137.

How many square metres of flooring were fitted?

6 Moshin bought eight garden chairs. When he got to the till he used a $10 voucher as part payment. His final bill was $56.

a Set this problem up as an equation, using c as the cost of one chair.

b Solve the equation to find the cost of one chair.

7 This diagram shows the traffic flow through a one-way system in a town centre.

Cars enter at A and at each junction the fractions show the proportion of cars that take each route.

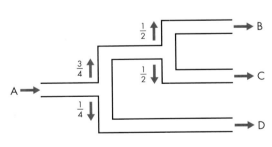

a 1200 cars enter at A. How many come out of each of the exits, B, C and D?

b If 300 cars exit at B, how many cars entered at A?

c If 500 cars exit at D, how many exit at B?

8 A rectangular room is 3 m longer than it is wide. The perimeter is 16 m.

Floor tiles cost 9 dollars per square metre. How much will it cost to cover the floor?

CORE

9 A boy is Y years old. His father is 25 years older than he is.
The sum of their ages is 31. How old is the boy?

10 Another boy is X years old. His sister is twice as old as he is. The sum of their ages is 27.
How old is the boy?

11 The diagram shows a square.

Find the value of x if the perimeter is 44 cm.

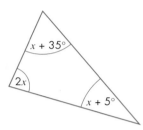

$(4x - 1)$

12 Max thought of a number. He then multiplied his number by 3. He added 4 to the answer.
He then doubled that answer to get a final value of 38. What number did he start with?

13 The angles of a triangle, in degrees, are $2x$, $x + 5$
and $x + 35$.

a Write down an equation to show that the angles
add up to 180 degrees.

b Solve your equation to find the value of x.

14 Five friends went for a meal in a restaurant. The bill was \$$x$.
They decided to add a \$10 tip and split the bill equally between them.

Each person paid \$9.50.

a Set up this problem as an equation.

b Solve the equation to work out the bill before the tip was added.

15 A teacher asked her class to find three angles of a triangle
that were consecutive even numbers.

Tammy wrote: $x + x + 2 + x + 4 = 180$
$$3x + 6 = 180$$
$$3x = 174$$
$$x = 58$$

So the angles are 58°, 60° and 62°.

The teacher then asked the class to find four angles of a quadrilateral that are
consecutive even numbers.

Can this be done? Explain your answer.

> **ADVICE AND TIPS**
>
> Do the same type of
> working as Tammy did for
> a triangle. Work out the
> value of x. What happens?

16 Maria has a large and a small bottle of cola. The large
bottle holds 50 cl more than the small bottle.

From the large bottle she fills four cups and has 18 cl
left over.

From the small bottle she fills three cups and has 1 cl
left over.

How much cola was there in each bottle?

> **ADVICE AND TIPS**
>
> Set up equations for both
> using x as the amount of
> cola in a cup. Put them
> equal but remember to
> add 50 to the small bottle
> equation to allow for the
> difference. Solve for x,
> then work out how much
> is in each bottle.

So far, all the equations you have worked with have been **linear equations**. Now you will look at equations that involve **quadratic expressions** such as $x^2 - 2x - 3$, which contain the square of the variable.

Solving the quadratic equation $x^2 + ax + b = 0$

To solve a **quadratic equation** such as $x^2 - 2x - 3 = 0$, you first must be able to factorise it. Work through Examples 8 to 10 below to see how this is done.

EXAMPLE 8

Solve the equation $x^2 + 6x + 5 = 0$.

This factorises into $(x + 5)(x + 1) = 0$.

The only way this expression can ever equal 0 is if the value of one of the expressions in the brackets is 0.

Hence either $(x + 5) = 0$ or $(x + 1) = 0$
$$\Rightarrow x + 5 = 0 \quad \text{or} \quad x + 1 = 0$$
$$\Rightarrow x \quad = -5 \text{ or } \quad x \quad = -1$$

So the solution is $x = -5$ or $x = -1$.

There are *two* possible values for x.

EXAMPLE 9

Solve the equation $x^2 + 3x - 10 = 0$.

This factorises into $(x + 5)(x - 2) = 0$.

Hence either $(x + 5) = 0$ or $(x - 2) = 0$
$$\Rightarrow x + 5 \quad = 0 \text{ or } x - 2 \quad = 0$$
$$\Rightarrow \quad x \quad = -5 \text{ or } x \quad = 2.$$

So the solution is $x = -5$ or $x = 2$.

EXAMPLE 10

Solve the equation $x^2 - 6x + 9 = 0$.

This factorises into $(x - 3)(x - 3) = 0$.

The equation has repeated roots.

That is: $(x - 3)^2 = 0$

Hence, there is only one solution, $x = 3$.

EXTENDED

EXERCISE 13E

Solve the equations in questions **1–12**.

1 $(x + 2)(x + 5) = 0$ **2** $(t + 3)(t + 1) = 0$ **3** $(a + 6)(a + 4) = 0$

4 $(x + 3)(x - 2) = 0$ **5** $(x + 1)(x - 3) = 0$ **6** $(t + 4)(t - 5) = 0$

7 $(x - 1)(x + 2) = 0$ **8** $(x - 2)(x + 5) = 0$ **9** $(a - 7)(a + 4) = 0$

10 $(x - 3)(x - 2) = 0$ **11** $(x - 1)(x - 5) = 0$ **12** $(a - 4)(a - 3) = 0$

First factorise, then solve the equations in questions **13–26**.

13 $x^2 + 5x + 4 = 0$ **14** $x^2 + 11x + 18 = 0$ **15** $x^2 - 6x + 8 = 0$

16 $x^2 - 8x + 15 = 0$ **17** $x^2 - 3x - 10 = 0$ **18** $x^2 - 2x - 15 = 0$

19 $t^2 + 4t - 12 = 0$ **20** $t^2 + 3t - 18 = 0$ **21** $x^2 - x - 2 = 0$

22 $x^2 + 4x + 4 = 0$ **23** $m^2 + 10m + 25 = 0$ **24** $t^2 - 8t + 16 = 0$

25 $t^2 + 8t + 12 = 0$ **26** $a^2 - 14a + 49 = 0$

27 A woman is x years old. Her husband is three years younger. The product of their ages is 550.

   a Set up a quadratic equation to represent this situation.

   b How old is the woman?

> **ADVICE AND TIPS**
>
> If one solution to a real-life problem is negative, reject it and only give the positive answer.

28 A rectangular field is 40 m longer than it is wide. The area is 48 000 square metres.

The farmer wants to place a fence all around the field.

How long will the fence be?

> **ADVICE AND TIPS**
>
> Let the width be x, set up a quadratic equation and solve it to get x.

First rearrange the equations in questions **29–37**, then solve them.

29 $x^2 + 10x = -24$ **30** $x^2 - 18x = -32$

31 $x^2 + 2x = 24$ **32** $x^2 + 3x = 54$

33 $t^2 + 7t = 30$ **34** $x^2 - 7x = 44$

35 $t^2 - t = 72$ **36** $x^2 = 17x - 72$

37 $x^2 + 1 = 2x$

> **ADVICE AND TIPS**
>
> You cannot solve a quadratic equation by factorisation unless it is in the form $x^2 + ax + b = 0$

38 A teacher asks her class to solve $x^2 - 3x = 4$.

This is Mario's answer.

$x^2 - 3x - 4 = 0$
$(x - 4)(x + 1) = 0$
Hence $x - 4 = 0$ or $x + 1 = 0$
$x = 4$ or -1

This is Sylvan's answer.

$x(x - 3) = 4$
Hence $x = 4$ or $x - 3 = 4 \Rightarrow x = -3 + 4 = -1$

When the teacher reads out the answer of $x = 4$ or -1, both students mark their work as correct.

Who used the correct method and what mistakes did the other student make?

Solving the general quadratic equation by factorisation

The general quadratic equation is of the form $ax^2 + bx + c = 0$ where a, b and c are positive or negative whole numbers. (It is easier to make sure that a is always positive.) Before you can solve any quadratic equation by factorisation, you must rearrange it to this form.

The method is similar to that used to solve equations of the form $x^2 + ax + b = 0$.
That is, you have to find two **factors** of $ax^2 + bx + c$ with a product of 0.

EXAMPLE 11

Solve these quadratic equations. **a** $12x^2 - 28x = -15$ **b** $30x^2 - 5x - 5 = 0$

a First, rearrange the equation to the general form.

$12x^2 - 28x + 15 = 0$

This factorises into $(2x - 3)(6x - 5) = 0$.

The only way this product can equal 0 is if the value of the expression in one of the brackets is 0. Hence:

either $2x - 3 = 0$ or $6x - 5 = 0$
$\Rightarrow 2x = 3$ or $6x = 5$
$\Rightarrow x = \frac{3}{2}$ or $x = \frac{5}{6}$

So the solution is $x = 1\frac{1}{2}$ or $x = \frac{5}{6}$

Note: It is almost always the case that if a solution is a fraction which is then changed into a rounded decimal number, the original equation cannot be evaluated exactly, using that decimal number. So it is preferable to leave the solution in its fraction form. This is called the *rational form*.

b This equation is already in the general form and it will factorise to
$(15x + 5)(2x - 1) = 0$ or $(3x + 1)(10x - 5) = 0$.

Look again at the equation. There is a common factor of 5 which can be taken out to give:

$5(6x^2 - x - 1 = 0)$

This is much easier to factorise to $5(3x + 1)(2x - 1) = 0$, which can be solved to give $x = -\frac{1}{3}$ or $x = \frac{1}{2}$.

Special cases

Sometimes the value of b or c may be zero. (Note that if a is zero the equation is no longer a quadratic equation but a linear equation.)

EXAMPLE 12

Solve these quadratic equations. **a** $3x^2 - 4 = 0$ **b** $4x^2 - 25 = 0$ **c** $6x^2 - x = 0$

a Rearrange to get $3x^2 = 4$.

Divide both sides by 3: $\qquad x^2 = \frac{4}{3}$

Take the square root on both sides: $x = \pm\sqrt{\dfrac{4}{3}} = \pm\dfrac{2}{\sqrt{3}} = \pm\dfrac{2\sqrt{3}}{3}$

Note: A square root can be positive or negative. The symbol $\pm$ s means that the square root has a positive and a negative value, *both* of which must be used in solving for x.

b You can use the method of part **a** or you can treat this as the **difference of two squares**. This can be factorised to $(2x - 5)(2x + 5) = 0$.

Each set of brackets can be put equal to zero.

$2x - 5 = 0 \implies x = +\frac{5}{2}$

$2x + 5 = 0 \implies x = -\frac{5}{2}$ $\qquad$ So the solution is $x = \pm\frac{5}{2}$.

c There is a common factor of x, so factorise as $x(6x - 1) = 0$.

There is only one set of brackets this time but each factor can be equal to zero, so $x = 0$ or $6x - 1 = 0$.

Hence, $x = 0$ or $\frac{1}{6}$.

EXERCISE 13F

Give your answers either in rational form or as mixed numbers.

1 Solve these equations.

a $3x^2 + 8x - 3 = 0$

b $6x^2 - 5x - 4 = 0$

c $5x^2 - 9x - 2 = 0$

d $4t^2 - 4t - 35 = 0$

e $18t^2 + 9t + 1 = 0$

f $3t^2 - 14t + 8 = 0$

g $6x^2 + 15x - 9 = 0$

h $12x^2 - 16x - 35 = 0$

i $15t^2 + 4t - 35 = 0$

j $28x^2 - 85x + 63 = 0$

k $24x^2 - 19x + 2 = 0$

l $16t^2 - 1 = 0$

m $4x^2 + 9x = 0$

n $25t^2 - 49 = 0$

o $9m^2 - 24m - 9 = 0$

> **ADVICE AND TIPS**
>
> Look out for the special cases where b or c is zero.

EXTENDED

2 Rearrange these equations into the general form and then solve them.

a $x^2 - x = 42$

b $8x(x + 1) = 30$

c $(x + 1)(x - 2) = 40$

d $13x^2 = 11 - 2x$

e $(x + 1)(x - 2) = 4$

f $10x^2 - x = 2$

g $8x^2 + 6x + 3 = 2x^2 + x + 2$

h $25x^2 = 10 - 45x$

i $8x - 16 - x^2 = 0$

j $(2x + 1)(5x + 2) = (2x - 2)(x - 2)$

k $5x + 5 = 30x^2 + 15x + 5$

l $2m^2 = 50$

m $6x^2 + 30 = 5 - 3x^2 - 30x$

n $4x^2 + 4x - 49 = 4x$

o $2t^2 - t = 15$

3 Here are three equations.

A: $(x - 1)^2 = 0$ B: $3x + 2 = 5$ C: $x^2 - 4x = 5$

a Give some mathematical fact that equations A and B have in common.

b Give a mathematical reason why equation B is different from equations A and C.

4 Pythagoras' theorem states that the sum of the squares of the two short sides of a right-angled triangle equals the square of the long side (hypotenuse).

A right-angled triangle has sides $5x - 1$, $2x + 3$ and $x + 1$ cm.

a Show that $20x^2 - 24x - 9 = 0$.

b Find the area of the triangle.

13.4 Solving quadratic equations by the quadratic formula   **E**

Many quadratic equations cannot be solved by factorisation because they do not have simple factors. Try to factorise, for example, $x^2 - 4x - 3 = 0$ or $3x^2 - 6x + 2 = 0$. You will find it is impossible.

One way to solve this type of equation is to use the **quadratic formula**. This formula can be used to solve *any* quadratic equation that can be solved (is **soluble**).

The solution of the equation $ax^2 + bx + c = 0$ is given by:

$$x = \frac{-b \pm \sqrt{b^2 - 4ac}}{2a}$$

where a and b are the **coefficients** of x^2 and x respectively and c is the **constant** term.

This is the quadratic formula, which you should memorise.

EXAMPLE 13

Solve $5x^2 - 11x - 4 = 0$, giving solutions correct to 2 decimal places.

Take the quadratic formula:

$$x = \frac{-b \pm \sqrt{b^2 - 4ac}}{2a}$$

and put $a = 5$, $b = -11$ and $c = -4$, which gives:

$$x = \frac{-(-11) \pm \sqrt{(-11)^2 - 4(5)(-4)}}{2(5)}$$

Note that the values for a, b and c have been put into the formula in brackets. This is to avoid mistakes in calculation. It is a very common mistake to get the sign of b wrong or to think that -11^2 is -121. Using brackets will help you do the calculation correctly.

$$x = \frac{11 \pm \sqrt{121 + 80}}{10} = \frac{11 \pm \sqrt{201}}{10}$$

$$\Rightarrow x = 2.52 \text{ or } -0.32$$

Note: The calculation has been done in stages. With a calculator it is possible just to work out the answer, but make sure you can use your calculator properly. Otherwise, break the calculation down. Remember the rule: 'if you try to do two things at once, you will probably get one of them wrong'.

Note: If you are asked to solve a quadratic equation to one or two decimal places, you can be sure that it can be solved *only* by the quadratic formula.

EXERCISE 13G

Use the quadratic formula to solve the equations in questions **1–15**. Give your answers to 2 decimal places.

ADVICE AND TIPS

Use brackets when substituting and do not try to work two things out at the same time.

1 $2x^2 + x - 8 = 0$

2 $3x^2 + 5x + 1 = 0$

3 $x^2 - x - 10 = 0$

4 $5x^2 + 2x - 1 = 0$

5 $7x^2 + 12x + 2 = 0$

6 $3x^2 + 11x + 9 = 0$

7 $4x^2 + 9x + 3 = 0$

8 $6x^2 + 22x + 19 = 0$

9 $x^2 + 3x - 6 = 0$

10 $3x^2 - 7x + 1 = 0$

11 $2x^2 + 11x + 4 = 0$

12 $4x^2 + 5x - 3 = 0$

13 $4x^2 - 9x + 4 = 0$

14 $7x^2 + 3x - 2 = 0$

15 $5x^2 - 10x + 1 = 0$

EXTENDED

16 A rectangular lawn is 2 m longer than it is wide.

The area of the lawn is 21 m². The gardener wants to edge the lawn with edging strips, which are sold in lengths of $1\frac{1}{2}$ m. How many will she need to buy?

17 Shaun is solving a quadratic equation, using the formula.

He correctly substitutes values for a, b and c to get:

$$x = \frac{3 \pm \sqrt{37}}{2}$$

What is the equation Shaun is trying to solve?

18 Hasan uses the quadratic formula to solve $4x^2 - 4x + 1 = 0$.

Miriam uses factorisation to solve $4x^2 - 4x + 1 = 0$.

They both find something unusual in their solutions.

Explain what this is, and why.

13.5 Solving quadratic equations by completing the square   E

Another method for solving quadratic equations is **completing the square**. You can use this method as an alternative to the quadratic formula.

You will remember that:

$$(x + a)^2 = x^2 + 2ax + a^2$$

which can be rearranged to give:

$$x^2 + 2ax = (x + a)^2 - a^2$$

This is the basic principle behind completing the square.

There are three basic steps in rewriting $x^2 + px + q$ in the form $(x + a)^2 + b$.

Step 1: Ignore q and just look at the first two terms, $x^2 + px$.

Step 2: Rewrite $x^2 + px$ as $\left(x + \dfrac{p}{2}\right)^2 - \left(\dfrac{p}{2}\right)^2$.

Step 3: Bring q back to get $x^2 + px + q = \left(x + \dfrac{p}{2}\right)^2 - \left(\dfrac{p}{2}\right)^2 + q$.

EXAMPLE 14

Rewrite each expression in the form $(x \pm a)^2 \pm b$.

a $x^2 + 6x - 7$

b $x^2 - 8x + 3$

a Ignore -7 for the moment.

Rewrite $x^2 + 6x$ as $(x + 3)^2 - 9$.

(Expand $(x + 3)^2 - 9 = x^2 + 6x + 9 - 9 = x^2 + 6x$. The 9 is subtracted to get rid of the constant term when the brackets are expanded.)

Now bring the -7 back, so $x^2 + 6x - 7 = (x + 3)^2 - 9 - 7$

Combine the constant terms to get the final answer:

$$x^2 + 6x - 7 = (x + 3)^2 - 16$$

b Ignore $+3$ for the moment.

Rewrite $x^2 - 8x$ as $(x - 4)^2 - 16$.

(Note that you still subtract $(-4)^2$, as $(-4)^2 = +16$.)

Now bring the $+3$ back, so $x^2 - 8x + 3 = (x - 4)^2 - 16 + 3$.

Combine the constant terms to get the final answer:

$$x^2 - 8x + 3 = (x - 4)^2 - 13$$

EXAMPLE 15

Rewrite $x^2 + 4x - 7$ in the form $(x + a)^2 - b$. Hence solve the equation $x^2 + 4x - 7 = 0$, giving your answers to 2 decimal places.

Note that:

$$x^2 + 4x = (x + 2)^2 - 4$$

So:

$$x^2 + 4x - 7 = (x + 2)^2 - 4 - 7 = (x + 2)^2 - 11$$

When $x^2 + 4x - 7 = 0$, you can rewrite the equations completing the square as:

$$(x + 2)^2 - 11 = 0$$

Rearranging gives $(x + 2)^2 = 11$.

Taking the **square root** of both sides gives:

$$x + 2 = \pm\sqrt{11}$$

$$\Rightarrow x = -2 \pm \sqrt{11}$$

This answer could be left like this, but you are asked to calculate it to 2 decimal places.

$$\Rightarrow x = 1.32 \text{ or } -5.32 \text{ (to 2 decimal places)}$$

To solve $ax^2 + bx + c = 0$ when a is not 1, start by dividing through by a.

EXAMPLE 16

Solve by completing the square.

$$2x^2 - 6x - 7 = 0$$

Divide by 2: $x^2 - 3x - 3.5 = 0$

$$x^2 - 3x = (x - 1.5)^2 - 2.25$$

So: $x^2 - 3x - 3.5 = (x - 1.5)^2 - 5.75$

When: $x^2 - 3x - 3.5 = 0$

then: $(x - 1.5)^2 = 5.75$

$$x - 1.5 = \pm\sqrt{5.75}$$

$$x = 1.5 \pm\sqrt{5.75}$$

$$x = 3.90 \text{ or } x = -0.90$$

EXERCISE 13H

EXTENDED

1 Write an equivalent expression in the form $(x \pm a)^2 - b$.

  a $x^2 + 4x$ **b** $x^2 + 14x$ **c** $x^2 - 6x$ **d** $x^2 + 6x$

  e $x^2 - 3x$ **f** $x^2 - 9x$ **g** $x^2 + 13x$ **h** $x^2 + 10x$

  i $x^2 + 8x$ **j** $x^2 - 2x$ **k** $x^2 + 2x$

2 Write an equivalent expression in the form $(x \pm a)^2 - b$.
   Question **1** will help with **a** to **h**.

  a $x^2 + 4x - 1$ **b** $x^2 + 14x - 5$ **c** $x^2 - 6x + 3$ **d** $x^2 + 6x + 7$

  e $x^2 - 3x - 1$ **f** $x^2 + 6x + 3$ **g** $x^2 - 9x + 10$ **h** $x^2 + 13x + 35$

  i $x^2 + 8x - 6$ **j** $x^2 + 2x - 1$ **k** $x^2 - 2x - 7$ **l** $x^2 + 2x - 9$

3 Solve each equation by completing the square. Leave a square root sign in your answer
   where appropriate. The answers to question **2** will help.

  a $x^2 + 4x - 1 = 0$ **b** $x^2 + 14x - 5 = 0$ **c** $x^2 - 6x + 3 = 0$

  d $x^2 + 6x + 7 = 0$ **e** $x^2 - 3x - 1 = 0$ **f** $x^2 + 6x + 3 = 0$

  g $x^2 - 9x + 10 = 0$ **h** $x^2 + 13x + 35 = 0$ **i** $x^2 + 8x - 6 = 0$

  j $x^2 + 2x - 1 = 0$ **k** $x^2 - 2x - 7 = 0$ **l** $x^2 + 2x - 9 = 0$

4 Solve by completing the square. Give your answers to 2 decimal places.

  a $x^2 + 2x - 5 = 0$ **b** $x^2 - 4x - 7 = 0$ **c** $x^2 + 2x - 9 = 0$

5 Solve these equations by completing the square. Leave your answer in square root form.

  a $2x^2 - 6x - 3 = 0$ **b** $4x^2 - 8x + 1 = 0$ **c** $2x^2 + 5x - 10 = 0$

  d $0.5x^2 - 7.5x + 8 = 0$

6 Ahmed rewrites the expression $x^2 + px + q$ by completing the square. He does this correctly and gets $(x - 7)^2 - 52$.

What are the values of p and q?

7 **a** Jorge writes the steps to solve $x^2 + 6x + 7 = 0$ by completing the square. He writes them on sticky notes. Unfortunately he drops the sticky notes and they get out of order. Try to put the notes in the correct order.

Add 2 to both sides	Subtract 3 from both sides	Write $x^2 + 6x + 7 = 0$ as $(x + 3)^2 - 2 = 0$	Take the square root of both sides

13.6 Simultaneous equations

A pair of **simultaneous linear equations** is exactly that — two linear equations for which you want the same solution, and which you therefore *solve together*. For example,

$x + y = 10$ has many solutions:

$$x = 2, y = 8 \qquad x = 4, y = 6 \qquad x = 5, y = 5 \ldots$$

and $2x + y = 14$ has many solutions:

$$x = 2, y = 10 \qquad x = 3, y = 8 \qquad x = 4, y = 6 \ldots$$

But only *one* solution, $x = 4$ and $y = 6$, satisfies both equations at the same time.

Elimination method

One way to solve simultaneous equations is by the *elimination method*. There are six steps in this method.

Step 1 is to balance the coefficients of one of the variables.

Step 2 is to **eliminate** this variable by adding or subtracting the equations.

Step 3 is to solve the resulting linear equation in the other variable.

Step 4 is to **substitute** the value found back into one of the previous equations.

Step 5 is to solve the resulting equation.

Step 6 is to check that the two values found satisfy the original equations.

EXAMPLE 17

Solve the equations: $6x + y = 15$ and $4x + y = 11$.

Label the equations so that the method can be clearly explained.

$$6x + y = 15 \qquad (1)$$
$$4x + y = 11 \qquad (2)$$

Step 1: Since the y-term in both equations has the same coefficient there is no need to balance them.

Step 2: Subtract one equation from the other. (Equation (1) minus equation (2) will give positive values.)

$$(1) - (2) \qquad\qquad 2x = 4$$

Step 3: Solve this equation: $\qquad x = 2$

Step 4: Substitute $x = 2$ into one of the original equations. (Usually it is best to the one with smallest numbers involved.)

So substitute into: $4x + y = 11$

which gives: $\qquad 8 + y = 11$

Step 5: Solve this equation: $\qquad y = 3$

Step 6: Test the solution in the original equations. So substitute $x = 2$ and $y = 3$ into $6x + y$, which gives $12 + 3 = 15$ and into $4x + y$, which gives $8 + 3 = 11$. These are correct, so you can confidently say the solution is $x = 2$ and $y = 3$.

EXAMPLE 18

Solve these equations.

$$5x + y = 22 \qquad (1)$$
$$2x - y = 6 \qquad (2)$$

Step 1: Both equations have the same y-coefficient but with *different* signs so there is no need to balance them.

Step 2: As the signs are different, *add* the two equations, to eliminate the y-terms.

$$(1) + (2) \qquad\qquad 7x = 28$$

Step 3: Solve this equation: $\qquad x = 4$

Step 4: Substitute $x = 4$ into one of the original equations, $5x + y = 22$,

which gives: $\qquad 20 + y = 22$

Step 5: Solve this equation: $\qquad y = 2$

Step 6: Test the solution by putting $x = 4$ and $y = 2$ into the original equations, $2x - y$, which gives $8 - 2 = 6$ and $5x + y$ which gives $20 + 2 = 22$. These are correct, so the solution is $x = 4$ and $y = 2$.

Substitution method

This is an alternative method. The method you use depends very much on the coefficients of the variables and the way that the equations are written in the first place. There are five steps in the substitution method.

Step 1 is to rearrange one of the equations into the form $y = \ldots$ or $x = \ldots$.

Step 2 is to substitute the right-hand side of this equation into the other equation in place of the variable on the left-hand side.

Step 3 is to expand and solve this equation.

Step 4 is to substitute the value into the $y = \ldots$ or $x = \ldots$ equation.

Step 5 is to check that the values work in both original equations.

EXAMPLE 19

Solve the simultaneous equations: $y = 2x + 3$, $3x + 4y = 1$.

Because the first equation is in the form $y = \ldots$ you can use the substitution method.

Again label the equations to help with explaining the method.

$$y = 2x + 3 \qquad (1)$$
$$3x + 4y = 1 \qquad (2)$$

Step 1: As equation (1) is in the form $y = \ldots$ there is no need to rearrange an equation.

Step 2: Substitute the right-hand side of equation (1) into equation (2) for the variable y.

$$3x + 4(2x + 3) = 1$$

Step 3: Expand and solve the equation. $\qquad 3x + 8x + 12 = 1$, $11x = -11$, $x = -1$

Step 4: Substitute $x = -1$ into $y = 2x + 3$: $\qquad y = -2 + 3 = 1$

Step 5: Test the values in $y = 2x + 3$, which gives $1 = -2 + 3$ and $3x + 4y = 1$, which gives $-3 + 4 = 1$. These are correct so the solution is $x = -1$ and $y = 1$.

EXERCISE 13I

1 Solve these simultaneous equations.

In question **1** parts **a** to **i** the coefficients of one of the variables are the same so there is no need to balance them. Subtract the equations when the identical terms have the same sign. Add the equations when the identical terms have opposite signs. In parts **j** to **l** use the substitution method.

a $4x + y = 17$
$2x + y = 9$

b $5x + 2y = 13$
$x + 2y = 9$

c $2x + y = 7$
$5x - y = 14$

CORE

d $3x + 2y = 11$

$2x - 2y = 14$

e $3x - 4y = 17$

$x - 4y = 3$

f $3x + 2y = 16$

$x - 2y = 4$

g $x + 3y = 9$

$x + y = 6$

h $2x + 5y = 16$

$2x + 3y = 8$

i $3x - y = 9$

$5x + y = 11$

j $2x + 5y = 37$

$y = 11 - 2x$

k $4x - 3y = 7$

$x = 13 - 3y$

l $4x - y = 17$

$x = 2 + y$

2 In this sequence, the next term is found by multiplying the previous term by a and then adding b. a and b are positive whole numbers.

3   14   47   ...   ...

a Explain why $3a + b = 14$.

b Set up another equation in a and b.

c Solve the equations to solve for a and b.

d Work out the next two terms in the sequence.

Balancing coefficients in one equation only

You could solve all the examples in Exercise 13I, question **1** by adding or subtracting the equations in each pair, or by substituting without rearranging. This does not always happen. The next examples show what to do when there are no identical terms, or when you need to rearrange.

EXAMPLE 20

Solve these equations. $3x + 2y = 18$ (1)

$2x - y = 5$ (2)

Step 1: Multiply equation (2) by 2. There are other ways to balance the coefficients but this is the easiest and leads to less work later. With practice, you will get used to which will be the best way to balance the coefficients.

$2 \times (2)$ $4x - 2y = 10$ (3)

Label this equation as number (3).

Be careful to multiply every term and not just the y-term. You could write:

$2 \times (2x - y = 5) \Rightarrow 4x - 2y = 10$ (3)

Step 2: As the signs of the y-terms are opposite, add the equations.

$(1) + (3)$ $7x = 28$

Be careful to add the correct equations. This is why labelling them is useful.

Step 3: Solve this equation: $x = 4$

Step 4: Substitute $x = 4$ into any equation, say $2x - y = 5$ $\Rightarrow$ $8 - y = 5$

Step 5: Solve this equation: $y = 3$

Step 6: Check: (1), $3 \times 4 + 2 \times 3 = 18$ and (2), $2 \times 4 - 3 = 5$, which are correct so the solution is $x = 4$ and $y = 3$.

EXAMPLE 21

Solve the simultaneous equations: $3x + y = 5$ (1)

$5x - 2y = 12$ (2)

Step 1: Multiply the first equation by 2: $6x + 2y = 10$ (3)

Step 2: Add (2) + (3): $11x = 22$

Step 3: Solve: $x = 2$

Step 4: Substitute back: $3 \times 2 + y = 5$

Step 5: Solve: $y = -1$

Step 6: Check: (1) $3 \times 2 - 1 = 5$ and (2) $5 \times 2 - 2 \times -1 = 10 + 2 = 12$, which are correct.

EXERCISE 13J

1 Solve parts **a** to **c** by the substitution method and the rest by first changing one of the equations in each pair to obtain identical terms for one unknown, and then adding or subtracting the equations to eliminate those terms.

a $5x + 2y = 4$
$4x - y = 11$

b $4x + 3y = 37$
$2x + y = 17$

c $x + 3y = 7$
$2x - y = 7$

d $2x + 3y = 19$
$6x + 2y = 22$

e $5x - 2y = 26$
$3x - y = 15$

f $10x - y = 3$
$3x + 2y = 17$

g $3x + 5y = 15$
$x + 3y = 7$

h $3x + 4y = 7$
$4x + 2y = 1$

i $5x - 2y = 24$
$3x + y = 21$

j $5x - 2y = 4$
$3x - 6y = 6$

k $2x + 3y = 13$
$4x + 7y = 31$

l $3x - 2y = 3$
$5x + 6y = 12$

2 **a** Francesca is solving the simultaneous equations $4x - 2y = 8$ and $2x - y = 4$.

She finds a solution of $x = 5$, $y = 6$ which works for both equations.

Explain why this is not a unique solution.

b Dimitri is solving the simultaneous equations $6x + 2y = 9$ and $3x + y = 7$.

Why is it impossible to find a solution that works for both equations?

CORE

Balancing coefficients in both equations

In some cases, you will need to change *both* equations to obtain identical terms. The next example shows you how to do this.

Note: The substitution method is not suitable for these types of equation as you end up with fractional terms.

EXAMPLE 22

Solve these equations. $\qquad 4x + 3y = 27 \qquad\qquad\qquad\qquad (1)$

$\qquad\qquad\qquad\qquad\qquad\qquad 5x - 2y = 5 \qquad\qquad\qquad\qquad (2)$

You need to change both equations to obtain identical terms in either x or y. However, you can see that if you make the y-coefficients the same, you will add the equations. Addition is always safer than subtraction, so this is obviously the better choice. Do this by multiplying the first equation by 2 (the y-coefficient of the second equation) and the second equation by 3 (the y-coefficient of the first equation).

Step 1: $\quad (1) \times 2$ or $2 \times (4x + 3y = 27) \qquad \Rightarrow \qquad 8x + 6y = 54 \quad (3)$

$\qquad\qquad (2) \times 3$ or $3 \times (5x - 2y = 5) \qquad \Rightarrow \qquad 15x - 6y = 15 \quad (4)$

Label the new equations (3) and (4).

Step 2: Eliminate one of the variables: $(3) + (4)$: $\qquad 23x = 69$

Step 3: Solve the equation: $\qquad\qquad\qquad\qquad\qquad x = 3$

Step 4: Substitute into equation (1): $\qquad\qquad 12 + 3y = 27$

Step 5: Solve the equation: $\qquad\qquad\qquad\qquad\qquad y = 5$

Step 6: Check: (1), $4 \times 3 + 3 \times 5 = 12 + 15 = 27$ and (2), $5 \times 3 - 2 \times 5 = 15 - 10 = 5$, which are correct so the solution is $x = 3$ and $y = 5$.

EXERCISE 13K

1 Solve these simultaneous equations.

a $2x + 5y = 15$
$\quad 3x - 2y = 13$

b $2x + 3y = 30$
$\quad 5x + 7y = 71$

c $2x - 3y = 15$
$\quad 5x + 7y = 52$

d $3x - 2y = 15$
$\quad 2x - 3y = 5$

e $5x - 3y = 14$
$\quad 4x - 5y = 6$

f $3x + 2y = 28$
$\quad 2x + 7y = 47$

g $2x + y = 4$
$\quad x - y = 5$

h $5x + 2y = 11$
$\quad 3x + 4y = 8$

i $x - 2y = 4$
$\quad 3x - y = -3$

j $3x + 2y = 2$
$\quad 2x + 6y = 13$

k $6x + 2y = 14$
$\quad 3x - 5y = 10$

l $2x + 4y = 15$
$\quad x + 5y = 21$

m $3x - y = 5$
$\quad x + 3y = -20$

n $3x - 4y = 4.5$
$\quad 2x + 2y = 10$

o $x - 5y = 15$
$\quad 3x - 7y = 17$

CORE

2 Here are four equations.

A: $5x + 2y = 1$

B: $4x + y = 9$

C: $3x - y = 5$

D: $3x + 2y = 3$

Here are four sets of (x, y) values.

$(1, -2)$, $(-1, 3)$, $(2, 1)$, $(3, -3)$

Match each pair of (x, y) values to a pair of equations.

> **ADVICE AND TIPS**
>
> You could solve each possible set of pairs but there are six to work out. Alternatively you can substitute values into the equations to see which work.

3 Find the area of the triangle enclosed by these three equations.

$y - x = 2$ $\qquad$ $x + y = 6$ $\qquad$ $3x + y = 6$

4 Find the area of the triangle enclosed by these three equations.

$x - 2y = 6$ $\qquad$ $x + 2y = 6$ $\qquad$ $x + y = 3$

> **ADVICE AND TIPS**
>
> Find the point of intersection of each pair of equations, plot the points on a grid and use any method to work out the area of the resulting triangle.

13.7 Solving inequalities

Ⓔ

Inequalities behave similarly to equations. You use the same rules to solve linear inequalities as you use for linear equations.

There are four inequality signs:

$<$ means 'less than'

$>$ means 'greater than'

$\leqslant$ means 'less than or equal to'

$\geqslant$ means 'greater than or equal to'.

Be careful. Never replace the inequality sign with an equals sign.

EXAMPLE 23

Solve $2x + 3 < 14$.

Rewrite this as: $\qquad$ $2x < 14 - 3$

$\qquad\qquad\qquad\qquad\quad$ $2x < 11$

Divide both sides by 2: $\quad$ $\dfrac{2x}{2} < \dfrac{11}{2}$

$\qquad\qquad\qquad\qquad\quad \Rightarrow x < 5.5$

This means that x can take any value below 5.5 but *not* the value 5.5.

EXAMPLE 24

Solve $\frac{x}{2} + 4 \geqslant 13$.

Solve as you would an equation but leave the inequality sign in place of the equals sign.

Subtract 4 from both sides: $\frac{x}{2} \geqslant 9$

Multiply both sides by 2: $x \geqslant 18$

This means that x can take any value above and including 18.

If you multiply or divide by a negative number when you are solving an inequality you must change the sign.

'less than' becomes 'more than'

'more than' becomes 'less than'

EXAMPLE 25

Solve the inequality $10 - 2x \geqslant 3$.

Subtract 10 from both sides: $-2x \geqslant -7$

Divide both sides by -2: $x \leqslant 3.5$

Note that you must reverse inequality signs when multiplying or dividing both sides by a negative number. So the inequality has changed in the last line.

You could solve example 25 in a different way:

$$10 - 2x \geqslant 3$$

Add $2x$ to both sides: $10 \geqslant 2x + 3$

Subtract 3 from both sides: $7 \geqslant 2x$

Divide both sides by 2: $3.5 \geqslant x$

The sign does not change this time.

$3.5 \geqslant x$ is equivalent to $x \leqslant 3.5$

EXERCISE 13L

1 Solve these linear inequalities.

a $x + 4 < 7$

b $t - 3 > 5$

c $p + 2 \geqslant 12$

d $2x - 3 < 7$

e $4y + 5 \leqslant 17$

f $3t - 4 > 11$

g $\frac{x}{2} + 4 < 7$

h $\frac{y}{5} + 3 \leqslant 6$

i $\frac{t}{3} - 2 \geqslant 4$

j $3(x - 2) < 15$

k $5(2x + 1) \leqslant 35$

l $2(4t - 3) \geqslant 34$

EXTENDED

215

2 Write down the largest integer value of x that satisfies each inequality.

a $x - 3 \leqslant 5$, where x is positive

b $x + 2 < 9$, where x is positive and even

c $3x - 11 < 40$, where x is a square number

d $5x - 8 \leqslant 15$, where x is positive and odd

e $2x + 1 < 19$, where x is positive and prime

3 Write down the smallest integer value of x that satisfies each inequality.

a $x - 2 \geqslant 9$, where x is positive

b $x - 2 > 13$, where x is positive and even

c $2x - 11 \geqslant 19$, where x is a square number

4 Ahmed went to town with $20 to buy two CDs. His bus fare was $3. The CDs were both the same price. When he reached home he still had some money in his pocket. What was the most each CD could cost?

ADVICE AND TIPS

Set up an inequality and solve it.

5 a Explain why you cannot make a triangle with three sticks of length 3 cm, 4 cm and 8 cm.

b Three sides of a triangle are x, $x + 2$ and 10 cm.

x is a whole number.

What is the smallest value x can take?

6 Five cards have inequalities and equations marked on them.

$$\boxed{x > 0} \quad \boxed{x < 3} \quad \boxed{x \geqslant 4} \quad \boxed{x = 2} \quad \boxed{x = 6}$$

The cards are shuffled, laid face down and then turned over, one at a time.

If the possible values on two consecutive cards have any numbers in common, then a point is scored.

If they do not have any numbers in common, then a point is deducted.

a The first two cards below score −1 because $x = 6$ and $x < 3$ have no numbers in common. Explain why the total for this combination scores 0.

$$\boxed{x = 6} \quad \boxed{x < 3} \quad \boxed{x > 0} \quad \boxed{x = 2} \quad \boxed{x \geqslant 4}$$

b What does this combination score?

$$\boxed{x > 0} \quad \boxed{x = 6} \quad \boxed{x \geqslant 4} \quad \boxed{x = 2} \quad \boxed{x < 3}$$

c Arrange the cards to give a maximum score of 4.

7 Solve these linear inequalities.

a $4x + 1 \geqslant 3x - 5$

b $5t - 3 \leqslant 2t + 5$

c $3y - 12 \leqslant y - 4$

d $2x + 3 \geqslant x + 1$

e $5w - 7 \leqslant 3w + 4$

f $2(4x - 1) \leqslant 3(x + 4)$

8 Solve these linear inequalities.

a $\dfrac{x + 4}{2} \leqslant 3$

b $\dfrac{x - 3}{5} > 7$

c $\dfrac{2x + 5}{3} < 6$

d $\dfrac{4x - 3}{5} \geqslant 5$

e $\dfrac{2t - 2}{7} > 4$

f $\dfrac{5y + 3}{5} \leqslant 2$

9 In this question n is always an integer.

a Find the largest possible value of n if $2n + 3 < 12$.

b Find the largest possible value of n if $\dfrac{n}{5} < 20$.

c Find the smallest possible value of n if $3(n - 7) \geqslant 10$.

d Find the smallest possible value of n if $\dfrac{6n - 2}{7} \geqslant 9$.

e Find the smallest possible value of n if $3n + 14 \leqslant 8n - 13$.

10 a If $20 - x > 4$, which of these numbers are possible values of x?

$-10 \qquad 0 \qquad 10 \qquad 20 \qquad 30$

b Solve the inequality $20 - x > 4$.

11 Solve these inequalities.

a $15 - x > 6$

b $18 - x \leqslant 7$

c $6 \geqslant 9 - x$

12 Solve these inequalities.

a $20 - 2x \leqslant 5$

b $3 - 4x \geqslant 11$

c $25 - 3x > 7$

d $2(6 - x) < 9$

e $\dfrac{10 - 2x}{5} \leqslant 4$

f $\dfrac{8 - 4x}{3} > 2$

Line graphs are used in many media, including newspapers and the textbooks of most of the subjects that you learn in school.

Graphs show the relationship between variables. Often one of these variables is time and the graph shows how the other variable changes over time.

For example, this graph on the right shows how the exchange rate between the dollar and the pound changed over six months in 2015.

Graphs like this make it easy to see what is happening to a variable – much easier than looking at lists of data. Here you can see instantly that the value of the pound, compared to US dollars, went up and down, but increased overall, over the five months.

A graph can show several variables to make it easier to compare them. The graph below shows data about a racing car going round a circuit. It compares the driver's acceleration and deceleration (speeding up and slowing down) with his steering. The green line is speed and the pink line is steering.

The graph gives the team engineers and trainers an instant picture of the way the driver goes round the course. It would be difficult to compare all this data in any other way.

Chapter

Graphs in practical situations

Topics   **Level**   **Key words**

 Conversion graphs   CORE   scale, estimate, conversion graph

 Travel graphs   CORE   average speed, distance–time graph

 Speed–time graphs   EXTENDED   speed–time, constant speed, distance travelled, rate of increase, acceleration, deceleration

In this chapter you will learn how to:

CORE	EXTENDED

• Interpret and use graphs in practical situations including travel graphs and conversion graphs. (C2.9 and E2.9)
 	
• Draw graphs from given data. (C2.9 and E2.9)
	• Apply the idea of rate of change to easy kinematics involving:
	  • distance–time and speed–time graphs
	  • acceleration and deceleration. (E2.9)
	
	• Calculate distance travelled as area under a linear speed–time graph. (E2.9)

Look at Examples 1 and 2, and make sure that you can understand the conversions. You need to be able to read these types of graph by finding a value on one axis and following it through to the other axis. Make sure you understand the **scales** on the axes to help you **estimate** the answers.

EXAMPLE 1

This is a **conversion graph** between litres and gallons.

a How many litres are there in 5 gallons?

b How many gallons are there in 15 litres?

From the graph you can see that:

a 5 gallons are approximately equivalent to 23 litres

b 15 litres are approximately equivalent to $3\frac{1}{4}$ gallons.

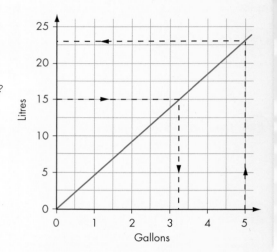

EXAMPLE 2

This is a graph of the charges made for units of electricity used in the home.

a How much will a customer who uses 500 units of electricity be charged?

b How many units of electricity will a customer who is charged $20 have used?

From the graph you can see that:

a a customer who uses 500 units of electricity will be charged $45

b a customer who is charged $20 will have used about 150 units.

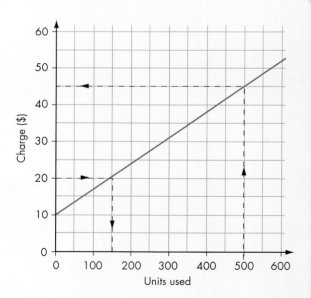

EXERCISE 14A

1 Mass can be measured in kilograms or pounds. This is a conversion graph between kilograms (kg) and pounds (lb).

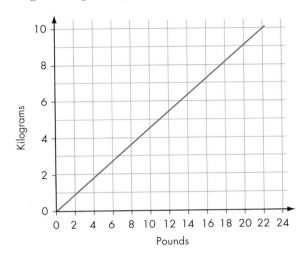

a Use the graph to make an approximate conversion of:

   i 18 lb to kilograms

   ii 5 lb to kilograms

   iii 4 kg to pounds

   iv 10 kg to pounds.

b Approximately how many pounds are equivalent to a total mass of 1 kg?

c Explain how you could use the graph to convert 48 lb to kilograms.

2 Distances can be measured in centimetres or inches. This is a conversion graph between inches (in) and centimetres (cm).

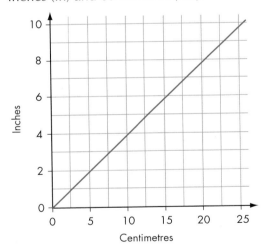

a Use the graph to make an approximate conversion of:

   i 4 in to centimetres

   ii 9 in to centimetres

   iii 5 cm to inches

   iv 22 cm to inches.

b Approximately how many centimetres are there in 1 inch?

c Explain how you could use the graph to convert 18 inches to centimetres.

3 This graph was produced to show approximately how much the British pound (£) is worth in Singapore dollars ($).

a Use the graph to make an approximate conversion of:

   i £100 to Singapore dollars

   ii £30 to Singapore dollars

   iii $150 to British pounds

   iv $250 to British pounds.

b Approximately how many Singapore dollars would you get for £1?

c What would happen to the conversion line on the graph if the pound is worth fewer Singapore dollars?

CORE

CORE

4 A hire firm hired out industrial heaters. They used this graph to approximate what the charges would be.

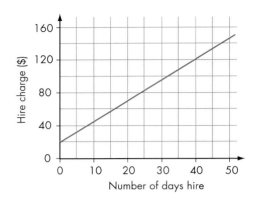

a Use the graph to find the approximate charge for hiring a heater for:

   i 40 days

   ii 25 days.

b Use the graph to find out how many days' hire you would get for a cost of:

   i $100

   ii $140.

5 A conference centre had this chart on the office wall so that the staff could see the approximate cost of a conference, based on the number of people attending it.

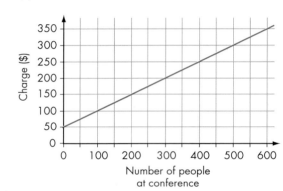

a Use the graph to find the approximate charge for:

   i 100 people

   ii 550 people.

b Use the graph to estimate how many people can attend a conference at the centre for a cost of:

   i $300

   ii $175.

6 A shopkeeper has to add sales tax to the prices of all goods he sells. He marked all goods with prices before tax was added and the sales assistant had to use this chart to convert these marked prices to selling prices.

a Use the chart to find the selling price of goods marked:

   i $60

   ii $25.

b What was the marked price if you bought something for:

   i $100

   ii $45?

7 In Europe temperatures are measured in Celsius and in the USA they are measured in Fahrenheit. Here is a conversion graph for the two scales.

a Use the graph to make an approximate conversion of:

i 35°C to Fahrenheit

ii 20°C to Fahrenheit

iii 50°F to Celsius

iv 90°F to Celsius.

b Water freezes at 0°C. What temperature is this in Fahrenheit?

8 I lost my fuel bill, but while talking to my friends I found out that:

Bill, who had used 850 units, was charged $57.50

Wendy, who had used 320 units, was charged $31

Rhanni, who had used 540 units, was charged $42.

a Plot the given information and draw a straight-line graph. Use a scale from 0 to 900 on the horizontal units axis, and from $0 to $60 on the vertical cost axis.

b Use your graph to find what I will be charged for 700 units.

9 Distances can be measured in kilometres or miles.

80 kilometres is approximately the same distance as 50 miles.

a Use this information to draw a conversion graph between kilometres and miles.

b Use the graph to convert 30 miles into kilometres.

c Use the graph to convert 25 kilometres into miles.

10 A candle gets shorter as it burns.

After 10 hours it has burnt down by 13 centimetres.

a Show this information on a graph.

b How much did the candle burn down in 7 hours?

c How long did it take to burn down by 5 centimetres?

11 This table shows how far a snail has moved after different periods of time.

Time in minutes	5	15	30
Distance in centimetres	13	39	78

a Draw a graph to show this information.

b How long did the snail take to move 60 centimetres?

As the name suggests, a travel graph gives information about how far someone or something has travelled over a given time period.

Travel graphs are sometimes called **distance–time** graphs.

You read a travel graph in a similar way to the conversion graphs you have just done. But you can also find the **average speed** from a distance–time graph, using the formula:

$$\text{average speed} = \frac{\text{total distance travelled}}{\text{total time taken}}$$

EXAMPLE 3

This distance–time graph represents a car journey from Murcia to Cartagena, a distance of 50 km, and back again.

a What can you say about points B, C and D?

b What can you say about the journey from D to F?

c Work out the average speed for each of the five stages of the journey.

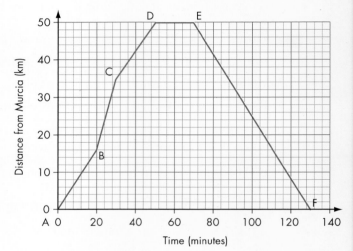

From the graph:

a B: After 20 minutes the car was 16 km away from Murcia.

C: After 30 minutes the car was 35 km away from Murcia.

D: After 50 minutes the car was 50 km away from Murcia, so at Cartagena.

b D–F: The car stayed at Cartagena for 20 minutes, and then took 60 minutes for the return journey.

c The average speeds over the five stages of the journey are worked out as follows.

A to B represents 16 km in 20 minutes.

20 minutes is $\frac{1}{3}$ of an hour, so multiply by 3 to give distance/hour.

Multiplying both numbers by 3 gives 48 km in 60 minutes, which is 48 km/h.

B to C represents 19 km in 10 minutes.

Multiplying both numbers by 6 gives 114 km in 60 minutes, which is 114 km/h.

C to D represents 15 km in 20 minutes.

Multiplying both numbers by 3 gives 45 km in 60 minutes, which is 45 km/h.

D to E represents a stop: speed is 0, no further distance travelled.

E to F represents the return journey of 50 km in 60 minutes, which is 50 km/h.

So, the return journey was at an average speed of 50 km/h.

EXERCISE 14B

1 Paulo was travelling in his car to a meeting. This distance–time graph illustrates his journey.

a How long after he set off did he:

  i stop for his break

  ii set off after his break

  iii get to his meeting place?

b At what average speed was he travelling:

  i over the first hour

  ii over the second hour

  iii for the last part of his journey?

c The meeting was scheduled to start at 10.30 am. What is the latest time he should have left home?

> **ADVICE AND TIPS**
>
> If a part of a journey takes 30 minutes, just double the distance to get the average speed.

2 Farid was travelling by car in Europe on his holiday.

This distance–time graph illustrates his journey.

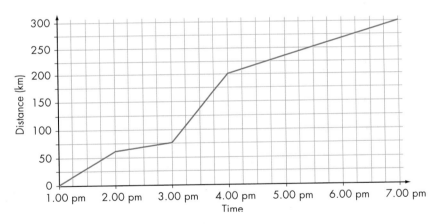

a His greatest speed was on the motorway.

  i How far did he travel on the motorway?

  ii What was his average speed on the motorway?

b i When did he travel the most slowly? ii What was his lowest average speed?

3 A small bus set off from Auzio to pick up Mikel and his family. It then went on to pick up Mikel's parents and grandparents. It then travelled further, dropping them all off at a hotel. The bus then went on a further 10 km to pick up another party and it took them back to Auzio. This distance–time graph illustrates the journey.

a How far from Auzio did Mikel's parents and grandparents live?

b How far from Auzio is the hotel at which they all stayed?

c What was the average speed of the bus on its way back to Auzio?

4 Reu and Yuto took part in a 5000 m race. It is illustrated in this graph.

a Yuto ran a steady race. What is his average speed in:

i metres per minute

ii kilometres per hour?

b Reu ran in spurts. What was his highest average speed?

c Who won the race and by how much?

5 Three friends, Patrick, Araf and Sean, ran a 1000 m race. The race is illustrated on the distance–time graph.

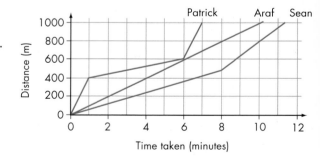

a Describe how each of them completed the race.

b **i** What is Araf's average speed in m/s?

ii What is this speed in km/h?

6 A walker sets off at 0900 from point P to walk along a trail at a steady pace of 6 km per hour.

90 minutes later, a cyclist sets off from P on the same trail at a steady pace of 15 km per hour.

a Draw a graph to illustrate the journeys of the walker and cyclist.

b At what time did the cyclist overtake the walker?

ADVICE AND TIPS

Mark a grid with a horizontal axis as time from 0900–1300, and the vertical axis as distance from 0 to 24. Draw lines for both walker and cyclist. Remember that the cyclist doesn't start until 1030.

7 Three school friends set off from school at the same time, 1545. They all lived 12 km away from the school. The distance–time graph illustrates their journeys.

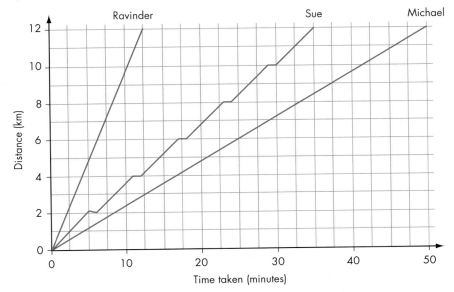

One of them went by bus, one cycled and one was taken by car.

a i Explain how you know that Sue used the bus.

ii Who went by car?

b At what times did each of them get home?

c i When the bus was moving, it covered 2 km in 5 minutes. What is this speed in kilometres per hour?

ii Overall, the bus covered 12 km in 35 minutes. What is this speed, in kilometres per hour?

iii How many stops did the bus make before Sue got off?

8 A girl walks at a steady speed of 1.5 metres/second for 20 seconds.

She stops for 20 seconds.

Then she walks at 2 metres/second for 10 seconds.

a How far has she walked altogether?

b Show her journey on a distance–time graph.

c What is her average speed for the whole journey?

9 A car drives along a motorway for 2 hours at a steady speed of 100 km/hour.

The driver has a break of 30 minutes.

She then drives back to where she started.

The whole trip takes 5 hours.

a Draw a distance–time graph to show the journey.

b What is the average speed for the second half of the journey?

CORE

10 A man walks to the top of a high hill.

He starts at 0900 and by 1000 he has travelled 5 km.

From 1000 to 1200 he covers another 8 km.

He takes an hour to complete the final 2 km to the top of the hill.

a What time did he reach the top?

b How far did he walk?

c Illustrate the journey with a distance–time graph.

d Find the average speed for each stage of the walk and for the whole journey.

14.3 Speed-time graphs

This is the distance–time graph for a cyclist.

The graph is a straight line. This shows that she is moving at constant speed.

$$\text{Speed} = \frac{\text{distance}}{\text{time}} = \frac{80}{20} = 4 \text{ m/s}$$

The distance she travels in
20 seconds = 4 × 20 = 80 metres.

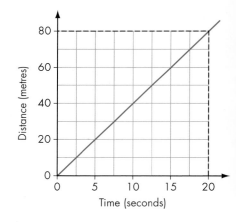

This is a **speed–time** graph of the same situation.

The horizontal line shows that she is moving at a **constant speed**.

The area between the line and the *x*-axis gives the **distance travelled**.

Area = 20 × 4 = 80 metres travelled in 20 seconds

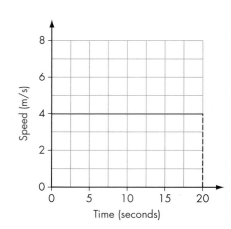

Here is a different speed–time graph for the cyclist.

In this case the speed increases uniformly from 4 m/s to 7 m/s in 10 seconds.

The **rate of increase** of speed is the **acceleration**.

Acceleration = $\dfrac{7-4}{20}$ = 0.15 m/s^2

As before, the area between the line and the x-axis gives the distance travelled.

Area = area of rectangle and area of triangle

$= 20 \times 4 + \frac{1}{2} \times 20 \times 3 = 110$ metres travelled

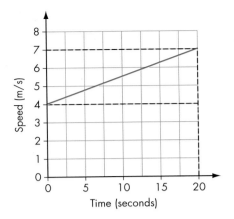

Note: if you know the formula for the area of a trapezium, you can use that here to give:

Area = $\frac{1}{2}(4 + 7) \times 20 = 110$ metres, as above.

EXAMPLE 4

This is a speed–time graph for a car over 15 seconds. Describe the car's journey.

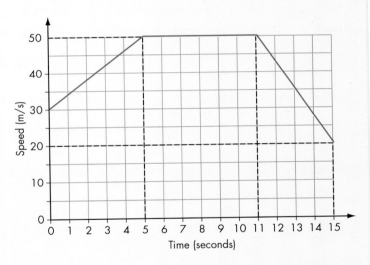

In the first five seconds the car accelerates steadily from 30 m/s to 50 m/s.

The acceleration is $\dfrac{50-30}{5} = 4$ m/s^2.

For the next six seconds the car travels at a constant speed of 50 m/s.

For the last four seconds the car slows down steadily from 50 m/s to 20 m/s.

Because the speed is decreasing you can say the car is **decelerating**.

The **deceleration** is $\dfrac{50-20}{4} = 7.5$ m/s^2.

The area under the line gives the distance travelled.

Distance travelled = area of trapezium (from 0 to 5 seconds) + area of rectangle (from 5 to 11 seconds) + area of trapezium (from 11 to 15 seconds).

$= 200 + 300 + 140$

$= 640$ metres.

EXERCISE 14C

1 The speed–time graph shows a car accelerating for 40 seconds and then decelerating.

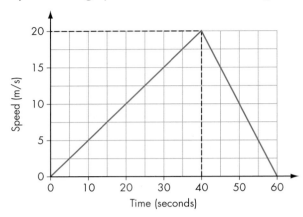

a What is the maximum speed of the car?

b What is the initial acceleration?

c What is the deceleration for the second part of the journey?

d What is the distance travelled?

e What is the average speed for the whole journey?

ADVICE AND TIPS

$$\text{Average speed} = \frac{\text{distance travelled}}{\text{time taken}}$$

2 The speed–time graph shows a train slowing down as it approaches a station.

a What is the deceleration of the train?

b What is the distance travelled?

3 The speed-time graph shows the speed of a boat over thirty seconds.

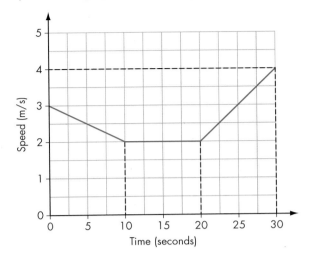

a What is the acceleration for the last ten seconds?

b What is the deceleration for the first ten seconds?

c What is the distance travelled?

d What is the average speed?

4 The graph illustrates a journey for a car.

a What is the initial acceleration?

b What is the final deceleration?

c What is the distance travelled?

d What is the average speed?

5 This table shows the speed of a cyclist as he descends a hill.

Time (s)12	0	5	10	15
Speed (m/s)	4	8		16

a Draw a speed–time graph for this journey.

b Find the acceleration of the cyclist.

c Find the distance travelled in the first ten seconds.

6 A car accelerates from 10 m/s at a rate of 2 m/s^2 for 8 seconds.

a What is the speed of the car after 8 seconds?

b Draw a speed–time graph to illustrate this.

c How far does the car travel in those 8 seconds?

7 This graph shows the journeys of two cars. The red car in this graph is travelling at a constant speed.

As the red car passes it the blue car starts from rest and accelerates for 15 seconds in the same direction as the red car.

The blue car then continues at a constant speed.

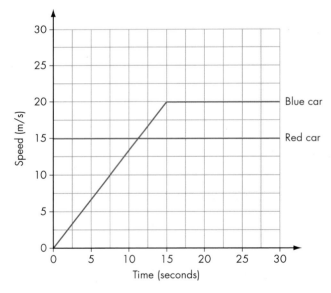

a What is the initial acceleration of the blue car?

b How far apart are the two cars after 30 seconds?

8 A car and a motorbike start at the same time and travel in the same direction.

Their speeds are shown on the speed–time graph.

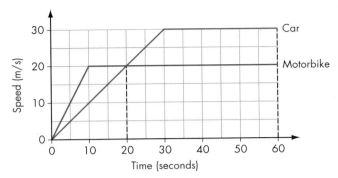

a What is the initial acceleration of the motorbike?

b i When are they travelling at the same speed?

  ii How far apart are they at that time?

c How far does the car travel when it is moving faster than the motorbike?

9 A truck decelerated uniformly from 20 m/s to rest. It travelled 150 m.

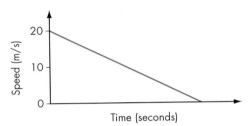

a How long did it take to come to rest?

b What was the rate of deceleration?

10 A cyclist started from rest and accelerated at 1.2m/s^2 for 5 seconds.

a What was the speed of the cyclist after 5 seconds?

b Sketch a speed–time graph for the 5 seconds.

c How far did the cyclist travel in the 5 seconds?

Why this chapter matters

The simplest type of graph is a straight-line graph. This shows that the value of one variable on the graph is always affected in a certain way by changes in the value of the other.

A graph gives you a good visual impression of the way two variables are related to each other.

This graph shows the results of an experiment that measured the voltage in an electrical circuit when different currents were flowing. The points are approximately in a straight line.

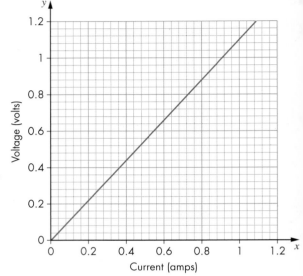

You could use this graph to find the voltage for any current you choose. For example, if you read up from 0.6 amps on the horizontal axis (the *x*-axis) to the graph line you find that the voltage on the vertical axis (the *y*-axis) is 0.66 volts and if the current is 0.8 amps the voltage is 0.88 volts.

You can see that there is a constant relationship between current and voltage: the current in amps is 1.1 × the voltage in volts. You can show this as an equation:

$y = 1.1x$ (voltage is on the *y*-axis and current is on the *x*-axis)

Then you can use this equation to find the voltage for any current you choose. You can also use it to find the current for any voltage. Once you know the relationship between variables, using the equation is quicker and easier than using the graph.

This connection between the geometry of graphs and the algebra of equations was made by a French mathematician, René Descartes in the 17th century CE. This picture shows him at work.

The coordinates in such graphs are called **Cartesian coordinates** (in Latin his name was *Cartesius*) and the *xy* grid they appear on is called the **Cartesian plane**.

The grid can be extended to include negative values as well. The graphs and equations shown on it are not always straight-line ones, as you will see in later chapters.

Straight-line graphs

Topics	Level	Key words
1 Drawing straight-line graphs	CORE	quadrant, coordinates, negative coordinates, Cartesian coordinates, equation of line, Cartesian plane, straight-line graph
2 The equation $y = mx + c$	CORE	slope, gradient, intercept, coefficient
3 Parallel lines	CORE	parallel
4 Points and lines	EXTENDED	midpoint, line segment
5 Perpendicular lines	EXTENDED	perpendicular

In this chapter you will learn how to:

CORE	EXTENDED
• Construct tables of values for functions of the form $ax + b$, $\pm x^2 + ax + b$, $(x \neq 0)$. where a and b are integral constants. (C2.10) • Draw and interpret such graphs. (C2.10) • Interpret and obtain the equation of a straight-line graph in the form $y = mx + c$. (C5.4 and E5.4) • Determine the equation of a straight line parallel to a given line. (C5.5 and E5.5) • Find the gradient of a straight line. (C5.2 and E5.2)	• Calculate the gradient of a straight line from the coordinates of two points on it. (E5.2) • Calculate the length and the coordinates of the midpoint of a straight line from the coordinates of its end points. (E5.3) • Find the gradient of a line perpendicular to a given line – find the gradient of parallel and perpendicular lines. (E5.6)

Using coordinates

A set of axes can form four sectors called **quadrants** but, so far, all the points you have read or plotted on graphs have been **coordinates** in the first quadrant (the top right section of a grid). The grid below shows you how to read and plot coordinates in all four quadrants and how to find the equations of vertical and horizontal lines. This involves using **negative coordinates**.

The coordinates of a point are given in the form (x, y), where x is the number along the x-axis and y is the number up the y-axis. These are sometimes called **Cartesian coordinates** after their inventor, René Descartes.

The grid is sometimes called the **Cartesian plane**.

The coordinates of the four points on the grid are:

A(2, 3)   B(−1, 2)   C(−3, −4)   D(1, −3)

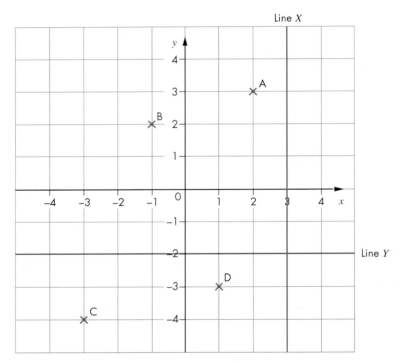

The x-coordinate of all the points on line X is 3.
So you can say the **equation of line** X is $x = 3$.

The y-coordinate of all the points on line Y is −2.
So you can say the equation of line Y is $y = -2$.

Note: The equation of the x-axis is $y = 0$ and the equation of the y-axis is $x = 0$.

Straight-line graphs

An equation of the form $y = mx + c$ where m and c are numbers will give a **straight-line graph**.

EXAMPLE 1

Draw the graph of $y = 4x - 5$ for values of x from 0 to 5. This is usually written as $0 \leqslant x \leqslant 5$.

Choose three values for x: these should be the highest and lowest x-values and one in between.

Work out the y-values by substituting the x-values into the equation.

Keep a record of your calculations in a table, as shown below.

x	0	3	5
y			

When $x = 0$, $y = 4(0) - 5 = -5$
This gives the point $(0, -5)$.

When $x = 3$, $y = 4(3) - 5 = 7$
This gives the point $(3, 7)$.

When $x = 5$, $y = 4(5) - 5 = 15$
This gives the point $(5, 15)$.

Hence your table is:

x	0	3	5
y	-5	7	15

You now have to decide the extent (range) of the axes. You can find this out by looking at the coordinates that you have so far.

The smallest x-value is 0, the largest is 5.
The smallest y-value is -5, the largest is 15.

Now draw the axes, plot the points and complete the graph.

It is nearly always a good idea to choose 0 as one of the x-values. You will often be given the range for the x-values and a set of axes to draw your graph.

EXERCISE 15A

Read through these hints before drawing the straight-line graphs in this exercise.

- Use the highest and lowest values of x given in the range.
- Do not pick x-values that are too close together, such as 1 and 2. Try to space them out so that you can draw a more accurate graph.
- Always label your graph with its equation. This is particularly important when you are drawing two graphs on the same set of axes.
- Create a table of values.

1 Draw the graph of $y = 3x + 4$ for x-values from 0 to 5 ($0 \leqslant x \leqslant 5$).

2 Draw the graph of $y = 2x - 5$ for $0 \leqslant x \leqslant 5$.

3 Draw the graph of $y = \dfrac{x}{2} - 3$ for $0 \leqslant x \leqslant 10$.

4 Draw the graph of $y = 3x + 5$ for $-3 \leqslant x \leqslant 3$.

5 Draw the graph of $y = \dfrac{x}{3} + 4$ for $-6 \leqslant x \leqslant 6$.

> **ADVICE AND TIPS**
>
> Complete the table of values first, then you will know the extent of the y-axis.

6 **a** On the same set of axes, draw the graphs of $y = 3x - 2$ and $y = 2x + 1$ for $0 \leqslant x \leqslant 5$.

   b At which point do the two lines intersect?

7 **a** On the same axes, draw the graphs of $y = 4x - 5$ and $y = 2x + 3$ for $0 \leqslant x \leqslant 5$.

   b At which point do the two lines intersect?

8 **a** On the same axes, draw the graphs of $y = \dfrac{x}{3} - 1$ and $y = \dfrac{x}{2} - 2$ for $0 \leqslant x \leqslant 12$.

   b At which point do the two lines intersect?

9 **a** On the same axes, draw the graphs of $y = 3x + 1$ and $y = 3x - 2$ for $0 \leqslant x \leqslant 4$.

   b Do the two lines intersect? If not, why not?

10 **a** Copy and complete the table to draw the graph of $x + y = 5$ for $0 \leqslant x \leqslant 5$.

x	0	1	2	3	4	5
y	5		3		1	

   b Now draw the graph of $x + y = 7$ for $0 \leqslant x \leqslant 7$.

11 A line has the equation $y = 1.5x + 3$.

   Decide whether each of these points is on the line or not.

   a (6, 12)   **b** (0, 4.5)

   c (2, 6)   **d** (10, 13)

   e (−2, 0)   **f** (−4, −6)

12 This is a graph of $y = 2x - 7$.

Use the graph to solve these equations.

a $2x - 7 = 5$

b $2x - 7 = 0$

c $2x - 7 = -3$

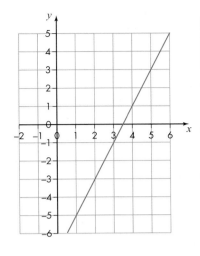

13 This is a graph of $y = 0.5x + 15$.

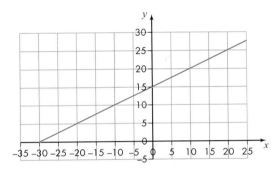

Use the graph to solve these equations.

a $0.5x + 15 = 25$ **b** $0.5x + 15 = 10$

15.2 The equation $y = mx + c$

Gradient

The **slope** of a line is called its **gradient**. The steeper the slope of the line, the larger the value of the gradient.

The graph shows the line with equation $y = 2x + 3$. You can measure the gradient of the line by drawing a right-angled triangle that has part of the line as its hypotenuse (sloping side). The gradient is then given by:

$$\text{gradient} = \frac{\text{distance measured up}}{\text{distance measured along}}$$

$$= \frac{6}{3}$$

$$= 2$$

In fact, if the equation of the line is $y = mx + c$, then m is the gradient of the line.

EXAMPLE 2

Show that the line with the equation $y = 0.5x - 1$ has a gradient of 0.5.

x	-2	0	2	4	6
y	-2	-1	0	1	2

This is a graph of the line.

From the shaded triangle, you can see that the gradient is:

$$\frac{3}{6} = 0.5.$$

What does the 'c' in $y = mx + c$ represent?

From the graphs above you can see it is the value of y, where the line crosses the y-axis.

$y = 2x + 3$ passes through (0, 3).

$y = 0.5x - 1$ passes through (0, -1).

Note that a line that slopes downwards from left to right has a negative gradient.

This line has a gradient of -1.

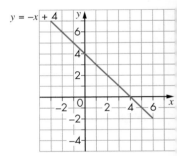

Summary

When a graph can be expressed in the form $y = mx + c$, the **coefficient** of x, which is m, is the **gradient**, and the constant term, which is c, is the **intercept** on the y-axis.

This means that if you know the gradient, m, of a line and its intercept, c, on the y-axis, you can write down the equation of the line immediately.

For example, if $m = 3$ and $c = -5$, the equation of the line is $y = 3x - 5$.

This gives a method of finding the equation of any line drawn on a pair of coordinate axes.

EXAMPLE 3

Find the equation of the line shown in diagram **A**.

A   B   C

First, find where the graph crosses the y-axis (diagram **B**).

So $c = 2$

Next, measure the gradient of the line (diagram **C**).

y-step $= 8$
x-step $= 2$
gradient $= 8 \div 2 = 4$

So $m = 4$

Finally, write down the equation of the line: $y = 4x + 2$.

EXERCISE 15B

1 You drew the graphs of the lines with these equations in Exercise 15A, questions 1–5.

In each case state the gradient.

Then check from your drawing that you are correct.

a $y = 3x + 4$

b $y = 2x - 5$

c $y = \dfrac{x}{2} - 3$

d $y = 3x + 5$

e $y = \dfrac{x}{3} + 4$

CORE

2 Give the equation of each of these lines, all of which have positive gradients. (Each square represents one unit.)

a

b

c

d

e

f

3 In each of these grids, there are two lines. (Each square represents one unit.)

a

b

c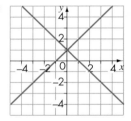

For each grid find the equation of each of the lines.

4 Give the equation of each of these lines, all of which have negative gradients. (Each square represents one unit.)

a

b

c

d

e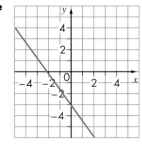

5 The line $y = 4x + c$ passes through $(1, 7)$.

    a Find the value of c.

    b Where does the line cross the y-axis?

6 The line $y = mx - 6$ passes through $(3, 6)$.

    a Find the value of m.

    b What is the gradient of the line?

7 Here are the equations of three lines.

$$y = 4x - 2$$
$$y = 3x - 4$$
$$y = 4x + 5$$

Which of these lines are parallel? Explain how you know.

15.3 Parallel lines

Consider the line with the equation $y = 0.5x + 1$.

The gradient is 0.5. Any line **parallel** to it will have the same gradient.

Examples are $y = 0.5x + 3$ and $y = 0.5x - 4$.
They all have 0.5 as the coefficient of x.

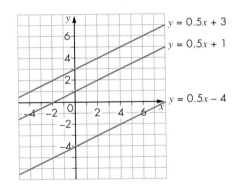

EXAMPLE 4

Find the equation of the line parallel to $y = 4x$ that passes through $(2, 5)$.

The line will have the equation $y = 4x + c$ for some value of c.

Substitute the coordinates of $(2, 5)$ into the equation.

$$5 = 4 \times 2 + c$$
$$5 = 8 + c$$
$$c = -3$$

So the equation is $y = 4x - 3$.

CORE

EXERCISE 15C

1 A line has the equation $y = 2x + 6$.

    a Show that $(3, 12)$ is on this line.

    b What is the gradient of the line?

    c Draw the graph of the line.

    d Find the equation of a parallel line passing through the origin.

    e Find the equation of a parallel line passing through $(0, 3)$.

2 A line has the equation $y = \dfrac{1}{4}x - 1$.

    a Find the coordinates of the points where the line crosses the y-axis and the x-axis.

    b Find the equation of a parallel line through $(8, 5)$.

3 A line has the equation $y = 8 - 2x$.

    a What is the gradient of this line?

    b Where does the line cross the x-axis?

    c Draw the line on a grid.

    d A parallel line passes through the origin. What is the equation?

    e Find the equation of a parallel line through $(5, 4)$.

4 **a** The line with equation $y = 5x + k$ passes through $(2, 11)$. Find the value of k.

    b A line that is parallel to the line in part a passes through $(4, 11)$.
    Find the equation of this line.

5 The two lines in this graph are parallel.

Find the equation of the lower line.

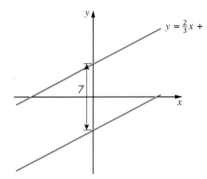

6 The three lines on this graph are parallel.

Find the equations of a and b.

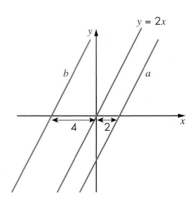

7 The equation of this line can be written as $y + 2x = 6$.

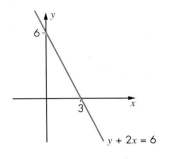

a Show that (1, 4) is on this line.

b Rearrange the formula in the form $y = mx + c$.

c What is the gradient of the line?

d Find the equation of a line that is parallel to this line, but passes through the origin.

15.4 Points and lines E

The equation of a line

If you know the coordinates of two points on a straight line you have enough information to find the equation of the line between them.

Consider the line through (–2, 2) and (4, 5).

Using the triangle shown:

$$\text{gradient} = \frac{\text{difference between } y\text{-coordinates}}{\text{difference between } x\text{-coordinates}}$$

$$= \frac{5 - 2}{4 - -2}$$

$$= \frac{3}{6}$$

$$= 0.5$$

So the equation must be $y = 0.5x + c$ for some value of c.

Use the coordinates of either of the points to find c.

Using $x = 4$ and $y = 5$:

$$5 = 0.5 \times 4 + c$$
$$5 = 2 + c$$
$$c = 3$$

So the equation is $y = 0.5x + 3$.

Midpoints

The **midpoint** of a **line segment** is the same distance from each end.

To find the coordinates of the midpoint, add the coordinates of the end points and divide by 2.

In the example shown:

$$\text{Midpoint is } \left(\frac{-2 + 4}{2}, \frac{2 + 5}{2} \right) = (1, 3.5)$$

The distance between two points

You can use Pythagoras' theorem (see Chapter 25 section 1) to find the distance between two points.

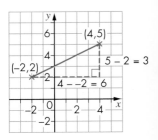

The lengths of the sides of the right-angled triangle shown are the differences between the x-coordinates and the y-coordinates of the end points.

In this case the lengths are 3 and 6.

Distance between points $= \sqrt{3^2 + 6^2} = \sqrt{45} = 6.71$ to 2 decimal places.

EXERCISE 15D

EXTENDED

1 Find the gradient of the line through each pair of points.

a (4, 0) and (6, 6)

b (0, 3) and (8, 7)

c (2, –2) and (4, 6)

d (1, 5) and (5, 1)

e (–4, 6) and (6, 1)

f (–5, –3) and (4, 3)

> **ADVICE AND TIPS**
>
> It will help to plot them on a grid.

2 Find the equation of the line joining each pair of points.

a (0, – 3) and (4, 5)

b (–4, 2) and (2, 5)

c (–1, –6) and (2, 6)

d (1, 5) and (4, –4)

3 Find the midpoints of the line segments joining the points in question 1.

4 A is (–3, 5), B is (1, 1) and C is (5, 9).

a Draw the triangle ABC on a coordinate grid.

b Find the equation of the straight line through A and C.

c Find the midpoint of AB.

d Find the equation of the straight line through the midpoints of AC and BC.

5 Find the length of the line segment joining each pair of points.

a (2, 2) and (6, 5)

b (–3, 2) and (9, 7)

c (1, 5) and (7, –3)

d (–6, –4) and (9, 4)

6 A circle is drawn with its centre at (2, 1) and radius 5.

Show that (–3, 1), (6, 4) and (5, –3) all lie on the circle.

7 Show that the triangle ABC in question 4 is an isosceles triangle.

Perpendicular lines

The lines in this diagram are **perpendicular** to each other.

This means that the angle between them is a right angle, 90°.

You can use the dotted triangles to work out the gradients.

The gradient of line A is $\frac{2}{3}$.

The gradient of line B is $-\frac{3}{2}$ or -1.5.

You can see that:

the gradient of line B = –(the reciprocal of line B)

You can also see that the product of the gradients is –1:

$$\frac{2}{3} \times -\frac{3}{2} = -1$$

The gradients of perpendicular lines usually have this property. The only exception is when the gradient of one of the lines is zero.

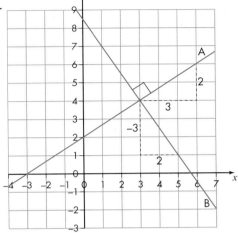

EXAMPLE 5

Find the equation of the line that is perpendicular to $2x + y = 3$ and passes through the point $(4, 6)$.

The given equation is equivalent to $y = -2x + 3$.

The gradient is –2.

The gradient of the line perpendicular to this line is $\frac{1}{2}$. (Find the reciprocal, change the sign).

The equation you want is $y = \frac{1}{2}x + c$ where c is a number.

Find the value of c by substituting the coordinates $(4, 6)$ in the equation.

$$6 = \frac{1}{2} \times 4 + c \qquad \div \qquad 6 = 2 + c \qquad \div \qquad c = 4$$

The equation of the perpendicular line is $y = \frac{1}{2}x + 4$.

You can check the answer to example 5 by drawing a diagram.

The lines are perpendicular.

 1 Work out the gradient of a line that is perpendicular to:

a line A

b line B

c line C.

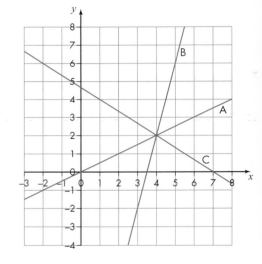

2 Here are the equations of two lines.

$$y = 2 + 5x \qquad 5y + x = 6$$

Show that the lines are perpendicular.

3 Work out the gradient of a line that is perpendicular to:

a $y = 3x$ **b** $y = \frac{1}{2}x - 15$ **c** $y = 3.5 - 0.4x$ **d** $y = 7.5x + 5.7$.

4 Work out the gradient of a line that is perpendicular to:

a $x + y + 5 = 0$ **b** $y + \frac{1}{2}x = 4$ **c** $3x + 4y = 5$ **d** $12y - 2x = 15$.

5 Find the equation of the line that is perpendicular to $y = 5x$ and passes through:

a the origin   **b** (0, 10)   **c** (10, 0).

EXTENDED

6 Find the equation of the line that is perpendicular to $5x + 3y = 30$ and passes through $(-20, -10)$.

7 Find the equation of the line that is perpendicular to AB and passes through C.

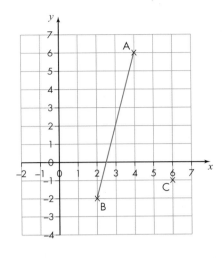

8 A is the point $(-4, 5)$ and B is the point $(6, 7)$.

The perpendicular bisector of AB is a line that is perpendicular to AB and passes through its mid-point.

Find the equation of the perpendicular bisector of AB.

9 This list includes the equations of two pairs of perpendicular lines.

  a $3x + 5y = 7$

  b $6x + 3y = 7$

  c $3y - 5x = 7$

  d $8x - 4y = 7$

  e $4y + 2x = 7$

Which is the odd one out?

10 The centre of a circle is the point $C(3, 2)$.

The circle passes through the point $P(-1, 4)$.

Work out the equation of the tangent to the circle at P.

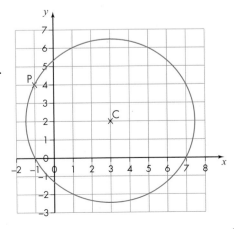

Why this chapter matters

There are many curves that can be seen in everyday life. Did you know that all these curves can be represented mathematically?

Below are a few examples of simple curves that you may have noticed. Can you think of others?

In mathematics, curves can take many shapes. These can be demonstrated using a cone, as shown on the right and below. If you make a cone out of modelling clay, you can see this for yourself. As you look at these curves, try to think of where you have seen them in your own life.

If you slice the cone parallel to the base, the shape you are left with is a circle.

If you slice the cone at an angle to the base, the shape you are left with is an ellipse.

If you slice the cone vertically, the shape you are left with is a hyperbola.

The curve that will be particularly important in this chapter is the parabola. Car headlights are shaped like parabolas.

If you slice the cone parallel to its side, the shape you are left with is a parabola.

All parabolas can be represented by **quadratic graphs**. In this chapter you will look at how to use **quadratic equations** to draw graphs that have this kind of curve.

The suspension cables on this bridge are also parabolas.

Chapter

16 Graphs of functions

Topics	Level	Key words
1 Quadratic graphs	**CORE**	quadratic graph, quadratic equation, parabola
2 Reciprocal graphs	**CORE**	reciprocal
3 More graphs	**EXTENDED**	cubic, exponential functions
4 Exponential graphs	**EXTENDED**	exponential growth, exponential decay
5 Estimating gradients	**EXTENDED**	gradient, tangent

In this chapter you will learn how to:

CORE	EXTENDED
• Construct tables of values for functions of the form $\pm x^2 + ax + b$, $\frac{a}{x}(x \neq 0)$ where a and b are integral constants. (C2.10) • Draw and interpret such graphs. (C2.10) • Solve quadratic equations approximately by graphical methods. (C2.10)	• Construct tables of values and draw graphs for functions of the form ax^n where a is a rational constant and $n = -2, -1, 0, 1, 2, 3$. (E2.10) • Construct simple sums of not more than three of the above and for functions of the form a^x where a is a positive integer. (E2.10) • Solve associated equations approximately by graphical methods. (E2.10) • Draw and interpret graphs representing exponential growth and decay problems. (E2.10) • Estimate gradients of curves by drawing tangents. (E2.11)

Quadratic graphs

A graph with a ∪ or ∩ shape is a quadratic graph.

A **quadratic graph** has an equation that involves a 'squared' term, such as x^2.

All of the following are **quadratic equations** and each would produce a quadratic graph.

$$y = x^2$$
$$y = x^2 + 5$$
$$y = x^2 - 3x$$
$$y = x^2 + 5x + 6$$
$$y = x^2 + 2x - 5$$

EXAMPLE 1

Draw the graph of $y = x^2$ for $-3 \leqslant x \leqslant 3$.

First make a table, as shown below.

x	−3	−2	−1	0	1	2	3
$y = x^2$	9	4	1	0	1	4	9

Now draw axes, with $-3 \leqslant x \leqslant 3$ and $0 \leqslant y \leqslant 9$, plot the points and join them to make a smooth curve.

This is the graph of $y = x^2$.
This type of graph is often referred to as a **parabola**.

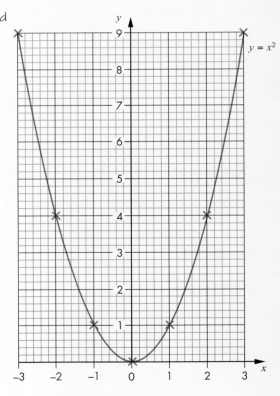

Here are some of the more common mistakes that you should try to avoid, when you are drawing a curve.

- When the points are too far apart, a curve tends to 'wobble'.

- Drawing curves in small sections leads to 'feathering'.

- The place where a curve should turn smoothly is drawn 'flat'.

- A curve is drawn through a point that, clearly, has been incorrectly plotted.

A quadratic curve drawn correctly will *always* be a smooth curve.

Here are some tips that will make it easier for you to draw smooth, curved graphs.

- If you are *right-handed*, turn your piece of paper or your exercise book round so that you draw from left to right. Your hand may be steadier this way than if you try to draw from right to left or away from your body.

  If you are *left-handed*, you may find drawing from right to left the more accurate way.

- Move your pencil over the points as a practice run without drawing the curve.

- Do one continuous curve and only stop at a plotted point.

- Use a *sharp* pencil and do not press too heavily, so that you may easily rub out mistakes.

EXAMPLE 2

a Draw the graph of $y = x^2 + 2x - 3$ for $-4 \leqslant x \leqslant 2$.

b Use your graph to find the value of y when $x = 1.6$.

c Use your graph to find solve the equation $x^2 + 2x - 3 = 1$.

a Draw a table like this, to help work out each step of the calculation.

x	−4	−3	−2	−1	0	1	2
x^2	16	9	4	1	0	1	4
$+2x$	−8	−6	−4	−2	0	2	4
−3	−3	−3	−3	−3	−3	−3	−3
$y = x^2 + 2x - 3$	5	0	−3	−4	−3	0	5

Generally, you do not need to work out all values in a table. If you use a calculator, you need only to work out the y-value. The other rows in the table are just working lines to break down the calculation.

b To find the corresponding y-value for any value of x, you start on the x-axis at that x-value, go up to the curve, across to the y-axis and read off the y-value. This procedure is marked on the graph with arrows.

Always show these arrows because even if you make a mistake and misread the scales, you have shown your method.

So when $x = 1.6$, $y = 2.8$.

c If $x^2 + 2x - 3 = 1$, this means that $y = 1$. This time, start at 1 on the y-axis and read off the two x-values that correspond to a y-value of 1.

Again, this procedure is marked on the graph with arrows.

So when $y = 1$, $x = -3.2$ or $x = 1.2$.

There are two possible values of x that make $x^2 + 2x - 3 = 1$.

Notice that the graph is symmetrical. The equation of the line of symmetry is $x = -1$, and the lowest point is $(-1, -4)$.

CORE

EXERCISE 16A

1 **a** Copy and complete the table for $y = x^2 + 2$.

x	−3	−2	−1	0	1	2	3
$y = x^2 + 2$	11			2		6	

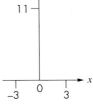

b Draw a graph of $y = x^2 + 2$ for $-3 \leqslant x \leqslant 3$.

2 **a** Copy and complete the table for $y = x^2 - 3x$ for $-3 \leqslant x \leqslant 5$. Use your table to plot the graph.

x	−3	−2	−1	0	1	2	3	4	5
x^2						4			
$-3x$						−6			
y						−2			

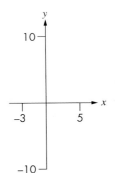

b Use your graph to find the value of y when $x = 3.5$.

c What are the coordinates of the lowest point on the graph?

d What is the equation of the line of symmetry?

e Use your graph to solve the equation $x^2 - 3x = 5$.

> **ADVICE AND TIPS**
>
> You may find you do not need the second and third rows.

3 **a** Copy and complete the table for the graph of $y = x^2 - 2x - 8$ for $-3 \leqslant x \leqslant 5$. Use your table to plot the graph.

x	−3	−2	−1	0	1	2	3	4	5
y					−8				

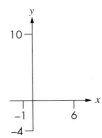

b Use your graph to find the value of y when $x = 0.5$.

c Use your graph to solve the equation $y = x^2 - 2x - 8 = 3$.

4 **a** Copy and complete the table for $y = x^2 - 5x + 4$ for $-1 \leqslant x \leqslant 6$. Use your table to plot the graph.

x	−1	0	1	2	3	4	5	6
y		4			−2			

b What are the coordinates of the lowest point on the graph?

c What is the equation of the line of symmetry?

d Use your graph to find the value of y when $x = -0.5$.

e Use your graph to solve the equation $x^2 - 5x + 4 = 3$.

CORE

5 a Copy and complete the table for $y = x^2 + 2x - 1$ for $-4 \leqslant x \leqslant 2$.
Use your table to plot the graph.

x	−4	−3	−2	−1	0	1	2
y	7						

b Use your graph to find the y-value when $x = -2.5$.

c Use your graph to solve the equation $x^2 + 2x - 1 = 1$.

6 a Copy and complete the table to draw the graph of $y = 12 - x^2$ for $-4 \leqslant x \leqslant 4$.

x	−4	−3	−2	−1	0	1	2	3	4
y				11				3	

b Use your graph to find the y-value when $x = 1.5$.

c Use your graph to solve the equation $12 - x^2 = 0$.

d Use your graph to solve the equation $12 - x^2 = 7$.

7 a Copy and complete the table to draw the graph of $y = x^2 + 4x$ for $-5 \leqslant x \leqslant 2$.

x	−5	−4	−3	−2	−1	0	1	2
x^2	25			4			1	
$+4x$	−20			−8			4	
y	5			−4			5	

b Where does the graph cross the x-axis?

c Use your graph to find the y-value when $x = -2.5$.

d Use your graph to solve the equation $x^2 + 4x = 3$.

8 a Copy and complete the table to draw the graph of $y = x^2 - 6x + 3$ for $-1 \leqslant x \leqslant 7$.

x	−1	0	1	2	3	4	5	6	7
y	10			−5			−2		

b Where does the graph cross the x-axis?

c Use your graph to find the y-value when $x = 3.5$.

d Use your graph to solve the equation $x^2 - 6x + 3 = 5$.

9 $y = 5x - x^2$

a Copy and complete this table of values.

x	−1	0	1	2	3	4	5	6
y								

b Draw a graph of $y = 5x - x^2$ for $-1 \leqslant x \leqslant 6$.

c What is the highest point on the graph?

d What is the equation of the line of symmetry on the graph?

e Solve the equation $5x - x^2 = 2$.

Reciprocal graphs

A **reciprocal** equation has the form $y = \frac{a}{x}$.

Examples of reciprocal equations are: $y = \frac{1}{x}$ $y = \frac{4}{x}$ $y = -\frac{3}{x}$

All reciprocal graphs have a similar shape and some symmetry properties.

EXAMPLE 3

Complete the table to draw the graph of $y = \frac{1}{x}$ for $-4 \leqslant x \leqslant 4$.

x	−4	−3	−2	−1	1	2	3	4
y								

Values are rounded to two decimal places, as it is unlikely that you could plot a value more accurately than this. The completed table looks like this.

x	−4	−3	−2	−1	1	2	3	4
y	−0.25	−0.33	0.5	−1	1	0.5	0.33	0.25

The graph plotted from these values is shown in **A**. This does not include much of the graph and does not show the properties of the reciprocal function. If you take x-values from −0.8 to 0.8 in steps of 0.2, you get the next table.

Note that you cannot use $x = 0$ since $\frac{1}{0}$ is infinity.

x	−0.8	−0.6	−0.4	−0.2	0.2	0.4	0.6	0.8
y	−1.25	−1.67	−2.5	−5	5	2.5	1.67	1.25

Plotting these points as well gives the graph in **B**.

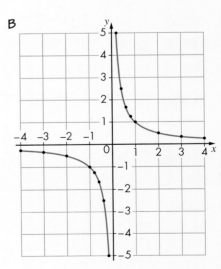

EXAMPLE 3 continued

The graph in **B** shows these properties.

- The lines $y = x$ and $y = -x$ are lines of symmetry.
- The closer x gets to zero, the nearer the graph gets to the y-axis.
- As x increases, the graph gets closer to the x-axis.

The graph never actually touches the axes, it just gets closer and closer to them.

These properties are true for *all reciprocal graphs*.

EXERCISE 16B

1 **a** Copy and complete the table to draw the graph of $y = \dfrac{2}{x}$ for $-4 \leqslant x \leqslant 4$.

x	0.2	0.4	0.5	0.8	1	1.5	2	3	4
y	10		4				1		0.5

b Use your graph to find the y-value when $x = 2.5$.

c Use your graph to solve the equation $\dfrac{2}{x} = 7$.

d Use your graph to solve the equation $\dfrac{2}{x} = -1.25$.

2 **a** Copy and complete the table to draw the graph of $y = \dfrac{5}{x}$ for $-20 \leqslant x \leqslant 20$.

x	0.2	0.4	0.5	1	2	5	10	15	20
y	25		10		2.5				0.25

b On the same axes, draw the line $y = x + 10$.

c Use your graph to solve the equation $\dfrac{5}{x} = x + 10$.

3 Draw a graph of $y = \dfrac{-2}{x}$ for $-4 \leqslant x \leqslant 4$.

Use a table of values like the one in question 1.

4 **a** Complete this table.

x	1	2	4	5	8	10	20
$\dfrac{20}{x}$							

b Draw a graph of $y = \dfrac{20}{x}$ for $-20 \leqslant x \leqslant 20$.

c On the same axes, draw the line with the equation $2x - 10$.

d Use your graph to solve the equation $\dfrac{20}{x} = 2x - 10$.

Cubic graphs

A **cubic** function or graph is one that contains a term in x^3. The following are examples of cubic graphs.

$$y = x^3 \qquad y = x^3 + 3x \qquad y = x^3 + x^2 + x + 1$$

The techniques used to draw them are exactly the same as those for quadratic and reciprocal graphs.

For example, here is a table of values and graph of $y = x^3 - x^2 - 4x + 4$

x	−3	−2.5	−2	−1.5	−1	−0.5	0	0.5	1	1.5	2	2.5	3
y	−20.00	−7.88	0.00	4.38	6.00	5.63	4.00	1.88	0.00	−0.88	0.00	3.38	10.00

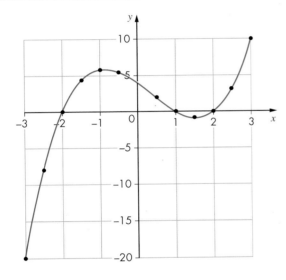

EXERCISE 16C

1 Draw the graph of $y = x^3$ for $-2 \leqslant x \leqslant 2$.

2 **a** Complete the table to draw the graph of $y = 0.5x^3$ for $-2.5 \leqslant x \leqslant 2.5$.

x	−2.5	−2	−1.5	−1	−0.5	0	0.5	1	1.5	2	2.5
y			−1.69			0.00					7.81

b Use your graph to solve the equation $0.5x^3 = 6$.

3 **a** Complete the table to draw the graph of $y = x^3 + 3$ for $-2.5 \leqslant x \leqslant 2.5$.

x	−2.5	−2	−1.5	−1	−0.5	0	0.5	1	1.5	2	2.5
y	−12.63			2.00		3.00	3.13			11.00	

b Use your graph to solve the equation $x^3 + 3 = 0$.

4 **a** Complete the table to draw the graph of $y = x^3 - 2x + 5$ for $-2 \leqslant x \leqslant 2$.

x	−2	−1.5	−1	−0.5	0	0.5	1	1.5	2
y	1.00	4.63			5.00	4.13			

b Use your graph to solve the equation $x^3 - 2x + 5 = 3$.

5 **a** Complete this table of values for $y = \dfrac{20}{x^2}$. Give the values to 2 decimal places.

x	1	2	3	4	5
y			2.22		

b Draw a graph of $y = \dfrac{20}{x^2}$ for $2 \leqslant x \leqslant 5$.

c Draw a graph of $y = \dfrac{20}{x^2}$ for $-5 \leqslant x \leqslant -2$.

6 **a** Complete this table of values for $y = \dfrac{100}{x^2} - 0.5x$.

x	4	5	6	7	8
y	4.25			−1.46	

b Draw a graph of $y = \dfrac{100}{x^2} - 0.5x$ for $4 \leqslant x \leqslant 8$.

c Use the graph to solve the equation $\dfrac{100}{x^2} - 0.5x = 0$.

7 This is a sphere.

The diameter is x cm.

The volume is $\dfrac{1}{6}\pi x^3$ cm^3.

a Draw a graph of $y = \dfrac{1}{6}\pi x^3$ for $6 \leqslant x \leqslant 10$.

Use a scale of 2 cm to 1 unit on the x-axis and 2 cm to 100 units on the y-axis.

b Use your graph to find the diameter of a sphere with a volume of 300 cm^3.

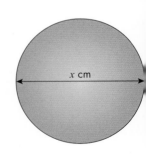

x cm

8 The volume of this cuboid is 1000 cm^3.

The top is a square of side x cm.

The height is y cm.

a Show that $y = \dfrac{1000}{x^2}$.

b Draw a graph of $y = \dfrac{1000}{x^2}$ for $5 \leqslant x \leqslant 10$.

Use a scale of 2 cm to 1 unit on the x-axis and 2 cm to 10 units on the y-axis

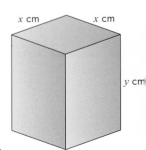

x cm   *x* cm

y cm

9 This is a cuboid. The end is a square of side x cm.

The length is 5 cm greater than the side of the square end.

The volume of the cuboid is $x^2(x + 5)$ cm³.

a Draw a graph of $y = x^2(x + 5)$ for $0 \leqslant x \leqslant 6$.

Use a scale of 2 cm to 1 unit on the x-axis and 2 cm to 100 units on the y-axis

b Use your graph to find the sides of the cuboid when the volume is 200 cm³.

16.4 Exponential graphs

There are 2000 monkeys in a forest. Scientists say that the numbers are increasing by 40% each year.

This is an example of **exponential growth**. The multiplier is 1.4.

After one year there will be 2000 × 1.4 = 2800 monkeys.

After two years there will be 2800 × 1.4 or 2000 × 1.4² = 3920 monkeys.

After x years there will be 2000×1.4^x monkeys.

> **ADVICE AND TIPS**
>
> Ignore the decimal fraction, you need only consider whole monkeys.

Here is a table of values.

x	0	1	2	3	4
y	2000	2800	3920	5488	7683

Check that 3920 × 1.4 = 5488 and 5488 × 1.4 = 7683.

A graph shows how the population of monkeys changes over time.

The graph shows that the population reaches 10 000 after nearly five years.

Graphs of exponential growth always have this shape.

As the value of x increases, the value of y increases and the graph gets steeper.

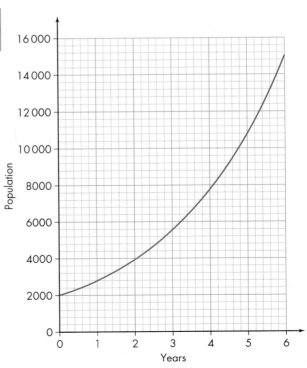

EXAMPLE 4

Atmospheric pressure is measure in units called hectopascals (hPa).

It decreases as height above sea level increases.

At sea level atmospheric pressure is 1000 hPa.

It decreases by 12% for every kilometre increase in height.

a Draw a graph to show how atmospheric pressure changes, up to a height of six kilometres.

b At what height is the atmospheric pressure half the value at sea level?

a The multiplier for a decrease of 12% is 0.88. $100\% - 12\% = 88\% = 0.88$

At a height of x km the pressure is 1000×0.8^x hPa.

Height above sea level x (km)	Pressure y (hPa)
0	1000
1	$1000 \times 0.88 = 880$
2	$1000 \times 0.88^2 = 774$
3	$1000 \times 0.88^3 = 681$
4	$681 \times 0.88 = 600$
5	$600 \times 0.88 = 528$
6	464

This graph shows the values.

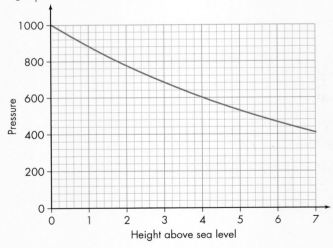

b Half the pressure at sea level is 500 mPa.

From the graph, when $y = 500$, $x = 5.4$.

The height is 5.4 km.

The graph in the example shows **exponential decay**.

As the value of x increases, the value of y decreases and the graph gets less steep.

EXERCISE 16D

1 A company buys some new machinery for $96 000.

The value of the machinery halves every year.

a Copy and complete this table.

Years	0	1	2	3	4
Value (thousand dollars)	96				

b Draw a graph to show how the value falls.

2 The price of a car is $25 000 when it is new.

The value falls by 20% per year.

a Show that the value after one year is $20 000.

b Copy and complete this table.

Years	0	1	2	3	4	5
Value ($)	25 000	20 000				

c Use the table to draw a graph showing the fall in value of the car.

d How long is it until the car is worth half its original value?

3 The population of a country in 2015 is 60 million.

The population increases by 15% every ten years.

a Copy and complete this table to show the predicted population.

Years	2015	2025	2035	2045	2055
Population (millions)	60	69			

b Draw a graph to show how the population changes.

c Estimate the year when the population will reach 100 million.

4 Mira has shares worth $500.

Their value increases by 30% each year.

a Copy and complete this table.

Years	0	1	2	3	4	5
Value ($)	500					1856

b Draw a graph to show how the value of the shares increases.

c How long does it take for the value of the shares to double?

5 There are 20 mice in a population.

Mice breed very quickly. The number of mice doubles every month.

a Draw a graph to show how the number of mice increases over six months.

b How long will it be until there are 200 mice?

6 Abram puts $400 in a bank.

After one year it is worth $480 dollars.

a What is the annual percentage increase?

b The value increases exponentially. Draw a graph to show how the value increases over four years.

c Use the graph to find the value after 2.5 years.

7 Marcus buys a picture for $5000.

The value is decreasing exponentially by 10% a year.

a Copy and complete this table to show the fall in value of the picture.

Years	0	1	2	3	4	5
Value ($)	5000		4050			2952

b Draw a graph to show how the value falls over eight years.

c How long does it take for the picture to halve in value?

8 This graph shows the rise in value of an investment.

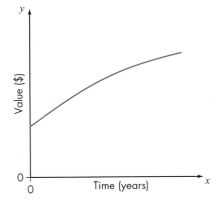

Mario says, 'The value increases all the time so this is exponential growth.'

Explain why Mario is not correct.

9 The graph shows the number of bacteria in a population that is growing exponentially.

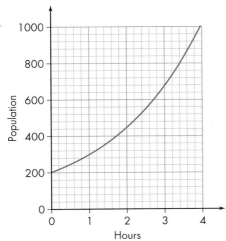

a What was the initial population?

b Work out the percentage growth of the population each hour.

c How long did the population take to triple in size?

10 Barak buys a motor bike for $1000.

The graph shows the fall in value of the bike.

a Show that the value halves every two years.

b Work out the percentage fall in the value of the bike every year.

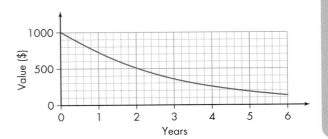

16.5 Estimating gradients   E

The **gradient** of a curve varies from point to point. At points on this curve to the left of A or to the right of B the gradient is positive.

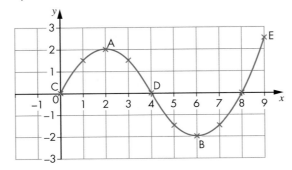

Between A and B the gradient is negative.

The curve is steepest near C, D or E and the gradient there will have the greatest magnitude.

You can estimate the gradient at any point on a curve by drawing a **tangent** at that point. This is a straight line that touches the curve at that point and has the same gradient. You can find the gradient of the tangent, which is a straight line, by drawing a triangle.

EXAMPLE 5

Find the gradient at the point P with the coordinates (7, −1.5).

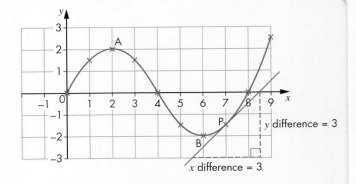

Draw a tangent at point P.

Draw a triangle and find the differences between the x-coordinates and y-coordinates.

$$\text{gradient} = \frac{\text{difference in } y}{\text{difference in } x} = \frac{3}{3} = 1$$

The gradient of the curve at P is 1.

Note that at A and B the gradient is 0.

EXERCISE 16E

1 The straight line is a tangent to the curve at the point P

Find the gradient of the curve at P.

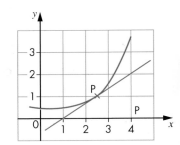

2 A and B have coordinates (−1, 1.5) and (3, 2).

Tangents to the curve have been drawn at A and B.

Calculate the gradient of the curve at A and at B.

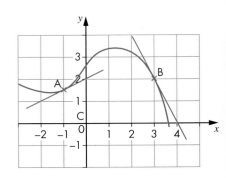

3 **a** Use this table of values to draw a graph of y for $0 \leqslant x \leqslant 4$.

x	0	1	2	3	4
y	2	1.5	2	3.5	6

b Draw a tangent to the curve at the point (3, 3.5).

c Estimate the gradient of the curve at the point (3, 3.5).

d At which point is the gradient of the curve 0?

4 **a** Copy and complete this table of values for $0.1x^3$.

x	0	0.5	1	1.5	2	2.5	3
$\mathbf{0.1}x^3$		0.01	0.1		0.8		

b Draw a graph of $y = 0.1x^3$ for $0 \leqslant x \leqslant 3$.

c Draw a tangent to the curve at (2, 0.8).

d Estimate the gradient of the curve at (2, 0.8).

5 **a** Copy and complete this table of values for $y = 2^x$.

x	−2	−1	0	1	2
y	0.25		1		

b Draw a graph of $y = 2^x$ for $-2 \leqslant x \leqslant 2$.

c Draw a tangent to the curve at (0, 1).

d Estimate the gradient of the curve at (0, 1).

6 **a** Copy and complete this table of values for $\dfrac{5}{x}$.

x	2	3	4	5	6
$\dfrac{5}{x}$			1.25		0.83

b Draw a graph of $y = \dfrac{5}{x}$ for $2 \leqslant x \leqslant 6$.

c Estimate the gradient of the curve at (4, 1.25).

Why this chapter matters

Patterns often appear in numbers. Prime numbers, square numbers and multiples all form patterns. Mathematical patterns also appear in nature.

There are many mathematical patterns that appear in nature. The most famous of these is probably the **Fibonacci** series.

1   1   2   3   5   8   13   21   …

This is formed by adding the two previous terms to get the next term.

The sequence was discovered by the Italian, Leonardo Fibonacci, in 1202, when he was investigating the breeding patterns of rabbits!

Since then, the pattern has been found in many other places in nature. The spirals found in a nautilus shell and in the seed heads of a sunflower plant also follow the Fibonacci series.

Fractals form another kind of pattern.

Fractals are geometric patterns that are continuously repeated on a smaller and smaller scale.

A good example of a fractal is this: start with an equilateral triangle and draw an equilateral triangle, a third the size of the original, on the middle of each side. Keep on repeating this and you will get an increasingly complex-looking shape.

The pattern shown here is called the Koch snowflake. It is named after the Swedish mathematician, Helge von Koch (1870–1924).

Fractals are commonly found in nature, a good example being the complex patterns found in plants, such as the leaves of a fern.

Number sequences

Topics	Level	Key words

1 Patterns in number sequences | CORE | sequence, term, difference, consecutive

2 The nth term of a sequence | CORE | coefficient, linear sequence, nth term, quadratic sequence, cubic sequence

3 General rules from patterns | CORE | rule, patterns

4 Further sequences | EXTENDED | exponential sequence

In this chapter you will learn how to:

CORE	EXTENDED
Calculate:	• Find the nth term of linear, quadratic, cubic and exponential sequences and simple combinations of these. (E2.7)
• Continue a given number sequence. (C2.7 and E2.7)	
• Recognise patterns in sequences and relationships between different sequences. (C2.7 and E2.7)	
• Find the nth term for linear sequences, and for simple quadratic and cubic sequences. (C2.7 and E2.7)	

A number **sequence** is an ordered set of numbers with a rule for finding every number in the sequence. The rule that takes you from one number to the next could be a simple addition or multiplication, but often it is more tricky than that. You always need to look very carefully at the pattern of a sequence.

Each number in a sequence is called a **term** and is in a certain position in the sequence.

Look at these sequences and their rules.

3, 6, 12, 24, … doubling the previous term each time … 48, 96, …

2, 5, 8, 11, … adding 3 to the previous term each time … 14, 17, …

1, 10, 100, 1000, … multiplying the previous term by 10 each time … 10 000, 100 000

80, 73, 66, 59, … subtracting 7 from the previous term each time … 52, 45, …

These are all quite straightforward once you have looked for the link from one term to the next (**consecutive** terms).

Differences

For some sequences you need to look at the **differences** between consecutive terms to determine the pattern.

EXAMPLE 1

Find the next two terms of the sequence 1, 3, 6, 10, 15, … .

Looking at the differences between consecutive terms:

```
1    3    6    10    15
  ↑    ↑    ↑     ↑
  2    3    4     5
```

So the sequence continues as follows.

```
1    3    6    10    15    21    28
  ↑    ↑    ↑     ↑ ⌐   ⌐  ⌐
  2    3    4     5    +6    +7
```

So the next two terms are 21 and 28.

The differences usually form a number sequence of their own, so you need to find the *sequence of the differences* before you can expand the original sequence.

EXERCISE 17A

1 Look at these number sequences. Write down the next three terms in each and explain how each sequence is formed.

a 1, 3, 5, 7, ...

b 2, 4, 6, 8, ...

c 5, 10, 20, 40, ...

d 1, 3, 9, 27, ...

e 4, 10, 16, 22, ...

f 3, 8, 13, 18, ...

g 2, 20, 200, 2000, ...

h 7, 10, 13, 16, ...

i 150, 141, 132, 123, ...

j 5, 15, 45, 135, ...

k 400, 200, 100, 50, ...

l 1, 5, 25, 125, ...

2 By considering the differences in these sequences, write down the next two terms in each one.

a 1, 2, 4, 7, 11, ...

b 1, 2, 5, 10, 17, ...

c 1, 3, 7, 13, 21, ...

d 1, 4, 10, 19, 31, ...

e 1, 9, 25, 49, 81, ...

f 1, 2, 7, 32, 157, ...

g 1, 3, 23, 223, 2223, ...

h 1, 2, 4, 5, 7, 8, 10, ...

i 2, 3, 5, 9, 17, ...

j 3, 8, 18, 33, 53, ...

3 Look at the sequences below. Find the rule for each sequence and write down its next three terms.

a 3, 6, 12, 24, ...

b 3, 9, 15, 21, 27, ...

c 128, 64, 32, 16, 8, ...

d 50, 47, 44, 41, ...

e 2, 5, 10, 17, 26, ...

f 5, 6, 8, 11, 15, 20, ...

g 5, 7, 8, 10, 11, 13, ...

h 40, 37, 34, 31, 28, ...

i 1, 3, 6, 10, 15, 21, ...

j 1, 2, 3, 4, ...

k 100, 20, 4, 0.8, ...

l 1, 0.5, 0.25, 0.125, ...

4 Look carefully at each number sequence.
Find the next two numbers in the sequence and try to explain the pattern.

a 1, 1, 2, 3, 5, 8, 13, ...

b 1, 4, 9, 16, 25, 36, ...

c 3, 4, 7, 11, 18, 29, ...

d 1, 8, 27, 64, 125, ...

> **ADVICE AND TIPS**
>
> These patterns do not go up by the same value each time so you will need to find another connection between the terms.

5 The sequence of triangular numbers builds up like this.

| 1 | 3 | 6 | 10 |

Find the next four triangular numbers.

CORE

6 The sequence of hexagonal numbers builds up like this.

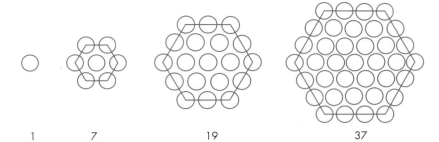

1   7   19   37

Find the next three hexagonal numbers.

7 The first term that these two sequences have in common is 17.

8, 11, 14, 17, 20, ...

1, 5, 9, 13, 17, ...

What are the next two terms that the two sequences have in common?

8 Here are the first few terms if two sequences.

2, 5, 8, 11, 14, ...

3, 6, 9, 12, 15, ...

Will the two sequences ever have a term in common? Yes or no?
Justify your answer.

9 Here are two sequences.

100, 95, 90, 85, …

3, 10, 17, 24, …

Write down all the numbers that will occur in **both** sequences.

17.2 The nth term of a sequence

Finding the rule

When using a number sequence, you sometimes need to know, say, its 50th term, or even a higher term in the sequence. To work out these terms, you need to find the rule that produces the sequence in its general form.

It may be helpful to look at the problem backwards. That is, take a rule and see how it produces a sequence. The rule is given for the general term, which is called the **nth term**.

EXAMPLE 2

The nth term of a sequence is $3n + 1$, where $n = 1, 2, 3, 4, 5, 6, \ldots$.
Write down the first five terms of the sequence.

Substituting $n = 1, 2, 3, 4, 5$ in turn:
$(3 \times 1 + 1), (3 \times 2 + 1), (3 \times 3 + 1), (3 \times 4 + 1), (3 \times 5 + 1), \ldots$
$4 7 10 13 16$

So the sequence is 4, 7, 10, 13, 16, ...

Notice that in Example 2 the difference between each term and the next is always 3, which is the **coefficient** of n (the number attached to n). Also, the constant term is the difference between the first term and the coefficient, that is, $4 - 3 = 1$.

EXERCISE 17B

1 Here are the nth terms of some sequences. Write down the first five terms of each sequence.

a $2n + 1$ for $n = 1, 2, 3, 4, 5$ **b** $3n - 2$ for $n = 1, 2, 3, 4, 5$

c $5n + 2$ for $n = 1, 2, 3, 4, 5$ **d** n^2 for $n = 1, 2, 3, 4, 5$

e $n^2 + 3$ for $n = 1, 2, 3, 4, 5$ **f** $20 - 2n$ for $n = 1, 2, 3, 4, 5$

2 Write down the first five terms of the sequence that has as its nth term:

a $n + 3$ **b** $3n - 1$ **c** $5n - 2$

d $n^2 - 1$ **e** $4n + 5$ **f** $45 - 3n$

> **ADVICE AND TIPS**
>
> Substitute numbers into the expressions until you can see how the sequence works.

3 The nth term of a sequence is $100 - 6n$.

a Work out the first four terms.

b Work out the first term that is less than zero.

4 Write down the first four terms of the sequence for which the nth term is:

a n^2 **b** $n^2 + 2$ **c** $2n^2$

d $4n^2 - 1$ **e** $200 - n^2$ **f** $\dfrac{n^2}{4}$

5 Write down the first four terms of a sequence for which the nth term is:

a n^3 **b** $n^3 + 1$ **c** $n^3 - 2$

d $3n^3 - 1$ **e** $\dfrac{n^3}{2}$ **f** $109 - n^3$

CORE

6 A haulage company uses this formula to calculate the cost of transporting n pallets.

For $n \leqslant 5$, the cost will be $\$(40n + 50)$

For $6 \leqslant n \leqslant 10$, the cost will be $\$(40n + 25)$

For $n \geqslant 11$, the cost will be $\$40n$

a How much will the company charge to transport 7 pallets?

b How much will the company charge to transport 15 pallets?

c A company is charged $170 for transporting pallets. How many pallets did they transport?

d Another haulage company uses the formula $\$50n$ to calculate the cost for transporting n pallets.

At what value of n do the two companies charge the same amount?

7 The nth term of a sequence is $3n + 7$.
The nth term of another sequence is $4n - 2$.

These two series have several terms in common but only one term that is common and has the same position in the sequence.

Without writing out the sequences, show how you can tell, using the expressions for the nth term, that this is the 9th term.

Finding the nth term of a linear sequence

In a **linear sequence** the *difference* between one term and the next is always the same.

For example:

2, 5, 8, 11, 14, ... difference of 3

The nth term of this sequence is given by $3n - 1$.

Here is another linear sequence.

5, 7, 9, 11, 13, ... difference of 2

The nth term of this sequence is given by $2n + 3$.

So, you can see that the nth term of a linear sequence is *always* of the form $An + b$, where:

- A, the coefficient of n, is the difference between each term and the next term (**consecutive** term)
- b is the difference between the first term and A.

EXAMPLE 3

From the sequence 5, 12, 19, 26, 33, ... , find:

a the nth term   **b** the 50th term.

a The difference between consecutive terms is 7. So the first part of the nth term is $7n$.
Subtract the difference, 7, from the first term, 5, which gives $5 - 7 = -2$.
So the nth term is given by $7n - 2$.

b Now find the 50th term by substituting $n = 50$ into the rule, $7n - 2$.
50th term $= 7 \times 50 - 2 = 350 - 2$
$= 348$

Quadratic and cubic sequences

The square numbers are:

1, 4, 9, 16, 25, 36, …

The nth square number is n^2.

So, for example, the 15th square number is $15^2 = 225$.

The cube numbers are:

1, 8, 27, 64, 125, …

The nth cube number is n^3.

So, for example, the 7th cube number is $7^3 = 7 \times 7 \times 7 = 343$.

Other sequences can be formed from square and cube numbers.

EXAMPLE 4

Find the nth term of this sequence.

5, 8, 13, 20, 29, …

The differences between terms are 3, 5, 7, 9, …

These are not the same, so it is *not* a linear sequence.

Compare the sequence with the square numbers.

5	8	13	20	29	…
1	4	9	16	25	…

You can see that each term is four more than a square number.

The nth term of this sequence is $n^2 + 4$.

A sequence that is based on square numbers is called a **quadratic sequence**.

EXAMPLE 5

Find the nth term of this sequence.

0, 7, 26, 63, 124

The differences are not equal. It is not linear.

Compare the terms with the square numbers and the cube numbers.

Sequence	0	7	26	63	124	…
Square numbers	1	4	9	16	25	…
Cube numbers	1	8	27	64	125	…

Each term is one less than a cube number.

The nth term of the sequence is $n^3 - 1$.

A sequence that is based on cube numbers is called a **cubic sequence**.

EXERCISE 17C

CORE

1 Find the next two terms and the nth term in each of these linear sequences.

ADVICE AND TIPS

Remember to look at the differences and the first term.

   a 3, 5, 7, 9, 11, ...   **b** 5, 9, 13, 17, 21, ...

   c 8, 13, 18, 23, 28, ...   **d** 2, 8, 14, 20, 26, ...

   e 5, 8, 11, 14, 17, ...   **f** 2, 9, 16, 23, 30, ...

   g 27, 25, 23, 21, 19, ... **h** 42, 38, 34, 30, 26, ... **i** 2, 5, 8, 11, 14, ...

   j 2, 12, 22, 32, ...   **k** 8, 12, 16, 20, ...   **l** 4, 9, 14, 19, 24, ...

2 Find the nth term and the 50th term in each of these linear sequences.

   a 4, 7, 10, 13, 16, ...   **b** 7, 9, 11, 13, 15, ...   **c** 3, 8, 13, 18, 23, ...

   d 1, 5, 9, 13, 17, ...   **e** 2, 10, 18, 26, ...   **f** 5, 6, 7, 8, 9, ...

   g 6, 11, 16, 21, 26, ...   **h** 3, 11, 19, 27, 35, ...   **i** 1, 4, 7, 10, 13, ...

   j 21, 24, 27, 30, 33, ... **k** 12, 19, 26, 33, 40, ... **l** 1, 9, 17, 25, 33, ...

3 For each sequence **a** to **j**, find:

   i the nth term   **ii** the 100th term.

     a 5, 9, 13, 17, 21, ...   **b** 3, 5, 7, 9, 11, 13, ...   **c** 4, 7, 10, 13, 16, ...

     d 8, 10, 12, 14, 16, ...   **e** 9, 13, 17, 21, ...   **f** 6, 11, 16, 21, ...

     g 0, 3, 6, 9, 12, ...   **h** 2, 8, 14, 20, 26, ...   **i** 7, 15, 23, 31, ...

     j 25, 27, 29, 31, ...

4 An online CD retail company uses this price chart. The company charges a standard basic price for a single CD, including postage and packing.

n	1	2	3	4	5	6	7	8	9	10	11	12	13	14	15
Charge ($)	10	18	26	34	42	49	57	65	73	81	88	96	104	112	120

   a Using the charges for 1 to 5 CDs, work out an expression for the nth term.

   b Using the charges for 6 to 10 CDs, work out an expression for the nth term.

   c Using the charges for 11 to 15 CDs, work out an expression for the nth term.

   d What is the basic charge for a CD?

5 Here are some quadratic sequences. Work out the nth term of each one.

   a 1, 4, 9, 16, 25, ...   **b** 3, 6, 11, 18, 27 ...   **c** 2, 8, 18, 32, 50, ... **d** 0, 3, 8, 15, 24

6 Here are some cubic sequences. Work out the nth term of each one.

   a 1, 8, 27, 64, ...   **b** 11, 18, 37, 74, ...

   c 0.5, 4, 13.5, 32, ...   **d** 10, 80, 270, 640, ...

7 Work out the *n*th term of each of these sequences.

a   6, 7, 8, 9, 10, …

b   6, 9, 14, 21, 30, …

c   6, 13, 32, 69, 130

d   6, 11, 16, 21, 26, …

e   5, 20, 45, 80, 125, …

f   5, 40, 135, 320, 625, …

17.3   General rules from patterns

Many problem-solving situations that you are likely to meet involve number sequences. So you need to be able to formulate general **rules** from given number **patterns**.

EXAMPLE 6

The diagram shows a pattern of squares building up.

a   How many squares will there be in the *n*th pattern?

b   Which pattern has 99 squares in it?

a   First, build up a table for the patterns.

Pattern number	1	2	3	4	5
Number of squares	1	3	5	7	9

Looking at the **difference** between consecutive patterns, you can see it is always two squares. So, use 2*n*.

Subtract the difference 2 from the first number, which gives $1 - 2 = -1$.

So the number of squares on the base of the *n*th pattern is $2n - 1$.

b   Now find *n* when $2n - 1 = 99$:

$$2n - 1 = 99$$
$$2n = 99 + 1 = 100$$
$$n = 100 \div 2 = 50$$

The pattern with 99 squares is the 50th.

When you are trying to find a general rule from a sequence of diagrams, always set up a table to connect the pattern number with the number of the variable (squares, matches, seats, etc.) for which you are trying to find the rule. Once you have set up the table, it is easy to find the *n*th term.

CORE

EXERCISE 17D

1 A pattern of squares is built up from matchsticks as shown.

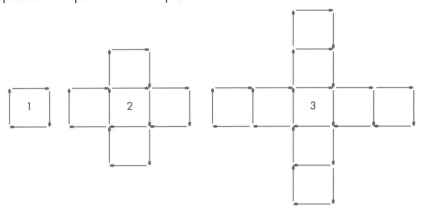

a Draw the fourth diagram.

b Copy and complete this table.

Pattern number	1	2	3	4
Number of squares	1	5	9	

c How many squares are in the *n*th diagram?

d How many squares are there in the 25th diagram?

e With 200 squares, which is the biggest diagram that could be made?

> **ADVICE AND TIPS**
>
> Write out the number sequences to help you see the patterns.

2 A pattern of triangles is built up from matchsticks.

1   2   3   4

a Draw the fifth set of triangles in this pattern.

b Copy and complete this table.

Pattern number	1	2	3	4	5
Number of matches	3				

c How many matchsticks are needed for the *n*th set of triangles?

d How many matchsticks are needed to make the 60th set of triangles?

e If there are only 100 matchsticks, which is the largest set of triangles that could be made?

3 A conference centre had tables each of which could sit six people.

1

When put together, the tables could seat people as shown.

2   3

a How many people could be seated at four tables put together this way?

b Copy and complete this table.

Number of tables	1	2	3	4
Number of seats	6			

c How many people could be seated at n tables put together in this way?

d At a conference, there were 50 people who wished to use the tables in this way.
How many tables would they need?

4 Prepacked fencing units come in the shape shown, made of four pieces of wood.

1

When you put them together in stages to make a fence, you also need joining pieces, so the fence will start to build up as shown below.

2   3

a How many pieces of wood would you have in a fence made up in:
  i five stages   **ii** n stages   **iii** 45 stages?

b I made a fence out of 124 pieces of wood. How many stages did I use?

CORE

5 Regular pentagons of side length 1 cm are joined together to make a pattern, as shown.

1   2   3   4

Copy this pattern and write down the perimeter of each shape. Put the results in a table.

a What is the perimeter of patterns like this made from:

   i six pentagons   **ii** n pentagons   **iii** 50 pentagons?

b What is the largest number of pentagons that can be put together like this to have a perimeter less than 1000 cm?

6 Lamp-posts are put at the end of every 100 m stretch of a motorway, as shown.

   1   2   3

a How many lamp-posts are needed for:

   i four 100 m stretches

   ii n 100 m stretches

   iii 8 km of this motorway.

b A new motorway is being built. The contractor has ordered 1598 lamp-posts. How long is this motorway?

7 A school dining hall had trapezium-shaped tables. Each table could seat five people, as shown here.

When the tables were joined together, as shown below, the individual tables could not seat as many people.

   1   2   3

a In this arrangement, how many could be seated if there were:

   i four tables

   ii n tables

   iii 13 tables?

b For an outside charity event, up to 200 people had to be seated. How many tables arranged like this did they need?

8 When setting out tins to make a display of a certain height, you need to know how many tins to start with at the bottom.

  a How many tins are needed on the bottom if you wish the display to be:

    i five tins high

    ii *n* tins high

    iii 18 tins high?

If the display is *n* tins high, the *total* number of tins is $T = \dfrac{n(n+1)}{2}$

  b Show that this formula gives the correct answer if $n = 2$ or $n = 3$.

  c How many tins will be needed to make a stack 10 tins high?

9 These are the instructions that were used to draw the patterns below.

For pattern 1, draw an equilateral triangle, mark the midpoints of each side and draw and colour in the equilateral triangle formed by these points.

For pattern 2, repeat this with the three white triangles remaining.

For pattern 3, repeat this with the nine white triangles.

    Pattern 1   Pattern 2   Pattern 3

The pattern is called a Sierpinski triangle and is one of the earliest examples of a fractal type pattern.

Pattern 4 is completed in the same way.

  a Copy and complete the table for patterns 1, 2 and 3.

Pattern	1	2	3	4
White triangles	3	9		
Coloured triangles	1	4		
Total	4			

  b Use the numbers in the first 3 columns of the table to help you complete column 4. Explain your method.

10 Thom is building three different patterns with matches.

He builds the patterns in steps.

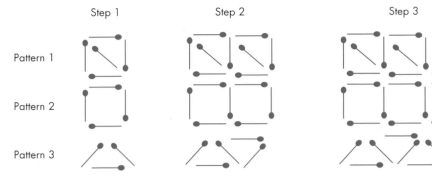

a Draw step 4 for each pattern.

b Complete this table for pattern 1.

Step	1	2	3	4
Number of matches	5			

c How many matches are needed for step n of pattern 1?

d How many matches are needed for step n of pattern 2?

e How many matches are needed for step n of pattern 3?

f What is the total number of matches needed for step n? Write your answer as simply as possible.

17.4 Further sequences

Look at these sequences.

6, 24, 96, 384, 1536, …

48, 24, 12, 6, 3, 1.5, …

In the first, you multiply each term by four to find the next one.

In the second, you multiply each term by 0.5 (or divide it by 2) to find the next term.

The nth term of the first sequence is 1.5×4^n.

Check: The first term is $1.5 \times 4 = 6$; the third term is $1.5 \times 4^3 = 1.5 \times 64 = 96$.

The nth term of the second sequence is 96×0.5^n.

Check: The first term is $96 \times 0.5 = 48$; the third term is $96 \times 0.5^3 = 96 \times 0.125 = 12$.

96×0.5^n can also be written as $\dfrac{96}{2^n}$. Check that this is correct.

These sequences are called **exponential sequences**. In these sequences, n is an exponent or power.

EXAMPLE 7

This is the start of an exponential sequence.

11, 12.1, 13.31, 14.641, ...

Work out the nth term.

The multiplier from one term to the next is always the same.

$\frac{12.1}{11} = 1.1 \qquad \frac{13.31}{12.1} = 1.1 \qquad \frac{14.641}{13.31} = 1.1$

The nth term is $a \times 1.1^n$ where a is a number.

The first term is 11 so $a \times 1.1 = 11 \rightarrow a = \frac{11}{1.1} = 10$.

The nth term is 10×1.1^n.

Sequences may be a combination of different types of sequence.

For example, look at this sequence.

3, 8, 15, 24, 35, ...

It is not linear, because the differences are not all the same.

It is not exponential because the multipliers are not all the same.

Compare the sequence with the square numbers.

n	1	2	3	4	5	...
Sequence	3	8	15	24	35	...
n^2	1	4	9	16	25	...
Difference between sequence and n^2	2	4	6	8	10	...

You can see that the difference is $2n$ each time.

The nth term is $n^2 + 2n$.

Check: If $n = 5$ then $n^2 + 2n = 25 + 10 = 35$ which is the fifth term.

EXTENDED

EXERCISE 17E

1 Find the next term in each of these exponential sequences.

   a 12, 36, 144, …

   b 13, 39, 117, 351, …

   c 2, 10, 50, 250, …

   d 20, 24, 28.8, 34.56, …

   e 4, 80, 1600, 32000, …

   f 240, 120, 60, 30, …

   g 1000, 200, 40, 8, …

   h 162, 108, 72, 48, …

2 Here are the nth terms of some sequences. In each case, find the first term and the fourth term.

   a 6^n

   b 8×2^n

   c 5×3^n

   d 40×0.5^n

   e $81 \times \left(\dfrac{2}{3}\right)^n$

3 Work out the nth term of each of these sequences.

   a 50, 100, 200, 400, 800, …

   b 3, 6, 12, 24, 48, …

   c 6, 18, 54, 162, ….

   d 120, 60, 30, 15, …

   e 45, 40.5, 36.45, 32.805, …

   f 80, 100, 125, …

4 Selina has \$5000 savings. She expects her savings to increase by 40% each year.

Years from now	0	1	2	3	4
Savings (\$)	5000	7000	9800		

   a Work out the two numbers missing from the table.

   b Work out an expression for her savings after n years.

5 Find the nth term of each of these sequences.

   a 2, 5, 10, 17, 26, …

   b 7, 10, 15, 22, 31, …

   c 0, 2, 6, 12, 20, …

   d 3, 12, 27, 48, 75, …

   e 1, 10, 25, 46, 73, …

   f 4, 14, 30, 52, 80, …

6 These are the first four triangular numbers.

  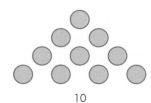

    1   3   6   10

   a Find the next two triangular numbers.

   b Find the nth term of this quadratic sequence.

      2, 6, 12, 20, 30, …

   c Use your answer from part **b** to find an expression for the nth triangular number.

   d Work out the 20th triangular number.

   e Show that 1275 is a triangular number.

7 Find the *n*th term of each of these sequences.

a 0, 7, 26, 63, 124, …

b 51, 58, 77, 114, 157, …

c 2, 10, 30, 68, 130, …

d 4, 14, 36, 76, 140, …

e 4, 32, 108, 256, 500, …

f 3, 30, 105, 252, 495, …

8 Here are some triangles made from oranges.

If these oranges are put in four layers they make a tetrahedron.

a Find the number of oranges in a tetrahedron with four layers.

The number of oranges in a tetrahedron with *n* layers is $\frac{1}{6}n(n + 1)(n + 2)$.

b Check that this expression gives the correct answer with $n = 4$.

c A shopkeeper has 100 oranges. He wants to display them in a tetrahedron. Work out the largest number of layers he can have.

d Show how to use the formula to add the first 20 triangular numbers.

9 Here is a sequence.

6, 16, 30, 48, 70, …

The *n*th term of this sequence is $an + bn^2$ where *a* and *b* are numbers.

a By looking at the first term, show that $a + b = 6$.

b Use the second term to find another equation involving *a* and *b*.

c Solve the equation from parts **a** and **b** simultaneously to find the *n*th term of the sequence.

Why this chapter matters

Indices are a useful way to write numbers. They show how different numbers are related to one another and they can make it easier to multiply or divide, or to compare the sizes of different numbers.

You probably already know about powers of numbers from chapters 5 and 9.

You use find powers of 10 when you write numbers in standard form, such as 3.7×10^6 or 8.92×10^{-5}.

The first is $3\,700\,000$ and the second is $0.000\,08\,92$.

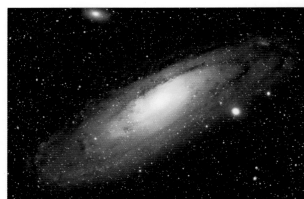

Using powers is a useful 'short cut'.
For example, the centre of the galaxy Andromeda is $24\,000\,000\,000\,000\,000\,000\,000$ km from our Sun.
It is much easier to write 24×10^{18}!

You will also find powers in squares and cubes such as 7^2 and 14^3.

Remember that $7^2 = 7 \times 7$ and $14^3 = 14 \times 14 \times 14$. This also saves time in writing out the numbers.

The small number that shows the power is called an index.

- In 7^2 the index is 2 and in 14^3 the index is 3.

- In 3.7×10^6 or 8.92×10^{-5} the indices are 6 and −5.

'Indices' is the plural of index. In the examples above 2, 3, 6 and −5 are indices.

In this chapter you will discover more useful ways to use indices. You might be surprised to find you can write fractions using indices. You can probably see how 8 can be written as a power of two (2^3) but it is not so obvious how $\frac{1}{8}$ can also be written as a power of two.

Warning

The word index has a number of other meanings in English. For example, you will find an index at the back of this book. In this chapter the word is always used to mean a power.

Chapter

Indices

Topics		Level	Key words
1	Using indices	**CORE**	index, indices, power, power 1, power 0
2	Negative indices	**CORE**	negative index, reciprocal
3	Multiplying and dividing with indices	**CORE**	
4	Fractional indices	**EXTENDED**	

In this chapter you will learn how to:

CORE	EXTENDED
• Understand the meaning and rules of indices. (C1.7 and E1.7) • Use and interpret positive, negative and zero indices. (C2.4 and E2.4) • Use the rules of indices. (C2.4 and E2.4)	• Use and interpret fractional indices, e.g. solve $32^x = 2$. (E2.4)

18.1 Using indices

An **index** is a convenient way of writing repetitive multiplications. The plural of index is **indices**.

The index tells you the number of times a number is multiplied by itself. For example:

$4^6 = 4 \times 4 \times 4 \times 4 \times 4 \times 4$ six lots of 4 multiplied together (call this '4 to the **power** 6')

$6^4 = 6 \times 6 \times 6 \times 6$ four lots of 6 multiplied together (call this '6 to the power 4')

$7^3 = 7 \times 7 \times 7$

$12^2 = 12 \times 12$

EXAMPLE 1

a Write each of these numbers out in full.

  i 4^3 **ii** 6^2 **iii** 7^5 **iv** 12^4

b Write each multiplication as a power.

  i $3 \times 3 \times 3 \times 3 \times 3 \times 3 \times 3 \times 3$

  ii $13 \times 13 \times 13 \times 13 \times 13$

  iii $7 \times 7 \times 7 \times 7$

  iv $5 \times 5 \times 5 \times 5 \times 5 \times 5 \times 5$

  a **i** $4^3 = 4 \times 4 \times 4$ **ii** $6^2 = 6 \times 6$

   iii $7^5 = 7 \times 7 \times 7 \times 7 \times 7$ **iv** $12^4 = 12 \times 12 \times 12 \times 12$

  b **i** $3 \times 3 \times 3 \times 3 \times 3 \times 3 \times 3 \times 3 = 3^8$

   ii $13 \times 13 \times 13 \times 13 \times 13 = 13^5$

   iii $7 \times 7 \times 7 \times 7 = 7^4$

   iv $5 \times 5 \times 5 \times 5 \times 5 \times 5 \times 5 = 5^7$

Working out powers on your calculator

The power button on your calculator will probably look like this $\boxed{x^\square}$.

To work out 5^7 on your calculator use the power key.

$5^7 = \boxed{5}\ \boxed{x^\square}\ \boxed{7} = 78\,125$

Two special powers

Power 1

Any number to the power 1 is the same as the number itself. This is always true so normally you do not write the power 1.

For example: $5^1 = 5$ $32^1 = 32$ $(-8)^1 = -8$

Power zero

Any number to the power 0 is equal to 1.

For example: $5^0 = 1$ $32^0 = 1$ $(-8)^0 = 1$

You can check these results on your calculator.

EXERCISE 18A

1 Write these expressions using index notation. Do not work them out yet.

a $2 \times 2 \times 2 \times 2$ **b** $3 \times 3 \times 3 \times 3 \times 3$

c 7×7 **d** $5 \times 5 \times 5$

e $10 \times 10 \times 10 \times 10 \times 10 \times 10 \times 10$ **f** $6 \times 6 \times 6 \times 6$

g 4 **h** $1 \times 1 \times 1 \times 1 \times 1 \times 1 \times 1$

i $0.5 \times 0.5 \times 0.5 \times 0.5$ **j** $100 \times 100 \times 100$

2 Write these power terms out in full. Do not work them out yet.

a 3^4 **b** 9^3 **c** 6^2 **d** 10^5 **e** 2^{10}

f 8^1 **g** 0.1^3 **h** 2.5^2 **i** 0.7^3 **j** 1000^2

3 Using the power key on your calculator (or another method), work out the values of the power terms in question **1**.

4 Using the power key on your calculator (or another method), work out the values of the power terms in question **2**.

5 A storage container is in the shape of a cube. The length of the container is 5 m.

To work out the volume of a cube, use the formula:

   volume = (length of edge)^3

Work out the total storage space in the container.

6 Write each number as a power of a different number.

The first one has been done for you.

a $32 = 2^5$ **b** 100 **c** 8 **d** 25

7 Without using a calculator, work out the values of these power terms.

a 2^0 **b** 4^1 **c** 5^0 **d** 1^9 **e** 1^{235}

8 The answers to question **7**, parts **d** and **e**, should tell you something special about powers of 1. What is it?

9 Write the answer to question **1**, part **j** as a power of 10.

10 Write the answer to question **2**, part **j** as a power of 10.

11 Using your calculator, or otherwise, work out the values of these power terms.

    a $(-1)^0$ **b** $(-1)^1$

    c $(-1)^2$ **d** $(-1)^4$

    e $(-1)^5$

12 Using your answers to question **11**, write down the answers to these power terms.

    a $(-1)^8$ **b** $(-1)^{11}$

    c $(-1)^{99}$ **d** $(-1)^{80}$

    e $(-1)^{126}$

13 The number 16 777 216 is a power of 2.

It is also a power of 4, a power of 8 and a power of 16.

Write the number as a power of 4, a power of 8 and a power of 16.

18.2   Negative indices

A **negative index** is a convenient way of writing the **reciprocal** of a number or term. (That is, one divided by that number or term.) For example:

$$x^{-a} = \frac{1}{x^a}$$

Here are some other examples.

$$5^{-2} = \frac{1}{5^2} \qquad\qquad 3^{-1} = \frac{1}{3} \qquad\qquad 5x^{-2} = \frac{5}{x^2}$$

EXAMPLE 2

Rewrite each number in the form 2^n.

    a 8   **b** $\dfrac{1}{4}$ **c** -32 **d** $-\dfrac{1}{64}$

    a $8 = 2 \times 2 \times 2 = 2^3$ **b** $\dfrac{1}{4} = \dfrac{1}{2^2} = 2^{-2}$

    c $-32 = -2^5$ **d** $-\dfrac{1}{64} = -\dfrac{1}{2^6} = -2^{-6}$

EXERCISE 18B

1 Write each of these in fraction form.

a 5^{-3} **b** 6^{-1} **c** 10^{-5} **d** 3^{-2} **e** 8^{-2}

f 9^{-1} **g** w^{-2} **h** t^{-1} **i** x^{-m} **j** $4m^{-3}$

ADVICE AND TIPS

If you move a power from top to bottom, or vice versa its sign changes. A negative power means a reciprocal: it does not mean the answer is negative.

2 Write each of these in negative index form.

a $\dfrac{1}{3^2}$ **b** $\dfrac{1}{5}$ **c** $\dfrac{1}{10^3}$ **d** $\dfrac{1}{m}$ **e** $\dfrac{1}{t^n}$

3 Change each expression into an index form of the type shown.

a All of the form 2^n

i 16 **ii** $\dfrac{1}{2}$ **iii** $\dfrac{1}{16}$ **iv** -8

b All of the form 10^n

i 1000 **ii** $\dfrac{1}{10}$ **iii** $\dfrac{1}{100}$ **iv** 1 million

c All of the form 5^n

i 125 **ii** $\dfrac{1}{5}$ **iii** $\dfrac{1}{25}$ **iv** $\dfrac{1}{625}$

d All of the form 3^n

i 9 **ii** $\dfrac{1}{27}$ **iii** $\dfrac{1}{81}$ **iv** -243

4 Rewrite each expression in fraction form.

a $5x^{-3}$ **b** $6t^{-1}$ **c** $7m^{-2}$ **d** $4q^{-4}$ **e** $10y^{-5}$

f $\frac{1}{2}x^{-3}$ **g** $\frac{1}{2}m^{-1}$ **h** $\frac{3}{4}t^{-4}$ **i** $\frac{4}{5}y^{-3}$ **j** $\frac{7}{8}x^{-5}$

5 Write each fraction in index form.

a $\dfrac{7}{x^3}$ **b** $\dfrac{10}{p}$ **c** $\dfrac{5}{t^2}$ **d** $\dfrac{8}{m^5}$ **e** $\dfrac{3}{y}$

6 Find the value of each number.

a $x = 5$

i x^2 **ii** x^{-3} **iii** $4x^{-1}$

b $t = 4$

i t^3 **ii** t^{-2} **iii** $5t^{-4}$

c $m = 2$

i m^3 **ii** m^{-5} **iii** $9m^{-1}$

d $w = 10$

i w^6 **ii** w^{-3} **iii** $25w^{-2}$

7 Two different numbers can be written in the form 2^n.

The sum of the numbers is 40.

What is the difference of the numbers?

8 x and y are integers.

$$x^2 - y^3 = 0$$

Work out possible values of x and y.

9 You are given that $8^7 = 2\,097\,152$.

Write down the value of 8^{-7}.

10 Put these in order from smallest to largest:

$$x^5 \qquad x^{-5} \qquad x^0$$

a when x is greater than 1

b when x is between 0 and 1

c when $x = -10$.

18.3 Multiplying and dividing with indices

When you *multiply* powers of the same number or variable, you *add* the indices. For example:

$$3^4 \times 3^5 = 3^{(4+5)} = 3^9$$
$$2^3 \times 2^4 \times 2^5 = 2^{12}$$
$$10^4 \times 10^{-2} = 10^2$$
$$10^{-3} \times 10^{-1} = 10^{-4}$$
$$a^x \times a^y = a^{(x+y)}$$

When you *divide* powers of the same number or variable, you *subtract* the indices. For example:

$$a^4 \div a^3 = a^{(4-3)} = a^1 = a$$
$$b^4 \div b^7 = b^{-3}$$
$$10^4 \div 10^{-2} = 10^6$$
$$10^{-2} \div 10^{-4} = 10^2$$
$$a^x \div a^y = a^{(x-y)}$$

When you *raise* a power to a further power, you *multiply* the indices. For example:

$(a^2)^3 = a^{2 \times 3} = a^6$

$(a^{-2})^4 = a^{-8}$

$(a^2)^6 = a^{12}$

$(a^x)^y = a^{xy}$

Here are some examples of different kinds of expressions that use powers.

$2a^2 \times 3a^4 = (2 \times 3) \times (a^2 \times a^4)$

$\qquad = 6 \times a^6 = 6a^6$

$4a^2b^3 \times 2ab^2 = (4 \times 2) \times (a^2 \times a) \times (b^3 \times b^2)$

$\qquad = 8a^3b^5$

$12a^5 \div 3a^2 = (12 \div 3) \times (a^5 \div a^2)$

$\qquad = 4a^3$

$(2a^2)^3 = (2)^3 \times (a^2)^3 = 8 \times a^6$

$\qquad = 8a^6$

EXERCISE 18C

1 Write these as single powers of 5.

  a $5^2 \times 5^2$ **b** 5×5^2 **c** $5^{-2} \times 5^4$

  d $5^6 \times 5^{-3}$ **e** $5^{-2} \times 5^{-3}$

2 Write these as single powers of 6.

  a $6^5 \div 6^2$ **b** $6^4 \div 6^4$ **c** $6^4 \div 6^{-2}$

  d $6^{-3} \div 6^4$ **e** $6^{-3} \div 6^{-5}$

3 Simplify these and write them as single powers of a.

  a $a^2 \times a$ **b** $a^3 \times a^2$ **c** $a^4 \times a^3$

  d $a^6 \div a^2$ **e** $a^3 \div a$ **f** $a^5 \div a^4$

4 **a** $a^x \times a^y = a^{10}$

    Write down a possible pair of values of x and y.

  b $a^x \div a^y = a^{10}$

    Write down a possible pair of values of x and y.

CORE

5 Write these as single powers of 4.

a $(4^2)^3$
b $(4^3)^5$
c $(4^1)^6$

d $(4^3)^{-2}$
e $(4^{-2})^{-3}$
f $(4^7)^0$

6 Simplify these expressions.

a $2a^2 \times 3a^3$
b $3a^4 \times 3a^{-2}$
c $(2a^2)^3$

d $-2a^2 \times 3a^2$
e $-4a^3 \times -2a^5$
f $-2a^4 \times 5a^{-7}$

7 Simplify these expressions.

a $6a^3 \div 2a^2$
b $12a^5 \div 3a^2$

c $15a^5 \div 5a$
d $18a^{-2} \div 3a^{-1}$

e $24a^5 \div 6a^{-2}$
f $30a \div 6a^5$

8 Simplify these expressions.

a $2a^2b^3 \times 4a^3b$

b $5a^2b^4 \times 2ab^{-3}$

c $6a^2b^3 \times 5a^{-4}b^{-5}$

d $12a^2b^4 \div 6ab$

e $24a^{-3}b^4 \div 3a^2b^{-3}$

9 Simplify these expressions.

a $\dfrac{6a^4b^3}{2ab}$

b $\dfrac{2a^2bc^2 \times 6abc^3}{4ab^2c}$

c $\dfrac{3abc \times 4a^3b^2c \times 6c^2}{9a^2bc}$

10 Write down **two** possible:

a multiplication questions with an answer of $12x^2y^5$

b division questions with an answer of $12x^2y^5$.

11 a, b and c are three different positive integers.

What is the smallest possible value of a^2b^3c?

12 Use the general rule for dividing powers of the same number, $\dfrac{a^x}{a^y} = a^{x-y}$, to prove that any number raised to the power zero is 1.

18.4 Fractional indices

Indices of the form $\frac{1}{n}$

Consider the problem $7^x \times 7^x = 7$. This can be written as:

$$7^{(x + x)} = 7$$
$$7^{2x} = 7^1 \implies 2x = 1 \implies x = \tfrac{1}{2}$$

If you now substitute $x = \tfrac{1}{2}$ back into the original equation, you see that:

$$7^{\frac{1}{2}} \times 7^{\frac{1}{2}} = 7$$

This makes $7^{\frac{1}{2}}$ the same as $\sqrt{7}$.

You can similarly show that $7^{\frac{1}{3}}$ is the same as $\sqrt[3]{7}$. And that, generally:

$$x^{\frac{1}{n}} = \sqrt[n]{x} \ (n\text{th root of } x)$$

So in summary:

Power $\tfrac{1}{2}$ is the same as positive square root.

Power $\tfrac{1}{3}$ is the same as cube root.

Power $\tfrac{1}{n}$ is the same as nth root.

For example:

$$49^{\frac{1}{2}} = \sqrt{49} = 7 \qquad 8^{\frac{1}{3}} = \sqrt[3]{8} = 2$$
$$10\,000^{\frac{1}{4}} = \sqrt[4]{10\,000} = 10 \qquad 36^{-\frac{1}{2}} = \frac{1}{\sqrt{36}} = \frac{1}{6}$$

If you have an expression in the form $\left(\dfrac{a}{b}\right)^n$ you can calculate it as $\dfrac{a^n}{b^n}$ and then write it as a fraction.

EXAMPLE 3

Write $\left(\dfrac{16}{25}\right)^{\frac{1}{2}}$ as a fraction.

You can find the power of the numerator and denominator separately.

$$\left(\frac{16}{25}\right)^{\frac{1}{2}} = \frac{16^{\frac{1}{2}}}{25^{\frac{1}{2}}}$$
$$= \frac{4}{5}$$

EXERCISE 18D

EXTENDED

1 Evaluate each number.

a $25^{\frac{1}{2}}$ b $100^{\frac{1}{2}}$ c $64^{\frac{1}{2}}$ d $81^{\frac{1}{2}}$ e $625^{\frac{1}{2}}$

f $27^{\frac{1}{3}}$ g $64^{\frac{1}{3}}$ h $1000^{\frac{1}{3}}$ i $125^{\frac{1}{3}}$ j $512^{\frac{1}{3}}$

k $144^{\frac{1}{2}}$ l $400^{\frac{1}{2}}$ m $625^{\frac{1}{4}}$ n $81^{\frac{1}{4}}$ o $100\,000^{\frac{1}{5}}$

p $729^{\frac{1}{6}}$ q $32^{\frac{1}{5}}$ r $1024^{\frac{1}{10}}$ s $1296^{\frac{1}{4}}$ t $216^{\frac{1}{3}}$

u $16^{-\frac{1}{2}}$ v $8^{-\frac{1}{3}}$ w $81^{-\frac{1}{4}}$ x $3125^{-\frac{1}{5}}$ y $1000\,000^{-\frac{1}{6}}$

2 Evaluate each number.

a $\left(\dfrac{25}{36}\right)^{\frac{1}{2}}$ b $\left(\dfrac{100}{36}\right)^{\frac{1}{2}}$ c $\left(\dfrac{64}{81}\right)^{\frac{1}{2}}$ d $\left(\dfrac{81}{25}\right)^{\frac{1}{2}}$ e $\left(\dfrac{25}{64}\right)^{\frac{1}{2}}$

f $\left(\dfrac{27}{125}\right)^{\frac{1}{3}}$ g $\left(\dfrac{8}{512}\right)^{\frac{1}{3}}$ h $\left(\dfrac{1000}{64}\right)^{\frac{1}{3}}$ i $\left(\dfrac{64}{125}\right)^{\frac{1}{3}}$ j $\left(\dfrac{512}{343}\right)^{\frac{1}{3}}$

3 Use the general rule for raising a power to another power to prove that $x^{\frac{1}{n}}$ is equivalent to $\sqrt[n]{x}$.

4 Which of these is the odd one out?

$16^{-\frac{1}{4}}$ $64^{-\frac{1}{2}}$ $8^{-\frac{1}{3}}$

Show how you decided.

5 Imagine that you are the teacher.

Write down how you would teach the class that $27^{-\frac{1}{3}}$ is equal to $\frac{1}{3}$.

6 $x^{-\frac{2}{3}} = y^{\frac{1}{3}}$

Find values for x and y that make this equation work.

7 Solve these equations.

a $2^x = 8$ b $8^x = 2$ c $4^x = 1$ d $16^x = 4$ e $100^x = 10$

f $81^x = 3$ g $16^x = 2$ h $125^x = 5$ i $1000^x = 10$ j $400^x = 20$

k $512^x = 8$ l $128^x = 2$

Indices of the form $\dfrac{a}{b}$

Here are two examples of this form.

$$t^{\frac{2}{3}} = t^{\frac{1}{3}} \times t^{\frac{1}{3}} = (\sqrt[3]{t})^2 \qquad\qquad 81^{\frac{3}{4}} = (\sqrt[4]{81})^3 = 3^3 = 27$$

If you have an expression of the form $\left(\dfrac{a}{b}\right)^{-n}$ you can invert it to calculate it as a fraction.

$$\left(\dfrac{a}{b}\right)^{-n} = \left(\dfrac{b}{a}\right)^{n}$$

EXAMPLE 4

Evaluate each number. **a** $16^{-\frac{1}{4}}$ **b** $32^{-\frac{4}{5}}$

When dealing with the negative index remember that it means reciprocal.

Work through problems like these one step at a time.

Step 1: Rewrite the calculation as a fraction by dealing with the negative power.
Step 2: Take the root of the base number given by the denominator of the fraction.
Step 3: Raise the result to the power given by the numerator of the fraction.
Step 4: Write out the answer as a fraction.

a **Step 1:** $16^{-\frac{1}{4}} = \left(\frac{1}{16}\right)^{\frac{1}{4}}$ **Step 2:** $16^{\frac{1}{4}} = \sqrt[4]{16} = 2$ **Step 3:** $2^1 = 2$ **Step 4:** $16^{-\frac{1}{4}} = \frac{1}{2}$

b **Step 1:** $32^{-\frac{4}{5}} = \left(\frac{1}{32}\right)^{\frac{4}{5}}$ **Step 2:** $32^{\frac{1}{5}} = \sqrt[5]{32} = 2$ **Step 3:** $2^4 = 16$ **Step 4:** $32^{-\frac{4}{5}} = \frac{1}{16}$

EXAMPLE 5

Write $\left(\frac{8}{27}\right)^{-\frac{2}{3}}$ as a fraction.

$$\left(\frac{8}{27}\right)^{-\frac{2}{3}} = \left(\frac{27}{8}\right)^{\frac{2}{3}}$$
$$= \frac{27^{\frac{2}{3}}}{8^{\frac{2}{3}}}$$
$$= \frac{(\sqrt[3]{27})^2}{(\sqrt[3]{8})^2} = \frac{3^2}{2^2} = \frac{9}{4}$$

EXERCISE 18E

1 Evaluate these.

a $32^{\frac{4}{5}}$ **b** $125^{\frac{2}{3}}$

c $1296^{\frac{3}{4}}$ **d** $243^{\frac{4}{5}}$

2 Rewrite these in index form.

a $\sqrt[3]{t^2}$ **b** $\sqrt[4]{m^3}$

c $\sqrt[5]{k^2}$ **d** $\sqrt{x^3}$

3 Evaluate these.

a $8^{\frac{2}{3}}$ **b** $27^{\frac{2}{3}}$

c $16^{\frac{3}{2}}$ **d** $625^{\frac{5}{4}}$

EXTENDED

4 Evaluate these.

a $25^{-\frac{1}{2}}$

b $36^{-\frac{1}{2}}$

c $16^{-\frac{1}{4}}$

d $81^{-\frac{1}{4}}$

e $16^{-\frac{1}{2}}$

f $8^{-\frac{1}{3}}$

g $32^{-\frac{1}{5}}$

h $27^{-\frac{1}{3}}$

5 Evaluate these.

a $25^{-\frac{3}{2}}$

b $36^{-\frac{3}{2}}$

c $16^{-\frac{3}{4}}$

d $81^{-\frac{3}{4}}$

e $64^{-\frac{4}{3}}$

f $8^{-\frac{2}{3}}$

g $32^{-\frac{2}{5}}$

h $27^{-\frac{2}{3}}$

6 Evaluate these.

a $100^{-\frac{5}{2}}$

b $144^{-\frac{1}{2}}$

c $125^{-\frac{2}{3}}$

d $9^{-\frac{3}{2}}$

e $4^{-\frac{5}{2}}$

f $64^{-\frac{5}{6}}$

g $27^{-\frac{4}{3}}$

h $169^{-\frac{1}{2}}$

7 Which of these is the odd one out?

$16^{-\frac{3}{4}}$ $\qquad$ $64^{-\frac{1}{2}}$ $\qquad$ $8^{-\frac{2}{3}}$

Show how you decided.

8 Imagine that you are the teacher.

Write down how you would teach the class that $27^{-\frac{2}{3}}$ is equal to $\frac{1}{9}$.

9 Write these as fractions.

a $\left(\frac{9}{4}\right)^{\frac{3}{2}}$

b $\left(\frac{27}{125}\right)^{\frac{2}{3}}$

c $\left(\frac{16}{9}\right)^{\frac{5}{2}}$

d $\left(\frac{4}{49}\right)^{\frac{3}{2}}$

e $\left(\frac{64}{27}\right)^{\frac{2}{3}}$

f $\left(\frac{16}{81}\right)^{\frac{3}{4}}$

g $\left(\frac{125}{64}\right)^{\frac{4}{3}}$

h $\left(\frac{64}{729}\right)^{\frac{5}{6}}$

10 Write these as fractions.

a $\left(\dfrac{3}{5}\right)^{-2}$ **b** $\left(\dfrac{4}{3}\right)^{-3}$

c $\left(\dfrac{9}{5}\right)^{-3}$ **d** $\left(\dfrac{2}{3}\right)^{-5}$

e $\left(\dfrac{9}{4}\right)^{-\frac{3}{2}}$ **f** $\left(\dfrac{4}{9}\right)^{-\frac{5}{2}}$

g $\left(\dfrac{8}{27}\right)^{-\frac{2}{3}}$ **h** $\left(\dfrac{49}{25}\right)^{-\frac{3}{2}}$

i $\left(\dfrac{125}{64}\right)^{-\frac{2}{3}}$ **j** $\left(\dfrac{25}{64}\right)^{-\frac{3}{2}}$

k $\left(\dfrac{16}{81}\right)^{-\frac{5}{4}}$ **l** $\left(\dfrac{2187}{128}\right)^{-\frac{5}{7}}$

11 Simplify these expressions.

a $x^{\frac{3}{2}} \times x^{\frac{5}{2}}$ **b** $x^{\frac{1}{2}} \times x^{-\frac{3}{2}}$

c $(8y^3)^{\frac{2}{3}}$ **d** $5x^{\frac{3}{2}} \div \frac{1}{2}x^{-\frac{1}{2}}$

e $4x^{\frac{1}{2}} \times 5x^{-\frac{3}{2}}$ **f** $\left(\dfrac{27}{y^3}\right)^{-\frac{1}{3}}$

12 Simplify these.

a $x^{\frac{1}{2}} \times x^{\frac{1}{2}}$

b $d^{-\frac{1}{2}} \times d^{-\frac{1}{2}}$

c $t^{\frac{1}{2}} \times t$

d $(x^{\frac{1}{2}})^4$

e $(y^2)^{\frac{1}{4}}$

f $a^{\frac{1}{2}} \times a^{\frac{3}{2}} \times a^2$

13 Simplify these.

a $x \div x^{\frac{1}{2}}$

b $y^{\frac{1}{2}} \div y^{1\frac{1}{2}}$

c $a^{\frac{1}{3}} \times a^{\frac{4}{3}}$

d $t^{-\frac{1}{2}} \times t^{-\frac{3}{2}}$

e $\dfrac{1}{d^{-2}}$

f $\dfrac{k^{\frac{1}{2}} \times k^{\frac{3}{2}}}{k^2}$

Why this chapter matters

In many real-life situations, variables are connected by a rule or relationship. It may be that as one variable increases the other increases. Alternatively, it may be that as one variable increases the other decreases.

In this chapter you will look at how quantities vary when they are related in some way.

As this plant gets older it becomes taller.

As the storm increases the number of sunbathers decreases.

As this car gets older it is worth less (and eventually it is worthless!).

As more songs are downloaded, there is less money left on the voucher.

Try to think of other variables that are connected in this way.

Chapter

19 Variation

Topics

 Direct variation

Level

EXTENDED

Key words

direct variation, direct proportion, constant of proportionality

 Inverse variation

EXTENDED

inverse variation, inverse proportion

In this chapter you will learn how to:

EXTENDED

- Express direct and inverse variation in algebraic terms and use this form of expression to find unknown quantities. (E2.8)

The term **direct variation** means the same as as **direct proportion**.

There is direct variation (or direct proportion) between two variables when one variable is a simple multiple of the other. That is, their ratio is a constant.

For example:

1 kilogram = 2.2 pounds   There is a multiplying factor of 2.2 between kilograms and pounds.

Area of a circle = πr^2 There is a multiplying factor of π between the area of a circle and the square of its radius.

Any question involving direct variation usually requires you first to find this multiplying factor (called the **constant of proportionality**), then to use it to solve a problem.

The symbol for variation or proportion is $\propto$.

So the statement 'Pay is directly proportional to time' can be mathematically written as:

pay $\propto$ *time*

which implies that:

pay $= k \times$ *time*

where k is the constant of proportionality.

There are four steps to be followed when you are using proportionality to solve problems.

Step 1: Set up the statement, using the proportionality symbol (you may use symbols to represent the variables).

Step 2: Set up the equation, using a constant of proportionality.

Step 3: Use given information to work out the value of the constant of proportionality.

Step 4: Substitute the value of the constant of proportionality into the equation and use this equation to find unknown values.

EXAMPLE 1

The cost of an item is directly proportional to the time spent making it. An item taking 6 hours to make costs \$30. Find:

a the cost of an item that takes 5 hours to make

b the length of time it takes to make an item costing \$40.

> **Step 1:** Let C dollars be the cost of making an item and t hours the time it takes.
>
> $$C \propto t$$
>
> **Step 2:** Setting up the equation gives:
>
> $$C = kt$$
>
> where k is the constant of proportionality.
>
> Note that you can 'replace' the proportionality sign $\propto$ with $= k$ to obtain the proportionality equation.
>
> **Step 3:** Since $C = 30$ when $t = 6$, then $30 = 6k$
>
> $$\Rightarrow \frac{30}{6} = k$$
>
> $$\Rightarrow k = 5$$
>
> **Step 4:** So the formula is $\quad C = 5t$.
>
> **a** When $t = 5 \qquad C = 5 \times 5 = 25$
>
>   So the cost is \$25.
>
> **b** When $C = 40 \qquad 40 = 5 \times t$
>
> $$\Rightarrow \frac{40}{5} = t \Rightarrow t = 8$$
>
> So the time spent making the item is 8 hours.

EXERCISE 19A

For questions **1** to **4**, first find the value of k, the constant of proportionality, and then the formula connecting the variables.

1 T is directly proportional to M. If $T = 20$ when $M = 4$, find the value of:

a T when $M = 3$

b M when $T = 10$.

2 W is directly proportional to F. If $W = 45$ when $F = 3$, find the value of:

a W when $F = 5$

b F when $W = 90$.

3 Q varies directly with P. If $Q = 100$ when $P = 2$, find the value of:

   a Q when $P = 3$

   b P when $Q = 300$.

4 X varies directly with Y. If $X = 17.5$ when $Y = 7$, find the value of:

   a X when $Y = 9$

   b Y when $X = 30$.

5 The distance covered by a train is directly proportional to the time taken for the journey. The train travels 105 kilometres in 3 hours.

   a What distance will the train cover in 5 hours?

   b How much time will it take for the train to cover 280 kilometres?

6 The cost of fuel delivered to your door is directly proportional to the mass received. When 250 kg is delivered, it costs 47.50 dollars.

   a How much will it cost to have 350 kg delivered?

   b How much would be delivered if the cost were 33.25 dollars?

7 The number of children who can play safely in a playground is directly proportional to the area of the playground. A playground with an area of 210 m^2 is safe for 60 children.

   a How many children can safely play in a playground of area 154 m^2?

   b A playgroup has 24 children. What is the smallest playground area in which they could safely play?

8 The number of spaces in a car park is directly proportional to the area of the car park.

   a A car park has 300 parking spaces in an area of 4500 m².

      It is decided to increase the area of the car park by 500 m² to make extra spaces.

      How many extra spaces will be made?

   b The old part of the car park is redesigned so that the original area has 10% more parking spaces.

      How many more spaces than in the original car park will there be altogether if the number of spaces in the new area is directly proportional to the number in the redesigned car park?

9 The number of passengers in a bus queue is directly proportional to the time that the person at the front of the queue has spent waiting.

Katya is the first to arrive at a bus stop. When she has been waiting 5 minutes the queue has 20 passengers.

A bus has room for 70 passengers.

How long had Katya been in the queue if the bus fills up from empty when it arrives and all passengers get on?

Direct proportions involving squares, cubes, square roots and cube roots

The process is the same as for a linear direct variation, as the next example shows.

EXAMPLE 2

The cost of a circular badge is directly proportional to the square of its radius. The cost of a badge with a radius of 2 cm is $0.68. Find:

a the cost of a badge of radius 2.4 cm   b the radius of a badge costing $1.53.

Step 1: Let C be the cost in dollars and r the radius of a badge in centimetres.
$$C \propto r^2$$

Step 2: Setting up the equation gives:
$$C = kr^2$$
where k is the constant of proportionality.

Step 3: $C = 0.68$ when $r = 2$. So:
$$0.68 = 4k$$
$$\Rightarrow \frac{0.68}{4} = k \Rightarrow k = 0.17$$

Step 4: So the formula is $C = 0.17r^2$.

a When $r = 2.4$ $C = 0.17 \times 2.4^2 = 0.98$ to 2 decimal places.
Rounding gives the cost as $0.98.

b When $C = 1.53$ $1.53 = 0.17r^2$
$$\Rightarrow \frac{1.53}{0.17} = 9 = r^2$$
$$\Rightarrow r = \sqrt{9} = 3$$

Hence, the radius is 3 cm.

EXERCISE 19B

For questions **1** to **6**, first find k, the constant of proportionality, and then the formula connecting the variables.

1 T is directly proportional to x^2. If $T = 36$ when $x = 3$, find the value of:

a T when $x = 5$

b x when $T = 400$.

2 W is directly proportional to M^2. If $W = 12$ when $M = 2$, find the value of:

a W when $M = 3$

b M when $W = 75$.

3 E varies directly with $\sqrt{C}$. If $E = 40$ when $C = 25$, find the value of:

   a E when $C = 49$ **b** C when $E = 10.4$.

4 X is directly proportional to $\sqrt{Y}$. If $X = 128$ when $Y = 16$, find the value of:

   a X when $Y = 36$ **b** Y when $X = 48$.

5 P is directly proportional to f^3. If $P = 400$ when $f = 10$, find the value of:

   a P when $f = 4$ **b** f when $P = 50$.

6 y is directly proportional to $\sqrt[3]{x}$. If $y = 100$ when $x = 125$, find the value of:

   a y when $x = 64$ **b** x when $y = 40$.

7 The cost of serving tea and biscuits varies directly with the square root of the number of people at the buffet. It costs $25 to serve tea and biscuits to 100 people.

   a How much will it cost to serve tea and biscuits to 400 people?

   b For a cost of $37.50, how many people could be served tea and biscuits?

8 In an experiment, the temperature, in °C, varied directly with the square of the pressure, in atmospheres (atm). The temperature was 20 °C when the pressure was 5 atm.

   a What will the temperature be at 2 atm?

   b What will the pressure be at 80 °C?

9 The mass, in grams, of ball bearings varies directly with the cube of the radius, measured in millimetres. A ball bearing of radius 4 mm has a mass of 115.2 g.

   a What will be the mass of a ball bearing of radius 6 mm?

   b A ball bearing has a mass of 48.6 g. What is its radius?

10 The energy, in joules (J), of a particle varies directly with the square of its speed, in m/s. A particle moving at 20 m/s has 50 J of energy.

   a How much energy has a particle moving at 4 m/s?

   b At what speed is a particle moving if it has 200 J of energy?

11 The cost, in dollars, of a trip varies directly with the square root of the number of miles travelled. The cost of a 100-mile trip is 35 dollars.

   a What is the cost of a 500-mile trip (to the nearest dollar)?

   b What is the distance of a trip costing 70 dollars?

12 A sculptor is making statues.

The amount of clay used is directly proportional to the cube of the height of the statue.

A statue is 10 cm tall and uses 500 cm³ of clay.

How much clay will a similar statue use if it is twice as tall?

13 The cost of making different-sized machines is proportional to the time taken.

A small machine costs $100 and takes two hours to make.

How much will a large machine cost that takes 5 hours to build?

14 The sketch graphs show each of these proportion statements.

a $y \propto x^2$

b $y \propto x$

c $y \propto \sqrt{x}$

Match each statement to the correct sketch.

15 Here are two tables.

Match each table to a graph in question **14**.

a

x	1	2	3
y	3	12	27

b

x	1	2	3
y	3	6	9

19.2 Inverse variation E

The term **inverse variation** means the same as **inverse proportion**.

There is inverse variation between two variables when one variable is directly proportional to the *reciprocal* of the other. That is, the product of the two variables is constant. So, as one variable increases, the other decreases.

For example, the faster you travel over a given distance, the less time it takes. So there is an inverse variation between speed and time. Speed is inversely proportional to time.

$$S \propto \frac{1}{T} \text{ and so } S = \frac{k}{T}$$

which can be written as $ST = k$.

EXAMPLE 3

M is inversely proportional to R. If $M = 9$ when $R = 4$, find the value of:

a M when $R = 2$ **b** R when $M = 3$.

Step 1: $M \propto \dfrac{1}{R}$

Step 2: Setting up the equation gives:

$$M = \dfrac{k}{R}$$

where k is the **constant of proportionality**.

Step 3: $M = 9$ when $R = 4$. So $9 = \dfrac{k}{4}$

$\Rightarrow 9 \times 4 = k \Rightarrow k = 36$

Step 4: The formula is $M = \dfrac{36}{R}$

a   When $R = 2$, then $M = \dfrac{36}{2} = 18$

b   When $M = 3$, then $3 = \dfrac{36}{R} \Rightarrow 3R = 36 \Rightarrow R = 12$

EXERCISE 19C

For questions **1** to **6**, first find the formula connecting the variables.

1 T is inversely proportional to m. If $T = 6$ when $m = 2$, find the value of:

     a T when $m = 4$ **b** m when $T = 4.8$.

2 W is inversely proportional to x. If $W = 5$ when $x = 12$, find the value of:

     a W when $x = 3$ **b** x when $W = 10$.

3 Q varies inversely with $(5 - t)$. If $Q = 8$ when $t = 3$, find the value of:

     a Q when $t = 10$ **b** t when $Q = 16$.

4 M varies inversely with t^2. If $M = 9$ when $t = 2$, find the value of:

     a M when $t = 3$ **b** t when $M = 1.44$.

5 W is inversely proportional to $\sqrt{T}$. If $W = 6$ when $T = 16$, find the value of:

     a W when $T = 25$ **b** T when $W = 2.4$.

6 y is inversely proportional to the cube of x. If $y = 4$ when $x = 2$, find the value of:

     a y when $x = 1$ **b** x when $y = \frac{1}{2}$.

EXTENDED

7 The grant available to a group of students was inversely proportional to the number of students. When 30 students needed a grant, they received $60 each.

  a What would the grant have been if 120 students had needed one?

  b If the grant had been $50 each, how many students would have received it?

8 While doing underwater tests in an ocean, scientists noticed that the temperature, in °C, was inversely proportional to the depth, in kilometres. When the temperature was 6 °C, the scientists were at a depth of 4 km.

  a What would the temperature have been at a depth of 8 km?

  b At what depth would they find the temperature at 2 °C?

9 A new engine had serious problems. The distance it went, in kilometres, without breaking down was inversely proportional to the square of its speed in metres per second (m/s). When the speed was 12 m/s, the engine lasted 3 km.

  a Find the distance covered before a breakdown, when the speed is 15 m/s.

  b On one test, the engine broke down after 6.75 km. What was the speed?

10 In a balloon it was noticed that the pressure, in atmospheres (atm), was inversely proportional to the square root of the height, in metres. When the balloon was at a height of 25 m, the pressure was 1.44 atm.

  a What was the pressure at a height of 9 m?

  b What would the height have been if the pressure was 0.72 atm?

11 The amount of waste from a firm, measured in tonnes per hour, is inversely proportional to the square root of the area of the filter beds that clean it, in square metres (m^2). The firm produces 1.25 tonnes of waste per hour, with filter beds of size 0.16 m^2.

  a The filter beds used to be only 0.01 m^2. How much waste did the firm produce then?

  b How much waste could be produced if the filter beds were 0.75 m^2?

12 Which statement is represented by the graph? Give a reason for your answer.

  A $y \propto x$

  B $y \propto \dfrac{1}{x}$

  C $y \propto \sqrt{x}$

13 In the table, y is inversely proportional to the cube root of x.

Complete the table, leaving your answers as fractions.

x	8	27	
y	1		$\frac{1}{2}$

Why this chapter matters

The theory of linear programming has been used by many companies to reduce their costs and increase productivity.

The theory of linear programming was developed at the start of the Second World War in 1939.

It was used to work out ways to supply armaments as efficiently as possible. It was such a powerful tool that the British and Americans did not want the Germans to know about it, so it was not made public until 1947.

George Dantzig was one of the inventors of linear programming. He came late to a lecture at University one day and saw two problems written on the blackboard. He copied them, thinking they were the homework assignment.

He solved both problems, but apologised to the lecturer later as he found them a little harder than the usual homework, so he took a few days to solve them and was late handing them in.

The lecturer was astonished. The problems he had written on the board were not homework but examples of 'impossible problems'. Not any more!

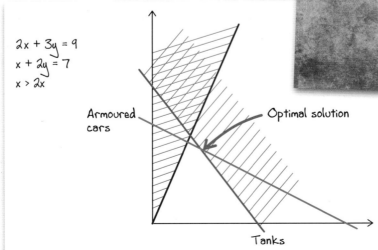

$2x + 3y = 9$
$x + 2y = 7$
$x > 2x$

Armoured cars

Optimal solution

Tanks

Chapter

20 Linear programming

Topics	Level	Key words
1 Graphical inequalities	EXTENDED	boundary, region, included, origin, dashed line, solid line
2 More than one inequality	EXTENDED	required
3 Linear programming	EXTENDED	linear programming

In this chapter you will learn how to:

EXTENDED

π

- Represent inequalities graphically. (E2.6)
- Use this representation in the solution of simple linear programming problems. (E2.6)

Graphical inequalities

A linear inequality can be plotted on a graph. The result is a **region** that lies on one side or the other of a straight line. You will recognise an inequality by the fact that it looks like an equation but instead of the equals sign it has an inequality sign: $<$, $>$, $\leq$, or $\geq$.

The following are examples of linear inequalities that can be represented on a graph.

$$y < 3 \qquad x > 7 \qquad -3 \leq y < 5 \qquad y \geq 2x + 3 \qquad 2x + 3y < 6 \qquad y \leq x$$

The method for graphing an inequality is to draw the **boundary** line that defines the inequality. This is found by replacing the inequality sign with an equals sign. When a strict inequality is stated ($<$ or $>$), the boundary line should be drawn as a *dashed* line to show that it is not included in the range of values. When $\leq$ or $\geq$ is used to state the inequality, the boundary line should be drawn as a *solid* line to show that the boundary is **included**.

After the boundary line has been drawn, shade the *unwanted region*.

To confirm on which side of the line the region lies, choose any point that is not on the boundary line and test it in the inequality. If it satisfies the inequality, that is the side required. If it doesn't, the other side is required.

Work through the six inequalities in this example to see how the procedure is applied.

ADVICE AND TIPS

You should shade the unwanted region of the graph, leaving the region that satisfies the inequality unshaded.

EXAMPLE 1

Show each inequality on a graph.

   a $y \leq 3$ **b** $x > 7$ **c** $-3 \leq y < 5$

   d $y \leq 2x + 3$ **e** $2x + 3y < 6$ **f** $y \leq x$

a Draw the line $y = 3$. Since the inequality is stated as $\leq$, the line is *solid*. Test a point that is not on the line. The **origin** is always a good choice if possible, as 0 is easy to test.

Putting 0 into the inequality gives $0 \leq 3$. The inequality is satisfied and so the region containing the origin is the side we want.

Shade the region on the other side of the line.

b Since the inequality is stated as $>$, the line is dashed. Draw the line $x = 7$.

Test the origin $(0, 0)$, which gives $0 > 7$. This is not true, so you want the other side of the line from the origin.

Shade the unwanted region as shown.

c Draw the lines $y = -3$ (solid for $\leqslant$) and $y = 5$ (dashed for $<$).

Test a point that is not on either line, say $(0, 0)$. Zero is between -3 and 5, so the required region lies between the lines.

Shade in the unwanted regions, outside these lines.

$-3 \leqslant y < 5$

d Draw the line $y = 2x + 3$. Since the inequality is stated as $\leqslant$, the line is solid.

Test a point that is not on the line, $(0, 0)$. Putting these x- and y-values in the inequality gives $0 \leqslant 2(0) + 3$, which is true. So the region that includes the origin is what you want.

Shade in the unwanted region on the other side of the line.

$y \leqslant 2x + 3$

e Draw the line $2x + 3y = 6$.

The easiest way is to find out where it crosses the axes.

If $x = 0$, $3y = 6 \Rightarrow y = 2$. Crosses y-axis at $(0, 2)$.

If $y = 0$, $2x = 6 \Rightarrow x = 3$. Crosses x-axis at $(3, 0)$.

Draw the line through these two points.

Since the inequality is stated as $<$, the line is dashed.

Test a point that is not on the line, say $(0, 0)$. Is it true that $2(0) + 3(0) < 6$? The answer is yes, so the origin is in the region that you want.

Shade in the unwanted region on the other side of the line.

$2x + 3y < 6$

f Draw the line $y = x$. Since the inequality is stated as $\leqslant$, the line is solid.

This time the origin is on the line, so pick any other point, say $(1, 3)$. Putting $x = 1$ and $y = 3$ in the inequality gives $3 \leqslant 1$. This is not true, so the point $(1, 3)$ is not in the region you want.

Shade in the region that includes $(1, 3)$, the unwanted region.

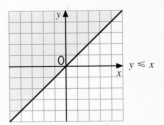

$y \leqslant x$

EXERCISE 20A

1 **a** Draw the line $x = 2$ (as a solid line).

    b Use shading to show the region defined by $x \leqslant 2$.

2 **a** Draw the line $y = -3$ (as a dashed line).

    b Use shading to show the region defined by $y > -3$.

3 **a** Draw the line $x = -2$ (as a solid line).

    b Draw the line $x = 1$ (as a solid line) on the same grid.

    c Use shading to show the region defined by $-2 \leqslant x \leqslant 1$.

4 **a** Draw the line $y = -1$ (as a dashed line).

    b Draw the line $y = 4$ (as a solid line) on the same grid.

    c Use shading to show the region defined by $-1 < y \leqslant 4$.

5 **a** On the same grid, draw the regions defined by these inequalities.

    i $-3 \leqslant x \leqslant 6$ **ii** $-4 < y \leqslant 5$

    b Are these points in the region defined by both inequalities?

    i $(2, 2)$ **ii** $(1, 5)$ **iii** $(-2, -4)$

6 **a** Draw the line $y = 2x - 1$ (as a dashed line).

    b Use shading to show the region defined by $y < 2x - 1$.

7 **a** Draw the line $3x - 4y = 12$ (as a solid line).

    b Use shading to show the region defined by $3x - 4y \leqslant 12$.

8 **a** Draw the line $y = \frac{1}{2}x + 3$ (as a solid line).

    b Use shading to show the region defined by $y \geqslant \frac{1}{2}x + 3$.

9 Use shading to show the region defined by $y < -3$.

10 **a** Draw the line $y = 3x - 4$ (as a solid line).

    b Draw the line $x + y = 10$ (as a solid line) on the same diagram.

    c Shade the diagram so that the region defined by $y \geqslant 3x - 4$ is left *unshaded*.

    d Shade the diagram so that the region defined by $x + y \leqslant 10$ is left *unshaded*.

    e Are these points in the region defined by both inequalities?

    i $(2, 1)$ **ii** $(2, 2)$ **iii** $(2, 3)$

When you have to show a region that satisfies more than one inequality, it is always clearer to *shade* the regions *not required*, so that the *required region* is left *blank*.

EXAMPLE 2

a On the same grid, show the regions that represent each inequality by shading the unwanted regions.

   i $x > 2$ **ii** $y \geqslant x$ **iii** $x + y < 8$

b Are these points in the region that satisfies all three inequalities?

   i $(3, 4)$ **ii** $(2, 6)$ **iii** $(3, 3)$

a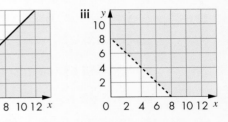

i The region $x > 2$ is shown unshaded in diagram **i**.
The boundary line is $x = 2$ (dashed).

ii The region $y \geqslant x$ is shown unshaded in diagram **ii**.
The boundary line is $y = x$ (solid).

iii The region $x + y < 8$ is shown unshaded in diagram **iii**.
The boundary line is $x + y = 8$ (dashed).
The regions have first been drawn separately so that each may be clearly seen. The diagram on the right shows all three regions on the same grid. The white triangular area defines the region that satisfies all three inequalities.

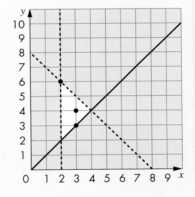

b **i** The point $(3, 4)$ is clearly within the region that satisfies all three inequalities.

   ii The point $(2, 6)$ is on the boundary lines $x = 2$ and $x + y = 8$. As these are dashed lines, they are not included in the region defined by all three inequalities. So, the point $(2, 6)$ is not in this region.

   iii The point $(3, 3)$ is on the boundary line $y = x$. As this is a solid line, it is included in the region defined by all three inequalities. So, the point $(3, 3)$ is included in this region.

EXTENDED

EXERCISE 20B

1 **a** Draw the line $y = x$ (as a solid line).

b Draw the line $2x + 5y = 10$ (as a solid line) on the same diagram.

c Draw the line $2x + y = 6$ (as a dashed line) on the same diagram.

> **ADVICE AND TIPS**
>
> Find the points where the line crosses each axis.

d Shade the diagram so that the region defined by $y \geqslant x$ is left *unshaded*.

e Shade the diagram so that the region defined by $2x + 5y \geqslant 10$ is left *unshaded*.

f Shade the diagram so that the region defined by $2x + y < 6$ is left *unshaded*.

g Are these points in the region defined by these inequalities?

   i (1, 1)   **ii** (2, 2)   **iii** (1, 3)

2 **a** On the same grid, draw the regions defined by the following inequalities. (Shade the diagram so that the overlapping region is left blank.)

   i $y > x - 3$ **ii** $3y + 4x \leqslant 24$ **iii** $x \geqslant 2$

b Are these points in the region defined by all three inequalities?

   i (1, 1)   **ii** (2, 2)   **iii** (3, 3)   **iv** (4, 4)

3 **a** On a graph draw the lines $y = x$, $x + y = 8$ and $y = 2$.

b Label the region R where $y \leqslant x$, $x + y \leqslant 8$ and $y \geqslant 2$. Shade the region that is *not* required.

4 **a** On a graph draw the lines $y = x - 4$, $y = 0.5x$ and $y = -x$.

b Show the region S where $y \geqslant x - 4$, $y \leqslant 0.5x$ and $y \geqslant -x$. Shade the region that is *not* required.

c What is the largest y-coordinate of a point in S?

d What is the smallest y-coordinate of a point in S?

e What is the largest value of $x + y$ for a point in S?

5 Explain how you would find which side of the line represents the inequality $y < x + 2$.

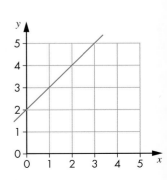

6 The region marked R is bounded by the lines $x + y = 3$, $y = \frac{1}{2}x + 3$ and $y = 5x - 15$.

   a What three inequalities are satisfied in region R?

   b What is the greatest value of $x + y$ in region R?

   c What is the greatest value of $x - y$ in region R?

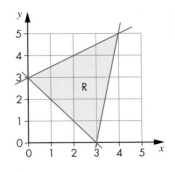

20.3 Linear programming E

There are practical situations that give rise to inequalities that can be shown on a graph. Solving problems in this way is called **linear programming**.

EXAMPLE 3

A boat trip costs $20 for adults and $10 for children.

Suppose there are x adults and y children on the trip.

a There must not be more than 10 people on the trip.
Explain why $x + y \leqslant 10$.

b The money taken for tickets must be at least $120.
Explain why $2x + y \geqslant 12$.

c Show the region satisfied by both these inequalities.

d What is the smallest number of adults that should be carried?

a The total number of passengers is 10.
The total must be 10 or less so $x + y \leqslant 10$.

b The total number of dollars taken is $20x + 10y$.
This must be at least 120 so $20x + 10y \geqslant 120$.
Divide both sides by 10 to get $2x + y \geqslant 12$.

c Negative numbers on the axes are not needed.
First draw the lines with equations $x + y = 10$ and $2x + y = 12$.
The first crosses the axes at $(10, 0)$ and $(0, 10)$.
The second crosses the axes at $(6, 0)$ and $(0, 12)$.
Shade the regions that are *not* required.
The solution must be in the unshaded region R.

d The smallest x-coordinate in this region is at $(2, 8)$.
x must be at least 2 so there must be at least 2 adults.

EXERCISE 20C

1 A man buys x cartons of milk and y bottles of water.

   a Explain what $x \geqslant 3$ means.

   b Write an inequality to show that he buys at least 2 bottles of water.

   c He cannot carry more than a total of 8 bottles and cartons. Write this as an inequality.

   d Draw a graph to show possible values of x and y.

2 There are x cars and y vans in a car park.

   a There are no more than 10 vehicles (cars and vans) in the car park.
      Write an inequality to show this.

   b Cars pay \$2 and vans pay \$5 to park. At least \$30 was paid in total.
      Write an inequality to show this.

   c Show the two inequalities on a grid.

   d What is the largest possible number of cars?

   e What is the smallest possible number of vans?

3 x girls and y boys are at a children's party.
There are more girls than boys.
The number of girls is less than twice the number of boys.
There are less than 12 children at the party.

   a Write down three inequalities to represent each of the statements above.

   b Show the region indicated by these inequalities.
      Shade the areas that are not included.

   c What is the largest possible number of girls at the party?

4 A house is home to x cats and y rabbits.
There are at least 6 animals.
There are no more than 12 animals.
There are more rabbits than cats.
There are at least two cats.

   a Write down four inequalities from these sentences.

   b Show the region that gives possible values of x and y by shading the area that is not required.

   c What is the smallest possible number of rabbits?

   d What is the largest possible number of cats?

5 At a concert, x people bought seats near the front for \$15 and y people bought seats near the back for \$10.

 a Ticket sales were at least \$600. Show that $3x + 2y \geqslant 120$.

 b No more than 50 tickets were sold. Write an inequality for this.

 c Show the region which represents these inequalities on a grid.

 d What is the largest possible number of back seats sold?

6 A company has a fleet of coaches in two sizes. There are x large coaches and y small coaches.

 a Write an inequality for each of these statements.

   i The total number of coaches must not exceed 15.

   ii There must be at least 5 coaches in total.

   iii The number of small coaches must not be less than half the number of large coaches.

 b Explain in words the meaning of $y \leqslant x + 2$.

 c Label the region R to show the four inequalities in parts **a** and **b**. Shade the area which is not required.

 d If there are 6 large coaches, how many small coaches could there be? Give all the possible values.

7 In a football league a team score 3 points for a win, 1 point for a draw and none for a loss.

 a If a team wins x matches and draws y matches, write down an expression for the total number of points gained.

 b Wayne knows these facts about his team's performance so far this season.
  ● They have gained at most 18 points.
  ● They have more wins than draws.
  ● They have at least two draws.

 Write an inequality, using x and y, for each of these three statements.

 c Show on a graph the region where the three inequalities are satisfied.

 d List all the possible numbers of wins and draws for Wayne's team.

Notation is important in mathematics. Try writing an equation or formula in words instead of symbols and you will see why notation makes things easier to understand.

If you drop a coin, how long does it take to reach the ground? That depends on the height you drop it from. We say that the time taken is a **function** of the height.

By dropping the coin from different heights and measuring the time it takes to fall, it would be possible to find a formula for the time in terms of the height.

Here are some other examples where one variable is a **function** of another.

- The cost of posting a parcel is a function of its mass.
- The time taken for a journey is a function of the distance travelled.
- The stopping distance of a car is a function of its speed.
- The cost of a second hand car is a function of its age.
- The time taken to download a computer file is a function of the size of the file.

The idea of an **inverse** occurs frequently in mathematics. It is not a difficult idea and it is a useful one. Putting a hat on and taking it off are inverse operations. Switching a light on and switching it off are also inverse operations.

Here are some examples of inverse operations in mathematics.

- Add 3 and subtract 3.
- Multiply by 5 and divide by 5.
- Rotate 90° clockwise and rotate 90° anticlockwise.
- Square a number and find the square root of the square number.

It is not always possible to find an inverse. Sadly the inverse of breaking a glass does not exist.

Chapter

21 Functions

Topics	Level	Key words
1 Function notation	EXTENDED	function
2 Inverse functions	EXTENDED	inverse
3 Composite functions	EXTENDED	composite
4 More about composite functions	EXTENDED	

In this chapter you will learn how to:

EXTENDED

- Use function notation, e.g. $f(x) = 3x - 5$, $f: x \rightarrow 3x - 5$ to describe simple functions. (E2.12)
- Find inverse functions $f^{-1}(x)$. (E2.12)
- Form composite functions as defined by $gf(x) = g(f(x))$. (E2.12)

You are familiar with equations written in x and y, such as $y = 3x - 4$ or $y = 2x^2 + 5x - 3$.

These equations are showing that y is a function of x. This means that the value of y depends on the value of x so that y changes when x changes.

Sometimes it is useful to use a different notation to show this. You could write the first equation above as $f(x) = 3x - 4$ and refer to it as 'function f'.

It is then easy to show the result of using different values for x. For example:

● 'the value of $f(x)$ when x is 5' can be written as $f(5)$.

So $f(5) = 3 \times 5 - 4 = 11$

● $f(1)$ means 'the value of $f(x)$ when x is 1'

So $f(1) = 3 \times 1 - 4 = -1$

and $f(-1) = -7$

If there are different functions in the same problem they can be represented by different letters, for example:

$g(x) = 2x^2 + 5x - 3$ or 'function g'.

Sometimes instead of $f(x) = 3x - 4$ you may see $f: x \rightarrow 3x - 4$. These two forms mean exactly the same thing.

EXERCISE 21A

1 $f(x) = 2x + 6$. Find the value of:

   a $f(3)$ **b** $f(10)$ **c** $f(\frac{1}{2})$ **d** $f(-4)$ **e** $f(-1.5)$.

2 $g(x) = \dfrac{x^2 + 1}{2}$. Find the value of:

   a $g(0)$ **b** $g(3)$ **c** $g(10)$ **d** $g(-2)$ **e** $g(-\frac{1}{2})$.

3 $f: x \rightarrow x^3 - 2x + 1$. Find the value of:

   a $f(2)$ **b** $f(-2)$ **c** $f(100)$ **d** $f(0)$ **e** $f(\frac{1}{2})$.

4 $g: x \rightarrow 2^x$. Find the value of:

   a $g(2)$ **b** $g(5)$ **c** $g(0)$ **d** $g(-1)$ **e** $g(-3)$.

5 $h(x) = \dfrac{x + 1}{x - 1}$. Find the value of:

   a $h(2)$ **b** $h(3)$ **c** $h(-1)$ **d** $h(0)$ **e** $h(1\frac{1}{2})$.

EXTENDED

6 f: $x \rightarrow 2x + 5$

    a If $f(a) = 20$, what is the value of a?

    b If $f(b) = 0$, what is the value of b?

    c If $f(c) = c$, what is the value of c?

7 $g(x) = \sqrt{x + 3}$

    a Find the value of $g(33)$.

    b If $g(a) = 10$, find the value of a.

    c If $g(b) = 2.5$, find the value of b.

8 $f(x) = 2x - 8$ and $g(x) = 10 - x$.

    a What is the value of x for which $f(x) = g(x)$?

    b Sketch the graphs of $y = f(x)$ and $y = g(x)$. At what point do they cross?

9 h: $x \rightarrow \dfrac{12}{x} + 1$ k: $x \rightarrow 2^x - 1$

    a Find the value of $h(6)$.

    b Find the value of $k(-1)$.

    c Solve the equation $h(x) = k(3)$.

    d Solve the equation $k(x) = h(-12)$.

21.2 Inverse functions

Suppose $f(x) = 2x + 6$.

Then $f(1) = 8$, $f(3) = 12$ and $f(-4) = -2$.

The **inverse** of f is the function that has the opposite effect and 'undoes' f. You write the inverse of f as f^{-1}.

Since f above means 'multiply by 2 and then add 6', the inverse will be 'subtract 6 and then divide by 2':

$$f^{-1}(x) = \frac{x - 6}{2}$$

So $f^{-1}(8) = 1$, $f^{-1}(12) = 3$ and $f^{-1}(-2) = -4$.

You can find the inverse by following these steps.

Step 1: Write $y = f(x)$

$$y = 2x + 6$$

Step 2: Rearrange to make x the subject.

$$y - 6 = 2x$$
$$\frac{y - 6}{2} = x$$
$$x = \frac{y - 6}{2}$$

Step 3: Replace y by x in the result.

$$f^{-1}(x) = \frac{x - 6}{2}$$

EXERCISE 21B

1 Find $f^{-1}(x)$ for each of these functions.

    a $f(x) = x + 7$ **b** $f(x) = 8x$ **c** $f(x) = \dfrac{x}{5}$ **d** $f(x) = x - 3$

2 $f: x \rightarrow \dfrac{x}{2} + 6$. Find the value of:

    a $f(4)$ **b** $f^{-1}(8)$ **c** $f(-2)$ **d** $f^{-1}(5)$.

3 Find $f^{-1}(x)$ for each of these functions.

    a $f(x) = \dfrac{x}{3} - 2$ **b** $f(x) = 4(x - 5)$ **c** $f(x) = \dfrac{x + 4}{5}$ **d** $f(x) = \dfrac{(3x - 6)}{5}$

    e $f(x) = 3(\dfrac{x}{2} + 4)$ **f** $f(x) = 4x^3$

4 $g(x) = \dfrac{2x + 5}{3}$. Find the value of:

    a $g^{-1}(3)$ **b** $g^{-1}(2)$ **c** $g^{-1}(0)$.

5 $f(x) = 10 - x$

    a Find an expression for $f^{-1}(x)$.

    b What do you notice about $f(x)$ and $f^{-1}(x)$?

6 Find $f^{-1}(x)$ for each of these functions.

    a $f(x) = \dfrac{8}{x}$ **b** $f(x) = \dfrac{20}{x} - 1$ **c** $f(x) = \dfrac{2}{x + 1}$

7 $f(x) = 2x - 4$

    a Find $f^{-1}(x)$.

    b On the same axes draw graphs of $y = f(x)$ and $y = f^{-1}(x)$.

    c Where do the lines cross?

EXTENDED

8 $f(x) = \dfrac{x+5}{2}$

Solve the equation $f(x) = f^{-1}(x)$.

9 $f^{-1}(x) = 3x - 2$

Find an expression for $f(x)$.

21.3   Composite functions

Consider a function given by $f(x) = 2x$.

So f means 'double it'.

$$f(2) = 4$$
$$f(3) = 6$$
$$f(5) = 10$$

and $f(-3) = -6$

Now suppose you have another function, g, and $g(x) = x - 3$.

So g means 'subtract 3'.

$$g(4) = 1$$
$$g(6) = 3$$
$$g(10) = 7$$

and $g(-6) = -9$

Now put those side by side to make a **composite** function.

$f(2) = 4$	and	$g(4) = 1$
$f(3) = 6$	and	$g(6) = 3$
$f(5) = 10$	and	$g(10) = 7$
$f(-3) = -6$	and	$g(-6) = -9$

In words they mean:

Start with 2, double it, subtract 3, the answer is 1.
Start with 3, double it, subtract 3, the answer is 3.
Start with 5, double it, subtract 3, the answer is 7.
Start with −3, double it, subtract 3, the answer is −9.

You can write that in symbols as:

$$gf(2) = 1$$
$$gf(3) = 3$$
$$gf(5) = 7$$
$$gf(-3) = -9$$

> **ADVICE AND TIPS**
>
> gf means 'first f, then g'.

EXAMPLE 1

$h(x) = x^2$ and $k(x) = x + 4$. Find:

a $kh(3)$ **b** $kh(-2)$ **c** $hk(5)$

a h means 'square it' and k means 'add 4'.

kh(3) means 'start with 3, square it, then add 4'.

So $kh(3) = 3^2 + 4 = 9 + 4 = 13$

b $kh(-2) = (-2)^2 + 4 = 4 + 4 = 8$

c hk(5) is the other way round.

It means 'start with 5, add 4, then square it'.

$hk(5) = (5 + 4)^2 = 9^2 = 81$

Look again at the functions f and g you considered earlier.

$f(x) = 2x$ and $g(x) = x - 3$

$gf(x)$ means 'first double x, then subtract 3'. So you can write:

$gf(x) = 2x - 3$

What about $fg(x)$?

That means 'first subtract 3, then double it'. So you can write:

$fg(x) = 2(x - 3)$

Looking at example 1, $h(x) = x^2$ and $k(x) = x + 4$.

Then $kh(x) = x^2 + 4$

and $hk(x) = (x + 4)^2$

EXERCISE 21C

1 $s(x) = x + 4$ and $t(x) = \dfrac{x}{2}$

    a Find the value of s(2) and ts(2).

    b Find the value of s(3) and ts(3).

    c Find the value of s(6) and ts(6).

    d Find an expression for ts(x).

    e Find the value of t(2) and st(2).

    f Find the value of t(3) and st(3).

    g Find the value of t(−10) and st(−10).

    h Find an expression for st(x).

2 $c(x) = x^3$ and $d(x) = 2x$.

    a Find the value of d(3) and cd(3).

    b Find the value of d(5) and cd(5).

    c Find an expression for cd(x).

    d Find the value of c(4) and dc(4).

    e Find an expression for dc(x).

EXTENDED

3 $r(x) = \sqrt{x}$ and $a(x) = 2x + 1$

    a Find the value of a(0), a(4) and a(12).   **b** Find the value of ra(0), ra(4) and ra(12).

    c Find an expression for ra(x).

4 $m(x) = 3x$

    a Find the value of m(2) and mm(2).   **b** Find the value of m(4) and mm(4).

    c Find an expression for mm(x).

5 $f(x) = 3x$ and $g(x) = x - 6$

    a Find an expression for fg(x).   **b** Find an expression for gf(x).

6 $a(x) = x + 4$ and $b(x) = x - 7$

    Show that ab(x) and ba(x) are identical.

21.4 More about composite functions

Suppose $f(x) = x^2 + 2$ and $g(x) = 2x - 3$.

Is it possible to find an expression for gf(x)?

Start with a value of x, say $x = 3$.

$f(3) = 3^2 + 2 = 11$

Now apply g to that answer.

$g(11) = 2 \times 11 - 3 = 19$

So   gf(3) = 19

gf(x) means 'start with x, apply f, and then apply g to the answer.'
If you start with x and 'apply f' you get $x^2 + 2$.

Now take that answer and 'apply g', double it and subtract 3.

$gf(x) = 2(x^2 + 2) - 3$

Now simplify that.

$gf(x) = 2(x^2 + 2) - 3$
$= 2x^2 + 4 - 3$
$= 2x^2 + 1$

So $gf(x) = 2x^2 + 1$

Check this works when x is 3.

$gf(3) = 2 \times 3^2 + 1$
$= 18 + 1$
$= 19$ as before.

EXERCISE 21D

1 $f(x) = \dfrac{x-3}{2}$ and $g(x) = 3x + 1$

Find the value of:

a $fg(3)$

b $gf(3)$

c $fg(6)$

d $gf(6)$.

2 $f(x) = x^2 - x$ and $g(x) = \dfrac{x}{2} + 3$

Find the value of:

a $fg(4)$

b $gf(4)$

c $fg(1)$

d $gf(1)$.

3 $f(x) = 2^x$ and $g(x) = 2x - 1$

Find the value of:

a $gf(2)$

b $fg(2)$

c $ff(3)$

d $gg(6)$.

4 $f(x) = 3x + 1$ and $g(x) = 2x - 2$

a Find an expression for $gf(x)$. Write your answer as simply as possible.

b Find an expression for $fg(x)$. Write your answer as simply as possible.

5 In each case find an expression for $fg(x)$. Write your answer as simply as possible.

a $f(x) = x^2$ and $g(x) = 3x + 4$

b $f(x) = 2x + 3$ and $g(x) = 3x - 4$

c $f(x) = \dfrac{x}{2} + 4$ and $g(x) = 4x - 2$

d $f(x) = 12 - x$ and $g(x) = 2x + 8$

6 $h(x) = 10 - x$ and $k(x) = 20 - x$

Find an expression for:

a $hk(x)$

b $kh(x)$

c $kk(x)$.

7 $h(x) = x^2$ and $k(x) = \dfrac{12}{x}$

Find an expression for:

a $hh(x)$

b $hk(x)$

c $kh(x)$

d $kk(x)$.

8 $m(x) = x^2 + 2x$ and $n(x) = 2x - 1$

a Find an expression for $mm(2)$.

b Show that $nn(x) = 4x - 3$.

c Show that $mn(x) = 4x^2 - 1$

9 $f(x) = \dfrac{1}{x - 4}$ and $g(x) = 3x + 1$.

a Find an expression for $fg(x)$.

b Find an expression for $gf(x)$, writing your answer as a single fraction.

10 $h(x) = \dfrac{x + 4}{2}$ and $k(x) = 3x - 5$

Find:

a $h^{-1}k(x)$

b $hh(x)$.

11 Suppose $f(x) = 0.5(x + 9)$.

a Show that $f(1) = 5$.

b Find the value of $f(5)$.

c Find $f(b)$ where b is the answer to part **b**.

d Continue in this way, using the last answer as the next value of x, to find the next six values.

e What is happening to the answers?

12 What happens in question 11 if you start with $f(25)$ instead of $f(1)$?

Note: drawings in questions are NOT TO SCALE unless otherwise specified.

- If the degree of accuracy is not specified in the question, and if the answer is not exact, give the answer to three significant figures. Give answers in degrees to one decimal place.

- For π, use either your calculator value or 3.142.

PAPER 1

1 Solve the equation.

$$\frac{n-8}{2} = 11$$ [2]

Paper 11 Q6 May/June 2014

2 Solve the following equations.

a $x + 9 = 16$ [1]

b $6y = 27$ [1]

Paper 11 Q5 Oct/Nov 2012

3 Rearrange this equation to make b the subject.

$$a = \frac{b}{5} - 9$$ [2]

Paper 11 Q9 May/June 2013

4 Here are the first four terms of a sequence.

4   11   18   25

Write down an expression for the nth term. [2]

Paper 11 Q10 May/June 2013

5 Simplify the following expression.

$$5a - 3b - 2a - b$$ [2]

Paper 11 Q7 Oct/Nov 2013

6 Solve the equation.

$$5 - 2x = 3x - 19$$ [2]

Paper 11 Q12 Oct/Nov 2013

7 Simplify the expression.

$$3j - 4k - 2 + 5j + k - 6$$ [2]

Paper 11 Q7 Oct/Nov 2012

8 **a** Factorise: $9y + 12$ [1]

b Expand: $a(a^2 - 7)$ [2]

Paper 11 Q13 Oct/Nov 2012

9 **a** Factorise completely.

$6ab - 24bc$ [2]

b Rearrange the following formula to make m the subject.

$$j = \frac{m}{n} - k$$ [2]

Paper 11 Q22 Oct/Nov 2013

10 **a** Multiply out the brackets.

$5(x + 3)$ [2][1]

b Factorise completely.

$12xy - 3x^2$ [2]

c Solve.

$5x - 24 = 51$ [2]

Paper 11 Q19 May/June 2013

11 **a** Simplify the expressions.

i $p^3 \times p^7$ **ii** $t^5 \div t^8$ [2]

b $(h^3)^k = h^{12}$

Find the value of k. [1]

Paper 11 Q16 May/June 2014

12 **a** Here are the first four terms of a sequence.

27   23   19   15

i Write down the next term in the sequence. [1]

ii Explain how you worked out your answer to part **a i**. [1]

b The nth term of a different sequence is $4n - 2$.

Write down the first three terms of this sequence. [1]

c Here are the first four terms of another sequence.

−1   2   5   8

Write down the nth term of this sequence. [2]

Paper 11 Q23 Oct/Nov 2013

PAPER 3

1 **a** **i** Complete the table of values for $y = \dfrac{8}{x}$, $x \neq 0$.

x	−8	−4	−2	−1		1	2	4	8
y		−2						2	

[3]

ii On a copy of this grid, draw the graph of $y = \dfrac{8}{x}$ for $-8 \leqslant x \leqslant -1$ and $1 \leqslant x \leqslant 8$.

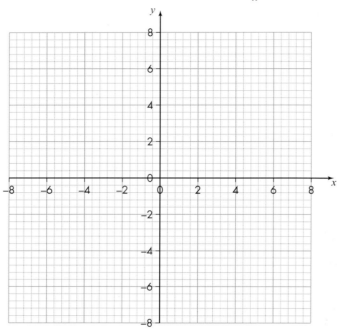

[4]

iii Write down the order of rotational symmetry of your graph. [1]

b **i** Complete this table of values for $y = 1.5x + 3$.

x	−6	−4	−2	0	2
y	−6	−3		3	

[2]

ii On your grid, draw the graph of $y = 1.5x + 3$. [1]

c Use your graph to solve the equation $\dfrac{8}{x} = 1.5x + 3$. [2]

d Write down the gradient of $y = 1.5x + 3$. [1]

Paper 31 Q6 May/Jun 2014

2 **a** The cost C, of a party for n people is calculated using the following formula.

$C = 130 + 4n$

i Calculate C when $n = 25$. [2]

ii Eurdley has a party that costs $1138. [2]

How many people is the party for?

b Solve the following equations.

i $3x = 27$ [1]

ii $8y - 4 = 24$ [2]

iii $4(5q - 2) = 72$ [3]

c Solve the simultaneous equations.

$6x + 8y = -31$

$14x - 5y = 46$ [4]

Paper 31 Q5 Oct/Nov 2013

3 **a** For these sequences, write down the next two terms and the rule for finding the next term.

i 84, 75, 66, 57, …, … [3]

ii 2, 6, 18, 54, …, … [3]

b For the sequence in part **a i:**

i write down an expression, in terms of n, for the nth term [2]

ii find the 21st term. [2]

Paper 31 Q9 May/Jun 2014

4 **a** In 2001 Arnold was x years old.

Ken is **34 years younger** than Arnold.

i Complete the table, in terms of x, for Arnold's and Ken's age.

	2001	2013
Arnold's age	x	
Ken's age		

[3]

ii In 2013 Arnold is **three** times as old as Ken.

Write down an equation in x and solve it. [4]

b Solve the simultaneous equations.

$3x + 2y = 18$

$2x - y = 19$ [3]

Paper 31 Q10 May/Jun 2013

PAPER 2

1 Solve the equation: $4x - 12 = 2(11 - 3x)$ [3]

Paper 22 Q10 Oct/Nov 2012

2 The equation of the line in the diagram is $y = 5 - x$.

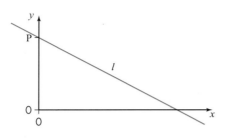

a The line cuts the y-axis at P. Write down the coordinates of P. [1]

b Write down the gradient of the line l. [1]

Paper 22 Q5 May/Jun 2014

3 The speed, v, of a wave is inversely proportional to the square root of the depth, d, of the water.

$v = 30$ when $d = 400$.

Find v when $d = 25$. [3]

Paper 22 Q11 Oct/Nov 2013

4 a $\left(\dfrac{3}{8}\right)^{\frac{3}{8}} \times \left(\dfrac{3}{8}\right)^{\frac{1}{8}} = p^{q}$

Find the values of p and q. [2]

b $5^{-3} + 5^{-4} = k \times 5^{-4}$

Find the value of k. [2]

Paper 22 Q14 Oct/Nov 2012

5 Solve the equation $\dfrac{3}{2x} + \dfrac{1}{x+1} = 0$. [3]

Paper 22 Q12 May/Jun 2014

6 Factorise completely.

a $4p^2q - 6pq^2$ [2]

b $u + 4t + ux + 4tx$ [2]

Paper 22 Q16 May/Jun 2014

7 Use the quadratic equation formula to solve $2x^2 + 7x - 3 = 0$.

Show all your working and give your answers correct to 2 decimal places. [4]

Paper 22 Q15 May/Jun 2013

8 Find the coordinates of the point of intersection of the two lines.

$2x - 7y = 2$

$4x + 5y = 42$ [3]

Paper 22 Q15 Oct/Nov 2013

9 Solve the inequality. $\dfrac{x}{2} + \dfrac{x-2}{3} < 5$ [3]

Paper 22 Q16 Oct/Nov 2013

10 Simplify. $\dfrac{x^2 + 6x - 7}{3x + 21}$ [4]

Paper 22 Q19 May/Jun 2014

11 $f(x) = 5x + 4$ $\qquad g(x) = \dfrac{1}{2x}, x \neq 0$ $\qquad h(x) = \left(\dfrac{1}{2}\right)^x$

Find:

a $fg(5)$ [2]

b $gg(x)$ in its simplest form [2]

c $f^{-1}(x)$ [2]

d the value of x when $h(x) = 8$. [2]

Paper 22 Q21 May/Jun 2013

12 32   25   18   11   4

These are the first five terms of a sequence. Find:

a the 6th term [1]

b the nth term [2]

c which term is equal to –322. [2]

Paper 22 Q20 May/Jun 2014

13 a The two lines $y = 2x + 8$ and $y = 2x - 12$ intersect the x-axis at P and Q.

Work out the distance PQ. [2]

b Write down the equation of the line with gradient –4 passing through (0, 5). [2]

c Find the equation of the line parallel to the line in part **b** passing through (5, 4). [3]

Paper 22 Q20 Oct/Nov 2012

PAPER 4

1 The table shows some values for the function $y = 11x - 2x^2 - 12$ for $1 \leqslant x \leqslant 4.5$.

x	1	1.5	2	2.5	3	3.5	4	4.5
y	−3		2	3	3			

a Complete the table of values. [3]

b On the grid below, draw the graph of $y = 11x - 2x^2 - 12$ for $1 \leqslant x \leqslant 4.5$.

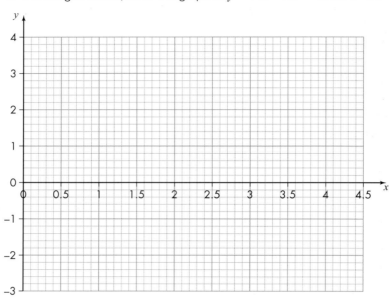

[4]

c By drawing a suitable line, use your graph to solve the equation

$11x - 2x^2 = 11$ for $1 \leqslant x \leqslant 4.5$. [2]

d The line $y = mx + 2$ is a tangent to the curve $y = 11x - 2x^2 - 12$ at the point P.

By drawing this tangent:

i find the coordinates of the point P [2]

ii work out the value of m. [2]

Paper 42 Q3 May/Jun 2013

2 a Write as a single fraction in its simplest form: $\dfrac{2x - 1}{2} - \dfrac{3x + 1}{5}$ [3]

b Expand and simplify: $(2x - 3)2 - 3x(x - 4)$ [4]

c i Factorise: $2x^2 + 5x - 3$ [2]

ii Simplify: $\dfrac{2x^2 + 5x - 3}{2x^2 - 18}$ [3]

Paper 42 Q3 Oct/Nov 2013

3 Paul buys a number of large sacks of fertiliser costing $x each.

He spends $27.

a Write down.in terms of x, an expression for the number of large sacks
which Paul buys. [1]

b Rula buys a number of small sacks of fertiliser.

Each small sack costs $2 less than a large sack.

Rula spends $25.

Write down, in terms of x, an expression for the number of small sacks
which Rula buys. [1]

c Rula buys 4 more sacks than Paul.

Write down an equation in x and show that it simplifies to $2x^2 - 3x - 27 = 0$. [4]

d Solve the equation $2x^2 - 3x - 27 = 0$. [3]

e Calculate the number of sacks that Paul buys. [1]

Paper 42 Q5 May/Jun 2013

4 $f(x) = \dfrac{1}{x}, x \neq 0$ $\qquad$ $g(x) = 1 - x$ $\qquad$ $h(x) = x^2 + 1$

a Find $fg\left(\dfrac{1}{2}\right)$ [2]

b Find $g{-1}(x)$, the inverse of $g(x)$. [1]

c Find $hg(x)$, giving your answer in its simplest form. [3]

d Find the value of x when $g(x) = 7$. [1]

e Solve the equation $h(x) = 3x$.

Show your working and give the answers correct to 2 decimal places. [4]

f A function $k(x)$ is its own inverse when $k^{-1}(x) = k(x)$.

For which of the functions $f(x)$, $g(x)$ and $h(x)$ is this true? [1]

Paper 42 Q10 May/Jun 2014

Why this chapter matters

Angles describe an amount of turn around a point. It is important to be able to measure them and understand their properties.

Why this chapter matters

In a regular polygon all the angles are the same size and all the sides are the same length. The shape of the polygon depends on how many angles and sides it has:

- a triangle has 3 sides and 3 angles
- a square has 4 sides and 4 angles
- a pentagon has 5 sides and 5 angles
- a hexagon has 6 sides and 6 angles.

The patterns below are made by putting together different regular polygons. How many different polygons can you see in them? Can you see any octagons (8-sided shapes) or decagons (10-sided shapes)?

  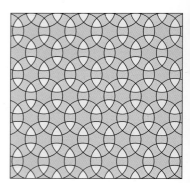

Some shapes fit together better than others because of the size of their angles.

Bees construct their hives from hexagon shapes, which can join together without gaps.

Squares and rectangles also fit together easily, which makes them an ideal shape to use in building.

In this chapter you will look at angles, the shapes they form, and their properties.

Chapter

22 Angle properties

Topics	Level	Key words
1 Angle facts	CORE	angles at a point, angles on straight lines, opposite angles, vertically opposite angles
2 Parallel lines	CORE	corresponding angles, alternate angles, allied angles, interior angles
3 Angles in a triangle	CORE	equilateral triangle, isosceles triangle, right-angled triangle
4 Angles in a quadrilateral	CORE	quadrilateral, parallelogram, rhombus, kite, trapezium
5 Regular polygons	CORE	polygon, regular polygon, external angles, pentagon, hexagon, octagon, square
6 Irregular polygons	EXTENDED	irregular polygon, heptagon, nonagon, decagon
7 Tangents and diameters	CORE	radius, tangent, point of contact, diameter, semi-circle
8 Angles in a circle	EXTENDED	arc, circumference, segment, subtended
9 Cyclic quadrilaterals	EXTENDED	cyclic quadrilateral, opposite segment, supplementary

In this chapter you will learn how to:

CORE

Calculate unknown angles using the following geometrical properties:

(a) angles at a point

(b) angles at a point on a straight line and intersecting straight lines

(c) angles formed within parallel lines

(d) angle properties of triangles and quadrilaterals

(e) angle properties of regular polygons

(f) angle in a semi-circle

(g) angle between tangent and radius of a circle. (C3.6 and E3.6)

EXTENDED

Use in addition the following geometrical properties:

(a) angle properties of irregular polygons

(b) angle at the centre of a circle is twice the angle at the circumference

(c) angles in the same segment are equal

(d) angles in opposite segments are supplementary; cyclic quadrilaterals. (E3.6)

Angles on a line

The **angles on a straight line** add up to 180°.

$a + b = 180°$

$c + d + e + f = 180°$

Draw an example for yourself (and measure all the angles) to show that the statement is true.

Angles at a point

The sum of the **angles at a point** is 360°. For example:

$a + b + c + d + e = 360°$

Again, check this for yourself by drawing an example and measuring the angles.

Sometimes equations can be used to solve angle problems.

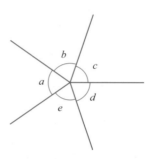

Opposite angles

Opposite angles are equal.

So $a = c$ and $b = d$.

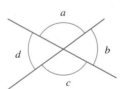

Sometimes opposite angles are called **vertically opposite angles**.

EXERCISE 22A

CORE

1 Calculate the value of *x* in each diagrams.

a

b

c

d

e

f

g

h

> **ADVICE AND TIPS**
>
> Never measure angles in questions like these. Diagrams in textbooks are not usually drawn accurately. Always calculate angles unless you are told to measure them.

2 Write down the value of *x* in each of these diagrams.

a

b

c

3 In the diagram, angle ABD is 45° and angle CBD is 125°.

Decide whether ABC is a straight line.

Write down how you decided.

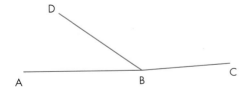

4 Calculate the value of x in each of these diagrams.

a

b

c

5 Calculate the value of x in each of these diagrams.

a

b

c

6 Calculate the value of x first and then calculate the value of y in each of these diagrams.

a

b

c

7 Shalini has a collection of tiles.
They are all equilateral triangles and are all the same size.

> **ADVICE AND TIPS**
>
> All the angles in an equilateral triangle are 60°.

She says that six of the tiles will fit together and leave no gaps.

Explain why Shalini is correct.

8 Work out the value of y in the diagram.

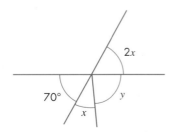

Angles in parallel lines

The arrowheads indicate that the lines are parallel and the line that crosses the parallel lines is called a transversal.

Notice that eight angles are formed.

Angles like these

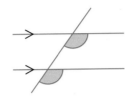

are called **corresponding angles** (Look for the letter F).

Corresponding angles are equal.

Angles like these

are called **alternate angles** (Look for the letter Z).

Alternate angles are equal.

Angles like these

are called **allied angles** or **interior angles** (Look for the letter C).

Allied angles add to 180°.

EXERCISE 22B

1 State the sizes of the lettered angles in each diagram.

a

b

c

d

e

f

CORE

2 State the sizes of the lettered angles in each diagram.

a

b

c

d

e

f

3 State the sizes of the lettered angles in these diagrams.

a

b

4 Calculate the values of x and y in these diagrams.

a

b

c

5 Calculate the values of x and y in these diagrams.

a

b

c

6 A company makes signs like the one below.

It has one line of symmetry.

ADVICE AND TIPS

Draw the line of symmetry on the shape first.

The company needs to know the size of the angle marked x on the diagram.

Work out the size of the angle labelled x.

7 In the diagram, AE is parallel to BD.

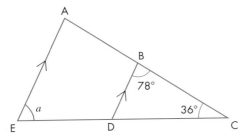

Work out the size of the angle labelled a.

8

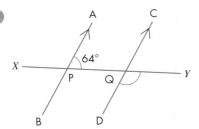

The line XY crosses the parallel lines AB and CD at P and Q.

a Work out the size of angle DQY. Give reasons for your answer.

b This is Vreni's solution.

> *Angle PQD = 64° (corresponding angles)*
> *So angle DQY = 124° (angles on a line = 190°)*

She has made a number of errors in her solution.

Write out the correct solution for the question.

CORE

CORE

9 Use the diagram to prove that the three angles in a triangle add up to 180°.

10 Prove that $p + q + r = 180°$.

22.3 Angles in a triangle

The three angles in any triangle add up to 180°.

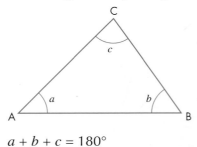

$a + b + c = 180°$

You can prove this by drawing a line through C parallel to AB.

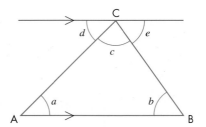

$a = d$ (alternate angles)

$b = e$ (alternate angles)

$d + e + c = 180°$ (angles on a straight line)

So $a + b + c = 180°$

Special triangles

Equilateral triangle

An **equilateral triangle** is a triangle with all its sides equal. Therefore, all three interior angles are 60°.

Isosceles triangle

An **isosceles triangle** is a triangle with two equal sides and, therefore, with two equal interior angles (at the foot of the equal sides).

Notice how to mark the equal sides and equal angles.

Right-angled triangle

A **right-angled triangle** has an interior angle of 90°.
$a + b = 90°$

EXERCISE 22C

1 Find the size of the angle marked with a letter in each of these triangles.

a

b

c

d

e

f

g

h

2 Do any of these sets of angles form the three angles of a triangle? Explain your answer.

   a 35°, 75°, 80°

   b 50°, 60°, 70°

   c 55°, 55°, 60°

   d 60°, 60°, 60°

   e 35°, 35°, 110°

   f 102°, 38°, 30°

CORE

3 In the triangle on the right, all the interior angles are the same.

   a What is the size of each angle?

   b What is the name of a special triangle like this?

   c What is special about the sides of this triangle?

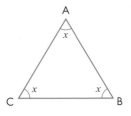

4 In the triangle on the right, two of the angles are the same.

   a Work out the size of the lettered angles.

   b What is the name of a special triangle like this?

   c What is special about the sides AC and AB of this triangle?

5 Find the size of the angle marked with a letter in each of these diagrams.

a

b

c

6 What is the special name for triangle DEF?

Show all your working to explain your answer.

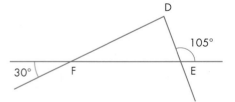

7 The diagram shows three intersecting straight lines.

Work out the values of a, b and c.

Give reasons for your answers.

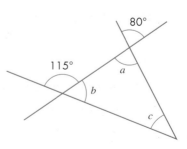

8 By using algebra, show that $x = a + b$.

Angles in a quadrilateral

The four angles in any **quadrilateral** add up to 360°.

$a + b + c + d = 360°$

You can check this by dividing the quadrilateral into two triangles.

The six angles of the triangles are the same as the four angles of the quadrilateral.

Sum of angles of a quadrilateral = 180° + 180° = 360°

EXAMPLE 1

Three angles of a quadrilateral are 125°, 130° and 60°.

Find the size of the fourth angle.

$125 + 130 + 60 + x = 360$

$315 + x = 360$

$x = 360 - 315$

$x = 45$

So the fourth angle is 45°.

Special quadrilaterals

A **parallelogram** has opposite sides that are parallel.

Its opposite sides are equal. Its diagonals bisect each other.
Its opposite angles are equal: that is, $\angle A = \angle C$ and $\angle B = \angle D$

A **rhombus** is a parallelogram with all its sides equal.

Its diagonals bisect each other at right angles.
Its diagonals also bisect the angles at the vertices.

A **kite** is a quadrilateral with two pairs of equal adjacent sides.

Its longer diagonal bisects its shorter diagonal at right angles.
The opposite angles between the sides of different lengths are equal.

A **trapezium** has two parallel sides.

The sum of the interior angles at the ends of each non-parallel side is 180°.

∠A + ∠D = 180° and ∠B + ∠C = 180°

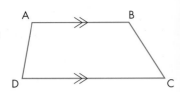

EXAMPLE 2

Find the size of the angles marked *x* and *y* in this parallelogram.

x = 55° (opposite angles are equal) and *y* = 125° (*x* + *y* = 180°)

EXERCISE 22D

1 Find the sizes of the lettered angles in these quadrilaterals.

a

b

c

d

e

f

2 Calculate the values of x and y in each of these quadrilaterals.

a

b

c

3 Find the value of x in each of these quadrilaterals and state what type of quadrilateral it could be. All angles are in degrees.

a A quadrilateral with angles $x + 10$, $x + 20$, $2x + 20$, $2x + 10$

b A quadrilateral with angles $x - 10$, $2x + 10$, $x - 10$, $2x + 10$

c A quadrilateral with angles $x - 10$, $2x$, $5x - 10$, $5x - 10$

d A quadrilateral with angles $4x + 10$, $5x - 10$, $3x + 30$, $2x + 50$

4 The diagram shows a parallelogram ABCD.

Work out the value of x, marked on the diagram.

5 Dani is making a kite and wants angle C to be half of angle A.

Work out the size of angles B and D.

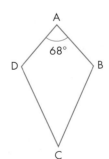

6 This quadrilateral is made from two isosceles triangles. They are both the same size.

Find the value of y in terms of x.

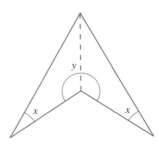

Regular polygons

Regular polygons

Here are five **regular polygons**.

| Square
4 sides | Pentagon
5 sides | Hexagon
6 sides | Octagon
8 sides | Decagon
10 sides |

A **polygon** is regular if all its interior angles are equal and all its sides are the same length.

A **square** is a regular four-sided shape that has an angle sum of 360°.

So, each angle is 360° ÷ 4 = 90°.

The angles of a regular polygon

Lines from the centre of a regular pentagon divide it into five isosceles triangles.

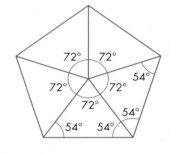

The angle at the centre is 360° so the angle of each isosceles triangle at the centre is:

$$360° ÷ 5 = 72°$$

The other angles in each triangle are identical so each one is:

$$(180 - 72) ÷ 2 = 54°$$

So each interior angle of a regular pentagon is:

$$2 × 54° = 108°$$

There is also an **exterior angle** at each vertex. It is:

$$180 - 108 = 72°$$

Notice that 72° = 360° ÷ 5.

General result

If a regular polygon has n sides each exterior angle is $\dfrac{360°}{n}$.

If you want to find the interior angle of a regular polygon it may be easier to find the exterior angle like this first. Then subtract it from 180° to find the interior angle.

Regular polygon	Number of sides	Exterior angle	Interior angle
Equilateral triangle	3	120°	60°
Square	4	90°	90°
Pentagon	5	72°	108°
Hexagon	6	60°	120°
Octagon	8	45°	135°

EXAMPLE 3

Calculate the size of the exterior and interior angle for a regular 12-sided polygon (a regular dodecagon).

Exterior angle $= \dfrac{360°}{12} = 30°$

interior angle $= 180° - 30° = 150°$

EXERCISE 22E

1 Each diagram shows an interior angle of a regular polygon. For each polygon, answer these questions.

  i What is the size of its exterior angle?

  ii How many sides does it have?

  iii What is the sum of its interior angles?

a 135°

b 160°

c 165°

d 144°

2 Each diagram shows an exterior angle of a regular polygon. For each polygon, answer these questions.

  i What is the size of its interior angle?

  ii How many sides does it have?

  iii What is the sum of its interior angles?

ADVICE AND TIPS

Remember that the angle sum is calculated as (number of sides − 2) × 180°.

a 8°

b 6°

c 24°

d 3°

3 None of these angles can be the interior angle of a regular polygon. Explain why.

a 173°

b 161°

c 169°

d 110°

CORE

4 None of these angles can be the exterior angle of a regular polygon. Explain why.

a 7°   b 26°   c 44°   d 13°

5 Draw a sketch of a regular octagon and join each vertex to the centre.

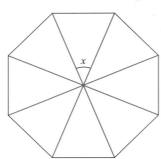

Calculate the value of the angle at the centre (marked *x*).

What connection does this have with the exterior angle?

Is this true for all regular polygons?

6 The diagram shows part of a regular polygon.

Each interior angle is 144°.

a What is the size of each exterior angle of the polygon?

b How many sides does the polygon have?

144°

22.6 Irregular polygons **E**

A polygon is regular if all its sides are the same length and all its angles are the same size. Any other polygon is **irregular**.

Interior angle

Exterior angle

The *exterior* angles of *any* polygon add up to 360°.

Interior angles

You can find the sum of the interior angles of any polygon by splitting it into triangles.

Quadrilateral

Two triangles

Pentagon

Three triangles

Hexagon

Four triangles

Heptagon

Five triangles

Since you already know that the angles in a triangle add up to 180°, you find the sum of the interior angles in a polygon by multiplying the number of triangles in the polygon by 180°, as shown in this table.

Shape	Name	Sum of interior angles
4-sided	**Quadrilateral**	2 × 180° = 360°
5-sided	**Pentagon**	3 × 180° = 540°
6-sided	**Hexagon**	4 × 180° = 720°
7-sided	**Heptagon**	5 × 180° = 900°
8-sided	**Octagon**	6 × 180° = 1080°
9-sided	**Nonagon**	7 × 180° = 1260°
10-sided	**Decagon**	8 × 180° = 1440°

As you can see from the table, for an n-sided polygon, the sum of the interior angles, S, is given by the formula:

$$S = 180(n - 2)°$$

Exterior angles

As in regular polygons the sum of all the exterior angles in an irregular polygon is 360° but their sizes may not be the same.

The size of any specific exterior angle = 180° – the size of its adjacent interior angle.

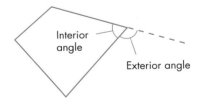

Interior angle

Exterior angle

EXAMPLE 4

Four angles of a pentagon are 100°.

How big is the fifth angle?

The interior angle sum of a pentagon is 3 × 180° = 540°.

Four angles add up to 400°.

The fifth angle must be 540 − 400 = 140°.

EXTENDED

EXERCISE 22F

1 Calculate the sum of the interior angles of polygons with these numbers of sides.

   a 10 sides   **b** 15 sides

   c 100 sides   **d** 45 sides

2 Find the number of sides of polygons with these interior angle sums.

   a 1260°   **b** 2340°

   c 18 000°   **d** 8640°

3 Calculate the size of the lettered angles in each of these polygons.

   a   **b**   **c**

4 Find the value of x in each of these polygons.

   a   **b**   **c**

5 What is the name of the regular polygon in which the interior angles are twice its exterior angles?

6 Wesley measured all the interior angles in a polygon. He added them up to make 991°, but he had missed out one angle.

   a What type of polygon did Wesley measure?

   b What is the size of the missing angle?

7 **a** In the triangle ABC, angle A is 42° and angle B is 67°.

   i Calculate the value of angle C.

   ii What is the value of the exterior angle at C?

   iii What connects the exterior angle at C with the sum of the angles at A and B?

   b Prove that any exterior angle of a triangle is equal to the sum of the two opposite interior angles.

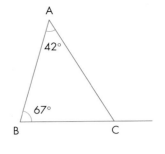

8 Two regular pentagons are placed together.

Work out the value of a.

9 A joiner is making tables so that the shape of each one is half a regular octagon, as shown in the diagram.

He needs to know the size of each angle on the top.

What are the sizes of the angles?

22.7 Tangents and diameters

A **tangent** is a straight line that touches a circle at one point only.

This point is called the **point of contact**.

A tangent to a circle is perpendicular to the **radius** drawn to the point of contact.

The radius OX is perpendicular to the tangent AB.

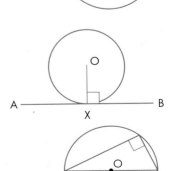

The **diameter** of a circle divides it into two **semi-circles**.

Every angle at the circumference of a semi-circle that is subtended by the diameter of the semi-circle is a right angle.

EXAMPLE 5

O is the centre of a circle. AP and BP are tangents.

Calculate the angle at P.

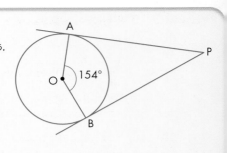

The angles at A and B are right angles.

OAPB is a quadrilateral so the interior angles add up to 360°.

Angle P is 360 − (90 + 90 + 154) = 26°.

EXERCISE 22G

CORE

1 In each diagram, TP and TQ are tangents to a circle with centre O. Find each value of x.

a

b

c

d

2 Each diagram shows a tangent to a circle with centre O. Find the value of x and y in each case.

a

b

c

d

3 In the diagram, O is the centre of the circle and AB is a tangent to the circle at C.

Explain why triangle BCD is isosceles.

Give reasons to justify your answer.

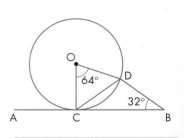

ADVICE AND TIPS

Look for isosceles triangles

Here are two more theorems you need to know about angles in circles.

If you draw lines from each end of an **arc** to the centre of a circle they form an angle at the centre. The arc has **subtended** an angle at the centre.

The angle at the centre of a circle is twice the angle at the **circumference** that is subtended by the same arc.

This diagram shows the angles subtended by arc AB.

$\angle AOB = 2 \times \angle ACB$

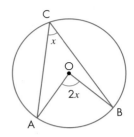

Angles subtended at the circumference in the same **segment** of a circle are equal.

Points C_1, C_2, C_3 and C_4 on the circumference are subtended by the same arc AB.

So $\angle AC_1B = \angle AC_2B = \angle AC_3B = \angle AC_4B$

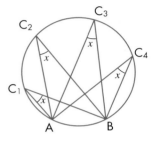

EXAMPLE 6

O is the centre of each circle. Find the sizes of angles marked *a* and *b* in each circle.

i

ii

i $a = 35°$ (angles in same segment)
    $b = 2 \times 35°$ (angle at centre = twice angle at circumference)
      $= 70°$

ii   With OP = OQ, triangle OPQ is isosceles and the sum of the angles in this triangle = 180°

    So $a + (2 \times 25°) = 180°$
              $a = 180° - (2 \times 25°)$
               $= 130°$
              $b = 130° \div 2$ (angle at centre = twice angle at circumference)
               $= 65°$

EXAMPLE 7

O is the centre of the circle. PQR is a straight line.

Find the size of the angle labelled a.

$\angle PQT = 180° - 72° = 108°$ (angles on straight line)

The reflex angle $\angle POT = 2 \times 108°$
(angle at centre = twice angle at circumference)
$$= 216°$$
$a + 216° = 360°$ (sum of angles around a point)
$$a = 360° - 216°$$
$$a = 144°$$

EXERCISE 22H

EXTENDED

1 Find the value of x in each of these circles with centre O.

a   b   c   d

e   f   g   h

2 Find the value of x in each of these circles with centre O.

a   b   c   d

e   f   g   h

3 In the diagram, O is the centre of the circle. Calculate these angles.

   a ∠ADB

   b ∠DBA

   c ∠CAD

4 In the diagram, O is the centre of the circle. Calculate these angles.

   a ∠EDF

   b ∠DEG

   c ∠EGF

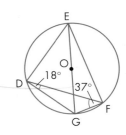

5 In the diagram XY is a diameter of the circle and ∠AZX is a.

Ben says that the value of a is 50°.

Give reasons to explain why he is wrong.

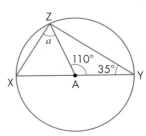

6 Find the values of x and y in each of these circles. O is the centre where shown.

a

b

c

d

e

f

7 In the diagram, O is the centre and AD is the diameter of the circle.

Find the value of x.

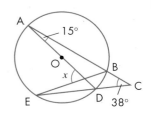

8 In the diagram, O is the centre of the circle and
∠CBD is x.

Show that the reflex ∠AOC is $2x$, giving reasons to
explain your answer.

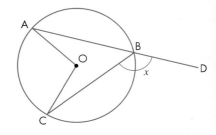

9 A, B, C and D are points on the circumference of
a circle with centre O.
Angle ABO is $x°$ and angle CBO is $y°$.

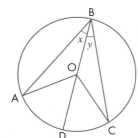

a State the value of angle BAO.

b State the value of angle AOD.

c Prove that the angle subtended by the chord AC at the centre
of a circle is twice the angle subtended at the circumference.

22.9 Cyclic quadrilaterals Ⓔ

There are *two* segments between points P and Q.

    a is the angle in one segment.

    b is the angle in the **opposite segment**.

Angles in opposite segments add up to 180°.
These angles (a and b) are **supplementary**.

    Proof: $a + b = \frac{1}{2}c + \frac{1}{2}d = \frac{1}{2}(c + d) = \frac{1}{2} \times 360 = 180°$

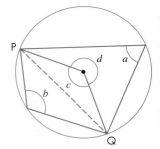

A quadrilateral whose four vertices lie on the circumference of a circle
is called a **cyclic quadrilateral**.

The sum of the opposite angles of a cyclic quadrilateral is 180°.

    $a + c = 180°$ and $b + d = 180°$

EXERCISE 22I

1 Find the sizes of the lettered angles in each of these circles.

a

b

c

d

e

f

g

h

2 Find the values of x and y in each of these circles. Where shown, O marks the centre of the circle.

a

b

c

d

e

f

g

h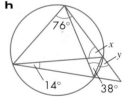

3 Find the values of x and y in each of these circles. Where shown, O marks the centre of the circle.

a

b

c

d

4 Find the values of x and y in each of these circles.

a   **b**   **c**   **d**

5 Find the values of the angles labelled with letters in each of these circles with centre O.

a   **b**   **c**   **d**

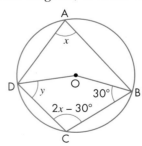

6 The cyclic quadrilateral PQRT has ∠ROQ equal to 38° where O is the centre of the circle. POT is a diameter and parallel to QR. Calculate these angles.

a ROT

b QRT

c QPT

7 In the diagram, O is the centre of the circle.

a Explain why $3x - 30° = 180°$.

b Work out the size of ∠CDO, labelled y on the diagram.

Give reasons in your working.

8 ABCD is a cyclic quadrilateral within a circle centre O and ∠AOC is $2x°$.

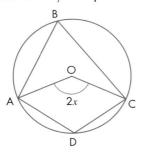

a Write down the size of ∠ABC.

b Write down the size of the reflex angle AOC.

c Prove that the sum of a pair of opposite angles of a cyclic quadrilateral is 180°.

9 In the diagram, ABCE is a parallelogram.

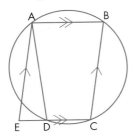

Prove that ∠AED = ∠ADE.

Give reasons in your working.

10 Two circles touch at D.

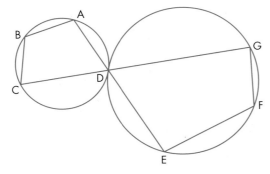

ADE and CDG are straight lines.

Explain why angles ABC and EFG must be equal in size.

Why this chapter matters

Thales of Miletus (624–547 BCE) was a Greek philosopher. Mathematicians believe he was the first person to use similar triangles to find the height of tall objects.

Thales discovered that, at a particular time of day, the height of an object and the length of its shadow were the same. He used this to calculate the height of the Egyptian pyramids.

You can apply the geometry of triangles to calculate heights. You can do this with an instrument called a clinometer. This means you can find the heights of trees, buildings and towers, mountains and other objects which are difficult to measure physically.

Astronomers use the geometry of triangles to measure the distance to nearby stars. They use the Earth's journey in its orbit around the Sun.

They measure the angle of the star twice, from the same point on Earth, but at opposite ends of its orbit. This gives them angle measurements at a known distance apart and from this triangle they can calculate the distance to the star.

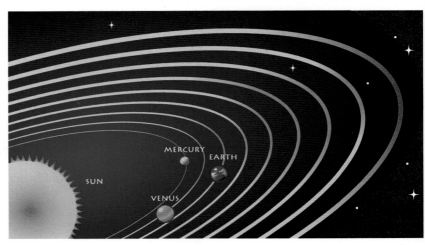

Telescopes and binoculars also use the geometry of triangles.

23 Geometrical terms and relationships

Topics	Level	Key words
1 Measuring and drawing angles	CORE	right angles, acute angles, obtuse angles, reflex angles, perpendicular, protractor
2 Bearings	CORE	bearing, three-figure bearing
3 Nets	CORE	net, cube, cuboid, prism, pyramid, vertex
4 Congruent shapes	CORE	congruent
5 Similar shapes	CORE	similar, enlargement, linear scale factor, corresponding angles, corresponding sides
6 Areas of similar triangles	EXTENDED	area scale factor
7 Areas and volumes of similar shapes	EXTENDED	solid shapes, volume scale factor

In this chapter you will learn how to:

CORE

- Use and interpret the geometrical terms: point, line, parallel, bearing, right angle, acute, obtuse and reflex angles, perpendicular, similarity and congruence. (C3.1 and E3.1)
- Calculate lengths of similar figures. (C3.4 and E3.4)
- Use and interpret vocabulary of triangles, quadrilaterals, circles, polygons and simple solid figures including nets. (C3.1 and E3.1)
- Calculate lengths of similar figures. (C4 and E4)
- Interpret and use three-figure bearings measured clockwise from the North (that is 000°–360°). (C6.1 and E6.1)

EXTENDED

- Use the relationships between areas of similar triangles, with corresponding results for similar figures. (E3.4)
- Extend to volumes and surface areas of similar solids. (E3.4)

A whole turn is divided into 360° or four **right angles** of 90° each.

An **acute angle** is less than one right angle.

An **obtuse angle** is between one and two right angles (90° and 180°).

A **reflex angle** is between two and four right angles (180° and 360°).

Two lines are **perpendicular** if the angle between them is 90°.

When you are using a **protractor**, it is important that you:

● place the centre of the protractor *exactly* on the corner (vertex) of the angle

● lay the baseline of the protractor *exactly* along one side of the angle.

You must follow these two steps to obtain an accurate value for the angle you are measuring.

You should already have discovered how easy it is to measure acute angles and obtuse angles, using the common semicircular protractor.

EXAMPLE 1

Measure the angles ABC, DEF and reflex GHI in the diagrams below.

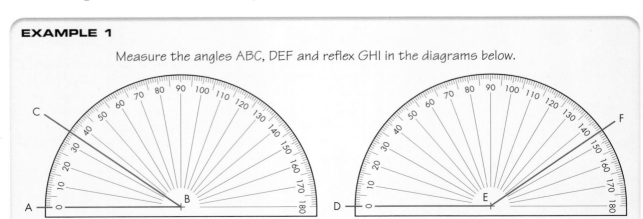

Acute angle ABC is 35° and obtuse angle DEF is 145°.

To measure reflex angles, such as angle GHI, it is easier to use a circular protractor if you have one.

Note the notation for angles.

Angle ABC, or ∠ABC, means the angle at B between the lines AB and BC.

Reflex angle GHI is 305°

EXERCISE 23A

1 Use a protractor to measure the size of each marked angle.

a

b

ADVICE AND TIPS

Check that your answer is sensible. If an angle is reflex, the answer must be over 180°.

c

d

ADVICE AND TIPS

A good way to measure a reflex angle is to measure the acute or obtuse angle and subtract it from 360°.

2 Use a protractor to draw angles of these sizes.

    a 30°   **b** 125°   **c** 90°   **d** 212°   **e** 324°   **f** 19°   **g** 171°

CORE

CORE

3 Find three pairs of perpendicular lines from the
following: AC, AD, AE, BE, CE, CF.

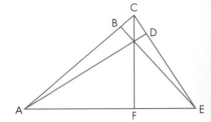

4 It is only safe to climb this ladder if the angle between
the ground and the ladder is between 72° and 78°.

Is it safe for Oliver to climb the ladder?

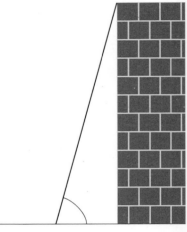

5 An obtuse angle is 10° more than an acute angle.
Write down a possible value for the size of the obtuse angle.

6 Use a ruler and a protractor to draw these triangles accurately. Then measure the
unmarked angle in each one.

a 60° 40° 7 cm

b 40° 6 cm

c 120° 35° 5 cm

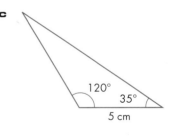

23.2 Bearings

The **bearing** of a point B from a point A is the angle through
which you turn *clockwise* as you change direction from
due north to the direction of B.

For example, in this diagram the bearing of B from A is 060°.

A bearing can have any value from 0° to 360°. It is usual to give all bearings as three figures. This is known as a **three-figure bearing**. So, in the example on the previous page, the bearing is written as 060°, using three figures. Here are three more examples.

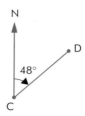

D is on a bearing of 048° from C

F is on a bearing of 110° from E

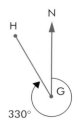

H is on a bearing of 330° from G

There are eight bearings that you should know. They are shown in the diagram.

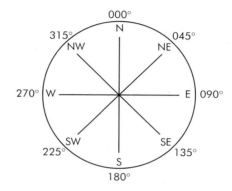

EXAMPLE 2

A, B and C are three towns.

Write down the bearing of B from A and the bearing of C from A.

The bearing of B from A is 070°.

The bearing of C from A is 360° − 115° = 245°.

Scale: 1 cm represents 10 km

EXERCISE 23B

1 Look at this map. By measuring angles, find the bearings of:

a T from D

b D from E

c M from D

d G from A

e M from G

f T from M.

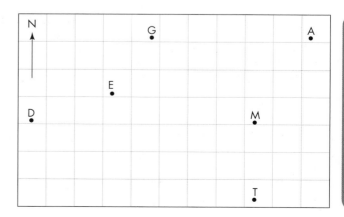

CORE

371

2 Draw sketches to illustrate these situations.

  a C is on a bearing of 170° from H.   **b** B is on a bearing of 310° from W.

3 A is due north from C. B is due east from A. B is on a bearing of 045° from C. Sketch the layout of the three points, A, B and C.

4 The Captain decided to sail his ship around the four sides of a square kilometre.

  a Assuming he started sailing due north, write down the next three bearings he would use in order to complete the square in a clockwise direction.

  b Assuming he started sailing on a bearing of 090°, write down the next three bearings he would use in order to complete the square in an anticlockwise direction.

5 The map shows a boat journey around an island, starting and finishing at S. On the map, 1 centimetre represents 10 kilometres. Measure the distance and bearing of each leg of the journey.

Copy and complete this table.

Leg	Actual distance	Bearing
1		
2		
3		
4		
5		

Scale: 1 cm = 10 km

6 The diagram shows a port P
and two harbours X and Y on
the coast.

a A fishing boat sails to X from P.

What is the three-figure
bearing of X from P?

b A yacht sails to Y from P.

What is the three-figure bearing of Y from P?

7 Draw diagrams to solve these problems.

a The three-figure bearing of A from B is 070°. Work out the three-figure bearing of
B from A.

b The three-figure bearing of P from Q is 145°. Work out the three-figure bearing of
Q from P.

c The three-figure bearing of X from Y is 324°. Work out the three-figure bearing of
Y from X.

8 The diagram shows the position of Kim's house H and the college C.

Scale: 1 cm represents 200 m

a Use the diagram to work out the actual distance from Kim's house to the college.

b Measure and write down the three-figure bearing of the college from Kim's house.

c The supermarket S is 600 m from Kim's house on a bearing of 150°.

Mark the position of S on a copy of the diagram.

9 Chen is flying a plane on a bearing of 072°.

He is told to fly due south towards an airport.

Through what angle does he need to turn?

10 A, B and C are three villages in a bay.

They lie on the vertices of a square.

The bearing of B from A is 030°.

Work out the bearing of A from C.

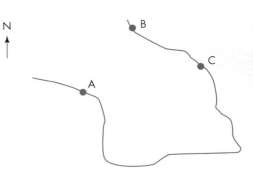

You should know these solid shapes:

A **cube** has six square faces.

A **cuboid** has rectangular faces.

A **prism** has a uniform cross-section.

Triangular prism   Hexagonal prism

A **pyramid** has a polygon-shaped base.
The other faces are triangles and meet
at the **vertex**.

Square-based pyramid   Pentagon-based pyramid

Each of these solid shapes can be made from a **net**.

A net is a flat shape which can be cut out and folded to make a solid shape.

EXAMPLE 3

Sketch the net for:

a cube   **b** square-based pyramid.

a This is a sketch of a net for a cube.

b This is a sketch of a net for a square-based pyramid.

EXERCISE 23C

1 Draw, on squared paper, an accurate net for each of these cuboids.

a 2 cm, 3 cm, 4 cm

b 3 cm, 4 cm, 5 cm

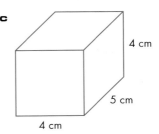

c 4 cm, 5 cm, 4 cm

2 Bashira is making an open box from card.

This is a sketch of the box.

Bashira has a piece of card that measures 15 cm by 21 cm. Can she make the box from this card?

4 cm, 3 cm, 7 cm

3 The shape on the right is a triangular prism. Its ends are isosceles triangles and its other faces are rectangles. Draw an accurate net for this prism. Use squared paper.

4 cm, 5 cm, 3 cm

4 Draw the nets of these shapes.

a

1 cm, 1 cm, 3 cm

Cuboid

b

4 cm, 4 cm, 4 cm, 4 cm, 4 cm

Square-based pyramid

c

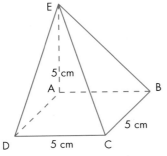

5 cm, 5 cm, 5 cm

Square-based pyramid, with point E directly above point A

d

3 cm, 4 cm, 6 cm

Right-angled triangular prism

CORE

CORE

5 Here is a net for a cube.

How many different nets can you draw for a cube?

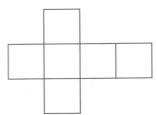

ADVICE AND TIPS

There are 11 altogether. How many can you find?

6 Which of these are nets for a square-based pyramid?

a

b

c

23.4 Congruent shapes

Two-dimensional shapes that are exactly the same size and shape as each other are **congruent**. For example, although they are in different positions, the triangles below are all congruent, because they are all exactly the same size and shape.

Congruent shapes fit exactly on top of each other. So, one way to see whether shapes are congruent is to trace one of them and check that it covers the other shapes exactly. For some of the shapes, you may have to turn your tracing paper over.

EXAMPLE 4

Which of these shapes is not congruent to the others?

a   b   c   d

Trace shape **a** and check whether it fits exactly on top of the others.

You should find that shape **b** is not congruent to the others.

EXERCISE 23D

1 State whether the shapes in each pair, **a** to **f** are congruent or not.

a

b

c

d

e

f

2 Which figure in each group, **a** to **c**, is not congruent to the other two?

a i   **ii**   **iii**

b i   **ii**   **iii**

c i   **ii**   **iii**

3 Draw a square PQRS. Draw in the diagonals PR and QS. Which triangles are congruent to each other?

4 Draw a rectangle EFGH. Draw in the diagonals EG and FH. Which triangles are congruent to each other?

5 Draw a parallelogram ABCD. Draw in the diagonals AC and BD. Which triangles are congruent to each other?

6 Draw an isosceles triangle ABC where AB = AC. Draw the line from A to the midpoint of BC. Which triangles are congruent to each other?

Two shapes are **similar** if one is an **enlargement** of the other.

Corresponding angles of similar shapes are equal.

These shapes are similar. Their corresponding angles are equal.

The ratio of their **corresponding sides** is 2 : 3.

The **linear scale factor** is $\frac{3}{2}$ because the lengths in shape B are $\frac{3}{2}$ of the corresponding lengths in shape A.

EXAMPLE 5

These two shapes are similar.

Find the values of x and y.

Look for two corresponding sides where you know the lengths.

AE corresponds to FJ and the lengths are 12 cm and 15 cm.

The scale factor is $\frac{15}{12} = 1.25$.

To find the value of x (GF) from the length given for BA:

$x = 1.25 \times AB$

$\quad = 1.25 \times 10$

$\quad = 12.5$

$x = 12.5$ cm

To find the value of y (CD) from the length of HI:

$HI = 1.25 \times y$

$9 = 1.25 \times y$

$y = \frac{9}{1.25}$

$\quad = 7.2$

$y = 7.2$ cm

EXERCISE 23E

1 These diagrams are drawn to scale. What is the linear scale factor of the enlargement in each case?

a

b

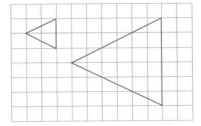

2 Are the shapes in each pair similar? If so, give the scale factor. If not, give a reason.

a

b

3 **a** Explain why these triangles are similar.

b Give the ratio of the sides.

c Which angle corresponds to angle C?

d Which side corresponds to side QP?

4 **a** Explain why these triangles are similar.

b Which angle corresponds to angle A?

c Which side corresponds to side AC?

5 **a** Explain why triangle ABC is similar to triangle AQR.

b Which angle corresponds to the angle at B?

c Which side of triangle AQR corresponds to side AC of triangle ABC? Your answers to question **4** may help you.

6 In the diagrams **a** to **d**, each pair of shapes are similar but not drawn to scale.

a Find the value of *x*.

b Find the values of *x* and *y*.

c Find the values of *x* and *y*.

d Calculate the length of QR.

7 a Explain why all squares are similar.

b Are all rectangles similar? Explain your answer.

8 Sean is standing next to a tree.

His height is 1.6 m and he casts a shadow that has a length of 2.4 m.

The tree casts a shadow that has a length of 7.8 m.

Use what you know about similar triangles to work out the height of the tree, *h*.

If two triangles have the same angles then they are similar.

Triangles ABC and DEF are similar.

Area ≈ 40 cm²

The linear scale factor is $\frac{18}{12} = 1.5$. The **area scale factor** is $1.5^2 = 2.25$.

If the area of triangle ABC is 40 cm² then the area of DEF is $40 \times 2.25 = 90$ cm².

If the linear scale factor is k, then the area scale factor is k^2.

EXERCISE 23F

1 These triangles are similar.

a What is the linear scale factor?

b The area of triangle ABC is 20 cm².

Calculate the area of triangle DEF.

2 These are equilateral triangles.

a Explain why they are similar.

b The area of the larger one is 60 cm² (to 1 significant figure)

Calculate the area of the smaller one.

3 The area of triangle ABC is 7 cm².

Calculate the area of triangle ADE.

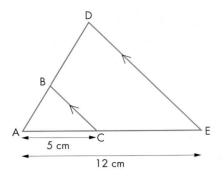

4 The area of triangle ABD is 25 cm².

Calculate the area of the trapezium CBDE.

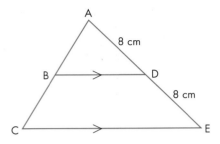

5 P is enlarged by a linear scale factor of 1.2 to make Q.

Q is enlarged by a linear scale factor of 1.2 to make R.

The area of Q is 100 cm².

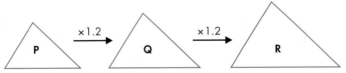

a What is the area of R?

b What is the area of P?

6 The area scale factor also applies to other similar shapes.

A and B are regular pentagons.

a Explain why they are similar.

b Calculate the area of B.

7 A, B and C are equilateral triangles.

B is four times the area of A.

a What is the linear scale factor?

b What is the value of x?

c C is twice the area of A.
Calculate the value of y.

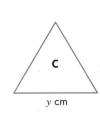

8 These shapes are similar.

The smaller one has an area of 210 cm^2.

Find the area of the larger one.

20 cm

26 cm

9 A photocopier has a setting for enlargements with a scale factor of 1.41%.

What is special about this particular value?

10 All circles are similar.

If a circle with a diameter of 8 cm has an area of 50.3 cm^2, what is the area of a circle with diameter 6 cm?

 23.7 Areas and volumes of similar shapes

You saw that if two shapes are similar and the linear scale factor is k then the area scale factor is k^2.

Two **solid shapes** are similar if corresponding lengths are in the same ratio and corresponding angles are equal. In that case the **volume scale factor** is k^3.

Generally, the relationship between similar shapes can be expressed as:

Length ratio $\quad x : y$ $\qquad$ Area ratio $\quad x^2 : y^2$ $\qquad$ Volume ratio $\quad x^3 : y^3$

EXAMPLE 6

A model yacht is made to a scale of $\frac{1}{20}$ of the size of the real yacht. The area of the sail of the model is 150 cm^2. What is the area of the sail of the real yacht?

At first, it may appear that you do not have enough information to solve this problem, but you can do it like this.

Linear scale factor = 1 : 20

Area scale factor $\quad$ = 1 : 400 (square of the linear scale factor)

Area of real sail $\quad$ = 400 × area of model sail

$\qquad\qquad\qquad$ = 400 × 150 cm^2

$\qquad\qquad\qquad$ = 60 000 cm^2 = 6 m^2

EXAMPLE 7

A bottle has a base radius of 4 cm, a height of 15 cm and a capacity of 650 cm^3.
A similar bottle has a base radius of 3 cm.

a What is the length ratio?

b What is the volume ratio?

c What is the volume of the smaller bottle?

a The length ratio is given by the ratio of the two radii, that is 4 : 3.

b The volume ratio is therefore $4^3 : 3^3 = 64 : 27$.

c Let v be the volume of the smaller bottle. Then the volume ratio is:

$$\frac{\text{volume of smaller bottle}}{\text{volume of larger bottle}} = \frac{v}{650} = \frac{27}{64}$$

$$\Rightarrow v = \frac{27 \times 650}{64} = 274 \text{ cm}^3 \text{ (3 significant figures)}$$

EXERCISE 23G

1 The length ratio between two similar solids is 2 : 5.

a What is the area ratio between the solids?

b What is the volume ratio between the solids?

2 The length ratio between two similar solids is 4 : 7.

a What is the area ratio between the solids?

b What is the volume ratio between the solids?

3 Copy and complete this table.

Linear scale factor	Linear ratio	Linear fraction	Area scale factor	Volume scale factor
2	1 : 2	$\frac{2}{1}$		
3				
$\frac{1}{4}$	4 : 1	$\frac{1}{4}$		$\frac{1}{64}$
			25	
				$\frac{1}{1000}$

4 A shape has an area of 15 cm^2. What is the area of a similar shape with lengths that are three times the corresponding lengths of the first shape?

5 A toy brick has a surface area of 14 cm^2. What would be the surface area of a similar toy brick with lengths that are:

a twice the corresponding lengths of the first brick

b three times the corresponding lengths of the first brick?

6 A rug has an area of 12 m². What area would be covered by rugs with lengths that are:

a twice the corresponding lengths of the first rug

b half the corresponding lengths of the first rug?

7 A brick has a volume of 300 cm³. What would be the volume of a similar brick whose lengths are:

a twice the corresponding lengths of the first brick

b three times the corresponding lengths of the first brick?

8 A tin of paint, 6 cm high, holds a half a litre of paint. How much paint would go into a similar tin which is 12 cm high?

9 A model statue is 10 cm high and has a volume of 100 cm³. The real statue is 2.4 m high. What is the volume of the real statue? Give your answer in m³.

10 A small tin of paint costs $0.75. What is the cost of a larger similar tin with height twice that of the smaller tin? Assume that the cost is based only on the volume of paint in the tin.

11 A small box of width 2 cm has a volume of 10 cm³. What is the width of a similar box with a volume of 80 cm³?

12 A cinema sells popcorn in two different-sized tubs that are similar in shape.

Show that it is true that the big tub is better value.

Popcorn

10 cm

20 cm

Small tub
$0.60

Large tub
$4

Better value if you buy the **Big** tub

13 The diameters of two ball bearings are given below.
Work out:

a the ratio of their radii

b the ratio of their surface areas

c the ratio of their volumes.

6 mm   8 mm

14 Cuboid A is similar to cuboid B.

The length of cuboid A is 10 cm and the length of cuboid B is 5 cm.

The volume of cuboid A is 720 cm³.

Zainab says that the volume of cuboid B must be 360 cm³.

Explain why she is wrong.

A

10 cm

B

5 cm

More complex problems on area and volume ratios

In some problems involving similar shapes, the length ratio is not given, so you need to start with the area ratio or the volume ratio. Then you will need to find the length ratio in order to proceed with the solution.

EXAMPLE 8

A manufacturer makes a range of clown hats that are all similar in shape. The smallest hat is 8 cm tall and uses 180 cm^2 of card. What will be the height of a hat made from 300 cm^2 of card?

The area ratio is 180 : 300

Therefore, the length ratio is $\sqrt{180} : \sqrt{300}$ (Do not calculate these yet.)

Let the height of the larger hat be H, then:

$$\frac{H}{8} = \frac{\sqrt{300}}{\sqrt{180}} = \sqrt{\frac{300}{180}}$$

$$\Rightarrow H = 8 \times \sqrt{\frac{300}{180}} = 10.3 \text{ cm (1 decimal place)}$$

EXAMPLE 9

Two similar tins hold respectively 1.5 litres and 2.5 litres of paint. The area of the label on the smaller tin is 85 cm^2. What is the area of the label on the larger tin?

The volume ratio is 1.5 : 2.5

Therefore, the length ratio is $\sqrt[3]{1.5} : \sqrt[3]{2.5}$ (Do not calculate these yet.)

So the area ratio is $(\sqrt[3]{1.5})^2 : (\sqrt[3]{2.5})^2$

Let the area of the label on the larger tin be A, then:

$$\frac{A}{85} = \frac{(\sqrt[3]{2.5})^2}{(\sqrt[3]{1.5})^2} = \left(\sqrt[3]{\frac{2.5}{1.5}}\right)^2$$

$$\Rightarrow A = 85 \times \left(\sqrt[3]{\frac{2.5}{1.5}}\right)^2 = 119 \text{ cm}^2 \text{ (3 significant figures)}$$

EXERCISE 23H

1 A firm produces three sizes of similar-shaped labels for its products. Their areas are 150 cm^2, 250 cm^2 and 400 cm^2.

The 250 cm^2 label fits around a can of height 8 cm. Find the heights of similar cans around which the other two labels would fit.

EXTENDED

2 A firm makes similar boxes in three different sizes: small, medium and large. These are the areas of their lids.

Small: 30 cm² Medium: 50 cm² Large: 75 cm²

The medium box is 5.5 cm high. Find the heights of the other two sizes.

3 A cone of height 8 cm can be made from a piece of card with an area of 140 cm². What is the height of a similar cone made from a similar piece of card with an area of 200 cm²?

4 It takes 5.6 litres of paint to paint a chimney that is 3 m high. What is the tallest similar chimney that can be painted with 8 litres of paint?

5 A piece of card, 1200 cm² in area, will make a tube 13 cm long. What is the length of a similar tube made from a similar piece of card with an area of 500 cm²?

6 If a television screen of area 220 cm² has a diagonal length of 21 cm, what is the diagonal length of a similar screen of area 350 cm²?

7 There are two similar bronze statues. One has a mass of 300 g, the other has a mass of 2 kg. The height of the smaller statue is 9 cm.

What is the height of the larger statue?

8 These are the sizes of the labels around three similar cans.

Small can: 24 cm² Medium can: 46 cm² Large can: 78 cm²

The medium size can is 6 cm tall with a mass of 380 g. Calculate:

a the heights of the other two sizes

b the masses of the other two sizes.

9 A statue has a mass of 840 kg. A similar statue was made out of the same material but is only two-fifths the height of the first one. What was the mass of the smaller statue?

10 A wooden model stands on a base of area 12 cm². A similar wooden model stands on a base of area 7.5 cm².

Calculate the mass of the smaller model if the larger one has a mass of 3.5 kg.

11 Stefan fills two similar jugs with orange juice.

The first jug holds 1.5 litres of juice and has a base diameter of 8 cm.

The second jug holds 2 litres of juice. Work out the base diameter of the second jug.

12 The total surface areas of two similar cuboids are 500 cm² and 800 cm².

If the width of one of the cuboids is 10 cm, calculate the two possible widths for the other cuboid.

13 The volumes of two similar cylinders are 256 cm³ and 864 cm³.

Which of these is the ratio of their surface areas?

a 2 : 3 **b** 4 : 9 **c** 8 : 27

Why this chapter matters

Engineers, town planners, architects, surveyors, builders and computer designers all need to work with great precision. Some projects involve working with very big lengths or distances; others involve very tiny ones. So how do they manage to work with these difficult measures?

The answer is that they all work with drawings drawn to scale. This allows them to represent lengths they cannot easily measure with standard equipment.

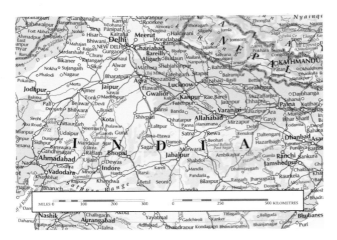

In a scale drawing, one length is used to represent another.

For example, a map cannot be drawn to the same size as the area it represents. The measurements are scaled down, to make a map of a size that can be conveniently used by drivers, tourists and walkers.

Architects use scale drawings to show views of a plannedhouse from different directions.

In computer design, people who design microchips need to scale up their drawings, as the dimensions they work with are so small. Use the internet to research more about how scale drawings are used.

24 Geometrical constructions

Topics	Level	Key words
1 Constructing shapes	CORE	construct, ruler, protractor, compasses, set square
2 Bisectors	CORE	bisect, angle bisector, line bisector, perpendicular bisector
3 Scale drawings	CORE	scale drawing
4 Drawing a locus	CORE	locus, loci, equidistant

In this chapter you will learn how to:

CORE

- Measure lines and angles. (C3.2 and E3.2)
- Construct a triangle given the three sides using ruler and pair of compasses only. (C3.2 and E3.2)
- Construct other simple geometrical figures from given data using protractors and set squares as necessary. (C3.2 and E3.2)
- Construct angle bisectors and perpendicular bisectors using straight edges and pair of compasses only. (C3.2 and E3.2)
- Read and make scale drawings. (C3.3 and E3.3)
- Use the following loci and the method of intersecting loci for sets of points in two dimensions which are:
  - at a given distance from a given point
  - at a given distance from a given straight line
  - equidistant from two given points
  - equidistant from two given intersecting straight lines. (C3.7 and E3.7)

You should be able to construct triangles and other polygons, using a **ruler**, a **protractor**, **compasses** and a **set square**.

When carrying out geometric constructions, always use a sharp pencil to give you thin, clear lines. The most important quality of your construction is its accuracy and requires fine, clean lines. Mark your points as small as you can make them, while ensuring they are clearly visible.

EXAMPLE 1

Construct a triangle with sides that are 5 cm, 4 cm and 6 cm long.

- **Step 1:** Draw the longest side as the base. In this case, the base will be 6 cm, which you draw along a ruler. (The diagrams in this example are drawn at half-size.)

- **Step 2:** Draw the second longest side, in this case the 5 cm side. Open the compasses to a radius of 5 cm (the length of the side), place the point on one end of the 6 cm line and draw a short faint arc, as shown here.

- **Step 3:** Draw the shortest side, in this case the 4 cm side. Open the compasses to a radius of 4 cm, place the point on the other end of the 6 cm line and draw a second short faint arc to intersect the first arc, as shown here.

- **Step 4:** Complete the triangle by joining each end of the base line to the point where the two arcs intersect.

Note: The arcs are construction lines and so you must leave them in, to show how you constructed the triangle.

4 cm   5 cm

6 cm

EXAMPLE 2

Make an accurate drawing of this trapezium.

● **Step 1:** Draw the line AB 8 cm long. Place the protractor on AB with the centre at A and mark 65°.

● **Step 2:** Draw a line from A through the 65° point. From A, using a pair of compasses, measure 6 cm along this line. Label this point D.

● **Step 3:** To draw a line through D parallel to AB, put one edge of a set square on AB and a ruler on the adjacent side of the set square.

● **Step 4:** Hold the ruler still and slide the set square up to D. Draw a line from D along the top of the set square.

● **Step 5:** Use a protractor to draw an angle of 72° at B and mark C where the lines cross.

EXERCISE 24A

CORE

1 Draw each triangle accurately and measure the sides and angles not given in the diagram.

ADVICE AND TIPS

Always make a sketch if one is not given in the question.

a

b

c

d

e

f

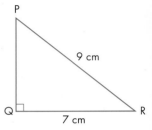

2 **a** Draw a triangle ABC, with AB = 7 cm, BC = 6 cm and AC = 5 cm.

ADVICE AND TIPS

Sketch the triangle first.

b Measure the sizes of ∠ABC, ∠BCA and ∠CAB.

3 Draw an isosceles triangle that has two sides of length 7 cm and the included angle of 50°. Measure the length of the base of the triangle.

4 Make an accurate drawing of this quadrilateral.

5 Make an accurate drawing of this diagram. Use a set square to help you draw the parallel lines.

6 Make an accurate drawing of this trapezium.

7 A triangle ABC has ∠ABC = 30°, AB = 6 cm and AC = 4 cm. There are two different triangles that can be drawn from this information.

What are the two different lengths that BC can be?

8 Construct an equilateral triangle of side length 5 cm. Measure the height of the triangle.

9 Construct a parallelogram with sides of length 5 cm and 8 cm and with an angle of 120° between them. Measure the height of the parallelogram.

10 A rope has 12 equally-spaced knots. It can be laid out to give a triangle, like this.

It will always be a right-angled triangle.

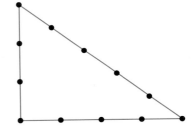

Here are two examples of such ropes.

 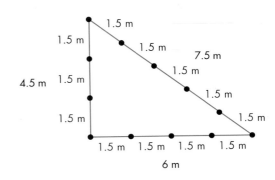

a Show, by constructing both of the above triangles (use a scale of 1 cm : 1 m), that each is a right-angled triangle.

b Choose a different triangle that you think might also be right-angled. Use the same knotted-rope idea to check.

11 Construct the triangle with the largest area that has a total perimeter of 12 cm.

12 Anil says that, as long as he knows all three angles of a triangle, he can draw it. Explain why Anil is wrong.

To **bisect** means to divide in half. So a bisector divides something into two equal parts.

- A **line bisector** divides a straight line into two equal lengths.
- An **angle bisector** is the straight line that divides an angle into two equal angles.

To construct a line bisector

It is usually more accurate to construct a line bisector than to measure its position (the midpoint of the line).

- **Step 1:** Here is a line to bisect.

- **Step 2:** Open your compasses to a radius of about three-quarters of the length of the line. Using each end of the line as a centre, and without changing the radius of your compasses, draw two intersecting arcs.

- **Step 3:** Join the two points where the arcs intersect. This line is the **perpendicular bisector** of the original line.

To construct an angle bisector

It is much more accurate to construct an angle bisector than to measure its position.

- **Step 1:** Here is an angle to bisect.

- **Step 2:** Open your compasses to any reasonable radius that is less than the length of the lines forming the angle. If in doubt, go for about 3 cm. With the vertex of the angle as centre, draw an arc through both lines.

- **Step 3:** With centres at the two points where this arc intersects the lines, draw two more arcs so that they intersect.

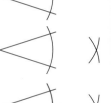

- **Step 4:** Join the point where these two arcs intersect to the vertex of the angle. This line is the angle bisector.

To construct a perpendicular from a point on a line (an angle of 90°)

This construction will produce a perpendicular from a point A on a line.

- Open your compasses to about 2 or 3 cm. With point A as centre, draw two short arcs to intersect the line at each side of the point.

- Now extend the radius of your compasses to about 4 cm. With centres at the two points at which the arcs intersect the line, draw two arcs to intersect at X above the line.

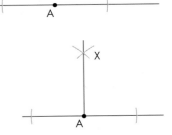

- Join AX.

  AX is perpendicular to the line.

Note: If you needed to construct a 90° angle at the end of a line, you would first have to extend the line.

You could be even more accurate by also drawing two arcs underneath the line, which would give three points in line.

To construct a perpendicular from a point to a line

This construction will produce a perpendicular from a point A to a line.

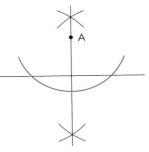

- With point A as centre, draw an arc that intersects the line at two points.

- With centres at these two points of intersection, draw two arcs to intersect each other both above and below the line.

- Join the two points where the arcs intersect. The resulting line passes through point A and is perpendicular to the line.

Note: When you are asked to *construct* a bisector of any type you must *only* use compasses, not a protractor.

1 Draw a line 7 cm long and bisect it. Check your accuracy by seeing if each half is 3.5 cm.

2 Draw a circle of about 4 cm radius.

Draw a triangle inside the circle so that the corners of the triangle touch the circle.

Bisect each side of the triangle.

The bisectors should all meet at the same point, which should be the centre of the circle.

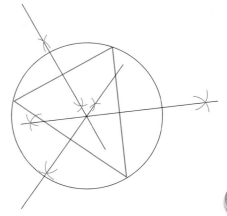

CORE

3 **a** Draw any triangle with sides that are between 5 cm and 10 cm long.

b On each side construct the line bisector.

All your line bisectors should intersect at the same point.

c Using this point as the centre, draw a circle that goes through every vertex of the triangle.

4 Repeat question **3** with a different triangle and check that you get a similar result.

5 **a** Draw this quadrilateral accurately.

b Construct the line bisector of each side. These all should intersect at the same point.

c Use this point as the centre of a circle that goes through each vertex of the quadrilateral. Draw this circle.

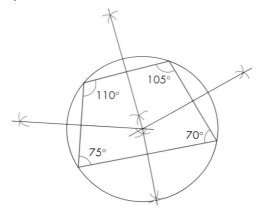

6 **a** Draw an angle of 50°.

b Construct the angle bisector.

c Check how accurate you have been by measuring each half. Both should be 25°.

7 Draw a circle with a radius of about 3 cm.

Draw a triangle so that the sides of the triangle are tangents to the circle.

Bisect each angle of the triangle.

The bisectors should all meet at the same point, which should be the centre of the circle.

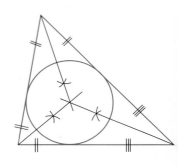

8 a Draw any triangle with sides that are between 5 cm and 10 cm long.

b At each angle construct the angle bisector.
All three bisectors should intersect at the same point.

c Use this point as the centre of a circle that just touches the sides of the triangle.

9 Repeat question **8** with a different triangle.

10 Draw a circle with radius about 4 cm.

Draw a quadrilateral, *not* a rectangle, inside the circle so that each vertex is on the circumference.

Construct the bisector of each side of the quadrilateral.

Where is the point where these bisectors all meet?

24.3 Scale drawings

A **scale drawing** is an accurate representation of a real object.

Scale drawings are usually smaller in size than the original objects. However, in certain cases, they are enlargements. Examples of these are drawings of miniature electronic circuits and very small watch movements.

You will generally be given the scale being used, for example, '1 cm represents 20 m'.

EXAMPLE 3

The diagram shows the front of a kennel.
1 cm on the diagram represents a measurement of 30 cm. Find:

a the actual width of the front

b the actual height of the doorway.

a The actual width of the front is:
4 cm × 30 = 120 cm

b The actual height of the doorway is:
1.5 cm × 30 = 45 cm

CORE

EXERCISE 24C

1 Look at this plan of a garden.

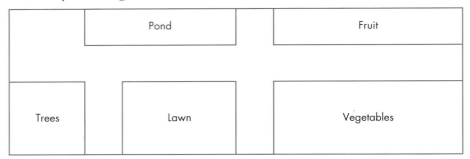

Scale: 1 cm represents 10 m

a State the actual dimensions of each plot of the garden.

b Calculate the actual area of each plot.

2 Below is a plan for a computer mouse mat.

Scale: 1 cm represents 6 cm

ADVICE AND TIPS

Remember to check the scale.

a How long is the actual mouse mat?

b How wide is the narrowest part of the mouse mat?

3 Below is a scale plan of the top of Ahmed's desk, in which 1 cm represents 10 cm.

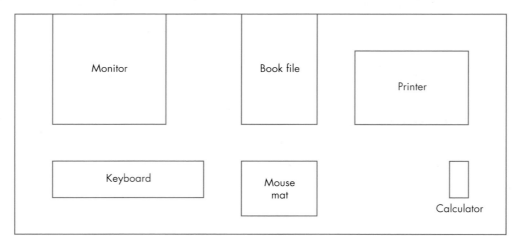

What are the actual dimensions of each of these objects?

a monitor   b keyboard   c mouse mat

d book file   e printer   f calculator

4 The diagram shows a sketch of a garden.

a Make an accurate scale drawing of the garden.

Use a scale of 1 cm to represent 2 m.

b Marie wants to plant flowers along the side marked *x* on the diagram. The flowers need to be planted 0.5 m apart. Use your scale drawing to work out how many plants she needs.

5 Look at the map below, which is drawn to a scale of 1 cm representing 2 km. Towns are shown with letters.

State these actual distances, correct to the nearest tenth of a kilometre.

a A to B   **b** B to C

c C to D   **d** D to E

e E to B   **f** B to D

6 This sketch shows the outline of a car park.

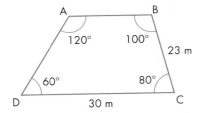

a Make a scale drawing in which 1 cm represents 5 m.

b What is the length of AB, in metres?

CORE

7 This map is drawn to a scale of 1 cm to 200 km.

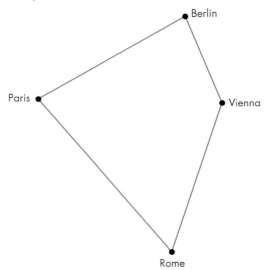

Find the distances between:

a Paris and Berlin

b Paris and Rome

c Rome and Vienna.

8 This is a scale drawing of the Great Beijing Wheel in China.

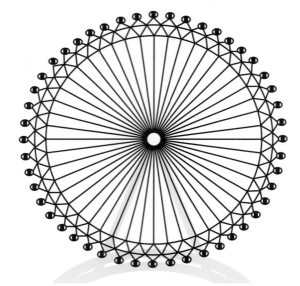

The height of the wheel is 210 m.

Which of these is the correct scale?

a 1 cm represents 30 cm

b 1 cm represents 7 m

c 1 cm represents 30 m

d 1 cm represents 300 m

A **locus** (plural **loci**) is the movement of a point according to a given rule.

EXAMPLE 4

A point P that moves so that it is always 5 cm from a fixed point A will have a locus that is a circle of radius 5 cm. Draw its path. Use a pair of compasses.

You can express this mathematically by saying the locus of the point P is such that AP = 5 cm.

EXAMPLE 5

Draw the locus of a point P that moves so that it is always the same distance from two fixed points A and B.

The locus is the perpendicular bisector of the line joining A and B.

You should be able to construct this with a pair of compasses and a straight edge.

You can express this mathematically by saying

the locus of the point P is such that AP = BP.

A point that is always the same distance from two points is **equidistant** from the two points.

EXAMPLE 6

Draw the locus of a point that is always 5 m from a long, straight wall.

The locus is a line parallel to the wall and 5 m from it.

5 m

EXAMPLE 7

Draw the locus of a point that moves so that it is always 5 cm from a line AB.

The locus is a racetrack shape around the line.

EXAMPLE 8

Draw the locus of points that are the same distance from line AB as from line AC.

The locus is the bisector of angle BAC.

You should be able to construct the bisector with compasses and a straight edge only.

CORE

EXERCISE 24D

1 A is a fixed point. Draw accurately the locus of the point P in each of these situations.

  a AP = 2 cm   **b** AP = 4 cm   **c** AP = 5 cm

2 A and B are two fixed points 5 cm apart.

  a Make an accurate drawing of A and B.

  b Using a straight edge and a pair of compasses only, show the points that are equidistant from A and B.

  c Draw the locus of points that are 4 cm from B.

  d Shade the region where points are closer to A than they are to B *and* less than 4 cm away from B.

3 **a** A horse is tethered in a field on a rope 4 m long. Make an accurate scale drawing of the area the horse can graze. Use 1 cm to represent 1 m.

  b The horse is still tethered by the same rope but there is now a long, straight fence running 2 m from the stake. Show the area that the horse can now graze.

4 ABCD is a square of side 4 cm. In each of these loci, the point P moves only inside the square. Draw the locus in each case.

  a AP = BP   **b** AP < BP   **c** AP = CP

  d CP < 4 cm   **e** CP > 2 cm   **f** CP > 5 cm

5 Draw triangle ABC.

The region R is defined as the set of points inside the triangle such that:

- they are closer to the line AB than the line AC
- they are closer to the point A than the point C.

Using a straight edge and compasses only, construct the region R.

6 ABCD is a rectangle.

Copy the diagram and draw the locus of all points outside the rectangle that are 2 cm from the edges of the rectangle.

7 **a** The diagram shows a park.

Make an accurate scale drawing, using 1 cm to represent 50 m.

b A path is the same distance from BC as from CD.

Use compasses and a straight edge only to show the path on your drawing.

c A second path is equidistant from B and C.

Use compasses and a straight edge only to construct this path.

d Shade the area of the park that is closer to BC than to CD *and* further from B than from C.

8 A broadcasting mast is 50 m away from a boundary fence.

On a scale drawing, where 1 cm represents 10 m, show the points that are 20 metres away from the fence and 40 metres away from the mast.

9 A square card ABCD of side 4 cm is placed against a ruler, as shown.

The card is rotated to the right about corner D until corner C touches the ruler. Then it is rotated to the right about corner C until corner B touches the ruler.

Make an accurate drawing to show the locus of corner A and the locus of corner B.

Why this chapter matters

How can you find the height of a mountain?

How do you draw an accurate map?

How can computers take an image and make it rotate so that you can view it from different directions?

How do Global Positioning Systems (GPS) work?

How can music be produced electronically?

The answer is by using the angles and sides of triangles and the connections between them. This important branch of mathematics is called trigonometry and is used in science, engineering, electronics and everyday life. This chapter gives a brief introduction to trigonometry.

The first major book of trigonometry was a written by an astronomer called Ptolemy, who lived in Alexandria, Egypt, over 1800 years ago.

It has tables of numbers, called 'trigonometric ratios', used in making calculations about the positions of stars and planets.

Trigonometry also helped Ptolemy to make a map of the world he knew. Today you no longer need to look up tables of values of trigonometric ratios because they are programmed into calculators and computers.

In the 19th century the French mathematician Jean Fourier showed how all musical sounds can be broken down into a combination of tones that can be described by trigonometry. His work makes it possible to imitate the sound of any instrument electronically.

25 Trigonometry

Topics		Level	Key words
1	Pythagoras' theorem	CORE	hypotenuse, Pythagoras' theorem
2	Trigonometric ratios	CORE	ratio, sine, cosine, tangent, opposite side, adjacent side
3	Calculating angles	CORE	inverse
4	Using sine, cosine and tangent functions	CORE	
5	Which ratio to use	CORE	
6	Angles of elevation and depression	EXTENDED	angle of elevation, angle of depression
7	Problems in three dimensions	EXTENDED	
8	Sine and cosine of obtuse angles	EXTENDED	obtuse angle
9	The sine rule and the cosine rule	EXTENDED	sine rule, cosine rule, included angle
10	Using sine to find the area of a triangle	EXTENDED	area sine rule

In this chapter you will learn how to:

CORE	EXTENDED
• Apply Pythagoras' theorem and the sine, cosine and tangent ratios for acute angles to the calculation of a side or of an angle of a right-angled triangle. (C6.2 and E6.2)	• Solve trigonometrical problems in two dimensions involving angles of elevation and depression. (E6.2) • Extend sine and cosine values to angles between 90° and 180°. (E6.2) • Solve problems using the sine and cosine rules for any triangle and the formula $\text{area of triangle} = \frac{1}{2}\,ab\sin C$. (E6.3) • Solve simple trigonometrical problems in three dimensions including angle between a line and a plane. (E6.4)

Pythagoras, who was a philosopher as well as a mathematician, was born in 580BCE in Greece. He later moved to Italy, where he established the Pythagorean Brotherhood, which was a secret society devoted to politics, mathematics and astronomy.

This is his famous theorem.

Consider squares being drawn on each side of a right-angled triangle, with sides 3 cm, 4 cm and 5 cm.

The longest side is called the **hypotenuse** and is always opposite the right angle.

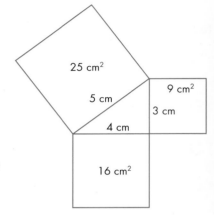

Pythagoras' theorem can then be stated as:

For any right-angled triangle, the area of the square drawn on the hypotenuse is equal to the sum of the areas of the squares drawn on the other two sides.

The usual description is:

In any right-angled triangle, the square of the hypotenuse is equal to the sum of the squares of the other two sides.

Pythagoras' theorem is more usually written as a formula:

$$c^2 = a^2 + b^2$$

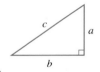

Remember that Pythagoras' theorem can only be used in right-angled triangles.

Finding the hypotenuse

EXAMPLE 1

Find the length of the hypotenuse, marked x on the diagram.

Using Pythagoras' theorem gives:
$$x^2 = 8^2 + 5.2^2 \text{ cm}^2$$
$$= 64 + 27.04 \text{ cm}^2$$
$$= 91.04 \text{ cm}^2$$

So $x = \sqrt{91.04} = 9.5$ cm (1 decimal place)

EXERCISE 25A

For each of the triangles in questions **1** to **9**, calculate the length of the hypotenuse, *x*, giving your answers to 1 decimal place.

1

9 cm
x
5 cm

2
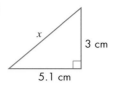
x
3 cm
5.1 cm

ADVICE AND TIPS

In these examples you are finding the hypotenuse. You need to add the squares of the two short sides in every case.

CORE

3

4.8 cm
7 cm
x

4

16 cm
x
13 cm

5

x
11 cm
15 cm

6

9 cm
15 cm
x

7

4 cm
3 cm
x

8

12 cm
x
5 cm

9

8 cm
x
6 cm

10 How does this diagram show that Pythagoras' theorem is true?

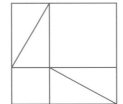

By rearranging the formula for Pythagoras' theorem, you can calculate the
length of one of the shorter sides.

$$c^2 = a^2 + b^2$$

So, $a^2 = c^2 - b^2$ or $b^2 = c^2 - a^2$

EXAMPLE 2

Find the length of the side labelled x.

x is one of the shorter sides.

So using Pythagoras' theorem gives:

$$x^2 = 15^2 - 11^2 \text{ cm}^2$$
$$= 225 - 121 \text{ cm}^2$$
$$= 104 \text{ cm}^2$$

So $x = \sqrt{104} = 10.2$ cm (1 decimal place)

EXERCISE 25B

1 For each of these triangles, calculate the length of the side
labelled x, giving your answers to 1 decimal place.

a

b

c

d

ADVICE AND TIPS

In these examples you are
finding a short side. You
need to subtract the
square of the other short
side from the square of the
hypotenuse in every case.

2 For each of these triangles, calculate the length labelled x, giving your answers to 1 decimal place.

a
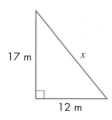
17 m
x
12 m

b

19 cm
11 cm
x

c

17 m
x
23 m

d

9 cm
x
8.5 cm

3 For each of these triangles, calculate the length marked x.

a

x
12 m
13 m

b

8 m
x
10 m

c

5 m
x
4 m

d

30 cm
x
40 cm

4 In question **3** you found sets of three numbers which satisfy $a^2 + b^2 = c^2$.

Can you find any more?

5 Calculate the value of x.

48.6 cm
23.4 cm
x cm

In trigonometry you will use three important **ratios** to calculate sides and angles: **sine**, **cosine** and **tangent**. These ratios are defined in terms of the sides of a right-angled triangle and an angle. The angle is often written as θ.

In a right-angled triangle:

- the side opposite the right angle is called the hypotenuse and is the longest side
- the side opposite the angle θ is called the **opposite side**
- the other side next to both the right angle and the angle θ is called the **adjacent side**.

The sine, cosine and tangent ratios for θ are defined as:

$$\text{sine } \theta = \frac{\text{Opposite}}{\text{Hypotenuse}} \qquad \text{cosine } \theta = \frac{\text{Adjacent}}{\text{Hypotenuse}} \qquad \text{tangent } \theta = \frac{\text{Opposite}}{\text{Adjacent}}$$

These ratios are usually abbreviated as:

$$\sin \theta = \frac{O}{H} \qquad \cos \theta = \frac{A}{H} \qquad \tan \theta = \frac{O}{A}$$

These abbreviated forms are also used on calculator keys.

Using your calculator

You will need to use a calculator to find trigonometric ratios.

Different calculators work in different ways, so make sure you know how to use your model.

Angles are not always measured in degrees. Sometimes radians or grads are used instead. You do not need to learn about those in your IGCSE course. Calculators can be set to operate in any of these three units, so make sure your calculator is operating in degrees.

Use your calculator to find the sine of 60 degrees.

You will probably press the keys $\boxed{\text{sin}}$ $\boxed{6}$ $\boxed{0}$ $\boxed{=}$ in that order, but it might be different on your calculator.

The answer should be 0.8660…

$3 \cos 57°$ is a short way of writing $3 \times \cos 57°$.

On most calculators you do not need to use the × button and you can just press the keys in the way it is written: $\boxed{3}$ $\boxed{\text{cos}}$ $\boxed{5}$ $\boxed{7}$ $\boxed{=}$

Check to see whether your calculator works this way.

The answer should be 1.63.

EXAMPLE 3

Find the value of 5.6 sin 30°.

This means 5.6 × sine of 30 degrees.

Remember that you may not need to press the × button.

5.6 sin 30° = 2.8

EXERCISE 25C

1 Find these values, rounding your answers to 3 significant figures.

   a sin 43°   **b** sin 56°   **c** sin 67.2°   **d** sin 90°

2 Find these values, rounding your answers to 3 significant figures.

   a cos 43°   **b** cos 56°   **c** cos 67.2°   **d** cos 90°

3 **a** **i** What is sin 35°?   **ii** What is cos 55°?

   b **i** What is sin 12°?   **ii** What is cos 78°?

   c **i** What is cos 67°?   **ii** What is sin 23°?

   d What connects the values in parts **a**, **b** and **c**?

   e Copy and complete these sentences.

     i sin 15° is the same as cos …

     ii cos 82° is the same as sin …

     iii sin x is the same as cos …

4 Use your calculator to work out the value of each ratio.

   a tan 43°   **b** tan 56°   **c** tan 67.2°   **d** tan 90°

   e tan 45°   **f** tan 20°   **g** tan 22°   **h** tan 0°

5 What is so different about tan compared with both sin and cos?

6 Use your calculator to work out the value of each ratio.

   a 4 sin 63°   **b** 7 tan 52°   **c** 5 tan 80°   **d** 9 cos 8°

7 Use your calculator to work out the values of the these ratios.

   a $\dfrac{5}{\sin 63°}$ **b** $\dfrac{6}{\cos 32°}$ **c** $\dfrac{3}{\tan 64°}$ **d** $\dfrac{7}{\tan 42°}$

CORE

CORE

8 Calculate sin x, cos x, and tan x for each triangle. Leave your answers as fractions.

a

3

x

5

4

b

x

12

13

5

c

25

7

x

24

25.3 **Calculating angles**

What angle has a cosine of 0.6? You can use a calculator to find out.

'The angle with a cosine of 0.6' is written as $\cos^{-1}$ 0.6 and is called the '**inverse** cosine of 0.6'.

Find out where $\cos^{-1}$ is on your calculator.

You will probably find it on the same key as cos, but you will need to press SHIFT or INV or 2ndF first.

Look to see if $\cos^{-1}$ is written above the cos key.

Check that $\cos^{-1}$ 0.6 = 53.1301… = 53.1° (1 decimal place)

Check that cos 53.1° = 0.600 (3 decimal places)

Check that you can find the inverse sine and the inverse tangent in the same way.

EXAMPLE 4

What angle has a sine of $\frac{3}{8}$?

You need to find $\sin^{-1} \frac{3}{8}$.

You could use the fraction button on your calculator or you could calculate $\sin^{-1} (3 \div 8)$.

If you use the fraction key you may not need brackets, or your calculator may put them in automatically.

Try to do it in both of these ways and then use whichever method you prefer.

The answer should be 22.0°.

EXERCISE 25D

Use your calculator to find the answers to these questions. Give your answers to 1 decimal place.

1 What angles have these sines?

    a 0.5   **b** 0.785   **c** 0.64   **d** 0.877   **e** 0.999   **f** 0.707

CORE

2 What angles have these cosines?

    a 0.5   **b** 0.64   **c** 0.999   **d** 0.707   **e** 0.2   **f** 0.7

3 What angles have these tangents?

    a 0.6   **b** 0.38   **c** 0.895   **d** 1.05   **e** 2.67   **f** 4.38

4 What happens when you try to find the angle with a sine of 1.2? What is the largest value of sine you can put into your calculator without getting an error when you ask for the inverse sine? What is the smallest?

5 **a** **i** What angle has a sine of 0.3? (Keep the answer in your calculator memory.)

    ii What angle has a cosine of 0.3?

    iii Add the two accurate answers of parts **i** and **ii** together.

  b Will you always get the same answer to this question, whatever number you start with?

25.4 Using sine, cosine and tangent functions

Sine function

Remember sine $\theta = \dfrac{\text{Opposite}}{\text{Hypotenuse}}$

You can use the **sine** ratio to calculate the lengths of sides and angles in right-angled triangles.

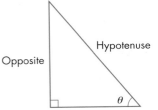

EXAMPLE 5

Find the size of angle θ, given that the opposite side is 7 cm and the hypotenuse is 10 cm.

Draw a diagram. (This is an essential step.)

From the information given, use sine.

$$\sin \theta = \frac{O}{H} = \frac{7}{10} = 0.7$$

What angle has a sine of 0.7?

To find out, use the inverse sine function on your calculator.

$$\sin^{-1} 0.7 = 44.4° \text{ (1 decimal place)}$$

EXAMPLE 6

Find the length of the side marked a in this triangle.

Side a is the opposite side, with 12 cm as the hypotenuse, so use sine.

$$\sin \theta = \frac{O}{H}$$

$$\sin 35° = \frac{a}{12}$$

So $a = 12 \sin 35° = 6.88$ cm (3 significant figures)

EXAMPLE 7

Find the length of the hypotenuse, h, in this triangle.

Note that although the angle is in the other corner, the opposite side is again given. So use sine.

$$\sin \theta = \frac{O}{H}$$

$$\sin 52° = \frac{8}{h}$$

So $h = \dfrac{8}{\sin 52°} = 10.2$ cm (3 significant figures)

EXERCISE 25E

1 Find the size of the angle marked x in each of these triangles.

a

b

c

2 Find the length of the side marked x in each of these triangles.

a

b

c

CORE

CORE

3 Find the length of the side marked x in each of these triangles.

a

b

c

4 Find the length of the side marked x in each of these triangles.

a

b

c

d

Cosine function

Remember cosine $\theta = \dfrac{\text{Adjacent}}{\text{Hypotenuse}}$

You can use the **cosine** ratio to calculate the lengths of sides and angles in right-angled triangles.

EXAMPLE 8

Find the size of angle θ, given that the adjacent side is 5 cm and the hypotenuse is 12 cm.

Draw a diagram. (This is an essential step.)

From the information given, use cosine.

$$\cos \theta = \frac{A}{H} = \frac{5}{12}$$

What angle has a cosine of $\frac{5}{12}$?

To find out, use the inverse cosine function on your calculator.

$\cos^{-1} = 65.4°$ (1 decimal place)

EXAMPLE 9

Find the length of the hypotenuse, h, in this triangle.

The adjacent side is given. So use cosine.

$$\cos \theta \quad = \frac{A}{H}$$

$$\cos 40° \quad = \frac{20}{h}$$

So $h = \dfrac{20}{\cos 40°} = 26.1$ cm (3 significant figures)

415

EXERCISE 25F

CORE

1 Find the size of the angle marked x in each of these triangles.

a

8 cm

5 cm

x

b

4 cm

x

1 cm

c

100 cm

x

160 cm

2 Find the length of the side marked x in each of these triangles.

a

8 cm

48°

x

b

36°

x

12 cm

c

11 cm

24°

x

d

52°

14 cm

x

3 Find the value of x in each of these triangles.

a

10 cm

56°

x

b

16 cm

x

11 cm

c

6.5 cm

49°

x

d

x

17 cm

13 cm

Tangent function

Remember tangent $\theta = \dfrac{\text{Opposite}}{\text{Adjacent}}$

You can use the **tangent** ratio to calculate the lengths of sides and angles in right-angled triangles.

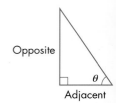

Opposite

θ

Adjacent

EXAMPLE 10

Find the length of the side marked x in this triangle.

The side marked x is the opposite side, with 9 cm as the adjacent side, so use tangent.

$\tan \theta \quad = \dfrac{O}{A}$

$\tan 62° \quad = \dfrac{x}{9}$

So $x = 9 \tan 62° = 16.9$ cm (3 significant figures)

62°

9 cm

x

EXERCISE 25G

1 Find the size of the angle marked x in each of these triangles.

a

6 cm

9 cm

b

x

20 cm

15 cm

c

35 cm

45 cm

x

2 Find the length of the side marked x in each of these triangles.

a

x

61°

5 cm

b

7 cm

43°

x

c

33°

11 cm

x

d

6 cm

34°

x

3 Find the value of x in each of these triangles.

a

x

63°

7 cm

b

9 cm

x

8 cm

c

x

52°

9 cm

d

x

4 cm

3.5 cm

Which ratio to use

The difficulty with any trigonometric problem is knowing which ratio to use to solve it.

These examples show you how to decide which ratio you need in any given situation.

EXAMPLE 11

Find the length of the side marked x in this triangle.

Step 1 Identify what information is given and what needs to be found. Namely, x is opposite the angle and 16 cm is the hypotenuse.

Step 2 Decide which ratio to use. Only one ratio uses opposite and hypotenuse: **sine**.

Step 3 Remember $\sin \theta = \dfrac{O}{H}$

Step 4 Put in the numbers and letters: $\sin 37° = \dfrac{x}{16}$

Step 5 Rearrange the equation and work out the answer:
$x = 16 \sin 37° = 9.629\,040\,371$ cm

Step 6 Give the answer to an appropriate degree of accuracy:
$x = 9.63$ cm (3 significant figures)

There is no need to write down every step as in Example 11. Step 1 can be done by marking the triangle. Steps 2 and 3 can be done in your head. Steps 4 to 6 are what you write down.

Remember that you must always show evidence of your working. Any reasonable attempt at identifying the sides and using a ratio will probably show that you undertand the method, but only if the fraction is the right way round.

The next examples are set out in a way that requires the *minimum* amount of working but gets *maximum* results.

EXAMPLE 12

Find the length of the side marked x in this triangle.

Mark on the triangle the side you know (H) and the side you want to find (A).

Recognise it is a **cosine** problem because you have A and H.

So $\cos 50° = \dfrac{x}{7}$

$x = 7 \cos 50° = 4.50$ cm (3 significant figures)

EXAMPLE 13

Find the size of the angle marked x in this triangle.

Mark on the triangle the sides you know.

Recognise it is a **tangent** problem because you have O and A.

So $\tan x = \dfrac{12}{7}$

$x = \tan^{-1} \dfrac{12}{7} = 59.7°$ (1 decimal place)

EXERCISE 25H

1 Find the length marked x in each of these triangles.

a

b

c

d

e

f

2 Find the size of the angle marked x in each of these triangles.

a

b

c

d

e

f
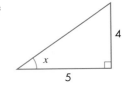

CORE

CORE

3 Find the value of the angle or length marked x in each of these triangles.

a

b

c

d

e

f

g

h

i

j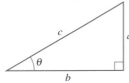

4 **a** How does this diagram show that $\tan \theta = \dfrac{\sin \theta}{\cos \theta}$?

b How does the diagram show that $(\sin \theta)^2 + (\cos \theta)^2 = 1$?

c Choose a value for θ and check the two results in parts **a** and **b** are true.

When you look *up* at an aircraft in the sky, the angle through which your line of sight turns, from looking straight ahead (the horizontal), is called the **angle of elevation**.

When you are standing on a high point and look *down* at a boat, the angle through which your line of sight turns, from looking straight ahead (the horizontal), is called the **angle of depression**.

EXAMPLE 14

From the top of a vertical cliff, 100 m high, Ali sees a boat out at sea. The angle of depression from Ali to the boat is 42°. How far from the base of the cliff is the boat?

The diagram of the situation is shown in figure **i**.

From this, you get the triangle shown in figure **ii**.

From figure **ii**, you see that this is a tangent problem.

So $\tan 42° = \dfrac{100}{x}$

$x = \dfrac{100}{\tan 42°} = 111$ m (3 significant figures)

EXERCISE 25I

In these questions, give any answers involving angles to the nearest degree.

1 Erik sees an aircraft in the sky. The aircraft is at a horizontal distance of 25 km from Erik. The angle of elevation is 22°. How high is the aircraft?

2 An aircraft is flying at an altitude of 4000 m and is 10 km from the airport. If a passenger can see the airport, what is the angle of depression?

EXTENDED

3 A man standing 200 m from the base of a television transmitter looks at the top of it and notices that the angle of elevation of the top is 65°. How high is the tower?

4 **a** From the top of a vertical cliff, 200 m high, a boat has an angle of depression of 52°. How far from the base of the cliff is the boat?

b The boat now sails away from the cliff so that the distance is doubled. Does that mean that the angle of depression is halved? Give a reason for your answer.

5 From a boat, the angle of elevation of the foot of a lighthouse on the edge of a cliff is 34°.

a If the cliff is 150 m high, how far from the base of the cliff is the boat?

b If the lighthouse is 50 m high, what would be the angle of elevation of the top of the lighthouse from the boat?

6 A bird flies from the top of a 12 m tall tree, at an angle of depression of 34°, to catch a worm on the ground.

a How far does the bird actually fly?

b How far was the worm from the base of the tree?

7 Sunil wants to work out the height of a building. He stands about 50 m away from the building. The angle of elevation from Sunil to the top of the building is about 15°. How tall is the building?

8 The top of a ski run is 100 m above the finishing line. The run is 300 m long. What is the angle of depression of the ski run?

9 Nessie and Cara are standing on opposite sides of a tree.

Nessie is 14 m away and the angle of elevation of the top of the tree is 30°.

Cara is 28 m away. She says the angle of elevation for her must be 15° because she is twice as far away.

Is she correct?

What do you think the angle of elevation is?

Problems in three dimensions (E)

To find the value of an angle or side in a three-dimensional figure you need to find a right-angled triangle in the figure. This triangle also has to include two known values that you can use in the calculation.

You must redraw this triangle separately as a plain, right-angled triangle. Add the known values and the unknown value you want to find. Then use the trigonometric ratios and Pythagoras' theorem to solve the problem.

EXAMPLE 15

A, B and C are three points at ground level. They are in the same horizontal plane. C is 50 km east of B. B is north of A. C is on a bearing of 050° from A.

An aircraft, flying east, passes over B and over C at the same height. When it passes over B, the angle of elevation from A is 12°. Find the angle of elevation of the aircraft from A when it is over C.

First, draw a diagram containing all the known information.

Next, use the right-angled triangle ABC to calculate AB and AC.

$$AB = \frac{50}{\tan 50°} = 41.95 \text{ km} \quad \text{(4 significant figures)}$$

$$AC = \frac{50}{\sin 50°} = 65.27 \text{ km} \quad \text{(4 significant figures)}$$

Then use the right-angled triangle ABX to calculate BX, and hence CY.

BX = 41.95 tan 12° = 8.917 km   (4 significant figures)

Finally, use the right-angled triangle ACY to calculate the required angle of elevation, θ.

$$\tan \theta = \frac{8.917}{65.27} = 0.1366$$

$$\Rightarrow \theta = \tan^{-1} 0.1366 = 7.8° \quad \text{(1 decimal place)}$$

Always write down intermediate working values to at least 4 significant figures, or use the answer on your calculator display to avoid inaccuracies in the final answer.

EXTENDED

EXERCISE 25J

1 A vertical flagpole AP stands at the corner of a rectangular courtyard ABCD.

Calculate the angle of elevation of P from C.

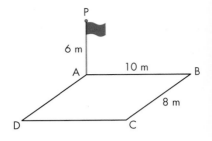

2 The diagram shows a pyramid. The base is a horizontal rectangle ABCD, 20 cm by 15 cm. The length of each sloping edge is 24 cm. The apex, V, is over the centre of the rectangular base. Calculate:

a the length of the diagonal AC

b the size of the angle VAC

c the height of the pyramid.

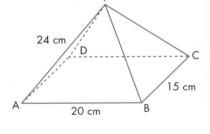

3 The diagram shows the roof of a building. The base ABCD is a horizontal rectangle 7 m by 4 m. The triangular ends are equilateral triangles. Each side of the roof is an isosceles trapezium. The length of the top of the roof, EF, is 5 m. Calculate:

a the length EM, where M is the midpoint of AB

b the size of angle EBC

c the size of the angle between EM and the base ABCD.

4 ABCD is a vertical rectangular plane. EDC is a horizontal triangular plane. Angle CDE = 90°, AB = 10 cm, BC = 4 cm and ED = 9 cm. Calculate:

a angle AED   **b** angle DEC

c EC   **d** angle BEC.

5 In the diagram, XABCD is a pyramid with a rectangular base.

Revina says that the angle between the edge XD and the base ABCD is 56.3°.

Work out the correct answer to show that Revina is wrong.

So far you have only used sines and cosines in right-angled triangles.
A calculator also gives the sine and cosine of **obtuse angles**.
Check that sin 115° = 0.906 and cos 115° = −0.423.

You cannot have a right-angled triangle with an obtuse angle
so how can you calculate sin and cos of an obtuse angle?

Imagine a rod OP lying on the x-axis as shown.

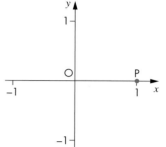

It rotates anticlockwise about the origin O.

ONP is a right-angled triangle.

The hypotenuse OP = 1

If the angle that OP makes with the x-axis is θ.

The adjacent side ON = OP × cos θ = 1 × cos θ = cos θ

So the x-coordinate of P is cos θ

Similarly:

The opposite side NP = OP × sin θ = 1 × sin θ = sin θ

So the y-coordinate of P is sin θ
Therefore the coordinates of P are (cos θ, sin θ).

Imagine that OP continues to rotate so that angle θ
becomes obtuse.
You can still define cos θ and sin θ as the coordinates of P.
You can see from the diagram that the x-coordinate for P
must now be negative.
The y-coordinate is still positive.

For example, if θ = 115°, the diagram looks like this:

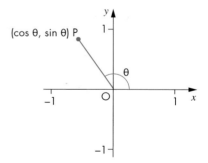

You can see that the angle adjoining θ on the x-axis must have
the same sine as θ. When adjacent angles add up to 180° their sines
are the same. So:

If angle θ is obtuse then sin θ = sin (180° − θ)

Their cosines have the same numeric value but the cosine of the
obtuse angle is negative:

If angle θ is obtuse cos θ = − cos (180° − θ)

For example, if θ = 115° then 180° − θ = 65°
Sin 115° = sin 65° = 0.906
Cos 115° = − cos 65° = − 0.423

EXERCISE 25K

Angle	10°	30°	50°	85°	90°	95°	130°	150°	170°
b Sine		0.5			1				

c Describe the symmetry of the graph.

d Find two examples from the table to show that if two angles add up to 180 degrees they have the same sine.

2 Find two angles that have a sine of 0.5.

3 Find two angles that have a sine of 0.72.

4 x is an obtuse angle and $\sin x = 0.84$. Find the value of x.

5 **a** Copy and complete this table.

Angle	15°	35°	60°	80°	90°	100°	120°	145°	165°
b Cosine									

c Describe the symmetry of the graph.

6 Find the size of each of these angles.

    a $\cos^{-1} 0.85$ **b** $\cos^{-1}(-0.85)$

    c $\cos^{-1}(-0.5)$ **d** $\cos^{-1} 0$

    e $\cos^{-1} 0.125$ **f** $\cos^{-1}(-0.125)$

7 Solve these equations where $0 \leqslant x \leqslant 180°$. Give your answers to the nearest degree. There may be more than one solution.

    a $\cos x = 0.6$ **b** $\cos x = -0.25$

    c $\sin x = \dfrac{3}{4}$ **d** $\sin x = 1$

    e $\cos x = 0$ **f** $\sin x = 0.95$

    g $\sin x = 2$ **h** $\sin x = \cos x$

25.9 The sine rule and the cosine rule

Any triangle has six measurements: three sides and three angles. To find any unknown angles or sides), you need to know at least three of the measurements. Any combination of three measurements – except that of all three angles – is enough to work out the rest.

When you need to find the value of a side or an angle in a triangle that contains no right angle, you can use one of two rules, depending on what you know about the triangle. These are the **sine rule** and the **cosine rule**.

The sine rule

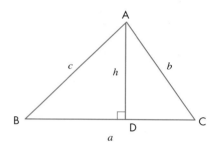

Take a triangle ABC and draw the perpendicular from A to the opposite side BC.

From right-angled triangle ADB, $h = c \sin B$

From right-angled triangle ADC, $h = b \sin C$

Therefore,

$$c \sin B = b \sin C$$

which can be rearranged to give:

$$\frac{c}{\sin C} = \frac{b}{\sin B}$$

By drawing a perpendicular from each of the other two vertices to the opposite side (or by algebraic symmetry), you see that:

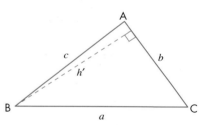

$$\frac{a}{\sin A} = \frac{c}{\sin C} \quad \text{and that} \quad \frac{a}{\sin A} = \frac{b}{\sin B}$$

These are usually combined in the form:

$$\frac{a}{\sin A} = \frac{b}{\sin B} = \frac{c}{\sin C}$$

which can be inverted to give:

$$\frac{\sin A}{a} = \frac{\sin B}{b} = \frac{\sin C}{c}$$

Remember, when using the sine rule: take each side in turn, divide it by the sine of the angle opposite and then equate the results.

Note:
- When you are calculating a *side*, use the rule with the *sides on top*.
- When you are calculating an *angle*, use the rule with the *sines on top*.

EXAMPLE 16

In triangle ABC, find the value of x.

Use the sine rule with sides on top, which gives:

$$\frac{x}{\sin 84°} = \frac{25}{\sin 47°}$$

$$\Rightarrow x = \frac{25 \sin 84°}{\sin 47°} = 34.0 \text{ cm} \quad (3 \text{ significant figures})$$

EXAMPLE 17

In triangle ABC, find the value of the acute angle x.

Use the sine rule with sines on top, which gives:

$$\frac{\sin x}{7} = \frac{\sin 40°}{6}$$

$$\Rightarrow \sin x = \frac{7 \sin 40°}{6} = 0.7499$$

$$\Rightarrow x = \sin^{-1} 0.7499 = 48.6° \quad (3 \text{ significant figures})$$

The sine rule works even if the triangle has an obtuse angle, because you can find the sine of an obtuse angle.

EXERCISE 25L

1 Find the length of the side labelled x in each of these triangles.

a

b

c

2 Find the size of the angle labelled x in each of these triangles.

a

b

c

EXTENDED

EXTENDED

3 To find the height of a tower standing on a small hill, Maria made some measurements (see diagram).

From a point B, the angle of elevation of C is 20°, the angle of elevation of A is 50° and the distance BC is 25 m.

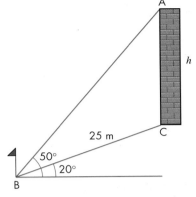

a Calculate these angles.

  i ABC

  ii BAC

b Using the sine rule and triangle ABC, calculate the height h of the tower.

4 Use the information on this sketch to calculate the width, w, of the river.

5 An old building is unsafe and is protected by a fence. A company is going to demolish the building and has to work out its height BD, marked h on the diagram.

Calculate the value of h, using the given information.

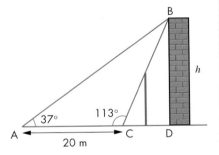

6 A mass is hung from a horizontal beam using two strings. The shorter string is 2.5 m long and makes an angle of 71° with the horizontal. The longer string makes an angle of 43° with the horizontal. What is the length of the longer string?

7 A rescue helicopter is based at an airfield at A.

It is sent out to rescue a man from a mountain at M, due north of A.

The helicopter then flies on a bearing of 145° to a hospital at H, as shown on the diagram.

Calculate the direct distance from the mountain to the hospital.

8 Triangle ABC has an obtuse angle at B.

Calculate the size of angle ABC.

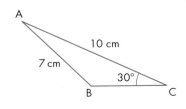

429

The cosine rule

Take the triangle, shown on the right, where D is the foot of the perpendicular to BC from A. The length of BD is x.

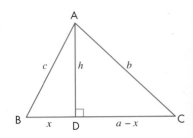

Using Pythagoras' theorem on triangle BDA:

$$h^2 = c^2 - x^2$$

Using Pythagoras' theorem on triangle ADC:

$$h^2 = b^2 - (a - x)^2$$

Therefore,

$$c^2 - x^2 = b^2 - (a - x)^2$$
$$c^2 - x^2 = b^2 - a^2 + 2ax - x^2$$
$$\Rightarrow c^2 = b^2 - a^2 + 2ax$$

From triangle BDA, $x = c \cos B$.

So:

$$c^2 = b^2 - a^2 + 2ac \cos B$$

Rearranging gives:

$$b^2 = a^2 + c^2 - 2ac \cos B$$

By algebraic symmetry:

$$a^2 = b^2 + c^2 - 2bc \cos A \quad \text{and} \quad c^2 = a^2 + b^2 - 2ab \cos C$$

This is the cosine rule, which can be best remembered from the diagram on the right, where:

$$a^2 = b^2 + c^2 - 2bc \cos A$$

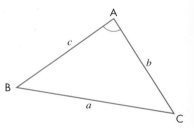

Note the symmetry of the rule and how the rule works using two adjacent sides and the angle between them (the **included angle**).

The formula can be rearranged to find any of the three angles.

$$\cos A = \frac{b^2 + c^2 - a^2}{2bc}$$

$$\cos B = \frac{a^2 + c^2 - b^2}{2ac}$$

$$\cos C = \frac{a^2 + b^2 - c^2}{2ab}$$

EXAMPLE 18

Find the value of x in this triangle.

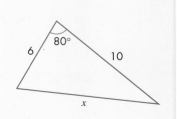

By the cosine rule:

$$x^2 = 6^2 + 10^2 - 2 \times 6 \times 10 \times \cos 80°$$
$$x^2 = 115.16$$
$$\Rightarrow x = 10.7 \text{ (3 significant figures)}$$

EXAMPLE 19

Find the value of x in this triangle.

By the cosine rule:

$$\cos x = \frac{5^2 + 7^2 - 8^2}{2 \times 5 \times 7} = 0.1428$$

$$\Rightarrow x = 81.8° \text{ (3 significant figures)}$$

It is possible to find the cosine of an angle that is greater than 90°. For example, cos 120° = –0.5.

EXERCISE 25M

EXTENDED

1 Find the length of the side marked x in each of these triangles.

a

b

c

2 Find the angle x in each of these triangles.

a

b

c

d Explain the significance of the answer to part **c**.

3 In triangle ABC, AB = 5 cm, BC = 6 cm and angle ABC = 55°. Find AC.

4 A triangle has two sides of length 40 cm and an angle of 110°. Work out the length of the third side of the triangle.

5 The diagram shows a trapezium ABCD.
AB = 6.7 cm, AD = 7.2 cm, CB = 9.3 cm and angle DAB = 100°.

Calculate:

a the length DB   **b** angle DBA

c angle DBC   **d** the length DC

6 A ship sails from a port on a bearing of 050° for 50 km then turns on a bearing of 150° for 40 km. A crewman is taken ill, so the ship drops anchor. What course and distance should a rescue helicopter from the port fly to reach the ship in the shortest possible time?

7 The three sides of a triangle are given as 3*a*, 5*a* and 7*a*. Calculate the smallest angle in the triangle.

8 Two ships, X and Y, leave a port at 9 am.

Ship X travels at an average speed of 20 km/h on a bearing of 075° from the port.

Ship Y travels at an average speed of 25 km/h on a bearing of 130° from the port.

Calculate the distance between the two ships at 11 am.

9 Calculate the size of the largest angle in the triangle ABC.

Choosing the correct rule

When finding the unknown sides and angles in a triangle, there are several situations that can occur.

Two sides and the included angle

1 Use the cosine rule to find the third side.

2 Use the sine rule to find either of the other angles.

3 Use the sum of the angles in a triangle to find the third angle.

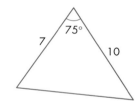

Two angles and a side

1 Use the sum of the angles in a triangle to find the third angle.

2, 3 Use the sine rule to find the other two sides.

Three sides

1 Use the cosine rule to find one angle.

2 Use the sine rule to find another angle.

3 Use the sum of the angles in a triangle to find the third angle.

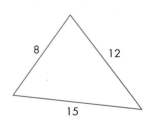

EXERCISE 25N

1 Find the length or angle x in each of these triangles.

a

b

c

d

e

f

g

h

i

2 The hands of a clock have lengths 3 cm and 5 cm. Find the distance between the tips of the hands at 4 o'clock.

3 A helicopter is seen hovering at a point which is in the same vertical plane as two towns, X and F, which are on the same level. Its distances from X and F are 8.5 km and 12 km respectively. The angle of elevation of the helicopter when observed from F is 43°. Calculate the distance between the two towns.

4 Triangle ABC has sides with lengths a, b and c, as shown in the diagram.

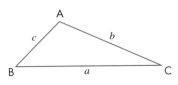

a What can you say about the angle BAC, if $b^2 + c^2 - a^2 = 0$?

b What can you say about the angle BAC, if $b^2 + c^2 - a^2 > 0$?

c What can you say about the angle BAC, if $b^2 + c^2 - a^2 < 0$?

5 The diagram shows a sketch of a field ABCD. A farmer wants to put a new fence round the perimeter of the field.

Calculate the perimeter of the field.

In triangle ABC, the vertical height is BD and the base is AC.

Let $BD = h$ and $AC = b$, then the **area** of the triangle is given by:

$$\tfrac{1}{2} \times AC \times BD = \tfrac{1}{2}bh$$

However, in triangle BCD:

$$h = BC \sin C = a \sin C$$

where $BC = a$.

Substituting into $\tfrac{1}{2}bh$ gives:

$$\tfrac{1}{2}b \times (a \sin C) = \tfrac{1}{2}ab \sin C$$

as the area of the triangle.

By taking the perpendicular from A to its opposite side BC, and the perpendicular from C to its opposite side AB, you can show that the area of the triangle is also given by:

$$\tfrac{1}{2}ac \sin B \quad \text{and} \quad \tfrac{1}{2}bc \sin A$$

Note the pattern: the area is given by the product of two sides multiplied by the sine of the included angle. This is the **area sine rule**. Starting from any of the three forms, you can use the sine rule to establish the other two.

EXAMPLE 20

Find the area of triangle ABC.

$$\text{Area} = \tfrac{1}{2}ab \sin C$$
$$\text{Area} = \tfrac{1}{2} \times 5 \times 7 \times \sin 38°$$
$$= 10.8 \text{ cm}^2 \text{ (3 significant figures)}$$

EXAMPLE 21

Find the area of triangle ABC.

You have all three sides but no angle. So first you must find an angle in order to apply the area sine rule.

Use the cosine rule to find angle C.

$$\cos C = \frac{a^2 + b^2 - c^2}{2ab}$$

$$= \frac{13^2 + 19^2 - 8^2}{2 \times 13 \times 19} = 0.9433$$

$$\Rightarrow C = \cos^{-1} 0.9433 = 19.4°$$

(Keep the exact value in your calculator memory.)

Now apply the area sine rule.

$$\tfrac{1}{2}ab \sin C = \tfrac{1}{2} \times 13 \times 19 \times \sin 19.4°$$
$$= 41.0 \text{ cm}^2 \text{ (3 significant figures)}$$

EXTENDED

EXERCISE 25O

1 Find the area of each of these triangles.

  a Triangle ABC with BC = 7 cm, AC = 8 cm and angle ACB = 59°

  b Triangle ABC with angle BAC = 86°, AC = 6.7 cm and AB = 8 cm

  c Triangle PQR with QR = 27 cm, PR = 19 cm and angle QRP = 109°

  d Triangle XYZ with XY = 231 cm, XZ = 191 cm and angle YXZ = 73°

  e Triangle LMN with LN = 63 cm, LM = 39 cm and angle NLM = 85°

2 The area of triangle ABC is 27 cm². If BC = 14 cm and angle BCA = 115°, find the length of AC.

3 The area of triangle LMN is 113 cm², LM = 16 cm and MN = 21 cm. Angle LMN is acute. Calculate these angles.

  a ∠LMN   **b** ∠MNL

4 A board is in the shape of a triangle with sides 60 cm, 70 cm and 80 cm. Find the area of the board.

5 Two circles, centres P and Q, have radii of 6 cm and 7 cm respectively. The circles intersect at X and Y. Given that PQ = 9 cm, find the area of triangle PXQ.

6 Sanjay is making a kite.
The diagram shows a sketch of his kite.

Calculate the area of the material required to make the kite.

7 The triangular area ABC is to be made a national park.

The bearing of B from A is 324°.
The bearing of C from A is 42°.

Calculate the area of the park.

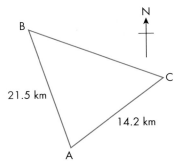

8 The diagram shows the dimensions of a four-sided field.

  a Show that the length of the diagonal BD is 66 metres to the nearest metre.

  b Calculate the size of angle C.

  c Calculate the area of the field.

Why this chapter matters

People have always needed to measure areas and volumes.

In everyday life, you will, for instance, need to find the area to work out how many tiles to buy to cover a floor; or you will need to find the volume to see how much water is needed to fill a swimming pool. You can do this quickly using formulae.

Measuring the world

From earliest times, farmers have wanted to know the area of their fields to see how many crops they could grow or animals they could support. When land is bought and sold, the cost depends on the area.

Volumes are important too. Volumes tell you how much space there is inside any structure. Whether it is a house, barn, aeroplane, car or office, the volume is important. In some countries there are regulations about the number of people who can use an office, based on the volume of the room.

Volumes of containers for liquids also need to be measured. Think, for example, of a car fuel tank, the water tank in a building, or a reservoir. It is important to be able to calculate the capacity of all these things.

So how do you measure areas and volumes? In this chapter, you will learn formulae that can be used to calculate areas and volumes of different shapes, based on a few measurements.

Many of these formulae were first worked out thousands of years ago. They are still in use today because they are important in everyday life.

The process of calculating areas and volumes using formulae is called **mensuration**.

Chapter

26 Mensuration

Topics		Level	Key words
1	Perimeter and area of a rectangle	CORE	length, width, perimeter, area
2	Area of a triangle	CORE	base, perpendicular height
3	Area of a parallelogram	CORE	parallelogram
4	Area of a trapezium	CORE	trapezium
5	Circumference and area of a circle	CORE	circumference, diameter, radius, π
6	Surface area and volume of a cuboid	CORE	volume, cuboid, surface area, litre
7	Volume of a prism	CORE	cross-section, prism
8	Volume and surface area of a cylinder	CORE	curved surface, cylinder
9	Arcs and sectors	EXTENDED	arc, sector, subtended
10	Volume of a pyramid	EXTENDED	pyramid, vertical height, vertex
11	Volume and surface area of a cone	EXTENDED	cone, vertical height, slant height
12	Volume and surface area of a sphere	EXTENDED	sphere

In this chapter you will learn how to:

CORE	EXTENDED
Carry out calculations involving:	**Solve problems involving:**
• the perimeter and area of a rectangle, triangle, parallelogram and trapezium and compound shapes derived from these (C4.2 and E4.2)	• the arc length and sector area as fractions of the circumference and area of a circle (E4.3)
• the circumference and area of a circle (C4.3 and E4.3)	• the surface area and volume of a sphere, pyramid and cone (given formulae for the sphere, pyramid and cone). (E4.4)
• the volume of a cuboid, prism and cylinder (C4.4 and E4.4)	
• the surface area of a cuboid and a cylinder (C4.4 and E4.4)	
• the areas and volumes of compound shapes (C4.5 and E4.5)	

The **perimeter** of a rectangle is the total distance around the outside.

length, *l*

width, *w*

$$\text{perimeter} = l + w + l + w$$
$$= 2(l + w)$$

The **area** of a rectangle is **length × width**.

$$\text{area} = lw$$

EXAMPLE 1

Calculate the area and perimeter of this rectangle.

11 cm

4 cm

Area of rectangle = length × width
$$= 11 \text{ cm} \times 4 \text{ cm}$$
$$= 44 \text{ cm}^2$$

Perimeter = 2 × 11 + 2 × 4
$$= 30 \text{ cm}$$

Some two-dimensional shapes are made up of two or more rectangles.

These shapes can be split into simpler shapes, which makes it easy to calculate their areas.

EXAMPLE 2

Find the area and perimeter of the shape shown on the right.

3 cm

2 cm

3 cm

10 cm

First, split the shape into two rectangles, A and B and find the missing lengths.

The perimeter is
$$3 + 2 + 7 + 3 + 10 + 5 = 30 \text{ cm}$$

area of A = 2 × 3 = 6 cm²

area of B = 10 × 3 = 30 cm²

The area of the shape is:

area of A + area of B = 6 + 30
$$= 36 \text{ cm}^2$$

3 cm

A

2 cm

10 – 3 = 7 cm

3 + 2 = 5 cm

B

3 cm

10 cm

EXERCISE 26A

1 Calculate the area and the perimeter for each of these rectangles.

a 7 cm, 5 cm   **b** 11 cm, 3 cm   **c** 15 cm, 3 cm

d 10 cm, 7 cm   **e** 8 cm, 7 cm   **f** 5 cm, 2 cm

2 Calculate the area and the perimeter for each of these rectangles.

a 8.2 cm, 6.5 cm   **b** 11.8 cm, 7.2 cm

3 A rectangular field is 150 m long and 45 m wide.

Fencing is needed to go all the way around the field.

The fencing is sold in 10-metre long pieces.

How many pieces are needed?

4 A soccer pitch is 160 m long and 70 m wide.

a Before a game, the players have to run about 1500 m to help them loosen up. How many times will they need to run round the perimeter of the pitch to do this?

b The groundsman waters the pitch at the rate of 100 m^2 per minute. How long will it take him to water the whole pitch?

5 What is the perimeter of a square with an area of 100 cm^2?

6 Which rectangle has the largest area? Which has the largest perimeter?

8 cm, 5 cm, **A**   11 cm, 4 cm, **B**   7 cm, 6 cm, **C**

Explain your answers.

CORE

7 Doubling the length and width of a rectangle doubles the area of the rectangle.

Is this statement:
- always true
- sometimes true
- never true?

Explain your answer.

8 Calculate the perimeter and area of each of these compound shapes below.

a

5 cm
2 cm
2 cm
10 cm

b

8 cm
4 cm
6 cm
4 cm

c

10 cm
7 cm
3 cm
10 cm

d

5 cm
3 cm
5 cm
5 cm

e
9 cm
2 cm
5 cm
5 cm

f

6 cm
3 cm
3 cm
11 cm

g

8 cm
4 cm
2 cm
8 cm

h

3 cm
9 cm
2 cm
5 cm

9 This compound shape is made from four rectangles that are all the same size.

Work out the area of the compound shape.

12 cm
9 cm

10 This shape is made from five squares that are all the same size.

It has an area of 80 cm².

Work out the perimeter of the shape.

26.2 Area of a triangle

The area of any triangle is given by the formula:

area = $\frac{1}{2}$ × **base** × **perpendicular height**

As an algebraic formula, this is written as:

$A = \frac{1}{2}bh$

This diagram shows why the area for a triangle is half the area of a rectangle with the same base and height.

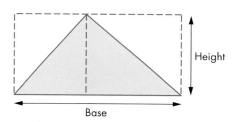

EXAMPLE 3

Calculate the area of this triangle.

4 cm

9 cm

Area = $\frac{1}{2}$ × 9 cm × 4 cm

   = $\frac{1}{2}$ × 36 cm²

   = 18 cm²

10·39.

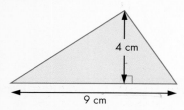

b   h =
Area 15/2 4 × 3 =
      12 cm²
      6 cm²

b   ?
.12 × R
124·7 =
62 cm²

EXAMPLE 4

Calculate the area of the shape shown below.

2 cm

7 cm   3 cm

This shape can be split into a rectangle (R) and a triangle (T).

2 cm   R   T

7 cm   3 cm

Area of the shape = area of R + area of T
$$= 7 \times 2 + \frac{1}{2} \times 2 \times 3$$
$$= 14 + 3$$
$$= 17 \text{ cm}^2$$

EXERCISE 26B

1 Calculate the area of each of these triangles.

a

7 cm

6 cm

b
3 cm

8 cm

c
7 cm

4 cm

d

10 cm   11 cm

e
12 cm
15 cm

f

14 cm
20 cm

2 Copy and complete the table for triangles **a** to **f**.

	Base	Perpendicular height	Area
a	8 cm	7 cm	
b		9 cm	36 cm²
c		5 cm	10 cm²
d	4 cm	3	6 cm²
e	6 cm		21 cm²
f	8 cm	11 cm	

CORE

3 This regular hexagon has an area of 48 cm².

What is the area of the square that surrounds the hexagon?

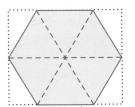

4 Find the area of each of these shapes.

a

> **ADVICE AND TIPS**
>
> Refer to Example 4 on how to find the area of a compound shape.

b

c

5 Find the area of each shaded shape.

a

> **ADVICE AND TIPS**
>
> Find the area of the outer shape and subtract the area of the inner shape.

b

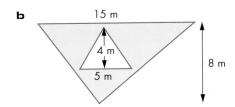

6 Write down the dimensions of two different-sized triangles that have the same area of 50 cm².

7 Which triangle is the odd one out? Give a reason for your answer.

a

b

c

A **parallelogram** can be changed into a rectangle by moving a triangle.

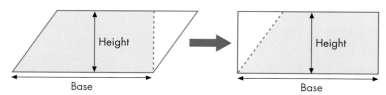

This shows that the area of the parallelogram is the area of a rectangle with the same base and height. The formula is:

area of a parallelogram = base × height

As an algebraic formula, this is written as:

$A = bh$

EXAMPLE 5

Find the area of this parallelogram.

Area = 8 cm × 6 cm
     = 48 cm²

EXERCISE 26C

1 Calculate the area of each parallelogram below.

a

b

c

d

e

f

2 Sandeep says that the area of this parallelogram is 30 cm².

Is she correct? Give a reason for your answer.

3 This shape is made from four parallelograms that are all the same size. The area of the shape is 120 cm².

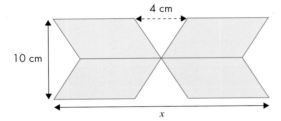

Work out the length marked x on the diagram.

4 This logo, made from two identical parallelograms, is cut from a sheet of card.

a Calculate the area of the logo.

b How many logos can be cut from a sheet of card that measures 1 m by 1 m?

26.4   Area of a trapezium

You can calculate the area of a **trapezium** by finding the average of the lengths of its parallel sides and multiplying this by the perpendicular height between them.

The area of a trapezium is given by this formula:

$$A = \tfrac{1}{2}(a + b)h$$

EXAMPLE 6

Find the area of the trapezium ABCD.

Area $= \frac{1}{2}(4 + 7) \times 3$

$\quad = \frac{1}{2} \times 11 \times 3$

$\quad = 16.5 \text{ cm}^2$

EXERCISE 26D

1 Copy and complete the table for the trapezia a–g.

	Parallel side 1	Parallel side 2	Perpendicular height	Area
a	8 cm	4 cm	5 cm	
b	10 cm	12 cm	7 cm	
c	7 cm	5 cm	4 cm	
d	5 cm	9 cm	6 cm	
e	3 cm	13 cm	5 cm	
f	4 cm	10 cm		42 cm^2
g	7 cm	8 cm		22.5 cm^2

> **ADVICE AND TIPS**
>
> Trapezia is the plural of trapezium.

2 Calculate the perimeter and the area of each trapezium.

a

b

c

> **ADVICE AND TIPS**
>
> Make sure you use the right measurement for the height. Sometimes you might be told the slant side length, which is not used for the area.

3 A trapezium has an area of 25 cm^2. Its vertical height is 5 cm.
Work out a possible pair of lengths for the two parallel sides.

4 Which of these shapes has the largest area?

a

b

c

5 Which of these shapes has the smallest area?

a

b

c

6 Which of these is the area of this trapezium?

a 45 cm^2 b 65 cm^2 c 70 cm^2

You must show your workings.

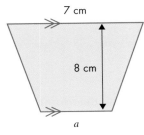

7 Work out the value of *a* so that the square and the trapezium have the same area.

8 The side of a ramp is a trapezium, as shown in the diagram. Calculate its area, giving your answer in square metres.

ADVICE AND TIPS

Change the height into metres first.

The perimeter of a circle is called the **circumference**.

You can calculate the circumference, C, of a circle by multiplying its **diameter**, d, by π.

The value of π is found on all scientific calculators, with $\pi = 3.141\,592\,654$, but if it is not on your calculator, then take $\pi = 3.142$.

The circumference of a circle is given by the formula:

circumference = π × diameter *or* $C = \pi d$

As the diameter is twice the **radius**, r, this formula can also be written as $C = 2\pi r$.

EXAMPLE 7

Calculate the circumference of the circle with a diameter of 4 cm.

Use the formula:

$C = \pi d$

$= \pi \times 4$

$= 12.6$ cm (rounded to 1 decimal place)

Remember The length of the radius of a circle is half the length of its diameter. So, when you are given a radius, in order to find a circumference you must first double the radius to get the diameter.

EXAMPLE 8

Calculate the diameter of a circle that has a circumference of 40 cm.

$C = \pi \times d$

$40 = \pi \times d$

$d = \dfrac{40}{\pi} = 12.7$ cm (rounded to 1 decimal place)

The area, A, of a circle is given by the formula:

area = π × radius^2 *or* $A = \pi \times r \times r$ *or* $A = \pi r^2$

Remember This formula uses the radius of a circle. So, when you are given the diameter of a circle, you must *halve* it to get the radius.

EXAMPLE 9

Calculate the area of a circle with a diameter of 12 cm.

First, halve the diameter to find the radius:

radius = 12 ÷ 2 = 6 cm

Then, find the area:

area = πr^2

$= \pi \times 6^2$

$= \pi \times 36$

$= 113.1$ cm^2 (rounded to 1 decimal place)

EXERCISE 26E

1 Calculate the circumference and area of each circle. Give your answers to 1 decimal place.

a 5 cm   b 3 cm   c 1.5 cm   d 4 cm

2 Calculate the circumference and area of each circle. Give your answers to 1 decimal place.

a 8 cm   b 5 cm   c 9.2 cm   d 4.7 cm

3 Calculate:

i the circumference

ii the area of each of these circles. Give your answers to 1 decimal place.

a 9 cm   b 22 cm   c 6.5 cm   d 28 cm

CORE

CORE

4 A circle has a circumference of 60 cm.

   a Calculate the diameter of the circle to 1 decimal place.

   b What is the radius of the circle to 1 decimal place?

   c Calculate the area of the circle to 1 decimal place.

5 Calculate the area of a circle with a circumference of 110 cm.

6 Calculate the perimeter and the area of each of these shapes. Give your answers to 1 decimal place.

a

12 cm

Semicircle

b

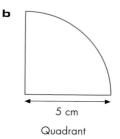

5 cm

Quadrant

7 Calculate the area of the shaded part of each of these diagrams.

a

6 m

2 m

b

5 cm

9 cm

c

4 cm

5 cm

> **ADVICE AND TIPS**
>
> In each diagram, subtract the area of the small circle from the area of the large circle.

8 The diagram shows a circular photograph frame.

Work out the area of the frame. Give your answer to 1 decimal place.

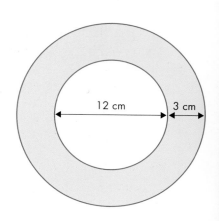

12 cm   3 cm

9 A square has sides of length a and a circle has radius r.

The area of the square is equal to the area of the circle.

Show that $r = \dfrac{a}{\sqrt{\pi}}$

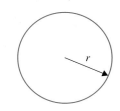

10 A circle fits exactly inside a square of sides 10 cm.

Calculate the area of the shaded region. Give your answer to 1 decimal place.

10 cm

10 cm

26.6 Surface area and volume of a cuboid

A cuboid is a box shape, all six faces of which are rectangles.

Height (h)

Length (l)

Width (w)

Every day you will see many examples of **cuboids**, such as food packets, DVD players – and even this book.

The **volume** of a cuboid is given by the formula:

volume = length × width × height *or* $V = l \times w \times h$ *or* $V = lwh$

You can claculate **surface area** of a cuboid by finding the total area of the six faces, which are rectangles. Notice that each pair of opposite rectangles have the same area. So, from the diagram above:

area of top and bottom rectangles = 2 × length × width = $2lw$

area of front and back rectangles = 2 × height × width = $2hw$

area of two side rectangles = 2 × height × length = $2hl$

Hence, the surface area of a cuboid is given by the formula:

surface area = $A = 2lw + 2hw + 2hl$

EXAMPLE 10

Calculate the volume and surface area of this cuboid.

3.5 cm

6 cm

4 cm

Volume = $V = lwh = 6 \times 4 \times 3.5 = 84$ cm^3

Surface area = $A = 2lw + 2hw + 2hl$

$= (2 \times 6 \times 4) + (2 \times 3.5 \times 4) + (2 \times 3.5 \times 6)$

$= 48 + 28 + 42 = 118$ cm^2

Note:

1 cm^3 = 1000 mm^3 and 1 m^3 = 1 000 000 cm^3

1000 cm^3 = 1 **litre**

1 m^3 = 1000 litres

EXERCISE 26F

1 Find **i** the volume and **ii** the surface area of each of these cuboids.

a

11 cm

3 cm

6 cm

b

5 cm

9 cm

2 cm

c

8 cm

15 cm

12 cm

d

21 cm

5 cm

5 cm

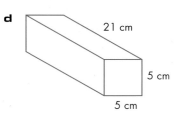

2 Find the capacity of a fish-tank with dimensions: length 40 cm, width 30 cm and height 20 cm. Give your answer in litres.

3 Find the volume of each cuboid

   a The area of the base is 40 cm^2 and the height is 4 cm.

   b The base has one side 10 cm and the other side 2 cm longer, and the height is 4 cm.

   c The area of the top is 25 cm^2 and the depth is 6 cm.

4 Calculate:

   i the volume

   ii the surface area of cubes with these edge lengths.

   a 4 cm   **b** 7 cm   **c** 10 mm   **d** 5 m   **e** 12 m

5 Safety regulations say that in a room where people sleep there should be at least 12 m^3 for each person. A dormitory is 20 m long, 13 m wide and 4 m high. What is the greatest number of people who can safely sleep in the dormitory?

6 Copy and complete the table for cuboids **a** to **e**.

	Length	Width	Height	Volume
a	8 cm	5 cm	4.5 cm	
b	12 cm	8 cm		480 cm^3
c	9 cm		5 cm	270 cm^3
d		7 cm	3.5 cm	245 cm^3
e	7.5 cm	5.4 cm	2 cm	

7 A tank contains 32 000 litres of water. The base of the tank measures 6.5 m by 3.1 m. Find the depth of water in the tank. Give your answer to one decimal place.

8 A room contains 168 m^3 of air. The height of the room is 3.5 m. What is the area of the floor?

9 What are the dimensions of cubes with these volumes?

   a 27 cm^3 **b** 125 m^3 **c** 8 mm^3 **d** 1.728 m^3

10 Calculate the volume of each of these shapes.

a   **b**

> **ADVICE AND TIPS**
>
> Split the solid into two separate cuboids and work out the dimensions of each of them from the information given.

11 A cuboid has volume of 125 cm^3 and a total surface area of 160 cm^2.

   Is it possible that this cuboid is a cube? Give a reason for your answer.

12 The volume of a cuboid is 1000 cm^3. What is the smallest surface area it could have?

A **prism** is a three dimensional shape that has the same **cross-section** running all the way through it.

Name:	Cuboid	Triangular prism	Cylinder	Cuboid	Hexagonal prism
Cross-section:	Rectangle	Triangle	Circle	Square	Hexagon

The volume of a prism is found by multiplying the area of its cross-section by the length of the prism (or height if the prism is stood on end).

That is, volume of prism = area of cross-section × length
or $V = Al$

EXAMPLE 11

Find the volume of the triangular prism.

The area of the triangular cross-section $A = \dfrac{5 \times 7}{2}$

$= 17.5 \ cm^2$

The volume is the area of its cross-section × length $= Al$

$= 17.5 \times 9$

$= 157.5 \ cm^3$

EXERCISE 26G

1 For each prism shown:

i calculate the area of the cross-section **ii** calculate the volume.

a

b

c

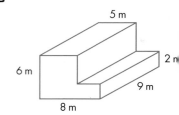

2 Calculate the volume of each of these prisms.

a

b

c

3 A swimming pool is 10 m wide and 25 m long.

It is 1.2 m deep at one end and 2.2 m deep at the other end. The floor slopes uniformly from one end to the other.

a Explain why the shape of the pool is a prism.

b The pool is filled with water at a rate of 2 m^3 per minute. How long will it take to fill the pool?

4 A building is in the shape of a prism with the dimensions shown in the diagram. Calculate the volume of air (in litres) inside the building.

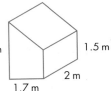

5 Each of these prisms has a uniform cross-section in the shape of a right-angled triangle.

a Find the volume of each prism.   **b** Find the total surface area of each prism.

i

ii

6 The top and bottom of the container shown here are the same size, both consisting of a rectangle, 4 cm by 9 cm, with a semi-circle at each end. The depth is 3 cm. Find the volume of the container.

7 In 2009 the sculptor Anish Kapoor exhibited a work called *Svayambh*. It was a block of red wax in the shape of a prism.

The cross-section was in the shape of an arched entrance.

It was 8 m long and weighed 30 tonnes. It slowly travelled through the galleries on a track.

Calculate the volume of wax used.

CORE

8 A horse trough is in the shape of a semi-circular prism, as shown.

What volume of water will the trough hold when it is filled to the top? Give your answer in litres.

9 The dimensions of the cross-section of a girder (in the shape of a prism), 2 m in length, are shown on the diagram. The girder is made of iron. 1 cm³ of iron weighs 79 g.

What is the mass of the girder?

10 Calculate the volume of this prism.

26.8 Volume and surface area of a cylinder

Volume

Since a **cylinder** is an example of a prism, you can calculate its **volume** by multiplying the area of one of its circular ends by the height.

That is,

volume = $\pi r^2 h$

where r is the radius of the cylinder and h is its height or length.

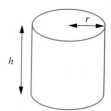

EXAMPLE 12

What is the volume of a cylinder with a radius of 5 cm and a height of 12 cm?

Volume = area of circular base × height

$= \pi r^2 h$

$= \pi \times 5^2 \times 12 \text{ cm}^3$

$= 942 \text{ cm}^3$ (3 significant figures)

Surface area

The total surface area of a cylinder is made up of the area of its **curved surface** plus the area of its two circular ends.

The curved surface area, when opened out, is a rectangle with length equal to the circumference of the circular end.

curved surface area = circumference of end × height of cylinder

$$= 2\pi rh \quad \text{or} \quad \pi dh$$

area of one end $= \pi r^2$

Therefore, total surface area $= 2\pi rh + 2\pi r^2 \quad$ or $\quad \pi dh + 2\pi r^2$

EXAMPLE 13

What is the total surface area of a cylinder with a radius of 15 cm and a height of 2.5 m?

First, you must change the dimensions to a common unit. Use centimetres in this case.

Total surface area $= \pi dh + 2\pi r^2$

$$= \pi \times 30 \times 250 + 2 \times \pi \times 15^2 \text{ cm}^2$$

$$= 23\,562 + 1414 \text{ cm}^2$$

$$= 24\,976 \text{ cm}^2$$

$$= 25\,000 \text{ cm}^2 \text{ (3 significant figures)}$$

EXERCISE 26H

1 For the cylinders below find:

    i the volume   **ii** the total surface area.

    Give your answers to 3 significant figures.

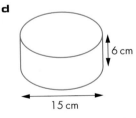

2 For each of these cylinder dimensions find:

    i the volume   **ii** the curved surface area.

    Give your answers in terms of π.

    a Base radius 3 cm and height 8 cm   **b** Base diameter 8 cm and height 7 cm

    c Base diameter 12 cm and height 5 cm   **d** Base radius of 10 m and length 6 m

CORE

CORE

3 A solid cylinder has a diameter of 8.4 cm and a height of 12.0 cm. Calculate the volume of the cylinder.

4 A cylindrical food can has a height of 10.5 cm and a diameter of 7.4 cm.

What can you say about the size of the paper label around the can?

5 A cylindrical container is 65 cm in diameter. Water is poured into the container until it is 1 m deep.

How much water is in the container? Give your answer in litres.

6 A drinks manufacturer plans a new drink in a can. The quantity in each can must be 330 ml.

Suggest a suitable height and diameter for the can. (You might like to look at the dimensions of a real drinks can.)

7 Wire is commonly made by putting hot metal through a hole in a plate.

What length of wire of diameter 1 mm can be made from a 1 cm cube of metal?

8 The engine size of a car is measured in litres. This tells you the total volume of its cylinders. Cylinders with the same volume can be long and thin or short and thick.

In a racing car, the diameter of a cylinder is twice its length. Suggest possible dimensions for a 0.4 litre racing car cylinder.

26.9 Arcs and sectors   E

A **sector** is part of a circle, bounded by two radii of the circle and one of the **arcs** formed by the intersections of these radii with the circumference.

The angle at O is the angle of the sector. The angle is **subtended** at the centre of the circle by the arc.

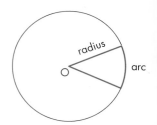

Length of an arc and area of a sector

A sector is a fraction of the whole circle, the size of the fraction being determined by the size of angle of the sector. The angle is often written as θ, a Greek letter pronounced *theta*. For example, the sector shown in the diagram represents the fraction $\dfrac{\theta}{360}$.

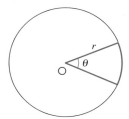

This applies to both its arc length and its area. Therefore:

$$\text{arc length} \quad = \frac{\theta}{360} \times 2\pi r \quad \text{or} \quad \frac{\theta}{360} \times \pi d$$

$$\text{sector area} \quad = \frac{\theta}{360} \times \pi r^2$$

EXAMPLE 14

Calculate the arc length and the area of the sector in the diagram.

The sector angle is 28° and the radius is 5 cm. Therefore:

$$\text{arc length} \quad = \frac{28}{360} \times \pi \times 2 \times 5 = 2.4 \text{ cm (1 decimal place)}$$

$$\text{sector area} \quad = \frac{28}{360} \times \pi \times 5^2 = 6.1 \text{ cm}^2 \text{ (1 decimal place)}$$

EXERCISE 26I

1 For each of these sectors, calculate: **i** the arc length **ii** the sector area.

a
40°
8 cm

b
95°
5 cm

c
78°
12 cm

d
130°
7 cm

2 Calculate the arc length and the area of a sector of which the arc subtends an angle of 60° at the centre of a circle with a diameter of 12 cm. Give your answer in terms of π.

3 Calculate the total perimeter of each of these sectors.

a
11 cm

b
22°
8.5 cm

4 Calculate the area of each of these sectors.

a
110°
7 cm

b
50°
8 cm

EXTENDED

5 O is the centre of a circle of radius 12.5 cm.

Calculate the length of the arc ACB.

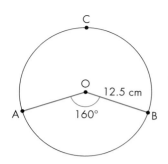

6 The diagram shows quarter of a circle. Calculate the area of the shaded shape, giving your answer in terms of π.

7 ABCD is a square of side length 8 cm. APC and AQC are arcs of the circles with centres D and B. Calculate the area of the shaded part.

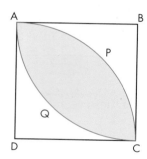

8 Find:

a the perimeter

b the area

of this shape.

26.10 Volume of a pyramid (E)

A **pyramid** is a three-dimensional shape with a base from which triangular faces rise to a common **vertex**. The base can be any polygon, but is usually a triangle, a rectangle or a square.

The volume of a pyramid is given by:

volume $= \frac{1}{3} \times$ base area $\times$ **vertical height**

$V = \frac{1}{3}Ah$

where A is the base area and h is the vertical height.

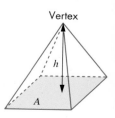

EXAMPLE 15

Calculate the volume of the pyramid.

Base area $= 5 \times 4 = 20$ cm^2

Volume $= \frac{1}{3} \times 20 \times 6 = 40$ cm^3

EXAMPLE 16

A pyramid, with a square base of side 8 cm, has a volume of 320 cm^3. What is the vertical height of the pyramid?

Let h be the vertical height of the pyramid. Then:

$$\text{volume} = \frac{1}{3} \times 64 \times h = 320 \text{ cm}^3$$

$$\frac{64h}{3} = 320 \text{ cm}^3$$

$$h = \frac{960}{64} \text{ cm}$$

$$h = 15 \text{ cm}$$

EXERCISE 26J

1 Calculate the volume of each of these pyramids, all with rectangular bases.

a

b

c

d

e

2 Calculate the volume of a pyramid that has a square base of side 9 cm and a vertical height of 10 cm.

EXTENDED

EXTENDED

3 Suppose you have six pyramids which have a height that is half the side of the square base.

   a Explain how they can fit together to make a cube.

   b How does this show that the formula for the volume of a pyramid is correct?

4 The glass pyramid outside the Louvre Museum in Paris was built in the 1980s. It is 20.6 m tall and the base is a square of side 35 m.

Suppose that instead of a pyramid, the building was a cuboid with the same square base, a flat roof and the same volume. How high would it have been?

5 Calculate the volume of each of these shapes.

a

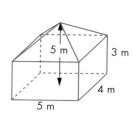

5 m
3 m
4 m
5 m

b

8 m
5 m
2.5 m
3 m

c

25 cm
20 cm
15 cm
10 cm

6 A crystal is in the form of two square-based pyramids joined at their bases (see diagram).

The crystal has a mass of 31.5 g.

What is the mass of 1 cm^3 of the substance?

7 cm
3 cm

7 A pyramid has a square base of side 6.4 cm. Its volume is 81.3 cm^3.

Calculate the height of the pyramid.

8 A pyramid has the same volume as a cube of side 10.0 cm.
The height of the pyramid is the same as the side of the square base.

Calculate the height of the pyramid.

9 The pyramid in the diagram has its top 5 cm cut off as shown.
The shape that is left is called a frustum. Calculate the volume of the frustum.

5 cm
3 cm
2 cm
10 cm
6 cm
9 cm

Volume and surface area of a cone

A **cone** can be treated as a pyramid with a circular base. Therefore, the formula for the volume of a cone is the same as that for a pyramid.

volume = $\frac{1}{3}$ × base area × vertical height

$V = \frac{1}{3}\pi r^2 h$

where r is the radius of the base and h is the **vertical height** of the cone.

The area of the curved surface of a cone is given by:

area of curved surface = π × radius × slant height

$S = \pi r l$

where l is the **slant height** of the cone.

So the total surface area of a cone is given by the are of the curved surface plus the area of its circular base.

$A = \pi r l + \pi r^2$

EXAMPLE 17

For the cone in the diagram, calculate:

i   its volume

ii   its total surface area.

Give your answers in terms of π.

i   The volume is given by $V = \frac{1}{3}\pi r^2 h$

$= \frac{1}{3} \times \pi \times 36 \times 8 = 96\pi$ cm^3

ii   The total surface area is given by $A = \pi r l + \pi r^2$

$= \pi \times 6 \times 10 + \pi \times 36 = 96\pi$ cm^2

EXERCISE 26K

1 For each cone, calculate:

i   its volume

ii   its total surface area.

Give your answers to 3 significant figures.

a

b

c

2 Find the total surface area of a cone with base radius 3 cm and slant height 5 cm. Give your answer in terms of π.

3 Calculate the volume of each of these shapes. Give your answers in terms of π.

a

b

4 You could work with a partner on this question.

A sector of a circle, as in the diagram, can be made into a cone (without a base) by sticking the two straight edges together.

a What would be the diameter of the base of the cone in this case?

b What is the diameter if the angle is changed to 180°?

c Investigate other angles.

5 A cone has the dimensions shown in the diagram.
Calculate the total surface area, leaving your answer in terms of π.

6 The slant height of a cone is equal to the base diameter. Show that the area of the curved surface is twice the area of the base.

7 The model shown on the right is made from aluminium.

What is the mass of the model, given that the density of aluminium is 2.7 g/cm³? (This means that 1 cm³ of aluminium has a mass of 2.7 g.)

8 A container in the shape of a cone, base radius 10 cm and vertical height 19 cm, is full of water. The water is poured into an empty cylinder of radius 15 cm. How high is the water in the cylinder?

The volume of a **sphere**, radius r, is given by:

$V = \frac{4}{3}\pi r^3$

Its surface area is given by:

$A = 4\pi r^2$

EXAMPLE 18

A sphere has a radius of 8 cm. Calculate:
i its volume
ii its surface area.

i The volume is given by:

$V = \frac{4}{3}\pi r^3$

$= \frac{4}{3} \times \pi \times 8^3 = \frac{2048}{3} \times \pi = 2140$ cm^3 (3 significant figures)

ii The surface area is given by:

$A = 4\pi r^2$

$= 4 \times \pi \times 8^2 = 256 \times \pi = 804$ cm^2 (3 significant figures)

EXERCISE 26L

1 Calculate the volume and surface area of each of these spheres. Give your answers in terms of π.

  a Radius 3 cm   **b** Radius 6 cm   **c** Diameter 20 cm

2 Calculate the volume and the surface area of a sphere with a diameter of 50 cm.

3 A sphere fits exactly into an open cubical box of side 25 cm. Calculate:

  a the surface area of the sphere   **b** the volume of the sphere.

4 A metal sphere of radius 15 cm is melted down and recast into a solid cylinder of radius 6 cm. Calculate the height of the cylinder.

5 Lead has a density of 11.35 g/cm^3. Calculate the maximum number of lead spheres of radius 1.5 mm that can be made from 1 kg of lead.

6 A sphere has a radius of 5.0 cm. A cone has a base radius of 8.0 cm. The sphere and the cone have the same volume.

Calculate the height of the cone.

7 A sphere of diameter 10 cm is carved out of a wooden block in the shape of a cube of side 10 cm. What percentage of the wood is wasted?

Why this chapter matters

If you look carefully, you will be able to spot symmetry all around you. It is present in the natural world and in objects made by humans. But, does it have a purpose and why do we need it?

Symmetry in nature, art and literature

Symmetry is everywhere you look in nature. Plants and animals have symmetrical body shapes and patterns. For example, if you divide a leaf in half, you will see that one half is the same shape as the other half.

Where is the symmetry in this butterfly, star fish and peacock? What effect does this symmetry have?

Pegasus.

This painting by a Dutch artist called M.C. Escher (1898–1972) uses line symmetry and rotational symmetry. Why do you think Escher used symmetry in his paintings?

Every tiger has its own unique pattern of stripes. These appear on the tiger's skin as well as its fur. What purpose do these symmetrical stripes serve?

Can you identify symmetry in the face of the tiger?

Symmetry in structures

St Peter's Basilica, Rome.

St Peter's Basilica, in the Vatican City in Rome, was started in 1506 and completed in 1626. It is a very symmetrical structure – see if you can identify all the symmetry that is present.

Why do you think that the designers of this building used symmetry?

These examples show some of the uses of symmetry in the world. Now think about where symmetry occurs in your own life – how important is it to you?

27 Symmetry

Topics		Level	Key words
1	Lines of symmetry	CORE	line of symmetry, mirror line
2	Rotational symmetry	CORE	rotational symmetry, order of rotational symmetry
3	Symmetry of special two-dimensional shapes	CORE	rectangle, square, parallelogram, kite, rhombus, trapezium, isosceles triangle, equilateral triangle
4	Symmetry of three-dimensional shapes	EXTENDED	plane of symmetry, axis of symmetry, cuboid, prism, pyramid, cone, cylinder, reflection
5	Symmetry in circles	EXTENDED	centre, chord, tangent, perpendicular bisector

In this chapter you will learn how to:

CORE	EXTENDED
<ul><li>Recognise rotational and line symmetry (including order of rotational symmetry) in two dimensions. (C3.5 and E3.5)</li><li>Recognise properties of triangles, quadrilaterals and circles directly related to their symmetries. (C3.5 and E3.5)</li></ul>	<ul><li>Recognise symmetry properties of the prism (including cylinder) and the pyramid (including cone). (E3.5)</li><li>Use the following symmetry properties of circles:<ul><li>equal chords are equidistant from the centre</li><li>the perpendicular bisector of a chord passes through the centre</li><li>tangents from an external point are equal in length (E3.5)</li></ul></li></ul>

27.1 Lines of symmetry

Many two-dimensional shapes have one or more lines of symmetry.

A **line of symmetry** is a line that can be drawn through a shape so that what can be seen on one side of the line is the mirror image of what is on the other side. This is why a line of symmetry is sometimes called a **mirror line**.

It is also the line along which a shape can be folded exactly onto itself.

EXAMPLE 1

Find the number of lines of symmetry for this cross.

There are four altogether.

EXERCISE 27A

1 Copy these shapes and draw on the lines of symmetry for each one. If it will help you, use tracing paper or a mirror to check your results.

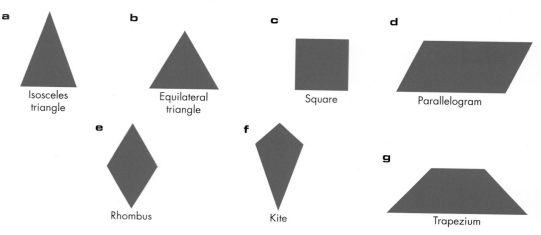

a Isosceles triangle b Equilateral triangle c Square d Parallelogram e Rhombus f Kite g Trapezium

2 **a** Find the number of lines of symmetry for each of these regular polygons.

i

Regular pentagon

ii

Regular hexagon

iii

Regular octagon

b How many lines of symmetry do you think a regular decagon has? (A decagon is a ten-sided polygon.)

3 Write down the number of lines of symmetry for each of these flags.

Austria   Canada   Iceland   Switzerland   Greece

4 These road signs all have lines of symmetry. Copy them and draw on the lines of symmetry for each one.

5 The animal and plant kingdoms are full of symmetry. Four examples are given below. State the number of lines of symmetry for each one.

a   **b**   **c**   **d**

Can you find other examples? Find suitable pictures, copy them and state the number of lines of symmetry each one has.

6 Copy this diagram.

On your copy, shade in four more squares so that the diagram has four lines of symmetry.

A two-dimensional shape has **rotational symmetry** if it can be rotated about a point to look exactly the same in a new position.

The **order of rotational symmetry** is the number of different positions in which the shape looks the same when it is rotated 360° about the point (that is, one complete turn).

The easiest way to find the order of rotational symmetry for any shape is to trace it and count the number of times that the shape looks the same as you turn the tracing paper through one complete turn.

EXAMPLE 2

Find the order of rotational symmetry for this shape.

First, hold the tracing paper on top of the shape and trace the shape. Then rotate the tracing paper and count how many times the tracing matches the original shape in one complete turn.

You will find three different positions.

So, the order of rotational symmetry for the shape is 3.

EXERCISE 27B

1 Copy these shapes and write below each one the order of rotational symmetry. If it will help you, use tracing paper.

a
Square

b
Rectangle

c
Parallelogram

d
Equilateral triangle

e
Regular hexagon

ADVICE AND TIPS

Remember a shape with rotational symmetry of order 1 has no rotational symmetry.

2 Find the order of rotational symmetry for each of these shapes.

a

b

c

d

e

CORE

3 These are Greek capital letters. Write down the order of rotational symmetry for each one.

ᵃ Φ   ᵇ H   ᶜ Z   ᵈ Θ   ᵉ Ξ

4 Here is a star pattern.

Inside the star there are two patterns that have rotational symmetry.

a What is the order of rotational symmetry of the whole star?

b What is the order of rotational symmetry of the two patterns inside the star?

5 Copy the grid on the right. On your copy, shade in four squares so that the shape has rotational symmetry of order 2.

27.3 Symmetry of special two-dimensional shapes

Some three and four-sided shapes have special names such as **isosceles triangle** or **parallelogram**. You need to know the symmetry properties of these shapes.

For example:

- An isosceles triangle has one line of symmetry and no rotational symmetry.

- A parallelogram has no lines of symmetry and rotational symmetry of order 2.

EXERCISE 27C

1 Draw diagrams to show all the lines of symmetry on:

a a **rectangle**   **b** a **kite**   **c** a **square**

d an **equilateral triangle**   **e** a **rhombus**.

CORE

2 **a** Which shape in question 1 has no rotational symmetry?

b Find the order of rotational symmetry for each of the others.

3 **a** What do you call a triangle with one line of symmetry?

b Can you draw a triangle with exactly two lines of symmetry?

4 What is the name for:

a a quadrilateral with no lines of symmetry and rotational symmetry of order 2

b a quadrilateral with rotational symmetry of order 4?

5 **a** Name two different quadrilaterals that have two lines of symmetry and rotational symmetry of order 2.

b Can you draw a quadrilateral that has two lines of symmetry but no rotational symmetry?

6 The dotted line is a line of symmetry.

a Which angles *must* be equal?

b Which sides *must* be equal?

c What is the name of this shape?

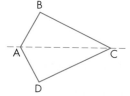

7 **a** What is the special name for a line of symmetry of a circle?

b How many lines of symmetry does a circle have?

c What is the order of rotational symmetry of a circle?

8 This shape has rotational symmetry of order 2.

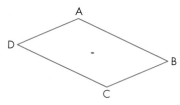

a Which angles *must* be equal?

b Which sides *must* be equal?

c What is the name of this shape?

9 If a **trapezium** has a line of symmetry, what can you say about its angles?

Symmetry of three-dimensional shapes

Three-dimensional shapes can be symmetrical in two ways.

A **cuboid** has a **plane of symmetry** that divides it into two halves. One is the **reflection** of the other half in a mirror on the plane of symmetry.

A cuboid has an **axis of symmetry**. It has rotational symmetry of order 2 about this axis.

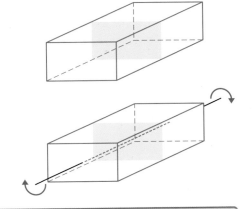

EXERCISE 27D

1 A cuboid has three planes of symmetry. Draw diagrams to show them.

2 A cuboid has three axes of symmetry.

 a Show them on a diagram.

 b What is the order of rotational symmetry about each axis?

3 This is an equilateral triangular **prism**.
It has *four* planes of symmetry.
One is shown in the diagram.

Draw diagrams to show the other three.

4 The diagram shows an equilateral prism.
AB and CD are axes of symmetry.

 a What is the order of rotational symmetry about each one?

 b The shape has two more axes of symmetry.
 Show them on a diagram.

5 The diagram shows an axis of symmetry for a square-based **pyramid**.

 a What is the order of rotational symmetry about the axis?

 b Does the pyramid have any other axes of symmetry?

6 Draw diagrams to show the planes of symmetry of a square-based pyramid.

How many are there?

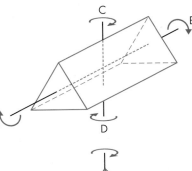

7 **a** Draw a diagram to show a plane of symmetry for a **cylinder**.

b One axis of symmetry for a cylinder is shown in this diagram. Draw a diagram to show a different one.

8 **a** How many axes of rotational symmetry does a **cone** have?

b How many planes of symmetry does a cone have?

9 The end of a prism is a regular hexagon.

a Show that the prism has seven planes of symmetry.

b How many axes of symmetry does it have?

27.5 Symmetry in circles   E

Here are some facts about circles it is useful to know:

1 If two **chords** are the same length, they are the same distance from the centre.

If O is the centre of the circle and AB and CD are equal in length, then OM = ON.

2 The **perpendicular bisector** of a chord passes through the centre of the circle.

In this diagram the **centre** must be on the broken line.

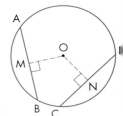

3 The two **tangents** from the point to a circle are equal in length.

PS and PT are the same length.

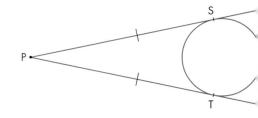

EXERCISE 27E

EXTENDED

1 a Draw a circle. Use a pair of compasses.

b Draw two chords, AB and CD.

c Use compasses and a ruler to construct the perpendicular bisector of AB.

d Construct the perpendicular bisector of CD.

e You should find that the perpendicular bisectors cross at the centre of the circle. Explain why this is the case.

f You could draw a circle by drawing round a circular object such as a plate or food can. How could you use chords to find the centre of the circle?

2 AB and CD are two chords of a circle that are the same length. O is the centre of the circle.

a What sort of triangle is AOB? Give a reason for your answer.

b Explain why triangles AOB and COD are congruent.

c If angle OAB is 65°, find the angle COD.

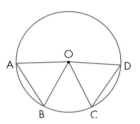

3 O is the centre of a circle. EF is a chord. M is the mid-point of EF.

a Show that triangles EOM and FOM are congruent.

b Explain why angle EMO is a right angle.

c If angle MOF is 72°, find angle MEO.

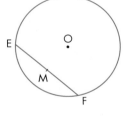

4 O is the centre of a circle. PX and PY are tangents.

a Explain why angle PXO must be 90°.

b Show that triangles XPO and YPO are congruent.

c Angle XPO = 17°. Calculate angle XOY.

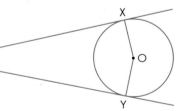

5 O is the centre of the circle. AB is a chord 6 cm long. Angle ACO is 90°. OC = 4 cm.

Calculate the radius of the circle.

6 O is the centre of the circle.

XZ and YZ are tangents.

Calculate the value of *a*.

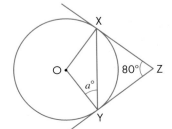

Why this chapter matters

> Vectors are used to represent any quantity that has both magnitude and direction. The velocity of a speeding car is its direction *and* its speed. Velocity is a vector.

To understand how a force acts on an object, you need to know the magnitude of the force and the direction in which it moves – that is, its vector.

In science, vectors are used to describe displacement, acceleration and momentum.

But are vectors used in real life? Yes! Here are some examples.

In the 1950s, a group of talented Brazilian footballers invented the swerving free kick. By kicking the ball in just the right place, they managed to make it curl around the wall of defending players and go into the goal. When a ball is in flight, it is acted upon by various forces which can be described by vectors.

Formula One teams always employ physicists and mathematicians to help them build the perfect racing car. Since vectors describe movements and forces, they are used as the basis of a car's design.

Pilots have to consider wind speed and direction when they plan to land an aircraft at an airport. Vectors are an important part of the computerised landing system.

Vectors play a key role in the design of aircraft wings, where an upward force or lift is needed to enable the aircraft to fly.

Vectors are used extensively in computer graphics. Software for animations uses the mathematics of vectors.

28 Vectors

Topics	Level	Key words
1 Introduction to vectors	**CORE**	magnitude, direction, vector, column, scalar, coordinate grid
2 Using vectors	**EXTENDED**	position vector
3 The magnitude of a vector	**EXTENDED**	magnitude, Pythagoras' theorem

In this chapter you will learn how to:

CORE	EXTENDED
• Describe a translation by using a vector represented by eg: $\begin{pmatrix} x \\ y \end{pmatrix}$, $\overrightarrow{AB}$ or **a**. (C7.1 and E7.1) • Add and subtract vectors. (C7.1 and E7.1) • Multiply a vector by a scalar. (C7.1 and E7.1)	• Calculate the magnitude of a vector $\begin{pmatrix} x \\ y \end{pmatrix}$ as $\sqrt{x^2 + y^2}$. (E7.3) • Represent vectors by directed line segments. (E7.3) • Use the sum and difference of two vectors to express given vectors in terms of two coplanar vectors. (E7.3) • Use position vectors. (E7.3)

A **vector** is something that has both **magnitude** and **direction** and can be represented by an arrow.

Examples are velocity, acceleration, force and momentum.

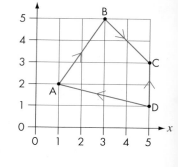

Vectors can be written down in several ways:

$\overrightarrow{AB}$ – giving the start and end points with an arrow over the top.

a – as a lower-case letter printed in **bold**. When you are writing vectors **a**, **b** ... by hand you cannot show them in bold type. Instead write them with a line underneath as <u>a</u>, <u>b</u> ... to show that they stand for a vector and not a number.

When vectors are drawn on a **coordinate grid** they can be represented by two numbers in brackets in a **column**.

The top number shows how far the line moves from one side to the other between its start and end points, and the bottom number shows how far it moves up or down.

For example, on this grid:

$\overrightarrow{AB} = \begin{pmatrix} 2 \\ 3 \end{pmatrix}$ means move 2 right and 3 upwards to get to point B from point A.

If a line moves either downwards or left the coordinate is negative. So:

$\overrightarrow{BC} = \begin{pmatrix} 2 \\ -2 \end{pmatrix}$ means move 2 right and 2 downwards to get to point C from point B.

$\overrightarrow{DA} = \begin{pmatrix} -4 \\ 1 \end{pmatrix}$ means move 4 left and 1 upwards to get to point A from point D.

$\overrightarrow{DC} = \begin{pmatrix} 0 \\ 2 \end{pmatrix}$ means move neither left or right, just upwards to get to point C from point D.

Notice that the line joining A and B can be written as AB and BA and these both have the same magnitude (in this case length).

But the vectors $\overrightarrow{AB}$ and $\overrightarrow{BA}$ are not the same because their directions are different:

$\overrightarrow{AB} = \begin{pmatrix} 2 \\ 3 \end{pmatrix}$ and $\overrightarrow{BA} = \begin{pmatrix} -2 \\ -3 \end{pmatrix}$

Therefore $\overrightarrow{BA} = -\overrightarrow{AB}$

ADVICE AND TIPS

Do not forget the arrow to indicate a vector.

The vectors on this grid show that if $\mathbf{a} = \begin{pmatrix} 1 \\ 2 \end{pmatrix}$ then $-\mathbf{a} = \begin{pmatrix} -1 \\ -2 \end{pmatrix}$.

$-\mathbf{a}$ is a vector with the same length (magnitude) but acting in the opposite direction as the vector **a**.

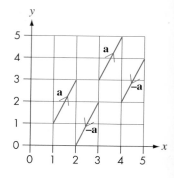

Adding, subtracting and multiplying vectors

Vectors are added together by placing them end to end.

On this grid **a** + **b** = **c**

or $\begin{pmatrix} 3 \\ 2 \end{pmatrix} + \begin{pmatrix} 1 \\ -3 \end{pmatrix} = \begin{pmatrix} 4 \\ -1 \end{pmatrix}$

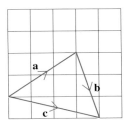

Notice that you add the top figures together (3 + 1 = 4) and the bottom figures together (2 + −3 = −1).

Vectors can be subtracted too.

a − **b** = **d**

a − **b** is the same as **a** + (−**b**)

$\begin{pmatrix} 3 \\ 2 \end{pmatrix} - \begin{pmatrix} 1 \\ -3 \end{pmatrix} = \begin{pmatrix} 3 \\ 2 \end{pmatrix} + \begin{pmatrix} -1 \\ 3 \end{pmatrix} = \begin{pmatrix} 2 \\ 5 \end{pmatrix}$

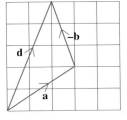

Notice that 3 − 1 = 2 and 2 − −3 = 5.

Vectors can be multiplied by a number.

If **a** = $\begin{pmatrix} 3 \\ 2 \end{pmatrix}$

$3\mathbf{a} = 3 \times \begin{pmatrix} 3 \\ 2 \end{pmatrix} = \begin{pmatrix} 9 \\ 6 \end{pmatrix}$

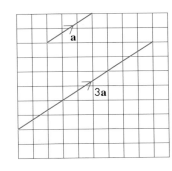

Notice that 3 × 3 = 9 and 3 × 2 = 6.

Here 3 is called a **scalar**, to distinguish it from a vector.

If k is a scalar, then $k\begin{pmatrix} x \\ y \end{pmatrix} = \begin{pmatrix} kx \\ ky \end{pmatrix}$.

EXERCISE 28A

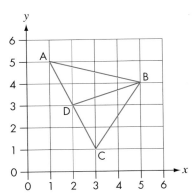

a Write these as column vectors.

  i $\overrightarrow{AB}$ **ii** $\overrightarrow{DB}$ **iii** $\overrightarrow{CB}$ **iv** $\overrightarrow{CA}$ **v** $\overrightarrow{AC}$ **vi** $\overrightarrow{DA}$

b Show that $\overrightarrow{AD} = \overrightarrow{DC}$. What does this tell you about the position of D on line AC?

CORE

CORE

2 F has the coordinates (4, 2), G has the coordinates (2, 6), M is the midpoint of FG and O is the origin.

    a Mark F, G and M on a coordinate grid.

       Write these as column vectors.

       i $\overrightarrow{FG}$ **ii** $\overrightarrow{GF}$ **iii** $\overrightarrow{OM}$ **iv** $\overrightarrow{MG}$ **v** $\overrightarrow{GO}$

3 A has the coordinates (3, 4).

$$\overrightarrow{AP} = \begin{pmatrix} 2 \\ 1 \end{pmatrix}, \overrightarrow{AQ} = \begin{pmatrix} 1 \\ -3 \end{pmatrix}, \overrightarrow{AR} = \begin{pmatrix} -2 \\ 0 \end{pmatrix}$$

Mark A, P, Q and R on a coordinate grid.

4 $\mathbf{a} = \begin{pmatrix} 2 \\ 3 \end{pmatrix}$ and $\mathbf{b} = \begin{pmatrix} 1 \\ -4 \end{pmatrix}$

Draw diagrams to show these vectors.

    a $\mathbf{a} + \mathbf{b}$ **b** $-\mathbf{b}$ **c** $\mathbf{a} - \mathbf{b}$ **d** $\mathbf{b} + \mathbf{a}$ **e** $\mathbf{b} - \mathbf{a}$ **f** $2\mathbf{b}$

5 $\mathbf{e} = \begin{pmatrix} -3 \\ -2 \end{pmatrix}$ and $\mathbf{f} = \begin{pmatrix} 2 \\ 4 \end{pmatrix}$

Work out these vectors.

    a $\mathbf{e} + \mathbf{f}$ **b** $3\mathbf{f}$ **c** $\mathbf{e} - \mathbf{f}$ **d** $\mathbf{f} - \mathbf{e}$ **e** $4\mathbf{e}$ **f** $2\mathbf{f} + \mathbf{e}$

6

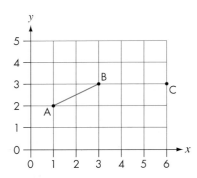

    a Copy this diagram.

    b $\overrightarrow{AE} = 2\overrightarrow{AB}$. Mark E on the grid.

    c $\overrightarrow{CD} = -2\overrightarrow{AB}$. Mark D on the grid.

    d $\overrightarrow{AB} = 2\overrightarrow{AM}$. Mark M on the grid.

    e $\overrightarrow{CN} = 2\overrightarrow{CB}$. Mark N on the grid.

    f If $\overrightarrow{DC} = k\overrightarrow{AM}$, what is the value of k?

This grid is made of identical parallelograms.

O is the origin.

The **position vector** of A is

$$\overrightarrow{OA} = \mathbf{a}$$

The **position vector** of B is

$$\overrightarrow{OB} = \mathbf{b}$$

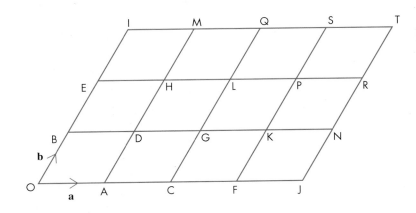

You can give the position vectors of the other points in terms of **a** and **b**. For example:

The position vector of G = $\overrightarrow{OG}$ = 2**a** + **b**

The position vector of S = $\overrightarrow{OS}$ = 3**a** + 3**b**

You can write other vectors in terms of **a** and **b**. For example:

$$\overrightarrow{CL} = 2\mathbf{b}$$
$$\overrightarrow{CP} = \overrightarrow{CL} + \overrightarrow{LP} = 2\mathbf{b} + \mathbf{a} \text{ or } \mathbf{a} + 2\mathbf{b}$$
$$\overrightarrow{CH} = \overrightarrow{CL} + \overrightarrow{LH} = 2\mathbf{b} + -\mathbf{a} = 2\mathbf{b} - \mathbf{a}$$

EXAMPLE 1

a Using the grid above, write down these vectors in terms of **a** and **b**.

i $\overrightarrow{BH}$ ii $\overrightarrow{HP}$ iii $\overrightarrow{GT}$

iv $\overrightarrow{TI}$ v $\overrightarrow{FH}$ vi $\overrightarrow{BQ}$

b What is the relationship between the following vectors?

i $\overrightarrow{BH}$ and $\overrightarrow{GT}$ ii $\overrightarrow{BQ}$ and $\overrightarrow{GT}$ iii $\overrightarrow{HP}$ and $\overrightarrow{TI}$

c Show that B, H and Q lie on the same straight line.

a i **a** + **b**   ii 2**a**   iii 2**a** + 2**b**   iv −4**a**   v −2**a** + 2**b**   vi 2**a** + 2**b**

b i $\overrightarrow{BH}$ and $\overrightarrow{GT}$ are parallel and $\overrightarrow{GT}$ is twice the length of $\overrightarrow{BH}$.

ii $\overrightarrow{BQ}$ and $\overrightarrow{GT}$ are equal.

iii $\overrightarrow{HP}$ and $\overrightarrow{TI}$ are in opposite directions and $\overrightarrow{TI}$ is twice the length of $\overrightarrow{HP}$.

c $\overrightarrow{BH}$ and $\overrightarrow{BQ}$ are parallel and start at the same point B. Therefore, B, H and Q must lie on the same straight line.

EXERCISE 28B

1 On this grid O is the origin, $\overrightarrow{OA}$ is **a** and $\overrightarrow{OB}$ is **b**.

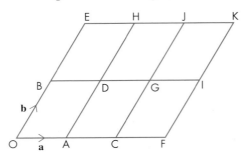

   a Name three other vectors equivalent to **a**.

   b Name three other vectors equivalent to **b**.

   c Name three vectors equivalent to –**a**.

   d Name three vectors equivalent to –**b**.

2 Using the same grid as in question **1**, write these vectors in terms of **a** and **b**.

   a the position vector of C   **b** the position vector of E

   c the position vector of K   **d** $\overrightarrow{OH}$

   e $\overrightarrow{AG}$ **f** $\overrightarrow{AK}$ **g** $\overrightarrow{BK}$

3 On the grid in question **1**, there are three vectors equivalent to $\overrightarrow{OG}$. Name all three.

4 On the grid in question **1**, there are three vectors that are three times the magnitude of $\overrightarrow{OA}$ and act in the same direction. Name all three.

5 Copy this grid.

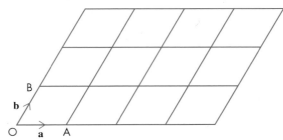

On your copy, mark the points C to G to show these vectors.

   a $\overrightarrow{OC} = 2\mathbf{a} + 3\mathbf{b}$ **b** $\overrightarrow{OD} = 2\mathbf{a} + \mathbf{b}$ **c** $\overrightarrow{OE} = 4\mathbf{a}$

   d $\overrightarrow{OF} = 4\mathbf{a} + 2\mathbf{b}$ **e** $\overrightarrow{OG} = \frac{1}{2}\mathbf{a} + 2\mathbf{b}$

6 On this grid, $\overrightarrow{OA}$ is **a** and $\overrightarrow{OB}$ is **b**.

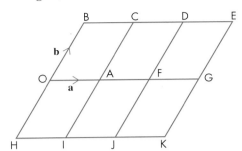

Write down vectors in terms of **a** and **b**.

a $\overrightarrow{OH}$ **b** $\overrightarrow{OK}$ **c** $\overrightarrow{OJ}$

d $\overrightarrow{OI}$ **e** $\overrightarrow{OC}$ **f** $\overrightarrow{CO}$

g $\overrightarrow{AK}$ **h** $\overrightarrow{DI}$ **i** $\overrightarrow{JE}$

j $\overrightarrow{AB}$ **k** $\overrightarrow{CK}$ **l** $\overrightarrow{DK}$

7 **a** On the grid in question **6**, there are two vectors that are twice the size of $\overrightarrow{AB}$ and act in the opposite direction. Name both of them.

b On the grid in question **6**, there are three vectors that are three times the size of $\overrightarrow{OA}$ and act in the opposite direction. Name all three.

8 Copy this grid.

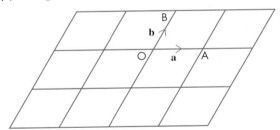

On your copy, mark the points C to P to show these vectors.

a $\overrightarrow{OC} = 2\mathbf{a} - \mathbf{b}$ **b** $\overrightarrow{OD} = 2\mathbf{a} + \mathbf{b}$

c $\overrightarrow{OE} = \mathbf{a} - 2\mathbf{b}$ **d** $\overrightarrow{OF} = \mathbf{b} - 2\mathbf{a}$

e $\overrightarrow{OG} = -\mathbf{a}$ **f** $\overrightarrow{OH} = -\mathbf{a} - 2\mathbf{b}$

g $\overrightarrow{OI} = 2\mathbf{a} - 2\mathbf{b}$ **h** $\overrightarrow{OJ} = -\mathbf{a} + \mathbf{b}$

i $\overrightarrow{OK} = -\mathbf{a} - \mathbf{b}$ **j** $\overrightarrow{OM} = -\mathbf{a} - \frac{3}{2}\mathbf{b}$

k $\overrightarrow{ON} = -\frac{1}{2}\mathbf{a} - 2\mathbf{b}$ **l** $\overrightarrow{OP} = \frac{3}{2}\mathbf{a} - \frac{3}{2}\mathbf{b}$

9 The diagram shows two sets of parallel lines. O is the origin.

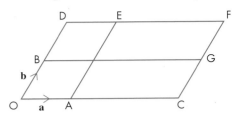

$\overrightarrow{OA} = \mathbf{a}$ and $\overrightarrow{OB} = \mathbf{b}$

$\overrightarrow{OC} = 3\overrightarrow{OA}$ and $\overrightarrow{OD} = 2\overrightarrow{OB}$

a Write down these vectors in terms of **a** and **b**.

   i $\overrightarrow{OF}$ **ii** $\overrightarrow{OG}$ **iii** $\overrightarrow{EG}$ **iv** $\overrightarrow{CE}$

b Write down two vectors that can be written as $3\mathbf{a} - \mathbf{b}$.

10 This grid shows the vectors $\overrightarrow{OA} = \mathbf{a}$ and $\overrightarrow{OB} = \mathbf{b}$. O is the origin.

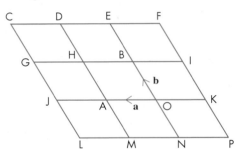

Write down the position vectors of these points.

a G   **b** F

c The midpoint of DH   **d** The centre of OAHB

e The centre of DCGH   **f** The centre of AMLJ

11 The diagram shows the vectors $\overrightarrow{OA} = \mathbf{a}$ and $\overrightarrow{OB} = \mathbf{b}$. M is the midpoint of AB.

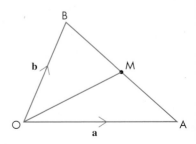

a **i** Work out the vector $\overrightarrow{AB}$ in terms of **a** and **b**.

   ii Work out the vector $\overrightarrow{AM}$.

   iii Explain why $\overrightarrow{OM} = \overrightarrow{OA} + \overrightarrow{AM}$.

   iv Using your answers to parts **ii** and **iii**, work out $\overrightarrow{OM}$ in terms of **a** and **b**.

b Copy the diagram and show on it the vector $\overrightarrow{OC}$ which is equal to $\mathbf{a} + \mathbf{b}$.

c Describe in geometrical terms the position of M in relation to O, A, B and C.

12 The diagram shows the vectors $\overrightarrow{OA} = \mathbf{a}$ and $\overrightarrow{OB} = \mathbf{b}$.
The point C divides the line AB in the ratio 1:2.

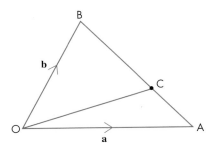

> **ADVICE AND TIPS**
>
> AC is $\frac{1}{3}$ the distance from A to B.

a i Work out the vector $\overrightarrow{AB}$.

   ii Work out the vector $\overrightarrow{AC}$.

  iii Work out the vector $\overrightarrow{OC}$ in terms of **a** and **b**.

b If C now divides the line AB in the ratio 1:3, write down the vector that represents $\overrightarrow{OC}$.

> **ADVICE AND TIPS**
>
> AC is now $\frac{1}{4}$ the distance from A to B.

13 The diagram shows the vectors $\overrightarrow{OA} = \mathbf{a}$ and $\overrightarrow{OB} = \mathbf{b}$.

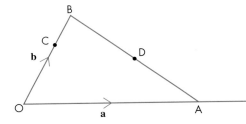

> **ADVICE AND TIPS**
>
> OC is $\frac{2}{3}$ the distance from O to B.

The point C divides OB in the ratio 2:1. The point E is such that $\overrightarrow{OE} = 2\overrightarrow{OA}$. D is the midpoint of AB.

a Write down (or work out) these vectors in terms of **a** and **b**.

  i $\overrightarrow{OC}$ **ii** $\overrightarrow{OD}$ **iii** $\overrightarrow{CO}$

b The vector $\overrightarrow{CD}$ can be written as $\overrightarrow{CD} = \overrightarrow{CO} + \overrightarrow{OD}$. Use this fact to work out $\overrightarrow{CD}$ in terms of **a** and **b**.

c Write down a similar rule to that in part **b** for the vector $\overrightarrow{DE}$. Use this rule to work out $\overrightarrow{DE}$ in terms of **a** and **b**.

d Explain why C, D and E lie on the same straight line.

The size or **magnitude** of a vector is represented by two vertical lines which stand for 'magnitude of' or 'length of', eg $|\overrightarrow{AB}|$ or $|\mathbf{a}|$.

If a vector is drawn on a rectangular coordinate grid you can use **Pythagoras' theorem** to calculate the magnitude.

For example:

if $\overrightarrow{AB} = \begin{pmatrix} 4 \\ -3 \end{pmatrix}$

then it can form the hypotenuse of a triangle with sides of lengths 3 and 4 as shown on the grid.

The square of the hypotenuse is equal to the sum of the squares of the other two sides so:

$$|\overrightarrow{AB}| = \sqrt{4^2 + 3^2}$$

$$= \sqrt{25}$$

$$= 5$$

In general, if $\mathbf{a} = \begin{pmatrix} x \\ y \end{pmatrix}$ then $|\mathbf{a}| = \sqrt{x^2 + y^2}$.

EXAMPLE 2

If A has a coordinates (3, −2) and B has coordinates (−3, 5), calculate $|\overrightarrow{AB}|$.

$$\overrightarrow{AB} = \begin{pmatrix} -6 \\ 7 \end{pmatrix}$$

$$|\overrightarrow{AB}| = \sqrt{(-6)^2 + 7^2}$$

$$= \sqrt{36 + 49}$$

$$= \sqrt{85}$$

$$= 9.22 \text{ to 2 decimal places}$$

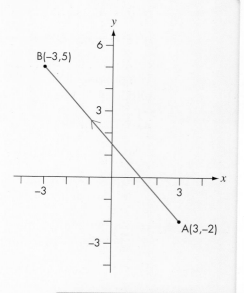

ADVICE AND TIPS

Remember that $(-6)^2 = 6^2$

EXTENDED

EXERCISE 28C

1 O is the origin.

P, Q and R have position vectors $\begin{pmatrix} 3 \\ 5 \end{pmatrix}$, $\begin{pmatrix} 6 \\ -2 \end{pmatrix}$, $\begin{pmatrix} 0 \\ -4 \end{pmatrix}$.

a Show O, P, Q and R on a diagram.

b Calculate $|\overrightarrow{OP}|$, $|\overrightarrow{OQ}|$ and $|\overrightarrow{OR}|$. You can leave square root signs in your answers.

c Calculate $|\overrightarrow{PQ}|$.

d Calculate $|\overrightarrow{QR}|$.

2 $\mathbf{a} = \begin{pmatrix} 6 \\ 8 \end{pmatrix}$ and $\mathbf{b} = \begin{pmatrix} 5 \\ -12 \end{pmatrix}$

a Calculate $|\mathbf{a}|$ and $|\mathbf{b}|$.

b Calculate $\mathbf{a} + \mathbf{b}$.

c Calculate $|\mathbf{a} + \mathbf{b}|$.

d Is it true that $|\mathbf{a} + \mathbf{b}| = |\mathbf{a}| + |\mathbf{b}|$? Give a reason for your answer.

e Calculate $|\mathbf{a} - \mathbf{b}|$.

f Calculate $|\mathbf{b} - \mathbf{a}|$.

g Is it always true that $|\mathbf{a} - \mathbf{b}| = |\mathbf{b} - \mathbf{a}|$? Give a reason for your answer.

3 A, B, C and D have position vectors $\begin{pmatrix} 4 \\ 4 \end{pmatrix}$, $\begin{pmatrix} 10 \\ 12 \end{pmatrix}$, $\begin{pmatrix} -4 \\ 10 \end{pmatrix}$ and $\begin{pmatrix} 4 \\ -6 \end{pmatrix}$.

a Calculate $|\overrightarrow{AB}|$, $|\overrightarrow{AC}|$ and $|\overrightarrow{AD}|$.

b Explain why B, C and D must lie on a circle with centre A and state the radius of the circle.

4 $\mathbf{c} = \begin{pmatrix} 1 \\ 4 \end{pmatrix}$ and $\mathbf{d} = \begin{pmatrix} -2 \\ 5 \end{pmatrix}$

Calculate:

a $|\mathbf{c}|$

b $|3\mathbf{d}|$

c $|2\mathbf{c} + \mathbf{d}|$

d $|4\mathbf{c} - 2\mathbf{d}|$.

How many sides does a strip of paper have? Two or one?

Take a strip of paper about 20 cm by 2 cm.

How many sides does it have? Easy! You can see that this has two sides, a topside and an underside. If you were to draw a line along one side of the strip, you would have one side with a line 20 cm long on it and one side blank.

Now mark the ends A and B, put a single twist in the strip of paper and tape (or glue) the two ends together, as shown.

How many sides does this strip of paper have now?

Take a pen and draw a line on the paper, starting at any point you like. Continue the line along the length of the paper – you will eventually come back to your starting point. Your strip has only one side now! There is no blank side.

You have transformed a two-sided piece of paper into a one-sided piece of paper.

This shape is called a Möbius strip. It is named after August Ferdinand Möbius, a 19th-century German mathematician. Möbius caused a revolution in geometry.

Möbius strips have a number of applications that use its property of one-sidedness, including conveyor belts in industry and in vacuum cleaners.

The Möbius strip has become the universal symbol of recycling. The symbol was created in 1970 by Gary Anderson at the University of Southern California, as part of a contest sponsored by a paper company.

The Möbius strip is a form of transformation. In this chapter, you will look at some other transformations of shapes.

Chapter

29 Transformations

Topics	Level	Key words
1 Translations	CORE	transformation, translation, vector
2 Reflections	CORE	reflection, object, image, mirror line
3 Further reflections	EXTENDED	
4 Rotations	CORE	rotation, centre of rotation, angle of rotation, clockwise, anticlockwise
5 Further rotations	EXTENDED	
6 Enlargements	CORE	scale factor, enlargement, centre of enlargement, ray methods, coordinate method
7 Further enlargements	EXTENDED	negative enlargement
8 Combined transformations	EXTENDED	

In this chapter you will learn how to:

CORE	EXTENDED
• Reflect simple plane figures in horizontal or vertical lines. (C7.2 and E7.2) • Construct given translations and enlargements (with positive and fractional scale factors) of simple plane figures. (C7.2) • Recognise and describe reflections, rotations, translations and enlargements (with positive and fractional scale factors). (C7.2) • Rotate simple plane figures about the origin, vertices or midpoints of edges of the figures, through multiples of 90°. (C7.2)	• Construct given translations and enlargements (including positive, negative and fractional scale factors) of simple plane figures. (E7.2) • Recognise and describe reflections, rotations, translations and enlargements (including positive, negative and fractional scale factors). (E7.2) • Rotate simple plane figures through multiples of 90°. (E7.2) • Use the following transformations of the plane: reflection (M); rotation (R); translation (T); enlargement (including fractional and negative scale factors) (E); and their combinations. (E7.5) • Identify and give precise descriptions of transformations connecting given figures. (E7.5)

A **transformation** changes the position or the size of a shape.

There are four basic ways of changing the position and size of two-dimensional shapes: a **translation**, a reflection, a rotation or an enlargement. All of these transformations, except enlargement, keep shapes congruent.

A translation is the 'movement' of a shape from one place to another without reflecting it or rotating it. It is sometimes called a glide, since the shape appears to glide from one place to another. Every point in the shape moves in the same direction and through the same distance.

You can describe translations by using **vectors**. A vector is represented by the combination of a horizontal shift and a vertical shift.

EXAMPLE 1

Use vectors to describe the translations of these triangles.

a A to B

b B to C

c C to D

d D to A

a The vector describing the translation from A to B is $\begin{pmatrix} 2 \\ 1 \end{pmatrix}$.

b The vector describing the translation from B to C is $\begin{pmatrix} 2 \\ 0 \end{pmatrix}$.

c The vector describing the translation from C to D is $\begin{pmatrix} -3 \\ 2 \end{pmatrix}$.

d The vector describing the translation from D to A is $\begin{pmatrix} -1 \\ -3 \end{pmatrix}$.

Note:

- The top number in the vector describes the horizontal movement. To the right +, to the left −.
- The bottom number in the vector describes the vertical movement. Upwards +, downwards −.
- These vectors are also called *direction vectors*.

EXERCISE 29A

1 Use vectors to describe these translations of the shapes on the grid below.

  a i A to B
    ii A to C
    iii A to D

  b i B to E
    ii B to F
    iii B to G

  c i C to A
    ii C to E
    iii C to G

  d i G to D
    ii F to G
    iii G to E

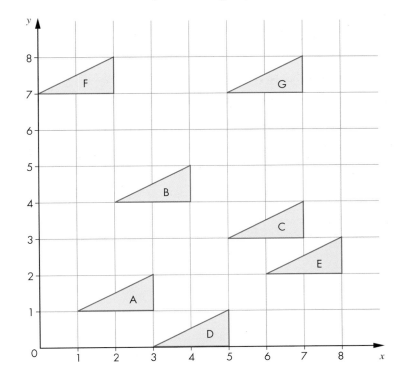

2 a Draw a set of coordinate axes and on it the triangle with coordinates A(1, 1), B(2, 1) and C(1, 3).

  b Draw the image of ABC after a translation with vector $\begin{pmatrix} 2 \\ 3 \end{pmatrix}$. Label this triangle P.

  c Draw the image of ABC after a translation with vector $\begin{pmatrix} -1 \\ 2 \end{pmatrix}$. Label this triangle Q.

  d Draw the image of ABC after a translation with vector $\begin{pmatrix} 3 \\ -2 \end{pmatrix}$. Label this triangle R.

  e Draw the image of ABC after a translation with vector $\begin{pmatrix} -2 \\ -4 \end{pmatrix}$. Label this triangle S.

3 Using your diagram from question **2**, use vectors to describe the translation that will move

  a P to Q   **b** Q to R   **c** R to S   **d** S to P

  e R to P   **f** S to Q   **g** R to Q   **h** P to S.

4 If a translation is given by:

$$\begin{pmatrix} x \\ y \end{pmatrix}$$

describe the translation that would take the image back to the original position.

CORE

5 A boat travels between three jetties X, Y and Z on a lake. Its journeys are described by direction vectors, with distance in kilometres.

The direction vector from X to Y is $\begin{pmatrix} 3 \\ -1 \end{pmatrix}$ and the direction vector from Y to Z is $\begin{pmatrix} -2 \\ -3 \end{pmatrix}$.

Using centimetre-squared paper and a scale of 1 cm : 1 km, draw a diagram to show journeys between X, Y and Z. Work out the direction vector for the journey from Z to X.

29.2 Reflections

A **reflection** transforms a shape so that it becomes a mirror image of itself.

EXAMPLE 2

Reflect this shape in the line provided.

The reflected shape looks like this.

Note: The reflection of each point in the original shape, called the **object**, is perpendicular to the mirror line. So if you 'fold' the whole diagram along the **mirror line**, the object will coincide with its reflection, called its **image**.

Object

Mirror line

Image

EXERCISE 29B

1 Copy the diagram.

On your copy, draw the reflection of the given triangle in each of these lines.

a $x = 2$

b $x = -1$

c $x = 3$

d $y = 2$

e $y = -1$

f y-axis

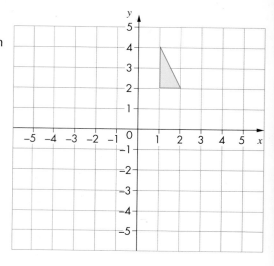

2 **a** Draw a pair of axes.
Label the *x*-axis from –5 to 5 and the *y*-axis from –5 to 5.

b Draw the triangle with coordinates A(1, 1), B(3, 1), C(4, 5).

c Reflect the triangle ABC in the *x*-axis. Label the image P.

d Reflect triangle P in the *y*-axis. Label the image Q.

e Reflect triangle Q in the *x*-axis. Label the image R.

f Describe the reflection that will move triangle ABC to triangle R.

3 **a** Draw a pair of axes.
Label the *x*-axis from –5 to +5 and the *y*-axis from –5 to +5.

b Reflect the points A(2, 1), B(5, 0), C(–3, 3), D(3, –2) in the *x*-axis.

c What do you notice about the values of the coordinates of the reflected points?

d What would the coordinates of the reflected point be if the point (*a*, *b*) were reflected in the *x*-axis?

4 **a** Draw a pair of axes.
Label the *x*-axis from –5 to +5 and the *y*-axis from –5 to +5.

b Reflect the points A(2, 1), B(0, 5), C(3, –2), D(–4, –3) in the *y*-axis.

c What do you notice about the values of the coordinates of the reflected points?

d What would the coordinates of the reflected point be if the point (*a*, *b*) were reflected in the *y*-axis?

5 By using the middle square as a starting square called ABCD, describe how to keep reflecting the square to obtain the final shape in the diagram.

6 Triangle A is drawn on a grid.

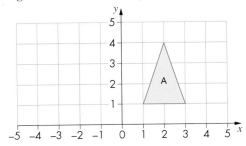

Triangle A is reflected to form a new triangle B.
The coordinates of triangle B are (–4, 4), (–3, 1) and (–5, 1).
Work out the equation of the mirror line.

You have been learning about reflections in horizontal or vertical lines.

You can reflect a shape in any line.

EXAMPLE 3

Draw the reflection of triangle *t* in the line with equation $y = x$.

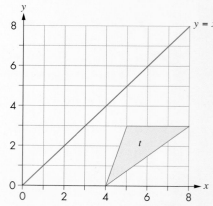

To find the image of each vertex of the triangle, draw lines perpendicular to the mirror.

Each vertex and its image are the same distance from the mirror but on opposite sides. Use the grid to help you find the new vertices.

Join the new vertices to draw the reflection M(*t*) of the triangle *t*.

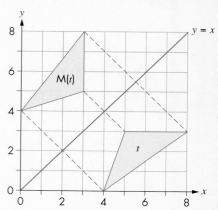

EXERCISE 29C

1 A designer used these instructions to create a design.

- Start with any rectangle ABCD.
- Reflect the rectangle ABCD in the line AC.
- Reflect the rectangle ABCD in the line BD.

Draw a rectangle and use the above to create a design.

2 Draw each of these triangles on squared paper, leaving plenty of space on the opposite side of the given mirror line. Then draw the reflection of each triangle.

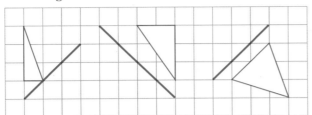

ADVICE AND TIPS

Turn the page around so that the mirror lines are vertical or horizontal.

3 a Draw a pair of axes and the lines $y = x$ and $y = -x$, as shown.

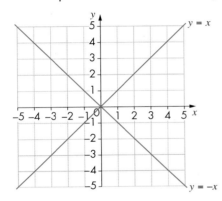

b Draw the triangle with coordinates A(2, 1), B(5, 1), C(5, 3).

c Draw the reflection of triangle ABC in the x-axis and label the image P.

d Draw the reflection of triangle P in the line $y = -x$ and label the image Q.

e Draw the reflection of triangle Q in the y-axis and label the image R.

f Draw the reflection of triangle R in the line $y = x$ and label the image S.

g Draw the reflection of triangle S in the x-axis and label the image T.

h Draw the reflection of triangle T in the line $y = -x$ and label the image U.

i Draw the reflection of triangle U in the y-axis and label the image W.

j What single reflection will move triangle W to triangle ABC?

4 Copy the diagram.

Reflect the triangle in these lines.

a $y = x$ **b** $x = 1$

c $y = -x$ **d** $y = -1$

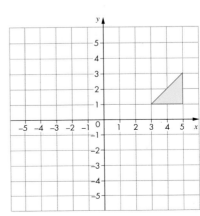

5 **a** Draw a pair of axes.
Label the x-axis from –5 to +5 and the y-axis from –5 to +5.

b Draw the line $y = x$.

c Reflect the points A(2, 1), B(5, 0), C(–3, 2), D(–2, –4) in the line $y = x$.

d What do you notice about the values of the coordinates of the reflected points?

e What would the coordinates of the reflected point be if the point (a, b) were reflected in the line $y = x$?

6 **a** Draw a pair of axes.
Label the x-axis from –5 to +5 and the y-axis from –5 to +5.

b Draw the line $y = -x$.

c Reflect the points A(2, 1), B(0, 5), C(3, –2), D(–4, –3) in the line $y = -x$.

d What do you notice about the values of the coordinates of the reflected points?

e What would the coordinates of the reflected point be if the point (a, b) were reflected in the line $y = -x$?

29.4 Rotations

A **rotation** transforms a shape to a new position by turning it about a fixed point called the **centre of rotation**.

EXAMPLE 4

Rotate each of these shapes as described.

a Rotation of 90° clockwise

b Rotation of 180°

The rotations look like this.

a Rotation of 90° clockwise

b Rotation of 180°

Note:

- The direction of turn or the **angle of rotation** is expressed as **clockwise** or **anticlockwise**.

- The position of the centre of rotation is always specified.

- The rotations 180° clockwise and 180° anticlockwise are the same.

The rotations that appear most frequently are 90° and 180°.

> **ADVICE AND TIPS**
>
> Use tracing paper to check rotations.

EXERCISE 29D

1 On squared paper, draw each of these shapes and its centre of rotation, leaving plenty of space all round the shape.

a Rotate each shape about its centre of rotation:

   i first by 90° clockwise (call the image A)

   ii then by 90° anticlockwise (call the image B).

b Describe, in each case, the rotation that would take:

   i A back to its original position   **ii** A to B.

2 A graphics designer came up with this routine for creating a design.

- Start with a triangle ABC.

- Reflect the triangle in the line AB.

- Rotate the whole shape about point C clockwise 90°, then a further clockwise 90°, then a further clockwise 90°.

From any triangle of your choice, create a design using the above routine.

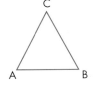

3 By using the middle square as a starting square, called ABCD, describe how to keep rotating the square to obtain the final shape in the diagram.

CORE

CORE

4 Copy the diagram. Rotate the given triangle by:

a 90° clockwise about (0, 0)

b 180° about (0, 0)

c 90° anticlockwise about (1, 4)

d 180° about (1, 3)

e 90° clockwise about (2, 2).

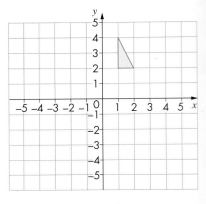

5 Give the centre and the angle for the rotations that will take:

a A onto B

b A onto C

c A onto D

d A onto E

6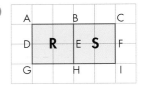

a A 180° rotation will take R onto S. Where is the centre of rotation?

b A 90° clockwise rotation will take R onto S. Where is the centre?

7 a Draw a pair of axes where both the x-values and y-values are from –5 to 5.

b Draw the triangle ABC, where A = (1, 2), B = (2, 4) and C = (4, 1).

c i Rotate triangle ABC 90° clockwise about the origin (0, 0) and label the image A′, B′, C′, where A′ is the image of A, etc.

ii Write down the coordinates of A′, B′, C′.

iii What connection is there between A, B, C and A′, B′, C′?

iv Will this connection always be so for a 90° clockwise rotation about the origin?

8 Repeat question **7**, but rotate triangle ABC through 180°.

9 Show that a reflection in the x-axis followed by a reflection in the y-axis is equivalent to a rotation of 180° about the origin.

EXERCISE 29E

1 Draw a set of x and y axes and label them both from 0 to 12.

Draw the triangle with vertices at (5,5), (7,5) and (7,8). Label it T.

  a Rotate T 180° about (4,5). Label the new triangle A.

  b Rotate T 90° clockwise about (7,4). Label the new triangle B.

  c Rotate T 90° anticlockwise about (7,9). Label the new triangle C.

  d What rotation will take triangle B onto triangle C?

2 **a** A 180° rotation will take square A onto square B.
Where is the centre of the rotation?

  b A 90° *clockwise* rotation will take A onto B.
Where is the centre of the rotation?

  c A 90° *anticlockwise* rotation will take A onto B.
Where is the centre of the rotation?

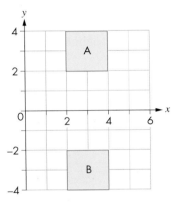

3 Give the centre and angle for the rotation of:

  a A onto B   **b** B onto C

  c C onto D   **d** D onto A.

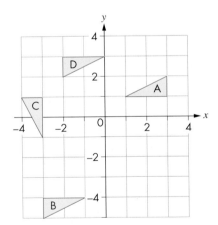

4 Show that a reflection in the line $y = x$ followed by a reflection in the line $y = -x$ is equivalent to a rotation of 180° about the origin.

5 **a** Draw a regular hexagon ABCDEF with centre O. The letters should go round the hexagon clockwise.

b Using O as the centre of rotation, describe a transformation that will result in the following movements.

i Triangle AOB to triangle BOC   **ii** Triangle AOB to triangle COD

iii Triangle AOB to triangle DOE   **iv** Triangle AOB to triangle EOF

c Describe the transformations that will move the rhombus ABCO to these positions.

i Rhombus BCDO   **ii** Rhombus DEFO

6 Triangle A, as shown on the grid, is rotated to form a new triangle B.

The coordinates of the vertices of B are (0, −2), (−3, −2) and (−3, −4).

Describe fully the rotation that maps triangle A onto triangle B.

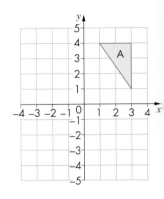

29.6   Enlargements

An **enlargement** changes the size of a shape to give a similar image. It always has a **centre of enlargement** and a **scale factor**. Every length of the enlarged shape will be:

original length × scale factor

The distance of each image point on the enlargement from the centre of enlargement will be:

distance of original point from centre of enlargement × scale factor

---

EXAMPLE 5

Enlarge triangle ABC by scale factor 3 about the centre of enlargement X.

This is the completed enlargement.

Note:

● Each length on the enlargement A'B'C' is three times the corresponding length on the original shape.

This means that the corresponding sides are in the same ratio:

AB : A'B' = AC : A'C' = BC : B'C' = 1 : 3

● The distance of any point on the enlargement from the centre of enlargement is three times the distance from the corresponding point on the original shape to the centre of enlargement.

There are two distinct ways to enlarge a shape: the **ray method** and the **coordinate method** (counting squares).

Ray method

This is the *only* way to construct an enlargement when the diagram is not on a grid.

EXAMPLE 6

Enlarge triangle ABC by scale factor 3 about the centre of enlargement X.

Notice that the rays have been drawn from the centre of enlargement to each vertex and beyond.

The distance from X to each vertex on triangle ABC is measured and multiplied by 3 to give the distance from X to each vertex A', B' and C' for the enlarged triangle A'B'C'.

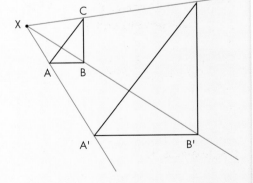

Once each image vertex has been found, the whole enlarged shape can then be drawn.

Check the measurements and see for yourself how the calculations have been done.

Notice again that the length of each side on the enlarged triangle is three times the length of the corresponding side on the original triangle.

Coordinate method

In this method, you use the coordinates of the vertices to 'count squares'.

EXAMPLE 7

Enlarge the triangle ABC by scale factor 3 from the centre of enlargement (1, 2).

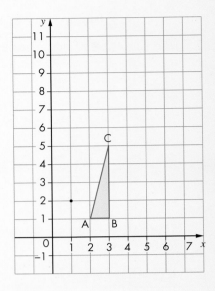

To find the coordinates of each image vertex, first work out the horizontal and vertical distances from each original vertex to the centre of enlargement.

Then multiply each of these distances by 3 to find the position of each image vertex.

For example, to find the coordinates of C' work out the distance from the centre of enlargement (1, 2) to the point C(3, 5).

horizontal distance = 2

vertical distance = 3

Make these 3 times longer to give:

new horizontal distance = 6

new vertical distance = 9

So the coordinates of C' are: (1 + 6, 2 + 9) = (7, 11)

Notice again that the length of each side is three times as long in the enlargement.

Negative enlargement

A **negative enlargement** produces an image shape on the opposite side of the centre of enlargement to the original shape.

EXAMPLE 8

Enlarge triangle ABC by scale factor −2, with the centre of enlargement at (1, 0).

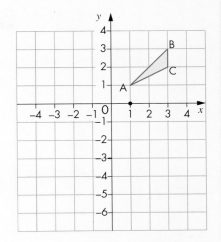

You can enlarge triangle ABC to give triangle A'B'C' by either the ray method or the coordinate method. You calculate the new lengths on the opposite side of the centre of enlargement to the original shape.

Notice how a negative scale factor also inverts the original shape.

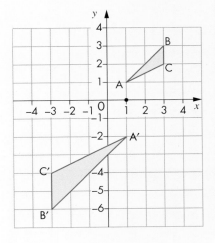

EXERCISE 29F

1 Copy each of these figures with its centre of enlargement. Then use the ray method to enlarge it by the given scale factor.

a

Scale factor 2

b

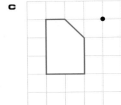

Scale factor 3

c

Scale factor 2

d

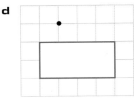

Scale factor 3

CORE

503

CORE

2 Copy each of these diagrams onto squared paper and enlarge it by scale factor 2, using the origin as the centre of enlargement.

a

b

c

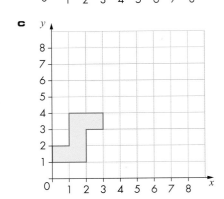

ADVICE AND TIPS

Even if you are using a counting square method, you can always check by using the ray method.

3 Enlarge each of these shapes by a scale factor of $\frac{1}{2}$ about its closest centre of enlargement.

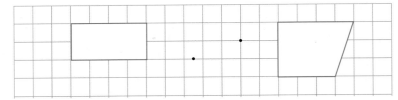

4 Copy this diagram onto squared paper.

a Enlarge the rectangle A by scale factor $\frac{1}{3}$ about the origin. Label the image B.

b Write down the ratio of the lengths of the sides of rectangle A to the lengths of the sides of rectangle B.

c Work out the ratio of the perimeter of rectangle A to the perimeter of rectangle B.

d Work out the ratio of the area of rectangle A to the area of rectangle B.

Further enlargements

Finding the centre of enlargement

You can find the centre of an enlargement by drawing rays.

EXAMPLE 9

E(a) is an enlargement of shape a.

Find the centre and the scale factor of the enlargement.

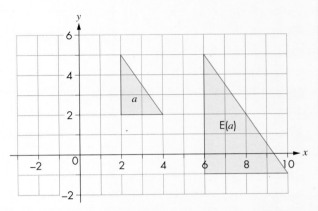

Draw rays through corresponding points on the object and image. They should all meet at one point. This is the centre of enlargement.

In this case the centre is (−2, 5).

To find the scale factor, find the ratio of corresponding sides.

Scale factor = $\dfrac{\text{height of E}(a)}{\text{height of } a}$

$= \dfrac{6}{3} = 2$

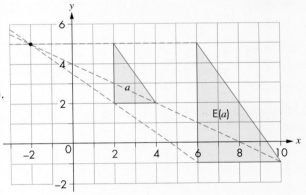

Fractional enlargement

Strange but true ... you can have an enlargement in mathematics that is actually smaller than the original shape!

EXAMPLE 10

Enlarge triangle ABC by a scale factor of $\frac{1}{2}$ about the centre of enlargement O.

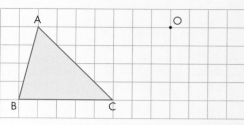

The enlargement of triangle ABC produces triangle A'B'C'.

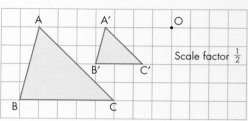

Scale factor $\frac{1}{2}$

EXTENDED

EXERCISE 29G

1 Copy each of these diagrams onto squared paper and enlarge it by scale factor 2, using the given centre of enlargement.

a

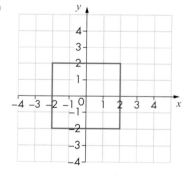

Centre of enlargement (–1, 1)

b

Centre of enlargement (–2, –3)

2 Copy this diagram onto squared paper.

Enlarge the triangle by scale factor –2 about the origin.

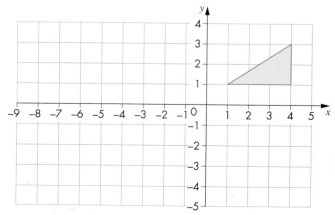

EXTENDED

3 Copy this diagram onto squared paper.

a Enlarge A by a scale factor of 3 about a centre (4, 5).

b Enlarge B by a scale factor $\frac{1}{2}$ about a centre (−1, −3).

c Enlarge B by scale factor $-\frac{1}{2}$ about a centre (−3, −1).

d What is the centre of enlargement and scale factor that maps B onto A?

e What is the centre of enlargement and scale factor that maps A onto B?

f What is the centre of enlargement and scale factor that maps the answer to part **b** to the answer to part **c**?

g What is the centre of enlargement and scale factor that maps the answer to part **c** to the answer to part **b**?

h What is the connection between the scale factors and the centres of enlargement in parts **d** and **e**, and in parts **f** and **g**?

4 Triangle A has vertices with coordinates (2, 1), (4, 1) and (4, 4).

Triangle B has vertices with coordinates (−5, 1), (−5, 7) and (−1, 7).

Describe fully the single transformation that maps triangle A onto triangle B.

29.8 **Combined transformations**

Sometimes you will need to use more than one **transformation** to produce the required image from the given object. In this exercise, you will revise the transformations you have met so far.

Remember, to describe:

- a **translation** fully, you need to use a vector
- a **reflection** fully, you need to use a mirror line
- a **rotation** fully, you need a centre of rotation, an angle of rotation and the direction of turn
- an **enlargement** fully, you need a centre of enlargement and a scale factor.

EXTENDED

EXERCISE 29H

1 The point P(3, 4) is reflected in the *x*-axis, then rotated by 90° clockwise about the origin. What are the coordinates of the image of P?

2 A point Q(5, 2) is rotated by 180° about the origin, then reflected in the *x*-axis.

  a What are the coordinates of the image of Q?

  b What single transformation would have taken point Q directly to the image point?

3 Describe fully the transformations that will map the shaded triangle onto each of the triangles A–F.

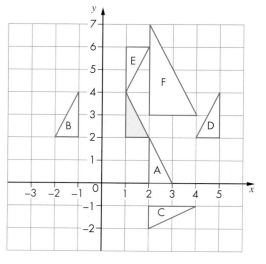

4 Describe fully the transformations that will result in the movement of:

  a T_1 to T_2

  b T_1 to T_6

  c T_2 to T_3

  d T_6 to T_2

  e T_6 to T_5

  f T_5 to T_4.

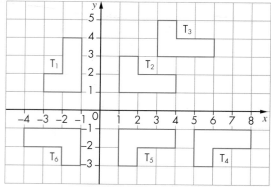

5 **a** Plot a triangle *t* with vertices (1, 1), (2, 1), (1, 3).

  b Reflect triangle *t* in the *y*-axis and label the image M(*t*).

  c Rotate triangle M(*t*) 90° anticlockwise about the origin and label the image RM(*t*).

  d Describe fully the transformation that will move triangle RM(*t*) back to triangle *t*.

6 Describe fully at least three different transformations that could move the square labelled S to the square labelled T.

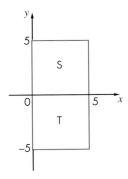

7 Copy the diagram onto squared paper.

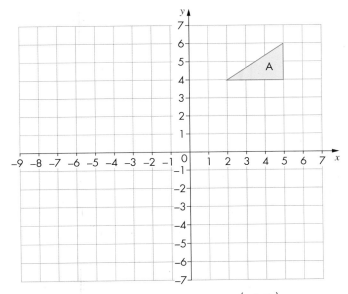

a Triangle A is translated by the vector $\begin{pmatrix} -1.5 \\ -3 \end{pmatrix}$ to give triangle B.

Triangle B is then enlarged by a scale factor -2 about the origin to give triangle C.

Draw triangles B and C on the diagram.

b Describe fully the single transformation that maps triangle C onto triangle A.

Why this chapter matters

Matrices are a simple way to store and process information systematically. A spreadsheet is a simple example of using matrices. The mathematics of matrices has been developed so that they can now be used in many different fields.

Encryption

The internet could not work without encryption – a way to code private information so that it remains secret. When you buy things online you want details of your bank account or credit card to be held securely. Matrices can be used to make encryption work effectively.

Marketing analysis

How can a company increase its profits? What would be the effect of changing the prices of products? How do sales vary over time? Matrices are a useful tool for market analysts to answer questions like these.

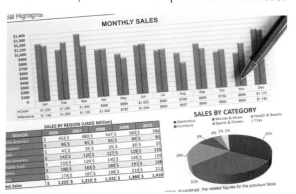

Black box analysis

All electrical devices have input and output currents and voltages. Matrices can tell you what outputs you will get for any given inputs. Electrical engineers call this 'black box analysis'.

Historical development

At first only mathematicians were interested in matrices. Then in the 1920s a German physicist called Werner Heisenberg found that they were the tool he needed to express his ideas about quantum theory. Since then more and more uses for matrices have been found in economics, engineering, architecture and other fields.

30 Matrices

Topics	Level	Key words
1 Introduction to matrices	EXTENDED	matrix, matrices, row, column, order, element, scalar quantity, matrix quantity, square matrix
2 Multiplying matrices	EXTENDED	
3 2 × 2 matrices	EXTENDED	zero matrix, identity matrix
4 The inverse of a 2 × 2 matrix	EXTENDED	determinant, inverse, singular, non-singular
5 Matrices and transformations	EXTENDED	image, position vector
6 Finding a matrix	EXTENDED	

In this chapter you will learn how to:

EXTENDED

- Display information in the form of a matrix of any order. (E7.4)
- Calculate the sum and product (where appropriate) of two matrices. (E7.4)
- Calculate the product of a matrix and a scalar quantity. (E7.4)
- Use the algebra of 2 × 2 matrices, including the zero and identity 2 × 2 matrices. (E7.4)
- Calculate the determinant $|\mathbf{A}|$ and inverse $|\mathbf{A}|^{-1}$ of a non-singular matrix $|\mathbf{A}|$. (E7.4)
- Describe transformations using coordinates and matrices (singular matrices are excluded). (E7.5)

A **matrix** is a rectangular array of numbers. The plural of matrix is **matrices**. Matrices can be used to store information.

For example, suppose there are two ferries A and B across a river.

On Monday:

- Ferry A carries 70 adults, 50 students and 42 children.
- Ferry B carries 52 adults, 65 students and 31 children.

Here is that information in a matrix:

$$\mathbf{M} = \begin{pmatrix} 70 & 50 & 42 \\ 52 & 65 & 30 \end{pmatrix}$$

This matrix has two **rows** and three **columns**. It is a 2 × 3 matrix (read this as '2 by 3').

2 x 3 is the **order** of the matrix. In the order of a matrix the rows come first, followed by columns.

The numbers in the matrix are the **elements** of the matrix.

Here is the matrix for ferry A and ferry B on Tuesday:

$$\mathbf{T} = \begin{pmatrix} 105 & 72 & 15 \\ 47 & 35 & 53 \end{pmatrix}$$

You can add these matrices to get the totals for the two days:

$$\mathbf{M} + \mathbf{T} = \begin{pmatrix} 70 & 50 & 42 \\ 52 & 65 & 30 \end{pmatrix} + \begin{pmatrix} 105 & 72 & 15 \\ 47 & 35 & 53 \end{pmatrix} + \begin{pmatrix} 175 & 122 & 57 \\ 99 & 100 & 83 \end{pmatrix}$$

To add matrices just add corresponding elements.

You can subtract matrices in a similar way.

You can only add and subtract matrices if they are the same order.

The ferries run five days a week. The ferry manager estimates the number of passengers in one week by multiplying the numbers on Monday by five:

$$5\mathbf{M} = 5 \times \begin{pmatrix} 70 & 50 & 42 \\ 52 & 65 & 30 \end{pmatrix} = \begin{pmatrix} 350 & 250 & 210 \\ 260 & 325 & 150 \end{pmatrix}$$

To multiply the matrix by any number, multiply every element by that number.

The multiplying number (that is 5 in the example above) is a **scalar quantity**. Quantities inside the matrix are called **matrix quantities**.

EXAMPLE 1

Here are three matrices.

$$A = \begin{pmatrix} 12 & -2 \\ 8 & 6 \\ -3 & 5 \end{pmatrix} \quad B = \begin{pmatrix} 8 & 6 & 3 \\ 2 & 5 & -8 \\ -10 & 12 & 7 \end{pmatrix} \quad C = \begin{pmatrix} 3 & -5 \\ -2 & 4 \\ 0 & 6 \end{pmatrix}$$

a Write down the order of each matrix.

b Find, if possible: $A + B$; $A + C$; $C - A$ and $4B$.

a The order of **A** is 3×2, **B** is 3×3 and **C** is 3×2.

b $A + B$ is impossible because **A** and **B** are different orders.

$$A + C = \begin{pmatrix} 15 & -7 \\ 6 & 10 \\ -3 & 11 \end{pmatrix}$$

$$C - A = \begin{pmatrix} 3 & -5 \\ -2 & 4 \\ 0 & 6 \end{pmatrix} - \begin{pmatrix} 12 & -2 \\ 8 & 6 \\ -3 & 5 \end{pmatrix} = \begin{pmatrix} -9 & -3 \\ -10 & -2 \\ 3 & 1 \end{pmatrix}$$

$$4B = \begin{pmatrix} 32 & 24 & 12 \\ 8 & 20 & -32 \\ -40 & 48 & 28 \end{pmatrix}$$

Note that **B** is called a **square matrix** because it has the same number of rows and columns.

EXERCISE 30A

1 In the 2012 Olympics, Japan won 7 gold medals, 14 silver and 17 bronze, a total of 38 medals. France won 34 medals: 11 gold, 11 silver and 12 bronze.

Put this information in a 4×2 matrix.

2 Write down the order of each of these matrices.

a $\begin{pmatrix} 5 & 6 & 0 & -5 & -3 \\ 4 & 2 & 12 & 13 & 0 \end{pmatrix}$

b $\begin{pmatrix} 6 & 7 & 5 \\ 13 & 0 & 0 \\ -4 & -6 & 1 \\ 8 & 2 & 10 \end{pmatrix}$

c $(13 \ 5 \ 8 \ 2)$

d $\begin{pmatrix} 3 \\ 5 \\ 8 \end{pmatrix}$

EXTENDED

EXTENDED

3 Calculate:

a $\begin{pmatrix} 2 & 8 \\ 4 & -1 \end{pmatrix} + \begin{pmatrix} 3 & -4 \\ -5 & 2 \end{pmatrix}$

b $\begin{pmatrix} 28 \\ 13 \\ 4 \end{pmatrix} + \begin{pmatrix} 15 \\ -11 \\ 0 \end{pmatrix}$

c $\begin{pmatrix} 3.5 & 6 & 7.5 \\ -1 & -2 & 0 \end{pmatrix} - \begin{pmatrix} 1.5 & 1.5 & 1 \\ 4 & 3.5 & 2 \end{pmatrix}$

d $\begin{pmatrix} 2 & 0 & -1 \\ 7 & 0 & -5 \end{pmatrix} + \begin{pmatrix} 0 & 4.5 & 7.5 \\ -1 & -5.5 & 3 \end{pmatrix}$

e $(3 \ 5 \ 7 \ 4) - (8 \ 1 \ 0 \ 5)$

4 $A = \begin{pmatrix} 4 & 2 \\ 10 & 3 \\ 1 & -1 \end{pmatrix}$ $B = \begin{pmatrix} 1 & 5 \\ 2 & 2 \\ 8 & 0 \end{pmatrix}$ $C = \begin{pmatrix} 2 & 10 & -1 \\ -4 & 2 & 0 \end{pmatrix}$ $D = \begin{pmatrix} 8 & 1 & -1 \\ 3 & 1 & 2 \end{pmatrix}$

Calculate these, where possible.

a $A + B$ b $B + C$ c $C + D$ d $A - B$ e $D - C$

f $3B$ g $\frac{1}{2}C$ h $3A - B$ i $2C + 3D$ j $3(C + D)$

5 $\begin{pmatrix} 6 & 2 & 8 & x \\ -3 & 4 & 1 & y \end{pmatrix} + \begin{pmatrix} 1 & 3 & 6 & 4 \\ -1 & -4 & w & 2 \end{pmatrix} = \begin{pmatrix} u & 5 & 14 & 2 \\ v & 0 & 0 & 6 \end{pmatrix}$

Find the values of u, v, w, x and y.

6 $A = \begin{pmatrix} 1 & 2 \\ 3 & 5 \end{pmatrix}$ and $B = \begin{pmatrix} 7 & 8 \\ 15 & 1 \end{pmatrix}$

a Work out matrix X if $2X + A = B$.

b Work out matrix Y if $4Y - A = B$.

30.2 Multiplying matrices   E

Look at the ferry example again.

The number of passengers on two different ferries on Monday is given by this matrix:

$$M = \begin{pmatrix} 70 & 50 & 42 \\ 52 & 65 & 30 \end{pmatrix}$$

The columns give the numbers of adults, students and children. The rows show ferries A and B.

The tickets cost $2 for adults, $1 for students and $0.50 for children. You can write them in a 3×1 matrix like this.

$$C = \begin{pmatrix} 2 \\ 1 \\ 0.5 \end{pmatrix}$$

To find out how much money each ferry took you can multiply **M** and **C** like this.

$$\mathbf{MC} = \begin{pmatrix} 70 & 50 & 42 \\ 52 & 65 & 30 \end{pmatrix} \times \begin{pmatrix} 2 \\ 1 \\ 0.5 \end{pmatrix} = \begin{pmatrix} 211 \\ 184 \end{pmatrix}$$

Ferry A took $211 and ferry B took $184.

When you multiply two matrices each *row* of the first matrix combines with each *column* of the second to give a single number.

In the example above, a 2×3 matrix multiplied by a 3×1 matrix gives a 2×1 answer.

A 2×4 matrix multiplied by a 4×1 matrix gives a 2×1 answer.

$$\begin{pmatrix} 1 & 2 & 3 & 4 \\ -2 & 0 & 1 & 2 \end{pmatrix} \begin{pmatrix} 5 \\ 6 \\ 1 \\ -3 \end{pmatrix} = \begin{pmatrix} 1 \times 5 + 2 \times 6 + 3 \times 1 + 4 \times -3 \\ -2 \times 5 + 0 \times 6 + 1 \times 1 + 2 \times -3 \end{pmatrix} = \begin{pmatrix} 8 \\ -15 \end{pmatrix}$$

A 2×4 matrix multiplied by a 4×2 matrix gives a 2×2 answer.

$$\begin{pmatrix} 1 & 2 & 3 & 4 \\ -2 & 0 & 1 & 2 \end{pmatrix} \begin{pmatrix} 5 & 2 \\ 6 & -1 \\ 1 & 3 \\ -3 & 1 \end{pmatrix} = \begin{pmatrix} 8 & 1 \times 2 + 2 \times -1 + 3 \times 3 + 4 \times 1 \\ -15 & -2 \times 2 + 0 \times -1 + 1 \times 3 + 2 \times 1 \end{pmatrix} = \begin{pmatrix} 8 & 13 \\ -15 & 1 \end{pmatrix}$$

You can only multiply two matrices if the number of *columns* in the first one is equal to the number of *rows* in the second.

$(m \times n \text{ matrix}) \times (n \times p \text{ matrix}) = m \times p \text{ matrix}$

EXERCISE 30B

1 Complete these matrix multiplications.

a $\begin{pmatrix} 2 & 1 \\ 3 & 2 \end{pmatrix} \begin{pmatrix} 4 & 5 \\ 1 & 2 \end{pmatrix} = \begin{pmatrix} 9 & 12 \\ 14 & \square \end{pmatrix}$

b $\begin{pmatrix} 2 & 1 & 2 \\ 3 & 2 & 4 \end{pmatrix} \begin{pmatrix} 3 \\ 5 \\ 2 \end{pmatrix} = \begin{pmatrix} 15 \\ \square \end{pmatrix}$

c $\begin{pmatrix} 2 & -1 & 3 \\ 1 & 5 & -2 \end{pmatrix} \begin{pmatrix} 3 & -2 \\ 1 & 3 \\ 0 & 1 \end{pmatrix} = \begin{pmatrix} 5 & \square \\ \square & 11 \end{pmatrix}$

d $\begin{pmatrix} 1 & 3 \\ 4 & 0 \end{pmatrix} \begin{pmatrix} 2 & -1 & 4 & 2 \\ 1 & 3 & 2 & 1 \end{pmatrix} = \begin{pmatrix} 5 & \square & \square & \square \\ 8 & \square & \square & \square \end{pmatrix}$

2 Multiply these matrices.

a $\begin{pmatrix} 3 & 1 & 2 & 2 \\ 0 & 2 & 1 & 3 \end{pmatrix} \begin{pmatrix} 1 \\ 0 \\ 2 \\ 1 \end{pmatrix}$

b $(3 \; 1 \; 5) \begin{pmatrix} 2 & -1 & 1 \\ 1 & 2 & 1 \\ 0 & 3 & -1 \end{pmatrix}$

c $\begin{pmatrix} 2 & 0 \\ 1 & 3 \end{pmatrix} \begin{pmatrix} 1 & 2 \\ 4 & 1 \end{pmatrix}$

d $\begin{pmatrix} -1 & 2 \\ 3 & 1 \end{pmatrix} \begin{pmatrix} 2 & 2 \\ 3 & 4 \end{pmatrix}$

e $\begin{pmatrix} 1 & 2 \\ 1 & 1 \end{pmatrix} \begin{pmatrix} 3 & 4 & 1 & 0 \\ 2 & 1 & 0 & 1 \end{pmatrix}$

EXTENDED

3 $A = \begin{pmatrix} 2 & 1 \\ 0 & 2 \end{pmatrix}$ $B = \begin{pmatrix} 3 & 1 & 2 \\ 1 & 2 & 3 \end{pmatrix}$ $C = \begin{pmatrix} 2 & 1 \\ 3 & 1 \end{pmatrix}$

  a Calculate: **i AB**   **ii AC**   **iii CB**   **iv CA**

  b Why is it impossible to calculate **BC**?

  c Is **AC** the same as **CA**?

4 A^2 means $A \times A$.

  a If $A = \begin{pmatrix} 3 & 1 \\ 1 & 2 \end{pmatrix}$ find A^2.

  b If $B = \begin{pmatrix} 0 & 2 \\ 1 & 5 \end{pmatrix}$ find B^2.

  c If $C = \begin{pmatrix} 1 & 2 \\ 2 & 1 \end{pmatrix}$ find C^2 and C^3.

5 **M** is a 2 × 5 matrix. **N** is a 5 × 2 matrix.

  a What is the order of **MN**?

  b What is the order of **NM**?

  c Is it possible to calculate M^2? Give a reason for your answer.

6 $\begin{pmatrix} 2 & 4 \\ 1 & 3 \end{pmatrix} \times \begin{pmatrix} 3 & x \\ y & 1 \end{pmatrix} = \begin{pmatrix} 14 & 0 \\ 9 & 1 \end{pmatrix}$

Find the values of x and y.

7 Calculate **AB** and **BA** if possible, or state that it is impossible.

  a $A = (1 \ 2 \ 4)$ $B = \begin{pmatrix} 4 \\ 2 \\ 1 \end{pmatrix}$ **b** $A = \begin{pmatrix} 2 & 0 \\ 1 & 5 \end{pmatrix}$ $B = \begin{pmatrix} 1 & 1 & 3 \\ 2 & 4 & 1 \end{pmatrix}$

  c $A = \begin{pmatrix} -1 & 1 \\ 2 & -2 \\ -3 & 3 \end{pmatrix}$ $B = \begin{pmatrix} 1 & 0 & 2 \\ 2 & 0 & -1 \end{pmatrix}$ **d** $A = \begin{pmatrix} 1 & 2 \\ 1 & 3 \\ 1 & 4 \end{pmatrix}$ $B = \begin{pmatrix} 8 & 2 \\ -1 & 5 \\ -2 & 6 \end{pmatrix}$

  e $A = \begin{pmatrix} 4 & -1 & -3 \\ -1 & 0 & 2 \end{pmatrix}$ $B = (1 \ 5)$ **f** $A = \begin{pmatrix} 5 & 4 \\ 6 & 5 \end{pmatrix}$ $B = \begin{pmatrix} 5 & -4 \\ -6 & 5 \end{pmatrix}$

8 **A** and **B** are two 2 × 2 matrices.

Is it always true, sometimes true or never true that **AB** = **BA**?

Give a reason for your answer.

2 × 2 matrices

In this section we shall be looking only at 2 × 2 matrices.

EXAMPLE 2

$$A = \begin{pmatrix} 2 & 1 \\ -2 & 3 \end{pmatrix} \text{ and } B = \begin{pmatrix} 4 & -1 \\ 1 & 2 \end{pmatrix}$$

Calculate **A + B**, **AB** and **BA**.

$$A + B = \begin{pmatrix} 2 & 1 \\ -2 & 3 \end{pmatrix} + \begin{pmatrix} 4 & -1 \\ 1 & 2 \end{pmatrix} = \begin{pmatrix} 2+4 & 1+-1 \\ -2+1 & 3+2 \end{pmatrix} = \begin{pmatrix} 6 & 0 \\ -1 & 5 \end{pmatrix}$$

$$AB = \begin{pmatrix} 2 & 1 \\ -2 & 3 \end{pmatrix}\begin{pmatrix} 4 & -1 \\ 1 & 2 \end{pmatrix} = \begin{pmatrix} 8+1 & -2+2 \\ -8+3 & 2+6 \end{pmatrix} = \begin{pmatrix} 9 & 0 \\ -5 & 8 \end{pmatrix}$$

$$BA = \begin{pmatrix} 4 & -1 \\ 1 & 2 \end{pmatrix}\begin{pmatrix} 2 & 1 \\ -2 & 3 \end{pmatrix} = \begin{pmatrix} 8+2 & 4+-3 \\ 2+-4 & 1+6 \end{pmatrix} = \begin{pmatrix} 10 & 1 \\ -2 & 7 \end{pmatrix}$$

Notice that **AB** ≠ **BA**

The zero matrix

Suppose $A = \begin{pmatrix} 2 & 1 \\ -2 & 3 \end{pmatrix}$ and $Z = \begin{pmatrix} 0 & 0 \\ 0 & 0 \end{pmatrix}$:

then $A + Z = \begin{pmatrix} 2 & 1 \\ -2 & 3 \end{pmatrix} + \begin{pmatrix} 0 & 0 \\ 0 & 0 \end{pmatrix} = \begin{pmatrix} 2 & 1 \\ -2 & 3 \end{pmatrix} = A$

and $AZ = \begin{pmatrix} 2 & 1 \\ -2 & 3 \end{pmatrix}\begin{pmatrix} 0 & 0 \\ 0 & 0 \end{pmatrix} = \begin{pmatrix} 0 & 0 \\ 0 & 0 \end{pmatrix} = \begin{pmatrix} 0 & 0 \\ 0 & 0 \end{pmatrix} = Z$

In fact if **M** is *any* 2 × 2 matrix:

then **M + Z = M**

and **MZ = Z**

You can see that Z is a bit like 0 (zero) when you add and multiply numbers. If a is any number:

then $a + 0 = a$

and $a \times 0 = 0$

For this reason $\begin{pmatrix} 0 & 0 \\ 0 & 0 \end{pmatrix}$ is called the 2 × 2 **zero matrix**.

The identity matrix

Suppose $I = \begin{pmatrix} 1 & 0 \\ 0 & 1 \end{pmatrix}$

If, as before, $A = \begin{pmatrix} 2 & 1 \\ -2 & 3 \end{pmatrix}$

then $AI = \begin{pmatrix} 2 & 1 \\ -2 & 3 \end{pmatrix}\begin{pmatrix} 1 & 0 \\ 0 & 1 \end{pmatrix} = \begin{pmatrix} 2+0 & 0+1 \\ -2+0 & 0+3 \end{pmatrix} = \begin{pmatrix} 2 & 1 \\ -2 & 3 \end{pmatrix} = A$

and $IA = \begin{pmatrix} 1 & 0 \\ 0 & 1 \end{pmatrix}\begin{pmatrix} 2 & 1 \\ -2 & 3 \end{pmatrix} = \begin{pmatrix} 2+0 & 1+0 \\ 0+-2 & 0+3 \end{pmatrix} = \begin{pmatrix} 2 & 1 \\ -2 & 3 \end{pmatrix} = A$

In fact, if **M** is *any* 2 × 2 matrix:

$$MI = IM = M$$

This is like multiplying any number (a) by 1:

$$a \times 1 = 1 \times a = a$$

$\begin{pmatrix} 1 & 0 \\ 0 & 1 \end{pmatrix}$ is called the 2 × 2 **identity matrix**.

This should remind you that if you multiply a 2 × 2 matrix by 1, the answer is identical to the matrix you started with.

EXERCISE 30C

In this exercise $Z = \begin{pmatrix} 0 & 0 \\ 0 & 0 \end{pmatrix}$ and $I = \begin{pmatrix} 1 & 0 \\ 0 & 1 \end{pmatrix}$.

1 $M = \begin{pmatrix} 4 & 1 \\ -2 & 1 \end{pmatrix}$ and $N = \begin{pmatrix} -2 & 2 \\ 3 & 1 \end{pmatrix}$

Calculate:

a $M + N$ **b** $N + M$

c MN **d** NM

e M^2 **f** N^2

2 $P = \begin{pmatrix} 3 & 4 \\ 5 & -2 \end{pmatrix}$ and $Q = \begin{pmatrix} 8 & -5 \\ 9 & 2 \end{pmatrix}$

Calculate:

a $P + Z$ **b** $Z + Q$

c PZ **d** IQ

e QI **f** ZQ

3 Calculate:

   a Z^2

   b I^2

4 $C = \begin{pmatrix} 4 & 1 \\ 1 & 2 \end{pmatrix}$

   Calculate:

   a $2C$ **b** C^2

   c $C + I$ **d** $(C + I)^2$

   e $C^2 + 2C$ **f** $C^2 + 2C + I$

5 $D = \begin{pmatrix} 3 & 1 \\ 8 & 2 \end{pmatrix}$

   Calculate:

   a $D - 2I$

   b $(D - 2I)^2$

6 $E = \begin{pmatrix} 2 & 4 \\ -3 & 2 \end{pmatrix}$

   Show that $(E - I)(E + I) = E^2 - I$

7 $R = \begin{pmatrix} 5 & 2 \\ 2 & 1 \end{pmatrix}$ and $S = \begin{pmatrix} 1 & -2 \\ -2 & 5 \end{pmatrix}$

   Calculate:

   a RS

   b SR

   c What do you notice about your answers to **a** and **b**?

8 $A = \begin{pmatrix} 4 & 2 \\ -2 & -1 \end{pmatrix}$ and $B = \begin{pmatrix} -1 & -2 \\ 2 & 4 \end{pmatrix}$

   Calculate:

   a AB

   b BA

   c What do you notice about your answers to **a** and **b**?

Suppose $\mathbf{M} = \begin{pmatrix} 3 & 5 \\ 1 & 2 \end{pmatrix}$ and $\mathbf{X} = \begin{pmatrix} 2 & -5 \\ -1 & 3 \end{pmatrix}$

then $\mathbf{MX} = \begin{pmatrix} 3 & 5 \\ 1 & 2 \end{pmatrix}\begin{pmatrix} 2 & -5 \\ -1 & 3 \end{pmatrix} = \begin{pmatrix} 1 & 0 \\ 0 & 1 \end{pmatrix} = \mathbf{I}$

and $\mathbf{XM} = \begin{pmatrix} 2 & -5 \\ -1 & 3 \end{pmatrix}\begin{pmatrix} 3 & 5 \\ 1 & 2 \end{pmatrix} = \begin{pmatrix} 1 & 0 \\ 0 & 1 \end{pmatrix} = \mathbf{I}$

So $\mathbf{MX} = \mathbf{XM} = \mathbf{I}$.

The matrix that can be multiplied by $\mathbf{M}$ to give the 2 × 2 identity matrix $\mathbf{I}$ is called the **inverse** of $\mathbf{M}$ and it is written as $\mathbf{M}^{-1}$.

You have seen that if $\mathbf{M} = \begin{pmatrix} 3 & 5 \\ 1 & 2 \end{pmatrix}$ then $\mathbf{M}^{-1} = \begin{pmatrix} 2 & -5 \\ -1 & 3 \end{pmatrix}$.

EXAMPLE 3

Show that if $\mathbf{M} = \begin{pmatrix} 4 & 1 \\ 2 & 3 \end{pmatrix}$ then $\mathbf{M}^{-1} = \begin{pmatrix} 0.3 & -0.1 \\ -0.2 & 0.4 \end{pmatrix}$.

$\mathbf{MM}^{-1} = \begin{pmatrix} 4 & 1 \\ 2 & 3 \end{pmatrix}\begin{pmatrix} 0.3 & -0.1 \\ -0.2 & 0.4 \end{pmatrix} = \begin{pmatrix} 1.2 + -0.2 & -0.4 + 0.4 \\ 0.6 + -0.6 & -0.2 + 1.2 \end{pmatrix} = \begin{pmatrix} 1 & 0 \\ 0 & 1 \end{pmatrix}$

and

$\mathbf{M}^{-1}\mathbf{M} = \begin{pmatrix} 0.3 & -0.1 \\ -0.2 & 0.4 \end{pmatrix}\begin{pmatrix} 4 & 1 \\ 2 & 3 \end{pmatrix} = \begin{pmatrix} 1.2 + -0.2 & 0.3 + -0.3 \\ -0.8 + 0.8 & -0.2 + 1.2 \end{pmatrix} = \begin{pmatrix} 1 & 0 \\ 0 & 1 \end{pmatrix}$

So $\mathbf{M}^{-1}$ is the inverse of $\mathbf{M}$.

The notation might remind you of numbers.

For example: $4^{-1} = \frac{1}{4}$.

then $4 \times 4^{-1} = 4 \times \frac{1}{4} = 1$

and $4^{-1} \times 4 = \frac{1}{4} \times 4 = 1$

This is similar to writing:

$\mathbf{M} \times \mathbf{M}^{-1} = \mathbf{I}$

or $\mathbf{M}^{-1} \times \mathbf{M} = \mathbf{I}$

Finding the inverse

How can you find the inverse of matrix $\mathbf{M} = \begin{pmatrix} a & b \\ c & d \end{pmatrix}$?

Take $\begin{pmatrix} 4 & 1 \\ 2 & 3 \end{pmatrix}$ as an example.

There are three steps.

Step 1: Find the **determinant** of M, $|\mathbf{M}| = ad - bc$

In this case $|\mathbf{M}| = 4 \times 3 - 1 \times 2 = 10$.

Step 2: Change the positions of a and d and the signs of b and c.

If $\mathbf{M} = \begin{pmatrix} 4 & 1 \\ 2 & 3 \end{pmatrix}$ you get $\begin{pmatrix} 3 & -1 \\ -2 & 4 \end{pmatrix}$

Step 3: Divide by the determinant.

$$\mathbf{M}^{-1} = \tfrac{1}{10}\begin{pmatrix} 3 & -1 \\ -2 & 4 \end{pmatrix} = \begin{pmatrix} 0.3 & -0.1 \\ -0.2 & 0.4 \end{pmatrix}$$

This is the matrix you saw in Example 3.

EXAMPLE 4

Find the inverse of $\mathbf{A} = \begin{pmatrix} 5 & 3 \\ -1 & 2 \end{pmatrix}$.

$|\mathbf{A}| = 5 \times 2 - 3 \times -1 = 13$

$$\mathbf{A}^{-1} = \tfrac{1}{13}\begin{pmatrix} 2 & -3 \\ 1 & 5 \end{pmatrix} = \begin{pmatrix} \tfrac{2}{13} & \tfrac{-3}{13} \\ \tfrac{1}{13} & \tfrac{5}{13} \end{pmatrix}$$

Check that $\mathbf{A}\mathbf{A}^{-1} = \mathbf{I}$ and $\mathbf{A}^{-1}\mathbf{A} = \mathbf{I}$

You can only carry out step 3 if $|\mathbf{M}| \neq 0$.

If $|\mathbf{M}| = 0$ then $\mathbf{M}$ is a **singular** matrix and $\mathbf{M}^{-1}$ does not exist.

For example, if $\mathbf{M} = \begin{pmatrix} 2 & 4 \\ 1 & 2 \end{pmatrix}$ then $|\mathbf{M}| = 4 - 4 = 0$ so $\mathbf{M}$ has no inverse.

If $|\mathbf{M}| \neq 0$ then the matrix $\mathbf{M}$ is **non-singular** and the inverse $\mathbf{M}^{-1}$ does exist.

EXERCISE 30D

1 $\mathbf{A} = \begin{pmatrix} 6 & 1 \\ 4 & 1 \end{pmatrix}$ and $\mathbf{B} = \begin{pmatrix} 4 & 1 \\ 4 & 2 \end{pmatrix}$ and $\mathbf{C} = \begin{pmatrix} 3 & -1 \\ -2 & 1 \end{pmatrix}$

  a Work out $|\mathbf{A}|$, $|\mathbf{B}|$, and $|\mathbf{C}|$.

  b Work out $\mathbf{A}^{-1}$, $\mathbf{B}^{-1}$ and $\mathbf{C}^{-1}$.

2 Show by multiplication that the inverse of $\begin{pmatrix} 10 & 11 \\ 9 & 10 \end{pmatrix}$ is $\begin{pmatrix} 10 & -11 \\ -9 & 10 \end{pmatrix}$.

3 Explain why the matrix $\begin{pmatrix} 6 & -3 \\ 2 & -1 \end{pmatrix}$ does not have an inverse.

4 State whether each of these matrices is singular or non-singular.

a $\begin{pmatrix} -4 & 8 \\ 2 & -4 \end{pmatrix}$ **b** $\begin{pmatrix} 5 & 5 \\ -3 & 3 \end{pmatrix}$ **c** $\begin{pmatrix} 2 & 5 \\ 4 & 10 \end{pmatrix}$ **d** $\begin{pmatrix} 1 & 5 \\ 1 & -5 \end{pmatrix}$

Find the inverse of each matrix, if it exists.

5 $\mathbf{M} = \begin{pmatrix} 1 & 3 \\ 4 & 2 \end{pmatrix}$, $\mathbf{N} = \begin{pmatrix} 5 & -4 \\ -2 & 3 \end{pmatrix}$, $\mathbf{S} = \begin{pmatrix} -1 & -2 \\ 4 & -3 \end{pmatrix}$ and $\mathbf{T} = \begin{pmatrix} 2 & 6 \\ -1 & -3 \end{pmatrix}$

a Work out $\mathbf{M} + \mathbf{N}$ and $|\mathbf{M} + \mathbf{N}|$.

b Is it true that $|\mathbf{M} + \mathbf{N}| = |\mathbf{M}| + |\mathbf{N}|$?

c One of the four matrices does not have an inverse. Which one?

6 Find the inverse of each of these matrices. Check each time that $\mathbf{MM}^{-1} = \mathbf{I}$.

a $\begin{pmatrix} 8 & 3 \\ 5 & 2 \end{pmatrix}$ **b** $\begin{pmatrix} 1 & -2 \\ 3 & -5 \end{pmatrix}$ **c** $\begin{pmatrix} 8 & 3 \\ 2 & 1 \end{pmatrix}$

d $\begin{pmatrix} 3 & 10 \\ -2 & -5 \end{pmatrix}$ **e** $\begin{pmatrix} 5 & -10 \\ -3 & 8 \end{pmatrix}$ **f** $\begin{pmatrix} 4 & 8 \\ 3 & 6 \end{pmatrix}$

7 The matrix $\begin{pmatrix} 6 & 10 \\ -3 & x \end{pmatrix}$ does not have an inverse. Find the value of x.

8 $\mathbf{A} = \begin{pmatrix} 20 & 9 \\ 11 & 5 \end{pmatrix}$ and $\mathbf{B} = \begin{pmatrix} 5 & -9 \\ -11 & 20 \end{pmatrix}$.

a Work out $\mathbf{AB}$ and $\mathbf{BA}$.

b What is $\mathbf{A}^{-1}$? What is $\mathbf{B}^{-1}$?

30.5 Matrices and transformations

2×2 matrices can represent transformations.

Suppose $\mathbf{M} = \begin{pmatrix} 0 & 1 \\ -1 & 0 \end{pmatrix}$

You can use it to transform the triangle shown on the grid here.

Write the **position vectors** of the vertices of the triangle in a matrix.

$$\begin{matrix} A & B & C \\ \begin{pmatrix} 2 & 5 & 5 \\ 1 & 1 & 3 \end{pmatrix} \end{matrix}$$

Multiply by **M**:

$$\begin{array}{cc} & A\ B\ C \\ \begin{pmatrix} 0 & 1 \\ -1 & 0 \end{pmatrix} & \begin{pmatrix} 2 & 5 & 5 \\ 1 & 1 & 3 \end{pmatrix} \end{array} = \begin{array}{c} A'\ B'\ C' \\ \begin{pmatrix} 1 & 1 & 3 \\ -2 & -5 & -5 \end{pmatrix} \end{array}$$

The new coordinates are shown on the grid to the right.

M represents a rotation of 90° clockwise around the origin.

Check that **M** has this effect on any position vector you choose.

If you know the 2 × 2 matrix that represents a transformation you can use it to calculate the coordinates of the **image** of any shape.

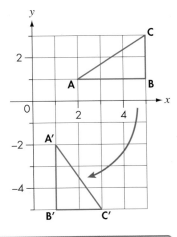

1 **a** Draw the quadrilateral with vertices at (–2, 0), (–5, 3), (–5, 5) and (–2, 6). Label it Q.

b Write the position vectors of the vertices in a 2 × 4 matrix.

c Multiply by the matrix $\mathbf{M} = \begin{pmatrix} 0 & 1 \\ -1 & 0 \end{pmatrix}$.

d Draw the image of Q under the transformation represented by **M** and check that it is a 90° clockwise rotation about the origin.

2 The triangle shown has vertices at (2, 3), (2, 5) and (5, 5).

a Transform the triangle, using the matrix $\mathbf{A} = \begin{pmatrix} 0 & 1 \\ 1 & 0 \end{pmatrix}$.

b Draw the triangle and its image. What transformation does **A** represent?

c Investigate the effect of matrix **A** on the rectangle with vertices at (1, 1), (4, 1), (4, –2) and (1, –2).

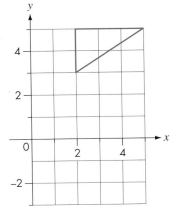

3 **a** Draw the triangle with vertices at (1, 0), (1, 4) and (3, 4).

b Transform the triangle, using the matrix $\begin{pmatrix} 0 & 1 \\ -1 & 0 \end{pmatrix}$.

c What transformation does $\begin{pmatrix} 0 & 1 \\ -1 & 0 \end{pmatrix}$ represent?

4 Use the triangle in question 3 and transform it using each of these matrices. Draw the result each time and describe fully the single transformation that the matrix represents.

a $\begin{pmatrix} -1 & 0 \\ 0 & 1 \end{pmatrix}$ **b** $\begin{pmatrix} -1 & 0 \\ 0 & -1 \end{pmatrix}$ **c** $\begin{pmatrix} 1 & 0 \\ 0 & -1 \end{pmatrix}$ **d** $\begin{pmatrix} 2 & 0 \\ 0 & 2 \end{pmatrix}$

e $\begin{pmatrix} -2 & 0 \\ 0 & -2 \end{pmatrix}$ **f** $\begin{pmatrix} 0 & -1 \\ 1 & 0 \end{pmatrix}$ **g** $\begin{pmatrix} 0 & 1 \\ 1 & 0 \end{pmatrix}$ **h** $\begin{pmatrix} 0 & -1 \\ -1 & 0 \end{pmatrix}$

EXTENDED

In the last exercise you saw that 2 × 2 matrices can represent reflections, rotations and enlargements.

If you want to find the matrix for a particular transformation, all you need to do is find what it does to the position vectors $\begin{pmatrix} 1 \\ 0 \end{pmatrix}$ and $\begin{pmatrix} 0 \\ 1 \end{pmatrix}$.

EXAMPLE 5

Find the matrix for a reflection in the line $y = -x$.

The image of $\begin{pmatrix} 1 \\ 0 \end{pmatrix}$ is $\begin{pmatrix} 0 \\ -1 \end{pmatrix}$.

The image of $\begin{pmatrix} 0 \\ 1 \end{pmatrix}$ is $\begin{pmatrix} -1 \\ 0 \end{pmatrix}$.

The matrix is $\begin{pmatrix} 0 & -1 \\ -1 & 0 \end{pmatrix}$.

The image of $\begin{pmatrix} 1 \\ 0 \end{pmatrix}$ | The image of $\begin{pmatrix} 0 \\ 1 \end{pmatrix}$

EXERCISE 30F

1. Draw any triangle and check that the matrix $\begin{pmatrix} 0 & -1 \\ -1 & 0 \end{pmatrix}$ in Example 5 does reflect it in the line $y = -x$.

2. Find the matrix which represents each transformation.

   a Reflection in the x-axis

   b Reflection in the y-axis

   c An enlargement, centre the origin, scale factor 4

   d Rotation of 90° clockwise about the origin

   e Rotation of 180° clockwise about the origin

3 Find the matrices which represent these transformations.

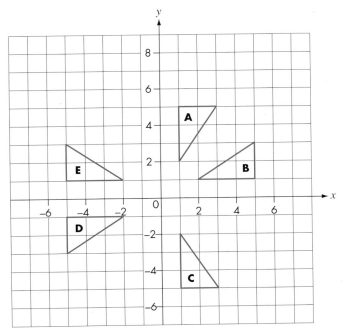

a **B to C**

b **B to E**

c **B to A**

d **E to C**

e **E to D**

4 a Find the matrix:

    i **C**, which represents a rotation of 90° clockwise about the origin

    ii **A**, which represents a rotation of 90° anticlockwise about the origin.

b Find $\mathbf{C}^2$. What transformation does it represent?

c Show that $\mathbf{C}^3 = \mathbf{A}$.

d Find $\mathbf{C}^4$ and state what matrix this is.

e Show that $\mathbf{C}^{-1} = \mathbf{A}$.

Examination questions

Note: drawings in questions are NOT TO SCALE unless otherwise specified.

- If the degree of accuracy is not specified in the question, and if the answer is not exact, give the answer to three significant figures. Give answers in degrees to one decimal place.
- For π, use either your calculator value or 3.142.

PAPER 1

1 Shade two more squares so that this pattern has rotational symmetry of order 2.

[1]

Paper 11 Q1 Oct/Nov 2012

2 $\mathbf{a} = \begin{pmatrix} 4 \\ -3 \end{pmatrix}$ $\mathbf{b} = \begin{pmatrix} -1 \\ 5 \end{pmatrix}$

Work out $\mathbf{a} - 2\mathbf{b}$.

[2]

Paper 11 Q7 May/Jun 2014

3

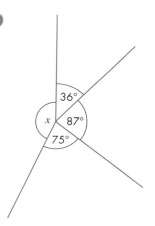

a Find the size of angle x. [1]

b What type of angle is x? [1]

Paper 11 Q3 Oct/Nov 2012

4 **a** The perimeter of a square is 28 mm.

Work out the length of one side of the square. [1]

b Calculate the volume of a cylinder with a radius 5.2 cm and height 15 cm. [2]

Paper 11 Q14 May/Jun 2013

5 The diagram shows a cuboid.

The volume of this cuboid is 720 cm³.

The width is 8 cm and the length is 15 cm.

Calculate h, the height of the cuboid. [2]

Paper 11 Q11 May/Jun 2014

6 Yim knows that one angle of an isosceles triangle is 48°.

He says that one of the other angles **must** be 66°.

Explain why Yim is wrong. [2]

Paper 11 Q13 Oct/Nov 2013

7 The diagram shows a circle inside a square.

The circumference of the circle touches all four sides of the square.

a Calculate the area of the circle when the side of the square is 15 cm. [2]

b Draw all the lines of symmetry on the diagram. [2]

Paper 11 Q21 May/Jun 2014

3

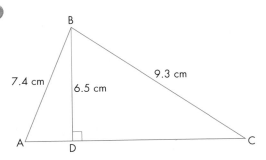

a Calculate AD. [3]

b Use trigonometry to calculate angle BCD. [2]

Paper 11 Q19 Oct/Nov 2012

PAPER 3

1 **a** In the diagram, ACE is a triangle.

B is a point on AC and D is a point on CE.

AE is parallel to BD, angle ACE = 70° and angle CBD = 40°.

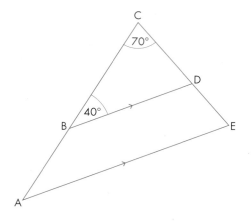

   i   Calculate angle BDC. [1]

   ii Write down the mathematical name of triangle BCD. [1]

   iii Calculate angle CAE. Give a reason for your answer. [2]

   iv Complete the following statement.

       Triangle ACE and triangle BCD are …. [1]

b In the diagram, A and B lie on a circle, centre O.

AC and BC are tangents to the circle and angle ACB = 55°.

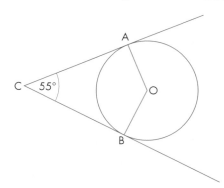

   i   Work out the reflex angle ACB. [1]

   ii Give a reason why angle OAC = angle OBC = 90°. [1]

   iii Work out angle AOB. [1]

   iv Write down the mathematical name of quadrilateral OACB. [1]

Paper 31 Q4 Oct/Nov 2012

2 **a** The diagram shows the positions of town A and town B, on a map.

    i Measure the length, in millimetres, of the line AB.   [1]

    ii Measure the bearing of town B from town A.   [1]

   b A triangular field has sides of length 550 m, 300 m and 400 m.

    i Construct the triangle, **using ruler and compasses only**.

    Use a scale of 1 cm to represent 50 m.

    The side of length 550 m should look like this.

550 m

        [3]

    ii By making a suitable measurement on your diagram, calculate the area of the field.

    Give your answer in square metres.   [3]

Paper 31 Q3 Oct/Nov 2013

3 The diagram shows two shapes, S and T, on a 1 cm² grid.

   P is the point (–2, 0).

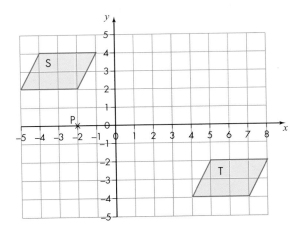

   a **i** Write down the mathematical name of shape S.   [1]

    ii How many lines of symmetry does shape S have?   [1]

   b Describe the single transformation that maps shape S onto shape T.   [2]

   c On a copy of the grid, draw the enlargement of shape S with scale factor 2 and centre P(–2, 0). You may need to extend the axes.

   Label the image E.   [2]

   a **i** Work out the area of shape S.   [2]

    ii How many shapes, identical to shape S, will fill shape E completely?   [1]

    iii Work out the area of shape E.   [1]

Paper 31 Q3 May/Jun 2014

PAPER 2

1 A hemisphere has a radius of 12 cm.

Calculate its volume. [2]

[The volume, V, of a sphere with radius r is $V = \frac{4}{3}\pi r^3$.]

Paper 22 Q8 Oct/Nov 2013

2 A car, 4.4 metres long, has a fuel tank that holds 65 litres of fuel when full.

The fuel tank of a mathematically similar model of the car holds 0.05 litres of fuel when full.

Calculate the length of the model car in centimetres. [3]

Paper 22 Q9 May/Jun 2013

3 OPQR is a trapezium with RQ parallel to OP and RQ = 2OP.

O is the origin, $\overrightarrow{OP}$ = **p** and $\overrightarrow{OR}$ = **r**.

M is the midpoint of PQ.

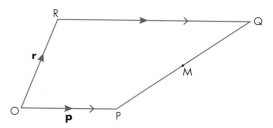

Find, in terms of **p** and **r**, in its simplest form:

a $\overrightarrow{PQ}$ [1]

b $\overrightarrow{OM}$, the position vector of M. [2]

Paper 22 Q14 May/Jun 2014

4 The exterior angle of a regular polygon is 36°.

What is the name of this polygon? [3]

Paper 22 Q9 Oct/Nov 2013

5 A water pipeline in Australia is a cylinder with radius 0.65 metres and length 85 kilometres.

85 km

0.65 m

Calculate the volume of water the pipeline contains when it is full.

Give your answer in cubic metres. [3]

Paper 22 Q8 Oct/Nov 2012

6 A, B, C, D and E are points on a circle.

Angle ABD = 58°, angle BAE = 85° and angle BDC = 19°.

BD and CA intersect at N.

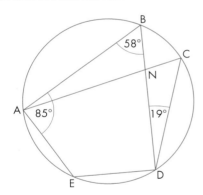

Calculate:

a angle BDE [1]

b angle AND. [2]

Paper 22 Q10 May/Jun 2013

EXTENDED

PAPER 4

1 The diagram shows a quadrilateral ABCD.

Angle BAD = 49° and angle ABD = 55°.

BD = 80 m, BC = 95 m and CD = 90 m.

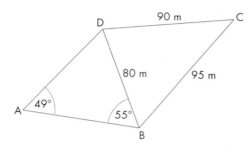

a Use the sine rule to calculate the length of AD. [3]

b Use the cosine rule to calculate angle BCD. [4]

c Calculate the area of the quadrilateral ABCD. [3]

d The quadrilateral represents a field.

Corn seeds are sown across the whole field at a cost of $3250 per hectare.

Calculate the cost of the corn seed used.

1 hectare = 10 000 m² [3]

Paper 42 Q3 May/Jun 2014

2 $A = \begin{pmatrix} 5 \\ 7 \end{pmatrix}$ $B = \begin{pmatrix} 6 \\ -4 \end{pmatrix}$ $C = \begin{pmatrix} 2 & 4 \\ 1 & 3 \end{pmatrix}$ $D = \begin{pmatrix} 2 & 9 \\ -1 & -3 \end{pmatrix}$

a Calculate the result of each of the following, if possible.

  i 3**A** [1]

  ii **AC** [1]

  iii **BA** [2]

  iv **C** + **D** [1]

  v **D**² [2]

b Calculate **C**⁻¹, the inverse of **C**. [2]

Paper 42 Q7 May/Jun 2013

3 **a** The diagram shows a prism of length 12 cm.

The cross-section is a regular hexagon of side 4 cm.

12 cm

4 cm

Calculate the total surface area of the prism. [4]

c Water flows through a cylindrical pipe of radius 0.74 cm.

It fills a 12 litre bucket in 4 minutes.

i Calculate the speed of the water through the pipe in centimetres per minute. [4]

ii When the 12 litre bucket is emptied into a circular pool, the water level rises by 5 **millimetres**.

Calculate the radius of the pool, correct to the nearest centimetre. [5]

Paper 42 Q9 May/Jun 2013

4 The total area of each of the following shapes is X.

The area of the shaded part of each shape is kX.

For each shape, find the value of k.

a

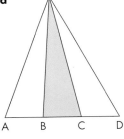

A   B   C   D

AB = BC = CD

b

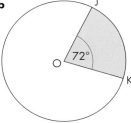

72°

O

J

K

Angle JOK = 72°

c

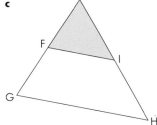

E

F

G

I

H

EF = FG and EI = IH

d

The shape is a
regular hexagon.

e

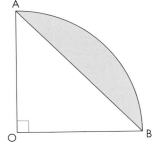

A

O

B

The diagram shows a sector of a circle centre O.
Angle AOB = 90°

[10]

Paper 42 Q11 May/Jun 2014

Why this chapter matters

Statistical graphs such as bar charts and line graphs are used in many areas of life from science to politics. They help us to analyse and interpret information.

One of the best ways to analyse information is to present it in a visual form. Some of the earliest types of statistical diagram were line graphs, bar charts and pie charts. They all show information in different ways.

Think about the owner of a bookshop.

He might use a graph like the one below to show how his sales go up and down over the year. Graphs like these are particularly good at showing trends in figures over time (see chapter 14).

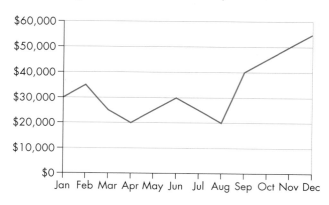

He might use a bar chart like the one on the right to show how many books they sell in different categories. Bar charts are very good at showing actual numbers.

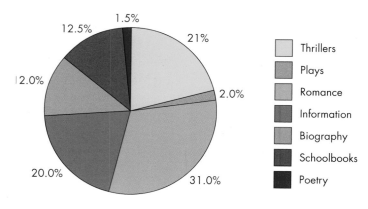

And he can get an idea of the percentage of different types of book he sells out of total sales by using a pie chart like the one on the left. Pie charts are good for analysing a whole (100%) by its parts.

This chapter introduces you to some of the most common forms of statistical representation. They fall into two groups – graphical diagrams such as bar charts and pie charts and quantitative diagrams such as frequency tables.

Statistical representation

Topics	Level	Key words
Frequency tables	CORE	tally chart, frequency, frequency table, classes, class interval, grouped frequency table
Pictograms	CORE	pictogram, symbol, key
Bar charts	CORE	bar chart, axis
Pie charts	CORE	pie chart, angle, sector
Scatter diagrams	CORE	scatter diagram, variables, correlation, positive correlation, negative correlation, zero correlation, line of best fit
Histograms	CORE	histogram
Histograms with bars of unequal width	EXTENDED	class frequency, frequency density

In this chapter you will learn how to:

CORE	EXTENDED

• Collect, classify and tabulate statistical data. (C9.1 and E9.1)
 • Read, interpret and draw simple inferences from tables and statistical diagrams. (C9.1 and E9.1)
 • Construct and read:
   • bar charts
   • pie charts
   • pictograms
   • simple frequency distributions
   • histograms with equal intervals and scatter diagrams. (C9.2 and E9.2)
 • Understand what is meant by positive, negative and zero correlation with reference to a scatter diagram. (C9.6 and E9.6)
 • Draw a straight line of best fit by eye. (C9.7 and E9.7)
	• Construct and read histograms with equal and unequal intervals (areas proportional to frequencies and vertical axis labelled 'frequency density'). (E9.2)

Frequency tables

Statistics is concerned with the collection and organisation of data, the representation of data on diagrams and the interpretation of data.

When you are collecting data for simple surveys, it is usual to use a **tally chart**. For example, data collection sheets are used to gather information about how people travel to work, how students spend their free time and the amount of time people spend watching TV.

It is easy to record the data by using tally marks, as shown in Example 1. Counting up the tally marks in each row of the chart gives the **frequency** of each category. By listing the frequencies in a column on the right-hand side of the chart, you can make a **frequency table** (see Example 1). Frequency tables are an important part of making statistical calculations.

EXAMPLE 1

Sandra wanted to find out about the ways in which students travelled to school. She carried out a survey. Her frequency table looked like this.

Method of travel	Tally	Frequency
Walk	HHT HHT HHT HHT HHT III	28
Car	HHT HHT II	12
Bus	HHT HHT HHT HHT III	23
Bicycle	HHT	5
Taxi	II	2

What does it tell you?

By adding together all the frequencies, you can see that 70 students took part in the survey. The frequencies also show you that more students travelled to school on foot than by any other method of transport.

Grouped data

Many surveys produce a lot of data that covers a wide range of values. In these cases, it is sensible to put the data into groups before attempting to compile a frequency table. These groups of data are called **classes** or **class intervals**.

Once the data has been grouped into classes, a **grouped frequency table** can be completed. The method is shown in Example 2.

EXAMPLE 2

These marks are for 36 students in a Year 10 mathematics examination.

31	49	52	79	40	29	66	71	73	19	51	47
81	67	40	52	20	84	65	73	60	54	60	59
25	89	21	91	84	77	18	37	55	41	72	38

a Construct a frequency table, using classes of 1–20, 21–40 and so on.

b What was the most frequent interval of marks?

a Draw the grid of the table shown below and put in the headings.

Next, list the classes, in order, in the column headed 'Marks'.

Using tally marks, indicate each student's score against the class to which it belongs. For example, 81, 84, 89 and 91 belong to the class 81–100, giving five tally marks, as shown below.

Finally, count the tally marks for each class and enter the result in the column headed 'Frequency'. The table is now complete.

Marks	Tally	Frequency
1–20	|||	3
21–40	||||| |||	8
41–60	||||| |||||		11
61–80	||||| ||||	9
81–100	|||||	5

b From the grouped frequency table, you can see that the highest number of students obtained a mark in the 41–60 interval.

EXERCISE 31A

1 Kurt kept a record of the number of goals scored by his local team in the last 20 matches. These are his results.

0 1 1 0 2 0 1 3 2 1

0 1 0 3 2 1 0 2 1 1

a Draw a frequency table for his data.   **b** Which was the most frequent score?

c How many goals were scored in total for the 20 matches?

2 Monique was doing a geography project on the weather. As part of her work, she kept a record of the daily midday temperatures in June.

Daily temperatures for June (°C)

15 18 19 21 23 22
20 23 22 24 24 25
26 26 20 19 19 20
18 18 19 17 16 15
16 16 17 18 20 22

a Copy and complete the grouped frequency table for her data.

b In which interval do the most temperatures lie?

c Describe what the weather was probably like throughout the month.

Temperature (°C)	Tally	Frequency
14–16		
17–19		
20–22		
23–25		
26–28		

3 In a game, Mitesh used a six-sided dice. He decided to keep a record of his scores to see whether the dice was fair. These are his scores.

2 4 2 6 1 5 4 3 3 2 3 6 2 1 3

5 4 3 4 2 1 6 5 1 6 4 1 2 3 4

a Draw a frequency table for his data.

b How many throws did Mitesh have during the game?

c Do you think the dice was a fair one? Explain why.

4 The data shows the heights, in centimetres, of a sample of 32 students.

172   158   160   175   180   167   159   180

167   166   178   184   179   156   165   166

184   175   170   165   164   172   154   186

167   172   170   181   157   165   152   164

a Draw a grouped frequency table for the data, using class intervals 151–155, 156–160, …

b In which interval do the most heights lie?

c Does this agree with a survey of the students in your class?

5 A student used a stopwatch to time how long it took her rabbit to find food left in its hutch.

This is her record, in seconds.

7	30	14	27	8	31	8	28	10	41	51	37	15	21	37	16	38
23	20	9	11	55	9	33	8	35	45	35	25	25	49	23	43	55
45	8	13	9	39	12	57	16	37	26	32	19	48	29	37		

Find the best way to put this data into a frequency chart to illustrate the length of time it took the rabbit to find the food.

6 A student was doing a survey to find the ages of people at a football competition.

He said that he would make a frequency table with the regions 15–20, 20–25, 25–30.

Explain what difficulty he could have with these class divisions.

Data collected from a survey can be presented in pictorial or diagrammatic form to help people to understand it more quickly. You see plenty of examples of this in newspapers and magazines and on TV, where every type of visual aid is used to communicate statistical information.

Pictograms

A **pictogram** is a frequency table in which frequency is represented by a repeated **symbol**. The symbol itself usually represents a number of items, as Example 4 on the next page shows. However, sometimes it is more sensible to let a symbol represent just a single unit, as in Example 3 below. The **key** tells you how many items are represented by a symbol.

EXAMPLE 3

The pictogram shows the number of phone calls made by Nurul from her mobile phone during a week.

Key 📱 represents 1 call

How many calls did Nurul make in the week?

From the pictogram, you can see that Nurul made a total of 27 calls.

Although pictograms can have great visual impact (particularly as used in advertising) and are easy to understand, they have a serious drawback. Apart from a half, fractions of a symbol cannot usually be drawn accurately and so frequencies are often represented only approximately by symbols.

Example 4 on the next page highlights this difficulty.

EXAMPLE 4

The pictogram shows the numbers of students who were late for school during a week.

Monday   ☺ ☺ ☺ ☺

Tuesday   ☺ ☺

Wednesday ☺ ☺ ☺

Thursday ☺ ☺ ☺

Friday   ☺ ☺ ☺ ☺ ☺

Key ☺ represents 5 students

How many students were late on:

a Monday   **b** Thursday?

Precisely how many students were late on Monday and Thursday respectively?

If you assume that each 'limb' of the symbol represents one student and its 'body' also represents one student, then the answers are:

a 19 students were late on Monday.   **b** 13 on Thursday.

EXERCISE 31B

1 The frequency table shows the numbers of cars parked in a shop's car park at various times of the day. Draw a pictogram to illustrate the data. Use a key of 1 symbol = 5 cars.

Time	9 am	11 am	1 am	3 am	5 am
Frequency	40	50	70	65	45

2 A milkman kept a record of how many pints of milk he delivered to 10 apartments on a particular morning. Draw a pictogram for the data. Use a key of 1 symbol = 1 pint.

Flat 1	Flat 2	Flat 3	Flat 4	Flat 5	Flat 6	Flat 7	Flat 8	Flat 9	Flat 10
2	3	1	2	4	3	2	1	5	1

3 The pictogram, taken from a Suntours brochure, shows the average daily hours of sunshine for five months in Tenerife.

a Write down the average daily hours of sunshine for each month.

b Give a reason why pictograms are useful in holiday brochures.

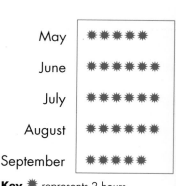

May   ✹✹✹✹✹

June   ✹✹✹✹✹✹

July   ✹✹✹✹✹✹

August   ✹✹✹✹✹✹

September ✹✹✹✹✹

Key ✹ represents 2 hours

4 The pictogram shows the amounts of money collected by six students after they had completed a sponsored walk for charity.

Anthony	$ $ $ $ $
Ben	$ $ $ $ $ $
Emma	$ $ $ $ $
Leanne	$ $ $ $
Reena	$ $ $ $ $ $
Simon	$ $ $ $ $ $ $

Key $ represents $5

a Who raised the most money?

b How much money was raised altogether by the six students?

c Robert also took part in the walk and raised $32. Why would it be difficult to include him on the pictogram?

5 A newspaper showed this pictogram about a family and the numbers of emails each family member received during one Sunday.

Key ⊠ represents 4 emails

		Frequency
Dad	⊠ ⊠ ⊠	
Mum	⊠ ▷	
Teenage son	⊠ ⊠ ⊠ ▷	
Teenage daughter		23
Young son		9

a How many emails did:

   i Dad receive   **ii** Mum receive   **iii** the teenage son receive?

b Copy and complete the pictogram.

c How many emails were received altogether?

31.3 Bar charts

A **bar chart** consists of a series of bars or blocks of the *same* width, drawn either vertically or horizontally from an **axis**.

The heights or lengths of the bars always represent *frequencies*.

Sometimes, the bars are separated by narrow gaps of equal width, which makes the chart easier to read.

EXAMPLE 5

The grouped frequency table shows the marks of 24 students in a test. Draw a bar chart for the data.

Marks	1–10	11–20	21–30	31–40	41–50
Frequency	2	3	5	8	6

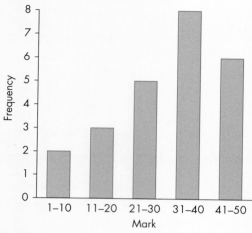

Note:

- Both axes are labelled.
- The class intervals are written under the middle of each bar.
- The bars are separated by equal spaces.

If you use a **dual bar chart**, it is easy to compare two sets of related data, as in Example 6.

EXAMPLE 6

This dual bar chart shows the average daily maximum temperatures for England and Turkey over a five-month period.

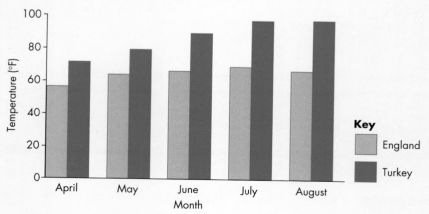

In which month was the *difference* between temperatures in England and Turkey the greatest?

The largest difference can be seen in August.

Note: You must always include a key to identify the two different sets of data.

EXERCISE 31C

1. For her survey on fitness, Samina asked a sample of people, as they left a sports centre, which activity they had taken part in. She then drew a bar chart to show her data.

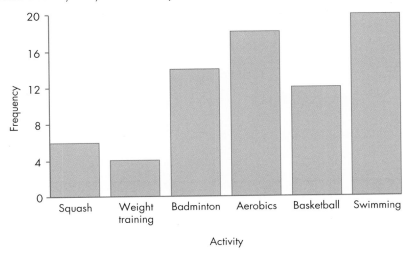

a Which was the most popular activity?

b How many people took part in Samina's survey?

2. The frequency table below shows the levels achieved by 100 students in their practice IGCSE examinations end of year tests.

Grade	F	E	D	C	B	A
Frequency	12	22	24	25	15	2

a Draw a suitable bar chart to illustrate the data.

b What fraction of the students achieved a grade C or grade B?

c Give one advantage of drawing a bar chart rather than a pictogram for this data.

3. This table shows the number of points Amir and Hasrul were each awarded in eight rounds of a general knowledge quiz.

Round	1	2	3	4	5	6	7	8
Amir	7	8	7	6	8	6	9	4
Hasrul	6	7	6	9	6	8	5	6

a Draw a dual bar chart to illustrate the data.

b Comment on how well each of them did in the quiz.

CORE

4 Mira did a survey about the time it took students in her class to get to school on a particular morning. She wrote down their times, correct to the nearest minute.

15 23 36 45   8 20 34 15 27 49

10 60   5 48 30 18 21   2 12 56

49 33 17 44 50 35 46 24 11 34

a Draw a grouped frequency table for Mira's data, using class intervals 1–10, 11–20, …

b Draw a bar chart to illustrate the data.

c What conclusions can Mira draw from the bar chart?

5 This table shows the number of accidents at a dangerous road junction over a six-year period.

Year	2005	2006	2007	2008	2009	2010
No. of accidents	6	8	7	9	6	4

a Draw a pictogram for the data.

b Draw a bar chart for the data.

c Which diagram would you use if you were going to argue that traffic lights should be installed at the junction? Explain why.

6 The diagram below shows the minimum and maximum temperatures for one day in August in five cities.

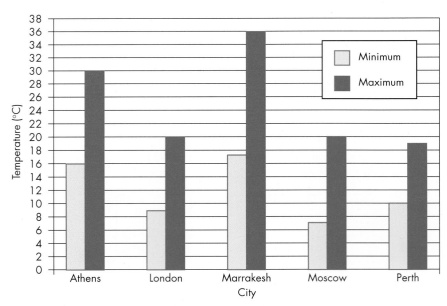

Lee says that the minimum temperature is always about half the maximum temperature for most cities.

Is Lee correct?

Give reasons to justify your answer.

Pie charts

Pictograms, bar charts and line graphs are easy to draw but they can be difficult to interpret when there is a big difference between the frequencies or there are only a few categories. In these cases, it is often more convenient to illustrate the data on a **pie chart**.

In a pie chart, the whole of the data is represented by a circle (the 'pie') and each category of it is represented by a **sector** of the circle (a 'slice of the pie'). The **angle** of each sector is proportional to the frequency of the category it represents.

So, a pie chart cannot show individual frequencies, as a bar chart can, for example. It can only show proportions.

Sometimes the pie chart will be marked off in equal sections rather than angles. In these cases, the numbers are always easy to work with.

EXAMPLE 7

20 people were surveyed about their preferred drink. Their replies are shown in the table.

Drink	Tea	Coffee	Milk	Cola
Frequency	6	7	4	3

Show the results on the pie chart given.

You can see that the pie chart has 10 equally-spaced divisions.

As there are 20 people, each division is worth two people. So the sector for tea will take up 3 of these divisions. In the same way, coffee will take up $3\frac{1}{2}$ divisions, milk will take up 2 divisions and cola will take up $1\frac{1}{2}$ divisions.

The finished pie chart will look like this.

Preferred drinks

Note:

- You should always label the sectors of the pie chart (use shading and a separate key if there is not enough space to write on the pie chart).

- Give your chart a title.

EXAMPLE 8

In a survey about holidays, 120 people were asked to state which type of transport they used on their last holiday. This table shows the results of the survey. Draw a pie chart to illustrate the data.

Type of transport	Train	Bus	Car	Ship	Plane
Frequency	24	12	59	11	14

You need to find the angle for the fraction of 360° that represents each type of transport. This is usually done in a table, as shown below.

Type of transport	Frequency	Calculation	Angle
Train	24	$\frac{24}{120} \times 360° = 72°$	72°
Bus	12	$\frac{12}{120} \times 360° = 36°$	36°
Car	59	$\frac{59}{120} \times 360° = 177°$	177°
Ship	11	$\frac{11}{120} \times 360° = 33°$	33°
Plane	14	$\frac{14}{120} \times 360° = 42°$	42°
Totals	120		360°

Draw the pie chart, using the calculated angle for each sector.

Note:

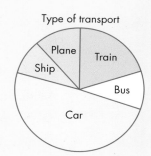

- Use the frequency total (120 in this case) to calculate each fraction.
- Check that the sum of all the angles is 360°.
- Label each sector.
- The angles or frequencies do not have to be shown on the pie chart.

EXERCISE 31D

1 Copy the diagram on the right and draw a pie chart to show each set of data.

a The favourite pets of 10 children

Pet	Bird	Cat	Rabbit
Frequency	4	5	1

b The makes of cars of 20 teachers

Make of car	Ford	Toyota	BMW	Nissan	Peugeot
Frequency	4	5	2	3	6

c The newspaper read by 40 office workers

Newspaper	*The Post*	*Today*	*The Mail*	*The Times*
Frequency	14	8	6	12

2 Draw a pie chart to represent each set of data.

ADVICE AND TIPS

Remember to complete a table as shown in the examples. Check that all angles add up to 360°.

a The numbers of children in 40 families

No. of children	0	1	2	3	4
Frequency	4	10	14	9	3

b How 90 students get to school

Journey to school	Walk	Car	Bus	Cycle
Frequency	4	10	14	9

3 Mariam asked 24 of her friends which sport they preferred to play. Her data is shown in this frequency table.

Sport	Rugby	Football	Tennis	Baseball	Basketball
Frequency	4	11	3	1	5

Illustrate her data in a pie chart.

4 Ameer wrote down the number of lessons he had per week in each subject on his school timetable.

Mathematics 5   English 5   Science 8   History 6
Geography 6   Arts 4   Sport 2

a How many lessons did Ameer have on his timetable?

b Draw a pie chart to show the data.

c Draw a bar chart to show the data.

d Which diagram better illustrates the data? Give a reason for your answer.

5 A market researcher asked 720 people which new brand of tinned beans they preferred. The results are given in the table.

A	248
B	264
C	152
D	56

a Draw a pie chart to illustrate the data.

b Why do you think pie charts are used to show this sort of information?

CORE

6 This pie chart shows the proportions of the different shoe sizes worn by 144 students in one year group in a school.

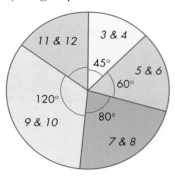

a What is the angle of the sector representing shoe sizes 11 and 12?

b How many students had a shoe size of 11 or 12?

7 The table below shows the numbers of candidates, at each grade, taking music examinations in Strings and Brass.

	Grades					Total number of candidates
	3	**4**	**5**	**6**	**7**	
Strings	300	980	1050	600	70	3000
Brass	250	360	300	120	70	1100

a Draw a pie chart to represent each of the two examinations.

b Compare the pie charts to decide which group of candidates, Strings or Brass, did better overall. Give reasons to justify your answer.

8 In a survey, a rail company asked passengers whether their service had improved.

What is the probability that a person picked at random from this survey answered "Don't know"?

A **scatter diagram** (also called a scattergraph or scattergram) is a method of comparing two **variables** by plotting their corresponding values on a graph. These values are usually taken from a table.

The variables are treated just like a set of (x, y) coordinates. This is shown in the scatter diagram on the right, in which the marks scored in a science test are plotted against the marks scored in a mathematics test.

This graph shows **positive correlation**. This means that students who get high marks in mathematics tests also tend to get high marks in science tests.

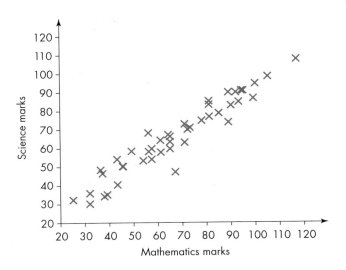

Correlation

There are different types of **correlation**. Here are three statements that may or may not be true.

- The taller people are, the wider their arm span is.
- The older a car is, the lower its value will be.
- The distance you live from your place of work will affect how much you earn.

These relationships could be tested by collecting data and plotting the points on a scatter diagram. For example, the first statement may give a scatter diagram like the one on the left below.

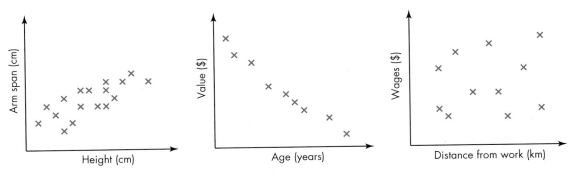

Positive correlation   **Negative correlation**   **Zero correlation**

This first diagram has **positive correlation** because, as one quantity increases, so does the other. From such a scatter diagram, you could say that the taller someone is, the wider their arm span.

Testing the second statement may give a scatter diagram like the middle one above. This has **negative correlation** because, as one quantity increases, the other quantity decreases. From such a scatter diagram, you could say that, as a car gets older, its value decreases.

Testing the third statement may give a scatter diagram like the one on the right, above. This scatter diagram has **zero correlation**. There is no relationship between the distance a person lives from their work and how much they earn.

EXAMPLE 9

The graphs below show the relationship between the temperature and the amount of ice-cream sold, and the relationship between the age of people and the amount of ice-cream they eat.

 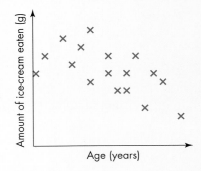

a Comment on the correlation of each graph.

b What does each graph tell you?

a The first graph has positive correlation and shows that, as the temperature increases, the amount of ice-cream sold increases.

b The second graph has negative correlation and shows that, as people get older, they eat less ice-cream.

Line of best fit

A **line of best fit** is a straight line that goes between all the points on a scatter diagram, passing as close as possible to all of them. You should try to have the same number of points on both sides of the line. Because you are drawing this line by eye, generous allowances are made around the correct answer. The line of best fit for the scatter diagram at the start of this section is shown below, left.

 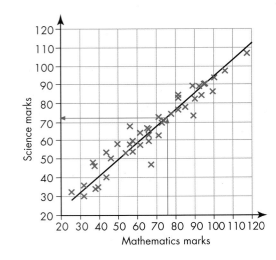

The line of best fit can be used to answer questions such as: 'A girl took the mathematics test and scored 75 marks but was ill for the science test. How many marks was she likely to have scored?'

You can find the answer by drawing a line up from 75 on the mathematics axis to the line of best fit and then drawing a line across to the science axis, as shown in the graph to the right of the graph showing the line of best fit. This gives 73, which is the mark she is likely to have scored in the science test.

EXERCISE 31E

1 Describe the correlation of each of these two graphs.

a

b

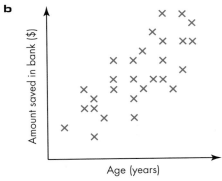

2 Explain what each graph in question 1 tells you.

3 The table below shows the results of a science experiment in which a ball is rolled along a desk top. The speed of the ball is measured at various points.

Distance from start (cm)	10	20	30	40	50	60	70	80
Speed (cm/s)	18	16	13	10	7	5	3	0

a Plot the data on a scatter diagram.

b Draw the line of best fit.

c If the ball's speed had been measured at 5 cm from the start, what is it likely to have been?

d Estimate how far the ball was from the start when its speed was 12 cm/s.

4 The heights, in centimetres, of 20 mothers and their 15-year-old daughters were measured. These are the results.

Mother	153	162	147	183	174	169	152	164	186	178
Daughter	145	155	142	167	167	151	145	152	163	168
Mother	175	173	158	168	181	173	166	162	180	156
Daughter	172	167	160	154	170	164	156	150	160	152

a Plot these results on a scatter diagram. Take the horizontal axis (the x-axis) for the mothers' heights from 140 to 200. Take the vertical axis (the y-axis) for the daughters' heights from 140 to 200.

b Is it true that the tall mothers have tall daughters?

5 The table shows the marks for ten students in their mathematics and geography examinations.

Student	Anna	Becky	Cath	Dema	Emma	Fatima	Greta	Hannah	Imogen	Sitara
Maths	145	155	142	167	167	151	145	152	163	168
Geog	175	173	158	168	181	173	166	162	180	156

a Plot the data on a scatter diagram. Take the x-axis for the mathematics scores and the y-axis for the geography scores.

b Draw the line of best fit.

c One of the students was ill when she took the geography examination. Which student was it most likely to be?

d If another student, Kate, was absent for the geography examination but scored 75 in mathematics, what mark would you expect her to have scored in geography?

e If another student, Lina, was absent for the mathematics examination but scored 65 in geography, what mark would you expect her to have scored in mathematics?

6 A teacher carried out a survey of 20 students from his class and asked them to say how many hours per week they spent playing sport and how many hours per week they spent watching TV. This table shows the results of the survey.

Student	1	2	3	4	5	6	7	8	9	10
Hours playing sport	12	3	5	15	11	0	9	7	6	12
Hours watching TV	18	26	24	16	19	27	12	13	17	14

Student	11	12	13	14	15	16	17	18	19	20
Hours playing sport	12	10	7	6	7	3	1	2	0	12
Hours watching TV	22	16	18	22	12	28	18	20	25	13

a Plot these results on a scatter diagram. Take the x-axis as the number of hours playing sport and the y-axis as the number of hours watching TV.

b If you knew that another student from the form watched 8 hours of TV a week, would you be able to predict how long they spent playing sport? Explain why.

7 The table shows the times taken and distances travelled by a taxi driver in 10 journeys on one day.

Distance (km)	1.6	8.3	5.2	6.6	4.8	7.2	3.9	5.8	8.8	5.4
Time (minutes)	3	17	11	13	9	15	8	11	16	10

a Draw a scatter diagram of this information, with time on the horizontal axis.

b Draw a line of best fit on your diagram.

c If a taxi journey takes 5 minutes, how far, in kilometres, would you expect the journey to have been?

d How much time would you expect a journey of 4 km to take?

8 Omar records the time taken, in hours, and the average speed, in kilometres per hour (km/h), for several different journeys.

Time (h)	0.5	0.8	1.1	1.3	1.6	1.75	2	2.4	2.6
Speed (km/h)	42	38	27	30	22	23	21	9	8

Estimate the average speed for a journey of 90 minutes.

9 Describe what you would expect the scatter graph to look like if someone said that it showed negative correlation.

31.6 Histograms

You should already be familiar with bar charts like the one on the right, in which the vertical axis represents frequency, and each bar has a label to show what it represents. (Sometimes it is more convenient to have the axes the other way round.)

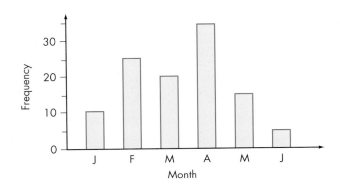

A **histogram** looks similar to a bar chart, but there are **three** fundamental differences.

- There are no gaps between the bars.

- The horizontal axis has a continuous scale.

- The area of each bar represents the class or group frequency of the bar.

Thia table shows times it takes people to walk to work.

Times, *t* minutes	$0 < t \leqslant 4$	$4 < t \leqslant 8$	$8 < t \leqslant 12$	$12 < t \leqslant 16$
Frequency	8	12	10	7

This **histogram** has been drawn from the table. The columns are *not* labelled $0 \leqslant t < 4$ and so on, as they would be on a bar chart. Instead there is a scale on the horizontal (time) axis. The first column is drawn between 0 and 4, the second between 4 and 8 and so on.

CORE

1 The table shows the range of heights of the girls in one year group in a school.

Height, *h* (cm)	$120 < h \leqslant 130$	$130 < h \leqslant 140$	$140 < h \leqslant 150$	$150 < h \leqslant 160$	$160 < h \leqslant 170$
Frequency	8	12	10	7	

Draw a histogram for this data.

2 A doctor was concerned at the length of time her patients had to wait to see her when they came to the morning clinic. Her survey gave these results.

Time, *m* (minutes)	$0 < m \leqslant 10$	$10 < m \leqslant 20$	$20 < m \leqslant 30$	$30 < m \leqslant 40$	$40 < m \leqslant 50$	$50 < m \leqslant 60$
Monday	5	8	17	9	7	4
Tuesday	9	8	16	3	2	1
Wednesday	7	6	18	2	1	1

a Draw a separate histogram for each day.

b On which day did patients tend to have to wait longer?

3 These are the prices of twenty second-hand cars, in dollars.

2590, 2650, 2650, 2790, 2850, 2925, 3090, 3125, 3125, 3420, 3595, 3740, 3750, 3920, 3945, 4095, 4150, 4200, 4750, 4785

a Copy and complete the frequency table to show the prices.

Prices (dollars)	2500–3000	3000–3500	3500–4000	4000–4500	4500–5000
Frequency					

b Illustrate the data on a histogram.

CORE

4 Boys and girls were given a simple task to complete. The times taken are shown below.

Times, t (minutes)	$0 < t \leqslant 4$	$4 < t \leqslant 8$	$8 < t \leqslant 12$	$12 < t \leqslant 16$	$16 < t \leqslant 20$
Boys	3	7	21	26	15
Girls	4	8	17	23	20

a Draw separate histograms to show the boys' times and the girls' times.

b How many boys and how many girls completed the task?

c What can you say about the longest time taken?

d Look at the two histograms and say whether you think boys or girls were better at the task. Give a reason for your answer.

5 These are the results of a survey of the masses of 50 girls and 50 boys of the same age.

Mass, k (kg)	$15 \leqslant k < 20$	$20 \leqslant k < 25$	$25 \leqslant k < 30$	$30 \leqslant k < 35$	$35 \leqslant k < 40$	$40 \leqslant k < 45$
Girls	4	6	12	14	6	8
Boys	1	4	16	10	15	4

a Show this data on two separate histograms.

b Children of this age who weigh less than 25kg are underweight. Shade the corresponding part of each histogram.

c Were more boys or girls underweight?

6 This histogram shows the times a group of employees took to travel to work one day.

a How many took more than 50 minutes?

b How many took between 20 and 50 minutes?

c What can you say about the shortest and longest journey times?

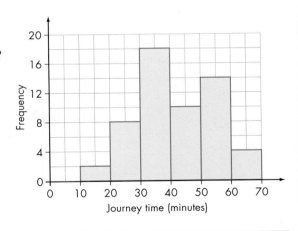

The company introduced flexible working so that employees could start work at different times. Here are the results of a survey of journey times after the change.

Journey time, t (minutes)	$0 \leqslant t < 10$	$10 \leqslant t < 20$	$20 \leqslant t < 30$	$30 \leqslant t < 40$	$40 \leqslant t < 50$	$50 \leqslant t < 50$	$60 \leqslant t < 70$
Frequency	2	4	14	12	14	5	5

d Show the new journey times on a histogram.

e Compare the two histograms. Have journey times been shortened? Justify your answer.

31.7 Histograms with bars of unequal width

Sometimes the data in a frequency distribution are grouped into classes with intervals that are different. In this case, the resulting histogram has bars of unequal widths.

The key fact to remember is that the area of a bar in a histogram represents the **class frequency** of the bar. So, in the case of an unequal-width histogram, you find the height to draw each bar by dividing its class frequency by its **class interval** width (bar width), which is the difference between the lower and upper bounds for each interval.

Conversely, given a histogram, you can find any of its class frequencies by multiplying the height of the corresponding bar by its width.

It is for this reason that the scale on the vertical axes of these histograms is always labelled 'frequency density', where:

$$\text{frequency density} = \frac{\text{frequency of class interval}}{\text{width of class interval}}$$

EXAMPLE 10

The heights of a group of girls were measured. The results were classified as shown in the table.

Height, h (cm)	$151 \leqslant h < 153$	$153 \leqslant h < 154$	$154 \leqslant h < 155$	$155 \leqslant h < 159$	$159 \leqslant h < 160$
Frequency	64	43	47	96	12

Draw a histogram to show the data.

It is convenient to write the table vertically and add two columns for class width and frequency density.

Calculate the class width by subtracting the lower class boundary from the upper class boundary. Calculate the frequency density by dividing the frequency by the class width.

Height, h (cm)	Frequency	Class width	Frequency density
$151 \leqslant h < 153$	64	2	32
$153 \leqslant h < 154$	43	1	43
$154 \leqslant h < 155$	47	1	47
$155 \leqslant h < 159$	96	4	24
$159 \leqslant h < 160$	12	1	12

The histogram can now be drawn.

The horizontal scale should be marked off as normal, from a value below the lowest value in the table to a value above the largest value in the table. In this case, mark the scale from 150 cm to 160 cm.

The vertical scale is always frequency density and is marked up to at least the largest frequency density in the table. In this case, 50 is a sensible value.

Each bar is drawn between the lower class interval and the upper class interval horizontally, and up to the frequency density vertically.

Now check that the area of each column is equal to the frequency.

151–153 is 2 × 32 = 64

153–154 is 1 × 43 = 43

and so on.

If the bars are of equal width, the frequency density and the frequency will be proportional. In that case you can use frequency on the vertical axis, as in section 31.6.

EXERCISE 31G

1 Draw histograms for these grouped frequency distributions.

a

Temperature, t (°C)	$8 \leqslant t < 10$	$10 \leqslant t < 12$	$12 \leqslant t < 15$	$15 \leqslant t < 17$	$17 \leqslant t < 20$	$20 \leqslant t < 24$
Frequency	5	13	18	4	3	6

b

Wage, w ($1000)	$6 \leqslant w < 10$	$10 \leqslant w < 12$	$12 \leqslant w < 16$	$16 \leqslant w < 24$
Frequency	16	54	60	24

c

Age, a (year)	$11 \leqslant a < 14$	$14 \leqslant a < 16$	$16 \leqslant a < 17$	$17 \leqslant w < 20$
Frequency	51	36	12	20

d

Pressure, p (mm)	$745 \leqslant p < 755$	$755 \leqslant p < 760$	$760 \leqslant p < 765$	$765 \leqslant p < 775$
Frequency	4	6	14	10

e

Time, t (min)	$0 \leqslant t < 8$	$8 \leqslant t < 12$	$12 \leqslant t < 16$	$16 \leqslant t < 20$
Frequency	72	84	54	36

EXTENDED

2 This information was gathered about the weekly pocket money given to 14-year-olds.

Pocket money, p ($)	$0 \leqslant p < 2$	$2 \leqslant p < 4$	$4 \leqslant p < 5$	$5 \leqslant p < 8$	$8 \leqslant p < 10$
Girls	8	15	22	12	4
Boys	6	11	25	15	6

Represent the information about the boys and girls on separate histograms.

3 The sales of the *Star* newspaper over 70 years are recorded in this table.

Years	1940–60	1961–80	1981–90	1991–2000	2001–05	2006–2010
Copies	62 000	68 000	71 000	75 000	63 000	52 000

Illustrate this information on a histogram. Take the class boundaries as 1940, 1960, 1980, 1990, 2000, 2005, 2010.

4 The Madrid trains were always late, so one month a survey was undertaken to find how many trains were late, and by how many minutes. The results are illustrated by this histogram.

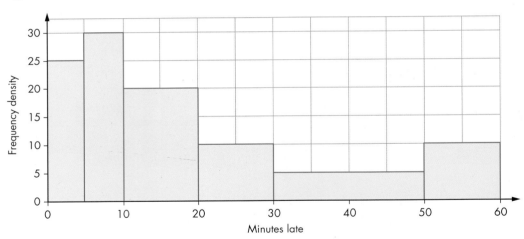

a How many trains were in the survey?

b How many trains were delayed for longer than 15 minutes?

5 For each of the frequency distributions illustrated in the histograms draw up the grouped frequency table.

a

b

c

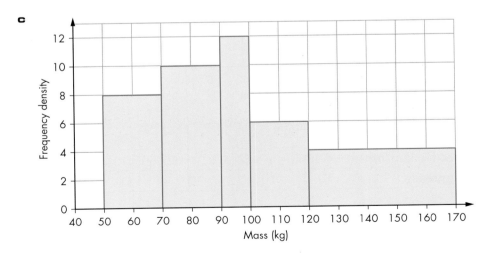

EXTENDED

6 All the patients in a hospital were asked how long it was since they last saw a doctor. The results are shown in the table.

Hours, h	$0 \leqslant h < 2$	$2 \leqslant h < 4$	$4 \leqslant h < 6$	$6 \leqslant h < 10$	$10 \leqslant h < 16$	$16 \leqslant h < 24$
Frequency	8	12	20	30	20	10

a Draw a histogram to illustrate the data.

b Estimate how many people waited more than 8 hours.

> **ADVICE AND TIPS**
>
> Find the area to the right of $h = 8$.

7 One summer, Albert monitored the mass of the tomatoes grown on each of his plants. His results are summarised in this table.

Mass, m (kg)	$6 \leqslant m < 10$	$10 \leqslant m < 12$	$12 \leqslant m < 16$	$16 \leqslant m < 20$	$20 \leqslant m < 25$
Frequency	8	15	28	16	10

a Draw a histogram for this distribution.

b Estimate how many plants produced more than 15 kg.

8 A survey was carried out to find the speeds of cars passing a particular point on a road. The histogram illustrates the results of the survey.

a Copy and complete this table.

Speed, v (km/h)	$0 < v \leqslant 40$	$40 < v \leqslant 50$	$50 < v \leqslant 60$	$60 < v \leqslant 70$	$70 < v \leqslant 80$	$80 < v \leqslant 100$
Frequency		10	40	110		

b Find the number of cars included in the survey.

9 The histogram shows the test scores for 320 students in a school.

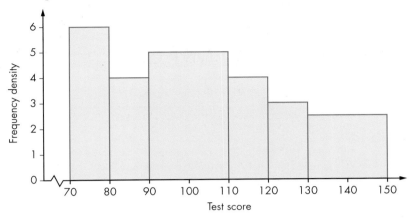

a How many students scored more than 120?

b The pass mark was 90. What percentage of students failed the test?

10 Adrienne and Bernice collected the same data about journey times but grouped it differently.

Here are Adrienne's figures.

Journey time (*t* minutes)	$0 \leqslant t < 5$	$5 \leqslant t < 10$	$10 \leqslant t < 15$	$15 \leqslant t < 20$	$20 \leqslant t < 25$	$25 \leqslant t < 30$
Frequency	10	15	30	12	6	3

Here are Bernice's figures.

Journey time (*t* minutes)	$0 \leqslant t < 10$	$10 \leqslant t < 15$	$15 \leqslant t < 30$
Frequency	25	30	21

a Draw a histogram for each set of figures. Use frequency density on the vertical axis each time.

b Describe any similarities or differences between the histograms.

EXTENDED

Why this chapter matters

The idea of 'average' is important in statistics. But there are several ways of working out an average, which have been developed over a long period of time.

Mean in Ancient India

There is a story about Rtuparna who was born in India around 5000 BCE. He wanted to estimate the amount of fruit on a tree:

- He counted how much fruit there was on one branch, then estimated the number of branches on the tree.

- He multiplied the estimated number of branches by the counted fruit on one branch.

He was amazed that the total was very close to the actual counted number of fruit when it was picked.

Rtuparna was one of the first to use arithmetic **mean**. The branch he chose was an average one representing all the branches. So the number of fruit on that branch would have been in the middle of the smallest and largest number of fruit on other branches on the tree.

Mode in Ancient Greece

This story comes from a war in Ancient Greece (431–404 BCE). It is about a battle between the Spartans and the Athenians.

The Athenians had to get over the Spartan Wall so they needed to work out its height. They started by counting the layers of bricks. This was done by hundreds of soldiers at the same time because many of them would get it wrong – but the majority would get it about right.

This is seen as an early use of the **mode**: the number of layers that occurred the most in the counting was taken as the one most likely to be correct.

They then had to guess the height of one brick and so calculate the total height of the wall. They could then make ladders long enough to reach the top of the wall.

The other average that you will use is the **median** (which finds the middle value), and there is no record of any use of this being used until the early 17th century.

These ancient examples demonstrate that you will not always work out the average in the same way – you must choose a method that is appropriate to the situation.

Chapter

32 Statistical measures

Topics	Level	Key words
1 The mode	CORE	average, mode, frequency, modal value
2 The median	CORE	median, middle value
3 The mean	CORE	average, mean
4 The range	CORE	range, spread, consistency
5 Which average to use	CORE	representative, appropriate, extreme values
6 Using frequency tables	CORE	frequency table
7 Grouped data	EXTENDED	grouped data, estimated, modal class, continuous data, discrete data
8 Cumulative frequency diagrams	EXTENDED	quartile, inter-quartile range, dispersion, cumulative frequency, lower quartile, upper quartile

In this chapter you will learn how to:

CORE	EXTENDED
• Calculate the mean, median, mode and range for individual and discrete data and distinguish between the purposes for which they are used. (C9.3 and E9.3)	• Calculate an estimate of the mean for grouped and continuous data. (E9.4) • Identify the modal class from a grouped frequency distribution. (E9.4) • Construct and use cumulative frequency diagrams. (E9.5) • Estimate and interpret the median, percentiles, quartiles and inter-quartile range. (E9.5)

The mode

Average is a term often used when describing or comparing sets of data, for example, average rainfall over a year or the average mark in a test for a group of students.

In each of the above examples, you are representing the whole set of many values (rainfall on every day of the year or the marks of all the students) by just a single, 'typical' value, which is called the average.

The idea of an average is extremely useful, because it enables you to compare one set of data with another set by comparing just two values – their averages.

There are several ways of expressing an average, but the most commonly used averages are the **mode**, the **median** and the **mean**.

The **mode** is the value that occurs the most in a set of data. That is, it is the value with the highest **frequency**.

The mode is a useful average because it is very easy to find and it can be applied to non-numerical data (qualitative data). For example, you could find the modal style of skirts sold in a particular month.

EXAMPLE 1

Suhail scored these numbers of goals in 12 football matches:

1 2 1 0 1 0 0 1 2 1 0 2

What is the mode of his scores?

The number that occurs most often in this list is 1. So, the mode is 1.

You can also say that the modal score or **modal value** is 1.

EXERCISE 32A

1. Find the mode for each set of data.

   a 3, 4, 7, 3, 2, 4, 5, 3, 4, 6, 8, 4, 2, 7

   b 47, 49, 45, 50, 47, 48, 51, 48, 51, 48, 52, 48

   c −1, 1, 0, −1, 2, −2, −2, −1, 0, 1, −1, 1, 0, −1, 2, −1, 2

   d $\frac{1}{2}, \frac{1}{4}, 1, \frac{1}{2}, \frac{3}{4}, \frac{1}{4}, 0, 1, \frac{3}{4}, \frac{1}{4}, 1, \frac{1}{4}, \frac{3}{4}, \frac{1}{4}, \frac{1}{2}$

   e 100, 10, 1000, 10, 100, 1000, 10, 1000, 100, 1000, 100, 10

   f 1.23, 3.21, 2.31, 3.21, 1.23, 3.12, 2.31, 1.32, 3.21, 2.31, 3.21

   ADVICE AND TIPS

   It helps to put the data in order or group all the same things together.

2. Find the mode for each set of data.

   a red, green, red, amber, green, red, amber, green, red, amber

   b rain, sun, cloud, sun, rain, fog, snow, rain, fog, sun, snow, sun

   c α, γ, α, β, γ, α, α, γ, β, α, β, γ, β, β, α, β, γ, β

   d ❋, ☆, ★, ★, ☆, ❋, ★, ☆, ★, ☆, ★, ❋, ✪, ☆, ★, ★, ☆

3 Halima did a survey to find the shoe sizes of students in her class. The bar chart illustrates her data.

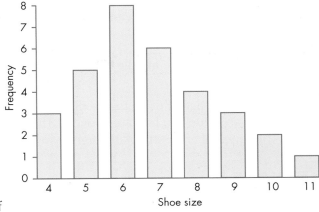

a How many students are there in Halima's class?

b What is the modal shoe size?

c Can you tell which are the boys' shoes sizes and which are the girls' shoe sizes?

d Halima then decided to draw a bar chart to show the shoe sizes of the boys and the girls separately. Do you think that the mode for the boys and the mode for the girls will be the same as the mode for the whole class? Explain your answer.

4 The frequency table shows the marks that a class obtained in a spelling test.

Mark	3	4	5	6	7	8	9	10
Frequency	1	2	6	5	5	4	3	4

a Write down the mode for their marks.

b Do you think this is a typical mark for the class? Explain your answer.

5 Explain why the mode is often referred to as the 'shopkeeper's average'.

6 This table shows the colours of eyes of the students in a class.

	Blue	Brown	Green
Boys	4	8	1
Girls	8	5	2

a How many students are there in the class?

b What is the modal eye colour for:

  i boys   **ii** girls   **iii** the whole class?

c After two students join the class the modal eye colour for the whole class is blue. Which of these statements is true?

- Both students had green eyes.
- Both students had brown eyes.
- Both students had blue eyes.
- You cannot tell what their eye colours were.

7 Here is a large set of raw data.

5 6 8 2 4 8 9 8 1 3 4 2 7 2 4 6 7 5 3 8
9 1 3 1 5 6 2 5 7 9 4 1 4 3 3 5 6 8 6 9
8 4 8 9 3 4 6 7 7 4 5 4 2 3 4 6 7 6 5 5

a What problems may occur if you attempted to find the mode by counting individual numbers?

b Explain a method that would make finding the mode more efficient and accurate.

c Use your method to find the mode of the data.

The median

The **median** is the **middle value** of a list of values when they are put in *order of size*, from lowest to highest.

The advantage of using the median as an average is that half the data values are below the median value and half are above it. Therefore, the average is only slightly affected by the presence of any particularly high or low values that are not typical of the data as a whole.

EXAMPLE 2

Find the median for this list of numbers.

2, 3, 5, 6, 1, 2, 3, 4, 5, 4, 6

Putting the list in numerical order gives:

1, 2, 2, 3, 3, **4**, 4, 5, 5, 6, 6

There are 11 numbers in the list, so the middle of the list is the 6th number. Therefore, the median is 4.

EXAMPLE 3

Find the median of the data shown in the frequency table.

Value	2	3	4	5	6	7
Frequency	2	4	6	7	8	3

First, add up the frequencies to find out how many pieces of data there are.

The total is 30 so the median value will be between the 15th and 16th values.

Now, add up the frequencies to give a running total, to find out where the 15th and 16th values are.

Value	2	3	4	5	6	7
Frequency	2	4	6	7	8	3
Running total	2	6	12	19	27	30

There are 12 data-values up to the value 4 and 19 up to the value 5.

Both the 15th and 16th values are 5, so the median is 5.

To find the median in a list of n values, written in order, use the rule:

$$\text{median} = \frac{n + 1}{2}\text{th value}$$

EXERCISE 32B

1 Find the median for each set of data.

a 7, 6, 2, 3, 1, 9, 5, 4, 8

b 26, 34, 45, 28, 27, 38, 40, 24, 27, 33, 32, 41, 38

c 4, 12, 7, 6, 10, 5, 11, 8, 14, 3, 2, 9

d 12, 16, 12, 32, 28, 24, 20, 28, 24, 32, 36, 16

e 10, 6, 0, 5, 7, 13, 11, 14, 6, 13, 15, 1, 4, 15

f −1, −8, 5, −3, 0, 1, −2, 4, 0, 2, −4, −3, 2

g 5.5, 5.05, 5.15, 5.2, 5.3, 5.35, 5.08, 5.9, 5.25

ADVICE AND TIPS

Remember to put the data in order before finding the median.

ADVICE AND TIPS

If there is an even number of pieces of data, the median will be halfway between the two middle values.

2 A group of 15 students had lunch in the school's cafeteria. Given below are the amounts that they spent.

$2.30, $2.20, $2, $2.50, $2.20, $3.50, $2.20, $2.25, $2.20, $2.30, $2.40, $2.20, $2.30, $2, $2.35

a Find the mode for the data.

b Find the median for the data.

c Which is the better average to use? Explain your answer.

3 **a** Find the median of 7, 4, 3, 8, 2, 6, 5, 2, 9, 8, 3.

b Without putting them in numerical order, write down the median for each of these sets.

i 17, 14, 13, 18, 12, 16, 15, 12, 19, 18, 13

ii 217, 214, 213, 218, 212, 216, 215, 212, 219, 218, 213

iii 12, 9, 8, 13, 7, 11, 10, 7, 14, 13, 8

iv 14, 8, 6, 16, 4, 12, 10, 4, 18, 16, 6

ADVICE AND TIPS

Look for a connection between the original data and the new data. For example, in **i**, the numbers are each 10 more than those in part **a**.

4 Given below are the age, height and mass of each of the seven players in a netball team.

	Ella	Linda	Pat	Marion	Amina	Martha	Elisa
Age (yr)	13	15	12	11	11	15	14
Height (cm)	161	165	162	158	154	168	169
Mass (kg)	41	42	37	32	35	42	40

a Find the median age of the team. Which player has the median age?

b Find the median height of the team. Which player has the median height?

c Find the median mass of the team. Which player has the median mass?

d Who would you choose as the average player in the team? Give a reason for your answer.

5 The table shows the numbers of sandwiches sold in a shop over 25 days.

Sandwiches sold	10	11	12	13	14	15	16
Frequency	2	3	6	4	3	4	3

a What is the modal number of sandwiches sold?

b What is the median number of sandwiches sold?

6 **a** Write down a list of nine numbers that has a median of 12.

b Write down a list of 10 numbers that has a median of 12.

c Write down a list of nine numbers that has a median of 12 and a mode of 8.

d Write down a list of 10 numbers that has a median of 12 and a mode of 8.

7 A list contains seven even numbers. The largest number is 24. The smallest number is half the largest. The mode is 14 and the median is 16. Two of the numbers add up to 42. What are the seven numbers?

8 Look at this list of numbers.

4, 4, 5, 8, 10, 11, 12, 15, 15, 16, 20

a Add four numbers to make the median 12.

b Add six numbers to make the median 12.

c What is the least number of numbers to add that will make the median 4?

9 Here are five payments.

$3, $5, $8, $100, $3000

Explain why the median is not a good average to use in this set of payments.

The **mean** of a set of data is the sum of all the values in the set divided by the total number of values in the set. That is:

$$\text{mean} = \frac{\text{sum of all values}}{\text{total number of values}}$$

This is what most people mean when they use the term **average**.

Another name for this average is the arithmetic **mean**.

The advantage of using the mean as an average is that it takes into account all the values in the set of data.

EXAMPLE 4

The ages of 11 players in a football squad are:

21, 23, 20, 27, 25, 24, 25, 30, 21, 22, 28

What is the mean age of the squad?

Sum of all the ages = 266

Total number in squad = 11

Therefore, mean age = $\frac{266}{11}$ = 24.1818… = 24.2 (1 decimal place)

EXERCISE 32C

1 Find the mean for each set of data.

a 7, 8, 3, 6, 7, 3, 8, 5, 4, 9

b 47, 3, 23, 19, 30, 22

c 42, 53, 47, 41, 37, 55, 40, 39, 44, 52

d 1.53, 1.51, 1.64, 1.55, 1.48, 1.62, 1.58, 1.65

e 1, 2, 0, 2, 5, 3, 1, 0, 1, 2, 3, 4

2 Calculate the mean for each set of data, giving your answer correct to 1 decimal place.

a 34, 56, 89, 34, 37, 56, 72, 60, 35, 66, 67

b 235, 256, 345, 267, 398, 456, 376, 307, 282

c 50, 70, 60, 50, 40, 80, 70, 60, 80, 40, 50, 40, 70

d 43.2, 56.5, 40.5, 37.9, 44.8, 49.7, 38.1, 41.6, 51.4

e 2, 3, 1, 0, 2, 5, 4, 3, 2, 0, 1, 3, 4, 5, 0, 3, 1, 2

CORE

CORE

3 The table shows the marks that 10 students obtained in mathematics, English and science in their examinations.

Student	Ahmed	Badru	Camille	Dayar	Evrim	Fatima	George	Helga	Imran	Josie
Maths	45	56	47	77	82	39	78	32	92	62
English	54	55	59	69	66	49	60	56	88	44
Science	62	58	48	41	80	56	72	40	81	52

  a Work out the mean mark for mathematics.

  b Work out the mean mark for English.

  c Work out the mean mark for science.

  d Which student obtained marks closest to the mean in all three subjects?

  e How many students were above the average mark in all three subjects?

4 Suni kept a record of the amount of time she spent on her homework over 10 days:

$\frac{1}{2}$ h, 20 min, 35 min, $\frac{1}{4}$ h, 1 h, $\frac{1}{2}$ h, $1\frac{1}{2}$ h, 40 min, $\frac{3}{4}$ h, 55 min

Calculate the mean time, in minutes, that Suni spent on her homework.

ADVICE AND TIPS

Convert all times to minutes, for example, $\frac{1}{4}$ h = 15 minutes.

5 These are the weekly wages of 10 people working in an office.

$350 $200 $180 $200 $350 $200 $240 $480 $300 $280

  a Find the modal wage.

  b Find the median wage.

  c Calculate the mean wage.

  d Which of the three averages best represents the office staff's wages? Give a reason for your answer.

ADVICE AND TIPS

Remember that the mean can be distorted by extreme values.

6 The ages of five people in a group of walkers are 38, 28, 30, 42 and 37.

  a Calculate the mean age of the group.

  b Steve, who is 41, joins the group. Calculate the new mean age of the group.

7 a Calculate the mean of 3, 7, 5, 8, 4, 6, 7, 8, 9 and 3.

  b Calculate the mean of 13, 17, 15, 18, 14, 16, 17, 18, 19 and 13. What do you notice?

  c Write down, without calculating, the mean for each of these sets of data.

   i 53, 57, 55, 58, 54, 56, 57, 58, 59, 53

   ii 103, 107, 105, 108, 104, 106, 107, 108, 109, 103

   iii 4, 8, 6, 9, 5, 7, 8, 9, 10, 4

ADVICE AND TIPS

Look for a connection between the original data and the new data. For example in **i** the numbers are 50 more.

8 Two families were in a competition.

Speed family		Roberts family	
Brian	aged 59	Frank	aged 64
Kath	aged 54	Marylin	aged 62
James	aged 34	David	aged 34
Helen	aged 34	James	aged 32
John	aged 30	Tom	aged 30
Joseph	aged 24	Helen	aged 30
Joy	aged 19	Evie	aged 16

Each family had to choose four members with a mean age of between 35 and 36.

Choose two teams, one from each family, that have this mean age between 35 and 36.

9 Asif had an average batting score of 35 runs. He had scored 315 runs in nine games of cricket.

What is the least number of runs he needs to score in the next match if he is to get a higher average score?

10 The mean age of a group of eight walkers is 42. Joanne joins the group and the mean age changes to 40. How old is Joanne?

32.4 The range

The **range** for a set of data is the highest value of the set minus the lowest value.

The range is *not* an average. It shows the **spread** of the data. You will use it when you are comparing two or more sets of similar data. You can also use it to comment on the **consistency** of two or more sets of data.

EXAMPLE 5

Rachel's marks in 10 mental arithmetic tests were 4, 4, 7, 6, 6, 5, 7, 6, 9 and 6.

Compare their marks. Adil's marks in the same tests were 6, 7, 6, 8, 5, 6, 5, 6, 5 and 6.

Rachel's mean mark is $60 \div 10 = 6$ and her range is $9 - 4 = 5$.

Adil's mean mark is $60 \div 10 = 6$ and his range is $8 - 5 = 3$.

Although the means are the same, Adil has a smaller range. This shows that Adil's results are more consistent.

EXERCISE 32D

CORE

1 Find the range for each set of data.

    a 3, 8, 7, 4, 5, 9, 10, 6, 7, 4

    b 62, 59, 81, 56, 70, 66, 82, 78, 62, 75

    c 1, 0, 4, 5, 3, 2, 5, 4, 2, 1, 0, 1, 4, 4

    d 3.5, 4.2, 5.5, 3.7, 3.2, 4.8, 5.6, 3.9, 5.5, 3.8

    e 2, –1, 0, 3, –1, –2, 1, –4, 2, 3, 0, 2, –2, 0, –3

2 The table shows the maximum and minimum temperatures at midday for five cities in England during a week in August.

	Birmingham	Leeds	London	Newcastle	Sheffield
Maximum temperature (°C)	28	25	26	27	24
Minimum temperature (°C)	23	22	24	20	21

    a Write down the range of the temperatures for each city.

    b What do the ranges tell you about the weather for England during the week?

3 Over a three-week period, a school sweet shop took these amounts.

	Monday	Tuesday	Wednesday	Thursday	Friday
Week 1	$32	$29	$36	$30	$28
Week 2	$34	$33	$25	$28	$20
Week 3	$35	$34	$31	$33	$32

    a Calculate the mean amount taken each week.

    b Find the range for each week.

    c What can you say about the total amounts taken for each of the three weeks?

4 In a golf tournament, the club coach had to choose either Maria or Fay to play in the first round. In the previous eight rounds, their scores were as follows.

    Maria's scores: 75, 92, 80, 73, 72, 88, 86, 90
    Fay's scores: 80, 87, 85, 76, 85, 79, 84, 88

    a Calculate the mean score for each golfer.

    b Find the range for each golfer.

    c Which golfer would you choose to play in the tournament? Explain why.

ADVICE AND TIPS

The best person to choose may not be the one with the biggest mean but could be the most consistent player.

5 Dan has a choice of two buses to get to school: Number 50 or Number 63. Over a month, he kept a record of the numbers of minutes each bus was late when it set off from his home bus stop.

No. 50: 4, 2, 0, 6, 4, 8, 8, 6, 3, 9

No. 63: 3, 4, 0, 10, 3, 5, 13, 1, 0, 1

a For each bus, calculate the mean number of minutes late.

b Find the range for each bus.

c Which bus would you advise Dan to catch? Give a reason for your answer.

6 The table gives the ages and heights of 10 children.

Name	Age (years)	Height (cm)
Evrim	9	121
Isaac	4	73
Lilla	8	93
Lewis	10	118
Evie	3	66
Badru	6	82
Oliver	4	78
Halima	2	69
Isambard	9	87
Chloe	7	82

a Chloe is having a party. She wants to invite as many children as possible but does not want the range of ages to be more than 5. Who will she invite?

b This is a sign at a theme park.

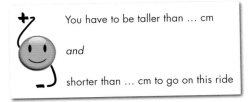

You have to be taller than ... cm

and

shorter than ... cm to go on this ride

Isaac is the shortest person who can go on the ride and Isambard is the tallest.

What are the smallest and largest missing values on the sign?

7 **a** The age range of a school quiz team is 20 years and the mean age is 34. Who would you expect to be in this team? Explain your answer.

b Another team has an average age of $15\frac{1}{2}$ and a range of 1. Who would you expect to be in this team? Explain your answer.

An average must be truly **representative** of a set of data. So, when you have to find an average, it is crucial to choose the **appropriate** type of average for this particular set of data.

If you use the wrong average, your results will be distorted and give misleading information.

This table, which compares the advantages and disadvantages of each type of average, will help you to make the correct decision.

	Mode	Median	Mean
Advantages	Very easy to find Not affected by **extreme values** Can be used for non-numerical data	Easy to find for ungrouped data Not affected by extreme values	Easy to find Uses all the values The total for a given number of values can be calculated from it
Disadvantages	Does not use all the values May not exist	Does not use all the values Often not understood	Extreme values can distort it Has to be calculated
Use for	Non-numerical data Finding the most likely value	Data with extreme values	Data with values that are spread in a balanced way

EXERCISE 32E

1 These are the ages of the members of a hockey team.

29 26 21 24 26 28 35 23 29 28 29

a Give:

i the modal age   **ii** the median age   **iii** the mean age.

b What is the range of the ages?

2 **a** For each set of data, find the mode, the median and the mean.

i 6, 10, 3, 4, 3, 6, 2, 9, 3, 4   **ii** 6, 8, 6, 10, 6, 9, 6, 10, 6, 8

iii 7, 4, 5, 3, 28, 8, 2, 4, 10, 9

b For each set of data, decide which average is the best one to use and give a reason.

3 These are the numbers of copies of *The Evening Star* sold on 12 consecutive evenings by a shop during a promotion exercise organised by the newspaper's publisher.

65 73 75 86 90 112 92 87 77 73 68 62

a Find the mode, the median and the mean for the sales.

b The shopkeeper had to report the average sale to the publisher after the promotion. Which of the three averages would you advise the shopkeeper to use? Explain why.

4 The mean age of a group of 10 young people is 15.

   a What do all their ages add up to?

   b What will be their mean age in five years' time?

5 Decide which average you would use for each statistic. Give a reason for your answer.

   a The average mark in an examination

   b The average pocket money for a group of 16-year-old students

   c The average shoe size for all the girls in one year at school

   d The average height for all the artistes on tour with a circus

   e The average hair colour for students in your school

   f The average mass of all newborn babies in a hospital's maternity ward

6 A pack of matches consisted of 12 boxes. The contents of each box were counted as:

   34   31   29   35   33   30   31   28   29   35   32   31

   On the box it stated 'Average contents 32 matches'. Is this correct?

7 Mr Brennan told each student their test mark and only gave the test statistics to the whole class. He gave the class the modal mark, the median mark and the mean mark.

   a Which average would tell a student whether they were in the top half or the bottom half of the class?

   b Which average tells the students nothing really?

   c Which average allows a student to gauge how well they have done compared with everyone else?

8 Three players were hoping to be chosen for the basketball team.

The table shows their scores in the last few games they played.

Tom	16, 10, 12, 10, 13, 8, 10
David	16, 8, 15, 25, 8
Mohaned	15, 2, 15, 3, 5

The teacher said they would be chosen by their best average score.

Which average would each boy want to be chosen by?

9 **a** Find five numbers that have *both* the properties below:

     ● a range of 5

     ● a mean of 5.

   b Find five numbers that have *all* the properties below:

     ● a range of 5

     ● a median of 5

     ● a mean of 5.

10 What is the average pay at a factory with 10 employees?

The boss said: '$43 295'

A worker said: '$18 210'

They were both correct.

Explain how this can be.

11 A list of nine numbers has a mean of 7.6. What number must be added to the list to give a new mean of 8?

12 A dance group of 17 people had a mean mass of 54.5 kg. To enter a competition there needed to be 18 people with an average mass of 54.4 kg or less. What is the maximum mass that the eighteenth person must have?

32.6 Using frequency tables

When you have gathered a lot of information, it is often convenient to put it together in a **frequency table**. From this table you can then find the values of the three averages and the range.

EXAMPLE 6

A survey was done on the number of people in each car leaving a shopping centre. The results are summarised in the table.

Number of people in each car	1	2	3	4	5	6
Frequency	45	198	121	76	52	13

For the number of people in a car, calculate:

a the mode **b** the median **c** the mean.

a The modal number of people in a car is easy to spot. It is the number with the largest frequency, which is 198. Hence, the modal number of people in a car is 2.

b You can find the median number of people in a car by working out where the middle of the set of numbers is located. First, add up frequencies to get the total number of cars surveyed, which comes to 505. Next, calculate the middle position.

$(505 + 1) \div 2 = 253$

Now add the frequencies across the table to find which group contains the 253rd item. The 243rd item is the end of the group with 2 in a car. Therefore, the 253rd item must be in the group with 3 in a car. Hence, the median number of people in a car is 3.

c To calculate the mean number of people in a car, multiply the number of people in the car by the frequency. This is best done in an extra column. Add these products to find the total number of people and divide by the total frequency (the number of cars surveyed).

Number in car	Frequency	Number in these cars
1	45	$1 \times 45 = 45$
2	198	$2 \times 198 = 396$
3	121	$3 \times 121 = 363$
4	76	$4 \times 76 = 304$
5	52	$5 \times 52 = 260$
6	13	$6 \times 13 = 78$
Totals	505	1446

Hence, the mean number of people in a car is $1446 \div 505 = 2.9$ (to 1 decimal place).

EXERCISE 32F

1 Find **i** the mode **ii** the median **iii** the mean from each frequency table below.

a A survey of the shoe sizes of all the boys in one year of a school gave these results.

Shoe size	4	5	6	7	8	9	10
Number of students	12	30	34	35	23	8	3

b A survey of the number of eggs laid by hens over a period of one week gave these results.

Number of eggs	0	1	2	3	4	5	6
Frequency	6	8	15	35	48	37	12

c This is a record of the number of babies born each week over one year in a small maternity unit.

Number of babies	0	1	2	3	4	5	6	7	8	9	10	11	12	13	14
Frequency	1	1	1	2	2	2	3	5	9	8	6	4	5	2	1

d A school did a survey on how many times in a week students arrived late at school. These are the findings.

Number of times late	0	1	2	3	4	5
Frequency	481	34	23	15	3	4

CORE

2 A survey of the number of children in each family of a school's intake gave these results.

Number of children	1	2	3	4	5
Frequency	214	328	97	26	3

a Assuming each child at the school is shown in the data, how many children are at the school?

b Calculate the mean number of children in a family.

c How many families have this mean number of children?

d How many families would consider themselves average from this survey?

3 A dentist kept records of how many teeth he extracted from his patients.

In 1989 he extracted 598 teeth from 271 patients.

In 1999 he extracted 332 teeth from 196 patients.

In 2009 he extracted 374 teeth from 288 patients.

a Calculate the average number of teeth taken from each patient in each year.

b Explain why you think the average number of teeth extracted falls each year.

4 One hundred cases of apples delivered to a supermarket were inspected and the numbers of bad apples were recorded.

Bad apples	0	1	2	3	4	5	6	7	8	9
Frequency	52	29	9	3	2	1	3	0	0	1

Give:

a the modal number of bad apples per case

b the mean number of bad apples per case.

5 Two dice are thrown together 60 times. The sums of the scores are shown below.

Score	2	3	4	5	6	7	8	9	10	11	12
Frequency	1	2	6	9	12	15	6	5	2	1	1

Find:

a the modal score

b the median score

c the mean score.

6 During a one-month period, the number of days off taken by 100 workers in a factory were noted as follows.

Number of days off	0	1	2	3	4
Number of workers	35	42	16	4	3

Calculate:

a the modal number of days off

b the median number of days off

c the mean number of days off.

7 Two friends often played golf together. They recorded the numbers of shots they made to get their balls into each hole over the last five games to compare who was more consistent and who was the better player. Their results were summarised in the table.

Number of shots	1	2	3	4	5	6	7	8	9
Roger	0	0	0	14	37	27	12	0	0
Brian	5	12	15	18	14	8	8	8	2

a What is the modal score for each player?

b What is the range of scores for each player?

c What is the median score for each player?

d What is the mean score for each player?

e Which player is the more consistent and why?

f Who would you say is the better player and why?

8 A tea stain on a newspaper removed four numbers from this frequency table of goals scored in 40 football matches one weekend.

Goals	0	1	2			5
Frequency	4	6	9			3

The mean number of goals scored is 2.4.

What could the missing four numbers be?

9 Manju made day trips to Mumbai frequently during a year.

The table shows how many days in a week she travelled.

Days	0	1	2	3	4	5
Frequency	17	2	4	13	15	1

Explain how you would find the median number of days Manju travelled in a week to Mumbai.

Sometimes the information you are given is grouped in some way (called **grouped data**), as in Example 7, which shows the range of weekly pocket money given to Year 12 students in a particular class.

Normally, grouped tables use **continuous data**, which is data that can have any value within a range of values, for example, height, mass, time, area and capacity. In these situations, the **mean** can only be **estimated** as you do not have all the information.

Discrete data is data that consists of separate numbers, for example, goals scored, marks in a test, number of children and shoe sizes.

In both cases, when using a grouped table to estimate the mean, first find the midpoint of the interval by adding the two end-values and then dividing by two.

EXAMPLE 7

Pocket money, p ($)	$0 < p < 1$	$1 < p < 2$	$2 < p < 3$	$3 < p < 4$	$4 < p < 5$
No. of students	2	5	5	9	15

a Write down the modal class.

b Calculate an estimate of the mean weekly pocket money.

a The modal class is easy to identify, since it is simply the one with the largest frequency. Here the modal class is $4 to $5.

b To estimate the mean, assume that each person in each class has the 'midpoint' amount, then build up the following table.

To find the midpoint value, add the two end-values together and then divide by two.

Pocket money, p ($$)	Frequency (f)	Midpoint (m)	$f \times m$
$0 < p \leqslant 1$	2	0.50	1.00
$1 < p \leqslant 2$	5	1.50	7.50
$2 < p \leqslant 3$	5	2.50	12.50
$3 < p \leqslant 4$	9	3.50	31.50
$4 < p \leqslant 5$	15	4.50	67.50
Totals	36		120

The estimated mean will be $120 ÷ 36 = $3.33 (rounded to the nearest cent).

If you had written 0.01–1.00, 1.01–2.00 and so on for the groups, then the midpoints would have been 0.505, 1.505 and so on. This would not have had a significant effect on the final answer as it is only an estimate.

Note that you *cannot* find the **median** or the range from a grouped table as you do not know the actual values.

EXERCISE 32G

1 For each table of values given below, find:

   i the modal group

   ii an estimate for the mean.

a

x	$0 < x \leqslant 10$	$10 < x \leqslant 20$	$20 < x \leqslant 30$	$30 < x \leqslant 40$	$40 < x \leqslant 50$
Frequency	4	6	11	17	9

b

y	$0 < y \leqslant 100$	$100 < y \leqslant 200$	$200 < y \leqslant 300$	$300 < y \leqslant 400$	$400 < y \leqslant 500$	$500 < x \leqslant 600$
Frequency	95	56	32	21	9	3

c

z	$0 < z \leqslant 5$	$5 < z \leqslant 10$	$10 < z \leqslant 15$	$15 < z \leqslant 20$
Frequency	16	27	19	13

d

Weeks	1–3	4–6	7–9	10–12	13–15
Frequency	5	8	14	10	7

2 Jason brought 100 pebbles back from the beach and found their masses, recording each mass to the nearest gram. His results are summarised in the table.

Mass, m (g)	$40 < m \leqslant 60$	$60 < m \leqslant 80$	$80 < m \leqslant 100$
Frequency	5	9	22

Mass, m (g)	$100 < m \leqslant 120$	$120 < m \leqslant 140$	$140 < m \leqslant 160$
Frequency	27	26	11

Find:

a the modal class of the pebbles

b an estimate of the total mass of all the pebbles

c an estimate of the mean mass of the pebbles.

3 A gardener measured the heights of all his daffodils to the nearest centimetre and summarised his results as follows.

Height (cm)	10–14	15–18	19–22	23–26	27–40
Frequency	21	57	65	52	12

a How many daffodils did the gardener have?

b What is the modal class of the daffodils?

c What is the estimated mean height of the daffodils?

4 A survey was created to see how quickly an emergency service got to cars that had broken down. The table summarises the results.

Time (min)	1–15	16–30	31–45	46–60	61–75	76–90	91–105
Frequency	2	23	48	31	27	18	11

  a How many calls were used in the survey?

  b Estimate the mean time taken per call.

  c Which average would the emergency service use for the average call-out time?

  d What percentage of calls do the emergency service get to within the hour?

5 One hundred light bulbs were tested by their manufacturer to see whether the average life-span of the manufacturer's bulbs was over 200 hours. The table summarises the results.

Life span, h (hours)	$150 < h \leq 175$	$175 < h \leq 200$	$200 < h \leq 225$	$225 < h \leq 250$	$250 < h \leq 275$
Frequency	24	45	18	10	3

  a What is the modal length of time a bulb lasts?

  b What percentage of bulbs last longer than 200 hours?

  c Estimate the mean life-span of the light bulbs.

  d Do you think the test shows that the average life-span is over 200 hours? Explain your answer fully.

6 Three shops each claimed to have the lowest average price increase over the year. The table summarises their price increases.

Price increase (p)	1–5	6–10	11–15	16–20	21–25	26–30	31–35
Soundbuy	4	10	14	23	19	8	2
Springfields	5	11	12	19	25	9	6
Setco	3	8	15	31	21	7	3

Using their average price increases, make a comparison of the supermarkets and say which one has the lowest price increases over the year. Remember to justify your answers.

7 The table shows the distances run, over a month, by an athlete who is training for a half-marathon.

Distance, d (km)	$0 < d \leq 5$	$5 < d \leq 10$	$10 < d \leq 15$	$15 < d \leq 20$	$20 < d \leq 25$
Frequency	3	8	13	5	2

A half-marathon is 21 kilometres. It is recommended that an athlete's daily average distance should be at least a third of the distance of the race for which they are training. Is this athlete doing enough training?

8 The table shows the points scored in a general-knowledge competition by all the players.

Points	0–9	10–19	20–29	30–39	40–49
Frequency	8	5	10	5	2

Balvir noticed that two frequencies were the wrong way round and that this made a difference of 1.7 to the arithmetic mean.

Which two frequencies were the wrong way round?

9 The profit made each week by a charity shop is shown in the table below.

Profit	$0–$500	$501–$1000	$1001–$1500	$1501–$2000
Frequency	15	26	8	3

Estimate the mean profit made each week.

10 The table shows the number of members of 100 football clubs.

Members	20–29	30–39	40–49	50–59	60–69
Frequency	16	34	27	18	5

Roger claims that the median number of members is 39.5.

Is he correct? Explain your answer.

32.8 Cumulative frequency diagrams

The **inter-quartile range** is a measure of the **dispersion** of a set of data. The advantage of the interquartile range is that it eliminates extreme values, and bases the measure of spread on the middle 50% of the data. This section will show you how to find the interquartile range and the median of a set of data by drawing a **cumulative frequency diagram**.

Look at the marks of 50 students in a mathematics test, which have been put into a grouped table. Note that it includes a column for the **cumulative frequency**, which you can find by adding each frequency to the sum of all preceding frequencies.

Mark	No. of students	Cumulative frequency
21 to 30	1	1
31 to 40	3	4
41 to 50	6	10
51 to 60	10	20
61 to 70	13	33
71 to 80	6	39
81 to 90	4	43
91 to 100	3	46
101 to 110	2	48
111 to 120	2	50

This data can then be used to plot a graph of the top value of each group against its cumulative frequency. The points to be plotted are (30, 1), (40, 7), (50, 13), (60, 21), etc., which will give the graph shown below. Note that the cumulative frequency is *always* the vertical axis.

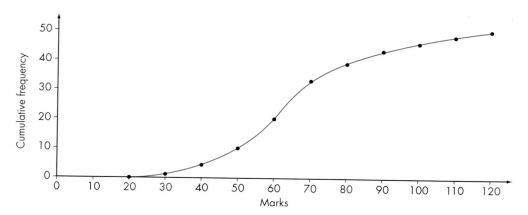

Also note that the scales on both axes are labelled at each graduation mark, in the usual way. *Do not* label the scales as shown below. It is *wrong*.

| 21–30 | 31–40 | 41–50 |

The plotted points can be joined by a freehand curve, to give a cumulative frequency diagram.

The median

The **median** is the middle item of data, once all the items have been put in order of size, from lowest to highest. So, if you have n items of data plotted as a cumulative frequency diagram, you can find the median from the middle value of the cumulative frequency, that is the $\frac{1}{2}n$th value.

But remember, if you want to find the median from a simple list of discrete data, you *must* use the $\frac{1}{2}(n + 1)$th value. The reason for the difference is that the cumulative frequency diagram treats the data as continuous, even for data such as examination marks, which are discrete. You can use the $\frac{1}{2}n$th value when working with cumulative frequency diagrams because you are only looking for an *estimate* of the median.

There are 50 values in the table on the previous page. To find the median:

- The middle value will be the 25th value.

- Draw a horizontal line from the 25th value to meet the graph.

- Now look down to the horizontal axis.

This will give an estimate of the median. In this example, the median is about 64 marks.

The inter-quartile range

By dividing the cumulative frequency into four parts, you can obtain **quartiles** and the inter-quartile range.

The **lower quartile** is the value one-quarter of the way up the cumulative frequency axis and is given by the $\frac{1}{4}n$th value.

The **upper quartile** is the value three-quarters of the way up the cumulative frequency axis and is given by the $\frac{3}{4}n$th value.

The inter-quartile range is the difference between the lower and upper quartiles.

These are illustrated on the graph below.

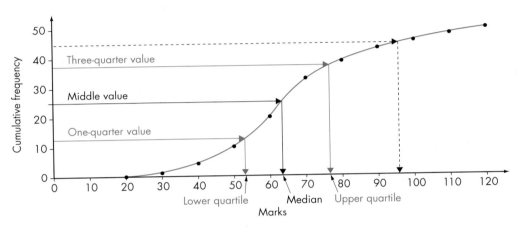

The quarter and three-quarter values out of 50 values are the 12.5th value and the 37.5th value. Draw lines across to the cumulative frequency curve from these values and down to the horizontal axis. These give the lower and upper quartiles. In this example, the lower quartile is 54 marks, the upper quartile is 77 marks and the inter-quartile range is $77 - 54 = 23$ marks.

As well as the median and the quartiles we can find percentiles. For example to find the 90th percentile:

90% of 50 is 45.

Draw a line across from a cumulative frequency of 45 and then down to the horizontal axis.

This shows that a mark of 95 is the 90th percentile.

So 90% of the students scored $\leqslant 95$ marks.

Other percentiles:

- The median is the 50th percentile.

- The upper quartile is the 75th percentile.

- The lower quartile is the 25th percentile.

EXAMPLE 8

This table shows the marks of 100 students in a mathematics test.

a Draw a cumulative frequency curve.

b Use your graph to find the median and the inter-quartile range.

c Students who score less than 44 have to have extra teaching. How many students will have to have extra teaching?

Mark, x	No. of students	Cumulative frequency
$21 \leqslant x \leqslant 30$	3	3
$31 \leqslant x \leqslant 40$	9	12
$41 \leqslant x \leqslant 50$	12	24
$51 \leqslant x \leqslant 60$	15	39
$61 \leqslant x \leqslant 70$	22	61
$71 \leqslant x \leqslant 80$	16	77
$81 \leqslant x \leqslant 90$	10	87
$91 \leqslant x \leqslant 100$	8	95
$101 \leqslant x \leqslant 110$	3	98
$111 \leqslant x \leqslant 120$	2	100

The groups are given in a different way to those in the table on page 575. You will meet several ways of giving groups (for example, 21–30, $20 < x \leqslant 30$, $21 < x < 30$) but the important thing to remember is to plot the top point of each group against the corresponding cumulative frequency.

a and **b** Draw the graph and add the lines for the median (50th value), lower and upper quartiles (25th and 75th values).

The required answers are read from the graph.

    Median = 65 marks
    Lower quartile = 51 marks
    Upper quartile = 79 marks
    Interquartile range = 79 − 51 = 28 marks

c At 44 on the mark axis, draw a perpendicular line to intersect the graph, and at the point of intersection draw a horizontal line across to the cumulative frequency axis, as shown. Number of students needing extra teaching is 18.

Note: An alternative way in which the table in Example 8 could have been set out is shown below. This arrangement has the advantage that the points to be plotted are taken straight from the last two columns. Decide which method you prefer.

Mark, x	No. of students	Less than	Cumulative frequency
$21 \leqslant x \leqslant 30$	3	30	3
$31 \leqslant x \leqslant 40$	9	40	12
$41 \leqslant x \leqslant 50$	12	50	24
$51 \leqslant x \leqslant 60$	15	60	39
$61 \leqslant x \leqslant 70$	22	70	61
$71 \leqslant x \leqslant 80$	16	80	77
$81 \leqslant x \leqslant 90$	10	90	87
$91 \leqslant x \leqslant 100$	8	100	95
$101 \leqslant x \leqslant 110$	3	110	98
$111 \leqslant x \leqslant 120$	2	120	100

EXERCISE 32H

1 A class of 30 students was asked to guess when one minute had passed. The table shows the results.

a Copy the table and complete a cumulative frequency column.

b Draw a cumulative frequency diagram.

c Use your diagram to estimate the median time and the inter-quartile range.

Time, x (seconds)	No. of students
$20 < x \leqslant 30$	1
$30 < x \leqslant 40$	3
$40 < x \leqslant 50$	6
$50 < x \leqslant 60$	12
$60 < x \leqslant 70$	3
$70 < x \leqslant 80$	3
$80 < x \leqslant 90$	2

2 A group of 50 pensioners was given the task in question **1**. The results are shown in the table.

a Copy the table and complete a cumulative frequency column.

b Draw a cumulative frequency diagram.

c Use your diagram to estimate the median time and the inter-quartile range.

d Which group, the students or the pensioners, was better at estimating time? Give a reason for your answer.

Time, x (seconds)	No. of pensioners
$10 < x \leqslant 20$	1
$20 < x \leqslant 30$	2
$30 < x \leqslant 40$	2
$40 < x \leqslant 50$	9
$50 < x \leqslant 60$	17
$60 < x \leqslant 70$	13
$70 < x \leqslant 80$	3
$80 < x \leqslant 90$	2
$90 < x \leqslant 100$	1

EXTENDED

3 The sizes of 360 senior schools are recorded in the table.

a Copy the table and complete a cumulative frequency column.

b Draw a cumulative frequency diagram.

c Use your diagram to estimate the median size of the schools and the inter-quartile range.

d Schools with fewer than 350 students are threatened with closure. About how many schools are threatened with closure?

e Use your graph to estimate the 90th percentile.

f Use your graph to estimate the 40th percentile.

No. of students	No. of schools
100–199	12
200–299	18
300–399	33
400–499	50
500–599	63
600–699	74
700–799	64
800–899	35
900–999	11

4 The temperature at a seaside town was recorded for 50 days. It was recorded to the nearest degree. The table shows the results.

a Copy the table and complete a cumulative frequency column.

b Draw a cumulative frequency diagram. Note that as the temperature is to the nearest degree the top values of the groups are 7.5 °C, 10.5 °C, 13.5 °C, 16.5 °C, etc.

Temperature (°C)	No. of days
5–7	2
8–10	3
11–13	5
14–16	6
17–19	6
20–22	9
23–25	8
26–28	6
29–31	5

c Use your diagram to estimate the median temperature and the inter-quartile range.

d Use your diagram to estimate the 10th percentile.

5 A game consists of throwing three darts and recording the total score. The results of the first 80 people to throw are recorded in the table.

a Draw a cumulative frequency diagram to show the data.

b Use your diagram to estimate the median score and the inter-quartile range.

Total score, x	No. of players
$1 \leqslant x \leqslant 20$	9
$21 \leqslant x \leqslant 40$	13
$41 \leqslant x \leqslant 60$	23
$61 \leqslant x \leqslant 80$	15
$81 \leqslant x \leqslant 100$	11
$101 \leqslant x \leqslant 120$	7
$121 \leqslant x \leqslant 140$	2

c People who score over 90 get a prize. About what percentage of the people get a prize?

6 One hundred children were asked to say how much pocket money they got in a week. The results are in the table.

Amount of pocket money (cents)	No. of children
51–100	6
101–150	10
151–200	20
201–250	28
251–300	18
301–350	11
351–400	5
401–450	2

a Copy the table and complete a cumulative frequency column.

b Draw a cumulative frequency diagram.

c Use your diagram to estimate the median amount of pocket money and the inter-quartile range.

d Estimate the 10th percentile and the 90th percentile.

7 Johan set his class an end-of-course test with two papers, A and B. He produced the cumulative frequency graphs, as shown.

a What is the median score for each paper?

b What is the inter-quartile range for each paper?

c Which is the harder paper? Explain how you know.

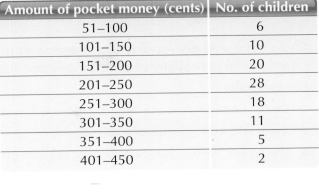

Johan wanted 80% of the students to pass each paper and 20% of the students to get a top grade in each paper.

d What marks for each paper give:

    i a pass   **ii** the top grade?

8 The lengths of time taken by 60 helpline telephone calls were recorded.

A cumulative frequency diagram of this data is shown here.

a Estimate the percentage of calls lasting more than 10 minutes.

b What is the 15th percentile for call lengths?

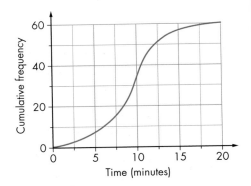

9 Byron was given a cumulative frequency diagram showing the marks obtained by students in a mental maths test.

He was told the top 10% were given the top grade.

How would he find the marks needed to gain this top award?

Why this chapter matters

Chance is a part of everyday life. Judgements are frequently made based on the probability of something happening.

For example:

- there is an 80% chance that my team will win the game tomorrow
- there is a 40% chance of rain tomorrow
- she has a 50–50 chance of having a baby girl
- there is a 10% chance of the bus being on time tonight.

In everyday life we talk about the probability of something happening. Two people might give different probabilities to the same events because of their different views. For example, some people might not agree that there is an 80% chance of your team winning the game. They might say that there is only a 70% chance of them winning tomorrow. A lot depends on what people believe or have experienced.

When people first started to predict the weather scientifically over 150 years ago, they used probabilities to do it. For example, meteorologists looked for three important indicators of rain:

- the number of nimbus clouds in the sky
- falling pressure on a barometer
- the direction of the wind and whether it was blowing from a part of the country with high rainfall.

If all three of these things occurred together rain would almost certainly follow soon.

Now, in the 21st century, probability theory is used to control the flow of traffic through road systems (below left) or the running of telephone exchanges (below right), and to look at patterns of the spread of infections.

33 Probability

Topics	Level	Key words
1 The probability scale	CORE	chance, outcome, event, probability, impossible, certain
2 Calculating probabilities	CORE	outcome, equally likely, probability fraction, random
3 Probability that an event will not happen	CORE	outcome
4 Probability in practice	CORE	experimental probability, trials, relative frequency
5 Combined events	EXTENDED	possibility diagram
6 Tree diagrams	EXTENDED	tree diagram

In this chapter you will learn how to:

CORE	EXTENDED
• Calculate the probability of a single event as either a fraction, decimal or percentage. (C8.1 and E8.1) • Understand and use the probability scale from 0 to 1. (C8.2 and E8.2) • Understand that: *the probability of an event occurring = 1 – the probability of the event not occurring.* (C8.3 and E8.3) • Understand relative frequency as an estimate of probability. (C8.4 and E8.4)	• Calculate the probability of simple combined events, using possibility diagrams and tree diagrams where appropriate. (E8.5)

Almost daily, you hear somebody talking about the probability of whether something will happen. They usually use words such as '**chance**', 'likelihood' or 'risk' rather than 'probability'. For example:

'What is the likelihood of rain tomorrow?'

'What chance does she have of winning the 100 metre sprint?'

'Is there a risk that his company will go bankrupt?'

You can give a value to the chance of any of these **outcomes** or **events** happening – and millions of others, as well. This value is called the **probability**.

It is true that some things are certain to happen and that some things cannot happen; that is, the chance of something happening can be anywhere between **impossible** and **certain**. This situation is represented on a sliding scale called the **probability scale**, as shown here.

Note: All probabilities lie somewhere in the range of **0** to **1**.

An outcome or an event that cannot happen (is impossible) has a probability of 0. For example, the probability that donkeys will fly is 0.

An outcome or an event that is certain to happen has a probability of 1. For example, the probability that the sun will rise tomorrow is 1.

EXAMPLE 1

Put arrows on the probability scale to show the probability of each of the outcomes of these events.

a You will get a head when throwing a coin.

b You will get a six when throwing a dice.

c You will have maths homework this week.

a This outcome is an even chance. (Commonly described as a fifty-fifty chance.)

b This outcome is fairly unlikely.

c This outcome is likely.

The arrows show the approximate probabilities on the probability scale.

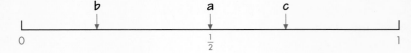

EXERCISE 33A

1 State whether each of these events is impossible, very unlikely, unlikely, even chance, likely, very likely or certain.

   a Someone in your class is left-handed.

   b You will live to be 100.

   c You get a score of seven when you throw a dice.

   d You will watch some TV this evening.

   e A new-born baby will be a girl.

2 Draw a probability scale and put an arrow to show the approximate probability of each of these events happening.

   a The next car you see will have been made in Japan.

   b A person in your class will have been born in the 20th century.

   c It will rain tomorrow.

   d In the next Olympic Games, someone will run the 1500 m race in 3 minutes.

   e During this week, you will have noodles with a meal.

3 **a** Draw a probability scale and mark an arrow to show the approximate probability of each of these events.

    A The next person to come into the room will be male.

    B The person sitting next to you in mathematics is over 16 years old.

    C Someone in the class will have a mobile phone.

   b What number on your scale corresponds to each arrow?

4 **a** Give two events of your own for which you think the probability of an outcome is:

    A impossible   B very unlikely

    C evens   D likely

    E certain.

   b Draw a probability scale numbered from 0 to 1 and put an arrow for each of your events.

   c What number on your scale corresponds to each arrow?

5 'The train was late yesterday so it is very likely that it will be late today.' Is this true?

In Exercise 33A, you may have had difficulty in knowing exactly where to put some of the arrows on the probability scale. It would have been easier for you if each result of the event could have been given a value, from 0 to 1, to represent the probability for that result.

For some events, this can be done by first finding all the possible results, or **outcomes**, for a particular event. For example, when you throw a coin there are two **equally likely** outcomes: it lands heads up or tails up. (The 'head' of a coin is the side which usually shows a head, the 'tail' is the side which shows the value of the coin.)

If you want to calculate the probability of getting a head, there is only one outcome that is possible. So, you can say that there is a 1 in 2, or 1 out of 2, chance of getting a head. This is usually given as a **probability fraction**, namely $\frac{1}{2}$. So, you would write the event as:

$$P(\text{head}) = \frac{1}{2}$$

Probabilities can also be written as decimals or percentages, so that:

$$P(\text{head}) = \frac{1}{2} \text{ or } 0.5 \text{ or } 50\%$$

The probability of an outcome is defined as:

$$P(\text{outcome}) = \frac{\text{number of ways the outcome can happen}}{\text{total number of possible outcomes}}$$

This definition always leads to a fraction, which should be cancelled to its simplest form.

Another probability term you will meet is at **random**. This means that the outcome cannot be predicted or affected by anyone.

EXAMPLE 2

The spinner shown here is spun and the score on the side on which it lands is recorded.

What is the probability that the score is:

a   2

b   odd

c   less than 5?

    a   There are two 2s out of six sides, so $P(2) = \frac{2}{6} = \frac{1}{3}$

    b   There are four odd numbers, so $P(\text{odd}) = \frac{4}{6} = \frac{2}{3}$

    c   All of the numbers are less than 5, so this is a certain event.

        $P(\text{less than 5}) = 1$

EXAMPLE 3

Bernice is always early, just on time or late for work.

The probability that she is early is 0.1, the probability she is just on time is 0.5.

What is the probability that she is late?

As all the possibilities are covered – that is 'early', 'on time' and 'late' – the total probability is 1. So:

P(early) + P(on time) = 0.1 + 0.5 = 0.6

So, the probability of Bernice being late is 1 − 0.6 = 0.4.

EXERCISE 33B

1 There are ten balls in a bag. One is red, two are blue, three are yellow and four are green. A ball is taken out without looking.

What is the probability that it is:

a red

b green

c green or yellow

d red or green

e white?

> **ADVICE AND TIPS**
>
> If an event is impossible, just write the probability as 0, not as a fraction such as $\frac{0}{6}$. If it is certain, write the probability as 1, not as a fraction such as $\frac{6}{6}$.

2 An 8-sided spinner has the numbers 1, 2, 3, 4, 5, 6, 7 and 8 on it. It is spun once.

What is the probability that the score is:

a 3

b more than 3

c an even number?

> **ADVICE AND TIPS**
>
> Remember to cancel the fractions if possible.

3 A bag contains only blue balls. If I take one out at random, what is the probability of each of these outcomes?

a I get a black ball.

b I get a blue ball.

4 Ten number cards with the numbers 1 to 10 inclusive are placed in a hat. Amir takes a number card out of the bag without looking. What is the probability that he draws:

a the number 7

b an even number

c a number greater than 6

d a number less than 3

e a number between 3 and 8?

5 A pencil case contains six red pens and five blue pens. Paulo takes out a pen without looking at it. What is the probability that he takes out:

a a red pen

b a blue pen

c a pen that is not blue?

6 A bag contains 50 balls. 10 are green, 15 are red and the rest are white. Galenia takes a ball from the bag at random. What is the probability that she takes:

a a green ball

b a white ball

c a ball that is not white

d a ball that is green or white?

7 There are 500 students in a school and 20 students in Ali's class. One person is chosen at random to welcome a special visitor.

What is the probability the person is in Ali's class?

8 Anton, Bianca, Charlie, Debbie and Elisabeth are in the same class. Their teacher wants two students to do a special job.

a Write down all the possible combinations of two people, for example, Anton and Bianca, Anton and Charlie. (There are 10 combinations altogether.)

b How many pairs give two boys?

c What is the probability of choosing two boys?

d How many pairs give a boy and a girl?

e What is the probability of choosing a boy and a girl?

f What is the probability of choosing two girls?

> **ADVICE AND TIPS**
>
> Try to be systematic when writing out all the pairs.

9 A bag contains 25 coloured balls. 12 are red, 7 are blue and the rest are green. Ravi takes a ball at random from the bag.

a Find:

    i P(he takes a red)   **ii** P(he takes a blue)   **iii** P(he takes a green).

b Add together the three probabilities. What do you notice?

c Explain your answer to part **b**.

10 The weather tomorrow will be sunny, cloudy or raining.

If P(sunny) = 40%, P(cloudy) = 25%, what is P(raining)?

11 At morning break, Priya has a choice of coffee, tea or hot chocolate.

If P(she chooses coffee) = 0.3 and P(she chooses hot chocolate) = 0.2, what is P(she chooses tea)?

12 The following information is known about the classes at Bradway School.

Year	Y1		Y2		Y3		Y4		Y5		Y6	
Class	P	Q	R	S	T	U	W	X	Y	Z	K	L
Girls	7	8	8	10	10	10	9	11	8	12	14	15
Boys	9	10	9	10	12	13	11	12	10	8	16	17

A class representative is chosen at random from each class.

Which class has the best chance of choosing a boy as the representative?

13 The teacher chooses, at random, a student to ring the school bell.

Tom says: 'It's even chances that the teacher chooses a boy or a girl.'

Explain why Tom might not be correct.

33.3 Probability that an event will not happen

In some questions in Exercise 33B, you were asked for the probability of something not happening. For example, in question 5 you were asked for the probability of picking a pen that is not blue. You could answer this because you knew how many pens there were in the case. However, sometimes you do not have this type of information.

The probability of throwing a six on a fair, six-sided dice is P(6) = $\frac{1}{6}$.

There are five **outcomes** that are not sixes: 1, 2, 3, 4, 5.

So, the probability of *not* throwing a six on a dice is:

$$P(\text{not a 6}) = \frac{5}{6}$$

Notice that:

$$P(6) = \frac{1}{6} \text{ and } P(\text{not a 6}) = \frac{5}{6}$$

So,

$$P(6) + P(\text{not a 6}) = 1$$

If you know that P(6) = $\frac{1}{6}$, then P(not a 6) is:

$$1 - \frac{1}{6} = \frac{5}{6}$$

So, if you know P(outcome happening), then:

$$P(\text{outcome not happening}) = 1 - P(\text{outcome happening})$$

EXAMPLE 4

A box of coloured pencils has 20 different pencils. There are four red pencils, five blue, one green, three yellow, two brown, one black, and four other colours.

A pencil is chosen at random. What is the probability that it is not red?

There are 4 red pencils out of 20.

The probability that a red is chosen is = $\frac{4}{20} = \frac{1}{5}$

The probability that a red is **not** chosen is $1 - \frac{1}{5} = \frac{4}{5}$

CORE

EXERCISE 33C

1 a The probability that a football team will win their next match is $\frac{1}{4}$. What is the probability that the team will not win?

b The probability that snow will fall during the winter holidays is 0.45. What is the probability that it will not snow?

c The probability that Paddy wins a game of chess is 0.7 and the probability that he draws the game is 0.1. What is the probability that he loses the game?

2 Look at Example 4.

What is the probability that the pencil is:

a not blue

b not yellow

c not black?

3 These letter cards are put into a bag.

a Lee takes a letter card at random.

   i What is the probability he takes a letter A?

   ii What is the probability he does not take a letter A?

b Ziad picks an M and keeps it. Tasnim now takes a letter from those remaining.

   i What is the probability she takes a letter A?

   ii What is the probability she does not take a letter A?

4 Hamzah is told: 'The chance of your winning this game is 0.3.'

Hamzah says: 'So I have a chance of 0.7 of losing.'

Explain why Hamzah might be wrong.

Probability in practice

Suppose you toss a coin many times and count how many times it lands with the 'heads' side showing.

The value of 'number of heads ÷ number of tosses' is called an **experimental probability**. As the number of **trials**, or experiments, increases, the value of the experimental probability gets closer to the theoretical probability, which in this case is $\frac{1}{2}$.

Experimental probability is also known as the **relative frequency** of an event. The relative frequency of an event is an estimate for the theoretical probability. It is given by:

$$\text{relative frequency of an outcome or event} = \frac{\text{frequency of the outcome or event}}{\text{total number of trials}}$$

EXAMPLE 5

The frequency table shows the speeds of 160 vehicles that pass a radar speed check on a fast road.

Speed (km/h)	20–29	30–39	40–49	50–59	60–69	70+
Frequency	14	23	28	35	52	8

a What is the experimental probability that a car is travelling faster than 70 km/h?

b If 500 vehicles pass the speed check, estimate how many will be travelling faster than 70 km/h.

a The experimental probability is the relative frequency, which is $\frac{8}{160} = \frac{1}{20}$.

b The number of vehicles travelling faster than 70 km/h will be $\frac{1}{20}$ of 500.

That is:

$500 \div 20 = 25$

EXERCISE 33D

1 Naseer throws a fair, six-sided dice and records the number of sixes that he gets after various numbers of throws. The table shows his results.

Number of throws	10	50	100	200	500	1000	2000
Number of sixes	2	4	10	21	74	163	329

a Calculate the experimental probability of scoring a 6 at each stage that Naseer recorded his results.

b How many ways can a dice land?

c How many of these ways give a 6?

d What is the theoretical probability of throwing a 6 with a dice?

e If Naseer threw the dice a total of 6000 times, how many sixes would you expect him to get?

2 Marie made a five-sided spinner, like the one shown in the diagram. She used it to play a board game with her friend Sarah.

The girls thought that the spinner was not very fair as it seemed to land on some numbers more than others. They threw the spinner 200 times and recorded the results. The results are shown in the table.

Side spinner lands on	1	2	3	4	5
Number of times	19	27	32	53	69

a Work out the experimental probability of each number.

b How many times would you expect each number to occur if the spinner is fair?

c Do you think that the spinner is fair? Give a reason for your answer.

3 A bottle contains 20 balls. The balls are either black or white.

Kenny conducts an experiment to see how many black balls there are in the bottle. He tips one ball at a time into a clear sealed tube at the end of the bottle. He records the number of black balls and tips them back into the bottle.

The results are shown in the table.

Number of samples	Number of black balls	Experimental probability
10	2	
100	25	
200	76	
500	210	
1000	385	
5000	1987	

a Copy the table and complete it by calculating the experimental probability of getting a black ball at each stage.

b Using this information, how many black balls do you think there are in the bottle?

4 Use a set of number cards from 1 to 10 and work with a partner. Take turns to choose a card and keep a record each time of what card you get, before returning it to the pack. Shuffle the cards each time and repeat the experiment 60 times. Put your results in a copy of this table.

Score	1	2	3	4	5	6	7	8	9	10
Total										

a How many times would you expect to get each number?

b Do you think you and your partner conducted this experiment fairly?

c Explain your answer to part **b**.

CORE

5 A four-sided dice has faces numbered 1, 2, 3 and 4. The score is the face on which it lands. Five students throw the dice to see if it is biased. They each throw it a different number of times. Their results are shown in the table.

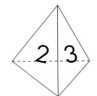

Student	Total number of throws	Score			
		1	2	3	4
Ali	20	7	6	3	4
Balvir	50	19	16	8	7
Caryl	250	102	76	42	30
Deema	80	25	25	12	18
Emma	150	61	46	26	17

a Which student will have the best set of results for finding the probability of each score? Why?

b Add up all the score columns and work out the relative frequency of each score. Give your answers to 2 decimal places.

c Is the dice biased? Explain your answer.

6 Andrew made an eight-sided spinner.

He tested it to see if it was fair.

He spun the spinner and recorded the results.

Unfortunately his little sister spilt something over his results table, so he could not see the middle part.

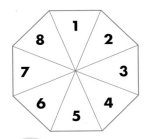

Number spinner lands on	1	2	3		6	7	8
Frequency	18	19	22		19	20	22

Assuming the spinner was a fair one, try to complete the missing parts of the table for Andrew.

7 At a computer factory, tests were carried out to see how many faulty computer chips were produced in one week.

	Monday	Tuesday	Wednesday	Thursday	Friday
Sample	850	630	1055	896	450
Number faulty	10	7	12	11	4

On which day was it most likely that the highest number of faulty computer chips were produced?

8 Steve tossed a coin 1000 times to see how many heads he got.

He said: 'If this is a fair coin, I should get 500 heads.'

Explain why he is wrong.

There are many situations where two events occur together. Four examples are given below.

Throwing two dice

Imagine that two dice, one red and one blue, are thrown.

The red dice can land with any one of six scores: 1, 2, 3, 4, 5 or 6.

The blue dice can also land with any one of six scores. This gives a total of 36 possible combinations.

These are shown in the left-hand diagram below, where combinations are given as (2, 3) and so on. The first number is the score on the blue dice and the second number is the score on the red dice.

The combination (2, 3) gives a total of 5. The total scores for all the combinations are shown in the diagram on the right-hand side. Diagrams that show all the outcomes of combined events are called **possibility diagrams**.

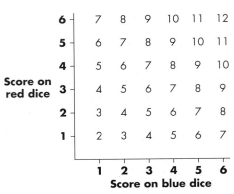

From the diagram on the right, you can see that there are two ways to get a score of 3. This gives a probability of scoring 3 as:

$$P(3) = \frac{2}{36} = \frac{1}{18}$$

From the diagram on the left, you can see that there are six ways to get a 'double'. This gives a probability of scoring a double as:

$$P(\text{double}) = \frac{6}{36} = \frac{1}{6}$$

Dice and coins

Throwing a dice and a coin

| | H | (1, H) | (2, H) | (3, H) | (4, H) | (5, H) | (6, H) |
| Outcome on coin | T | (1, T) | (2, T) | (3, T) | (4, T) | (5, T) | (6, T) |

Score on dice

Hence: P (head and an even number) = 3 ways out of 12 = $\frac{3}{12} = \frac{1}{4}$

EXTENDED

EXERCISE 33E

1 To answer these questions, use the diagram on the opposite page for all the possible scores when two fair dice are thrown together.

    a What is the most likely score?

    b Which two scores are least likely?

    c Write down the probabilities of throwing all the scores from 2 to 12.

    d What is the probability of a score that is:

        i bigger than 10   **ii** between 3 and 7

        iii even   **iv** a square number

        v a prime number   **vi** a triangular number?

2 Use the diagram on the opposite page that shows the outcomes when two fair, six-sided dice are thrown together as coordinates. What is the probability that:

    a the score is an even 'double'

    b at least one of the dice shows 2

    c the score on one dice is twice the score on the other dice

    d at least one of the dice shows a multiple of 3?

3 Use the diagram on the opposite page that shows the outcomes when two fair, six-sided dice are thrown together as coordinates. What is the probability that:

    a both dice show a 6

    b at least one of the dice will show a 6

    c exactly one dice shows a 6?

4 The diagram shows the scores for the event 'the difference between the scores when two fair, six-sided dice are thrown'. Copy and complete the diagram.

For the event described above, what is the probability of a difference of:

    a 1   **b** 0

    c 4   **d** 6

    e an odd number?

5 When two fair coins are thrown together, what is the probability of scoring:

    a two heads   **b** a head and a tail

    c at least one tail   **d** no tails?

Use a diagram of the outcomes when two coins are thrown together.

6 Two five-sided spinners are spun together and the total score of the faces that they land on is worked out. Copy and complete the possibility diagram shown.

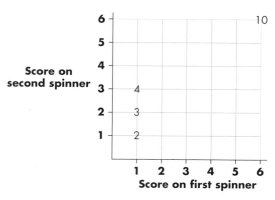

Score on second spinner

Score on first spinner

a What is the most likely score?

b When two five-sided spinners are spun together, what is the probability that:

   i the total score is 5

   ii the total score is an even number

   iii the score is a 'double'

   iv the total score is less than 7?

7 Two eight-sided spinners showing the numbers 1 to 8 were thrown at the same time.

   a Draw a possibility diagram to show the product of the two scores.

   b What is the probability that the product of the two spinners is an even square number?

8 Isaac rolls two dice and multiplies the numbers together to give their product. He wants to know the probability of rolling two dice that will give him a product between 19 and 35.

Explain why a possibility diagram will help him.

9 Nic went to a garden centre to buy some roses.

She found they came in six different colours – white, red, orange, yellow, pink and copper.

She also found they came in five different sizes – dwarf, small, medium, large climbing and rambling.

   a Draw a possibility diagram to show all the options.

   b She buys a random rose for Auntie Janet. Auntie Janet only likes red and pink roses and she does not like climbing or rambling roses.

   What is the probability that Nic has bought for Auntie Janet a rose:

   i that she likes

   ii that she does not like?

Imagine you have to draw two cards from this pack of six cards, but you must replace the first card before you select the second card.

You can tackle problems like this by using a **tree diagram**.

When you pick the first card, there are three possible outcomes: a square, a triangle or a circle. For a single event:

$$P(\text{square}) = \frac{3}{6} \qquad P(\text{triangle}) = \frac{2}{6} \qquad P(\text{circle}) = \frac{1}{6}$$

You can show this by depicting each event as a branch and writing its probability on the branch.

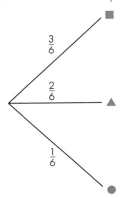

The diagram can then be extended to take into account a second choice. Because the first card has been replaced, you can still pick a square, a triangle or a circle. This is true no matter what is chosen the first time. You can demonstrate this by adding three more branches to the 'squares' branch in the diagram.

Here is the complete tree diagram.

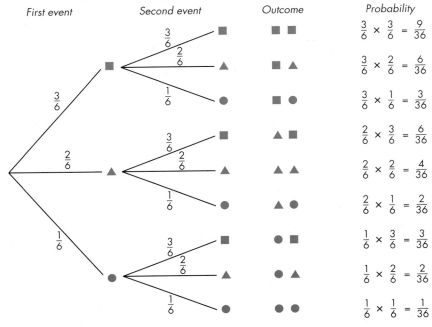

The probability of any outcome is calculated by multiplying all the probabilities on its branches. For instance:

$$P(\text{two squares}) = \frac{3}{6} \times \frac{3}{6} = \frac{9}{36}$$

$$P(\text{triangle followed by circle}) = \frac{2}{6} \times \frac{1}{6} = \frac{2}{36}$$

EXAMPLE 6

Using the tree diagram on the previous page, what is the probability of obtaining:

a two triangles

b a circle followed by a triangle

c a square and a triangle, in any order

d two circles

e two shapes that are the same?

a $P(\text{two triangles}) = \frac{4}{36}$

b $P(\text{circle followed by triangle}) = \frac{2}{36}$

c There are two places in the outcome column that have a square and a triangle. These are the second and fourth rows. The probability of each is $\frac{1}{6}$. Their combined probability is given by the addition rule.

$$P(\text{square and triangle, in any order}) = \frac{6}{36} + \frac{6}{36} = \frac{12}{36}$$

d $P(\text{two circles}) = \frac{1}{36}$

e There are three places in the outcome column that have two shapes the same. These are the first, fifth and last rows. The probabilities are respectively $\frac{9}{36}, \frac{4}{36}$ and $\frac{1}{36}$. Their combined probability is given by the addition rule.

$$P(\text{two shapes the same}) = \frac{9}{36} + \frac{4}{36} + \frac{1}{36} = \frac{14}{36}$$

EXERCISE 33F

1 A coin is tossed twice. Copy and complete this tree diagram to show all the possible outcomes.

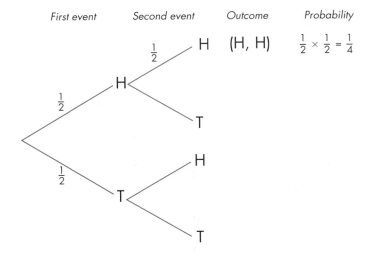

Use your tree diagram to work out the probability of getting:

a two heads

b a head and a tail

c at least one tail.

2 A bag contains lots of counters. There are equal numbers of green, gold, black and red counters. There are no other colours.

A counter is taken out at random. It is replaced. Then a second counter is taken.

a What is the probability that the first counter is gold?

b What is the probability that the second counter is gold?

c Draw a tree diagram to show the outcome of each counter being gold or not. Use it to work out the probability that:

   i both counters are gold

   ii at least one counter will be gold.

EXTENDED

3 On my way to work, I drive through two sets of road works with traffic lights that only show green or red. I know that the probability of the first set being green is $\frac{1}{3}$ and the probability of the second set being green is $\frac{1}{2}$.

a What is the probability that the first set of lights will be red?

b What is the probability that the second set of lights will be red?

c Copy and complete this tree diagram to show the possible outcomes when passing through both sets of lights.

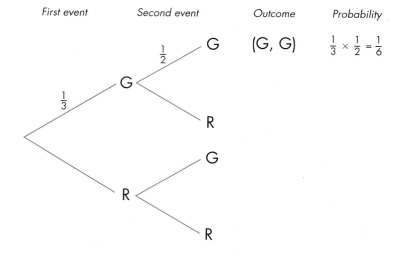

d Using the tree diagram, what is the probability that:

   i I do not get held up at either set of lights

   ii I get held up at exactly one set of lights

   iii I get held up at least once.

e Over a school term I make 90 journeys to work. On how many days can I expect to get two green lights?

4 Six out of every 10 cars in Britain are made in other countries.

a What is the probability that any car will be British made?

b Two cars can be seen approaching in the distance. Draw a tree diagram to work out the probability of each of these outcomes.

   i Both cars will be British made.

   ii One car will be British and the other car will be made elsewhere.

5 Three coins are tossed. Copy and complete the tree diagram and use it to answer the questions.

| First event | Second event | Third event | Outcome | Probability |

$(H, H, H) \quad \frac{1}{2} \times \frac{1}{2} \times \frac{1}{2} = \frac{1}{8}$

If a coin is tossed three times, what is the probability of scoring:

a three heads

b two heads and a tail

c at least one tail?

6 Thomas has to take a three-part language examination paper.

- The first part is speaking. He has a 0.4 chance of passing this part.
- The second is listening. He has a 0.5 chance of passing this part.
- The third part is writing. He has a 0.7 chance of passing this part.

a Draw a tree diagram covering three events, where the first event is passing or failing the speaking part of the examination, the second event is passing or failing the listening part and the third event is passing or failing the writing part.

b If he passes all three parts, his father will give him $20. What is the probability that he gets the money?

c If he passes two parts only, he can resit the other part. What is the chance he will have to resit?

d If he fails all three parts, he will be thrown off the course. What is the chance he is thrown off the course?

7 In a group of 10 girls, six like the pop group Smudge and four like the pop group Mirage. Two girls are to be chosen for a pop quiz.

   a What is the probability that the first girl chosen will be a Smudge fan?

   b The first girl chosen is a Smudge fan. Explain why the probability that the second girl chosen is a Smudge fan is $\frac{5}{9}$.

   c Copy and complete this tree diagram to show the probabilities when two girls are chosen.

   d Use your tree diagram to work out the probability that:

      i both girls chosen will like Smudge

      ii both girls chosen will like the same group

      iii the girls chosen will like different groups.

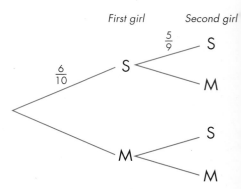

8 There are five white eggs and one brown egg in an egg box. Kate decides to make a two-egg omelette. She takes each egg from the box without looking at its colour.

   a What is the probability that the first egg taken is brown?

   b The first egg taken is brown. What is the probability that the second egg taken will be brown?

   c Copy this tree diagram and write the probabilities on the branches.

   d What is the probability that Kate gets an omelette made from:

      i two white eggs

      ii one white and one brown egg

      iii two brown eggs?

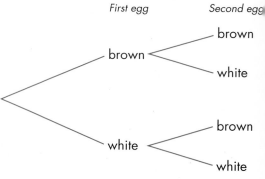

9 A box contains 10 red and 15 yellow balls. One is taken out and not replaced. Another is taken out.

   a The first ball taken out is red. What is the probability that the second ball is:

      i red

      ii yellow?

   b The first ball taken out is yellow. What is the probability that the second ball is:

      i red

      ii yellow?

10 A fruit bowl contains six oranges and eight lemons. Kevin takes two pieces of fruit at random.

    a If the first piece is an orange, what is the probability that the second is:

        i an orange

        ii a lemon?

    b What is the probability that:

        i both are oranges

        ii both are lemons?

11 A bag contains three black balls and seven red balls. A ball is taken out and not replaced. This is repeated twice. What is the probability that:

    a all three are black

    b exactly two are black

    c exactly one is black

    d none are black?

12 On my way to work, I pass two sets of traffic lights. The probability that the first is green is $\frac{1}{3}$. If the first is green, the probability that the second is green is $\frac{1}{3}$. If the first is red, the probability that the second is green is $\frac{2}{3}$. What is the probability that:

    a both are green

    b none are green

    c exactly one is green

    d at least one is green?

13 An engineering test is in two parts, a written test and a practical test. 90% of those who take the written test pass. When a person passes the written test, the probability that he or she will also pass the practical test is 60%. When a person fails the written test, the probability that he or she will pass the practical test is 20%.

    a What is the probability that someone passes both tests?

    b What is the probability that someone passes one test?

    c What is the probability that someone fails both tests?

    d What is the combined probability of the answers to parts **a**, **b** and **c**?

Note: drawings in questions are NOT TO SCALE unless otherwise specified.

- If the degree of accuracy is not specified in the question, and if the answer is not exact, give the answer to three significant figures. Give answers in degrees to one decimal place.

- For π, use either your calculator value or 3.142.

PAPER 1

1 The table shows the average monthly temperature (°C) for Fairbanks, Alaska.

Month	Jan	Feb	Mar	Apr	May	Jun	Jul	Aug	Sep	Oct	Nov	Dec
Temperature (°C)	−23.4	−19.8	−11.7	−0.8	9.2	15.4	16.9	13.8	7.5	−5.8	−21.4	−21.8

a Find:

   i the difference between the highest and the lowest temperatures [1]

   ii the median. [2]

b A month is chosen at random from the table.

Find the probability that its average temperature is below zero. [1]

Paper 11 Q19 May/Jun2014

2 One of the six letters is taken at random.

a Write down the probability that the letter is S. [1]

b The letter is replaced and again a letter is taken at random. This is repeated 600 times.

How many times would you expect the letter to be S? [1]

Paper 11 Q14 Oct/Nov 2013

3 8 15 7 8 7 15 4 13 4 3 10 2 9 4 5

a Write down the mode. [1]

b Work out the median. [2]

Paper 11 Q16 Oct/Nov 2013

4 The probability of Sachin's team winning any match is 0.45.

a Write down the probability of Sachin's team **not** winning any match. [1]

b In a season there are 40 matches.

How many matches should Sachin's team expect to win in a season? [2]

Paper 11 Q12 May/Jun 2013

5 The scatter diagram shows the rainfall and
the average temperature in a city for
the month of June, over a period of 10 years.

  a What type of correlation does the scatter diagram show? [1]

  b Describe the relationship between the rainfall and the average temperature. [1]

Paper 11 Q12 May/Jun 2014

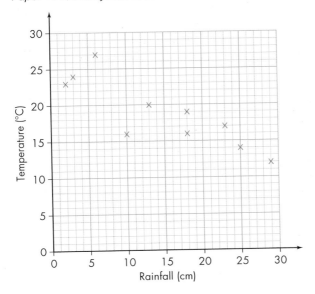

6 Marco throws a six-sided dice 27 times.

The bar chart shows his results.

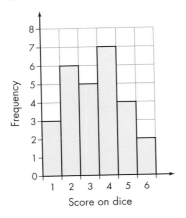

  a Write down the mode. [1]

  b Work out the probability that Marco throws a number less than 5. [2]

  c Calculate the mean. [3]

Paper 11 Q20 May/Jun 2013

CORE

1 **a** Amir asks his friends how many hours they spent playing sport last weekend.

His results are shown in the table below.

Number of hours	0	1	2	3	4	5
Frequency	6	2	3	1	2	1

  i Write down the mode.   [1]

  ii Find the median.   [1]

  iii Calculate the mean.   [3]

  iv On the grid, draw a bar chart to show the information given in the table.

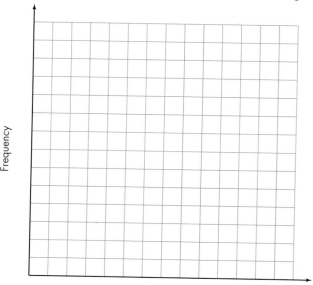

Frequency

Number of hours

          [4]

b Amir also asks 15 friends which was their favourite sport.

His results are shown in the table below.

Football	4
Cricket	5
Basketball	2
Badminton	4

Amir picks one of these friends at random.

Write down the probability that his friend's favourite sport is:

  i cricket   [1]

  ii **not** football   [1]

  iii basketball or badminton.   [1]

Paper 31 Q3 Oct/Nov 2012

2 The table shows a summary of the types of employment for 90 people.

Employment	Frequency	Pie chart sector angle
Retail	18	72°
Leisure industry	12	48°
Public service	35	
Others	25	

a i Copy and complete the table. [2]

ii Copy and complete the pie chart and label the sectors.

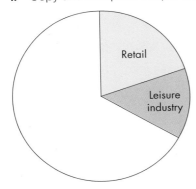

[2]

b Here are the ages of the people working in the leisure industry.

16   17   19   23   23   24   27   31   33   40   45   56

i Work out the range. [1]

ii Calculate the mean. [2]

iii Sabrina wants to interview someone working in the leisure industry.

She chooses one person at random.

Write down the probability that the person chosen is under 30 years old. [1]

Paper 31 Q4 May/June 2013

PAPER 2

1 Bruce plays a game of golf.

His score for each of the 18 holes is shown below.

2 3 4 5 4 6 2 3 4

4 5 3 4 3 5 4 4 4

The information is to be shown in a pie chart.

Calculate the sector angle for a score of 4. [2]

Paper 22 Q4 Oct/Nov 2013

2 During one day 48 people visited a museum.

The length of time each person spent in the museum was recorded.

The results are shown on the cumulative frequency diagram.

Work out:

a the median [1]

b the 20th percentile [2]

c the interquartile-range [2]

d the probability that a person chosen at random spends 2 hours or less in the museum. [2]

Paper 22 Q20 Oct/Nov 2013

3 The heights, in metres, of 200 trees in a park are measured.

Height h (m)	$2 < h \leqslant 6$	$6 < h \leqslant 10$	$10 < h \leqslant 13$	$13 < h \leqslant 17$	$17 < h \leqslant 19$	$19 < h \leqslant 20$
Frequency	23	47	45	38	32	15

   a Find which interval contains the median height. [1]

   b Calculate an estimate of the mean height. [4]

   c Complete the cumulative frequency table for the information given in the table above. [2]

Height h (m)	$2 < h \leqslant 6$	$h \leqslant 10$	$h \leqslant 13$	$h \leqslant 17$	$h \leqslant 19$	$h \leqslant 20$
Cumulative frequency	23					

Paper 22 Q20 May/Jun 2013

4 The owner of a small café records the average air temperature and the number of hot drinks he sells each week.

Air temperature (°C)	18	23	19	23	24	25
Number of hot drinks sold	12	8	13	10	9	7

   a On the grid, draw a scatter diagram to show this information. [2]

   b What type of correlation does your scatter diagram show? [1]

   c Draw a line of best fit on the grid. [1]

Paper 17 Q12 May/Jun 2013

PAPER 4

1 **a** The histogram shows some information about the masses (m grams) of 39 apples.

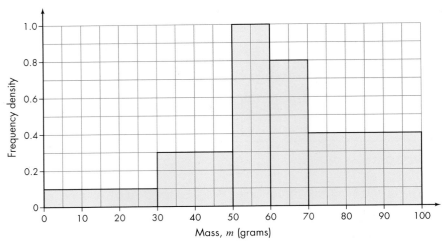

i Show that there are 12 apples in the interval $70 < m \leqslant 100$. [1]

ii Calculate an estimate of the mean mass of the 39 apples. [5]

b The mean mass of 20 oranges is 70 g.

One orange is eaten.

The mean mass of the remaining oranges is 70.5 g.

Find the mass of the orange that was eaten. [3]

Paper 42 Q7 May/Jun 2014

2 E N L A R G E M E N T

Prettie picks a card at random from the 11 cards above and does not replace it.

She then picks a second card at random and does not replace it.

a Find the probability that she picks:

i the letter L and then the letter G [2]

ii the letter E twice [2]

iii two letters that are the same. [2]

b Prettie now picks a third card at random.

Find the probability that the three letters:

i are all the same [2]

ii **do not** include a letter E [2]

iii include exactly two letters that are the same. [5]

Paper 42 Q6 Oct/Nov 2013

3 **In this question, give all your answers as fractions.**

When Ivan goes to school in winter, the probability that he wears a hat is $\frac{5}{8}$.

If he wears a hat, the probability that he wears a scarf is $\frac{2}{3}$.

If he does not wear a hat, the probability that he wears a scarf is $\frac{1}{6}$.

a Complete the tree diagram.

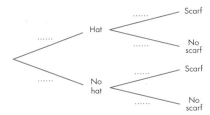

[3]

b Find the probability that Ivan:

i does not wear a hat and does not wear a scarf [2]

ii wears a hat but does not wear a scarf [2]

iii wears a hat or scarf but not both. [2]

c If Ivan wears a hat and a scarf, the probability that he wears gloves is $\frac{7}{10}$.

Calculate the probability that Ivan does **not** wear all three of hat, scarf and gloves. [3]

Paper 42 Q8 May/Jun 2013

Answers to Chapter 1

1.1　Square numbers

Exercise 1A

1　36, 49, 64, 81, 100, 121, 144, 169, 196, 225, 256, 289, 324, 361, 400

2　4, 9, 16, 25, 36, 49

3　**a** 50, 65, 82　　**b** 98, 128, 162
　c 51, 66, 83　　**d** 48, 63, 80

4　**a** 25, 169, 625, 1681
　b Answers in each row are the same

5　**a** 13　　**b** 5　　**c** 9　　**d** 11　　**e** 20

6　36 and 49

Exercise 1B

1　**a** 12, 24, 36　　**b** 20, 40, 60　　**c** 15, 30, 45
　d 18, 36, 54　　**e** 35, 70, 105

2

	Square number	Factor of 70
Even number	16	10
Multiple of 7	49	35

3　4761 (69^2)

4　24 seconds

5　30 seconds

6　**a** 12　　**b** 9　　**c** 6　　**d** 13
　e 15　　**f** 16　　**g** 10　　**h** 17

1.2　Multiples of whole numbers

Exercise 1C

1　**a** 3, 6, 9, 12, 15　　　　**b** 7, 14, 21, 28, 35
　c 9, 18, 27, 36, 45　　**d** 11, 22, 33, 44, 55
　e 16, 32, 48, 64, 80

2　**a** 72, 132, 216　**b** 161, 91　**c** 72, 102, 132, 78, 216

3　**a** 98　**b** 99　**c** 96　**d** 95　**e** 98　**f** 96

4　4 or 5 (as 2, 10 and 20 are not realistic answers)

5　**a** 18　　　　**b** 28　　　　**c** 15

6　5 numbers: 18, 36, 54, 72, 90

1.3　Factors of whole numbers

Exercise 1D

1　**a** 1, 2, 5, 10　　　　　　**b** 1, 2, 4, 7, 14, 28
　c 1, 2, 3, 6, 9, 18　　　**d** 1, 17
　e 1, 5, 25　　　　　　　**f** 1, 2, 4, 5, 8, 10, 20, 40
　g 1, 2, 3, 5, 6, 10, 15, 30　**h** 1, 3, 5, 9, 15, 45
　i 1, 2, 3, 4, 6, 8, 12, 24　**j** 1, 2, 4, 8, 16

2　**a** 55　　**b** 67　　**c** 29
　d 39　　**e** 65　　**f** 80

3　**a** 2　　**b** 2　　**c** 3　　**d** 5　　**e** 3
　f 3　　**g** 7　　**h** 5　　**i** 10　　**j** 11

4　5

Exercise 1E

1　**a** 168　　**b** 105　　**c** 84　　**d** 84
　e 96　　**f** 54　　**g** 75　　**h** 144

2　**a** 8　　**b** 7　　**c** 4　　**d** 14
　e 4　　**f** 9　　**g** 5　　**h** 4
　i 3　　**j** 16　　**k** 5　　**l** 18

3　18 and 24

1.4　Prime numbers

Exercise 1F

1　23 and 29

2　97

3　All these numbers are not prime.

4　3, 5, 7

5　Only if all 31 bars are in a single row, as 31 is a prime number and its only factors are 1 and 31.

1.5　Prime factorisation

Exercise 1G

1　**a** 36　　**b** 105　　**c** 250　　**d** 816
　e 714　　**f** 1715　　**g** 1089　　**h** 1352

2　**a** $2 \times 3^2 \times 5$　　**b** $2^3 \times 19$　　　**c** 2^6
　d $2 \times 3 \times 5 \times 11$　**e** 17^2　**f** $2^5 \times 5^2$
　g It is a prime number and cannot be factorised.
　h $7 \times 11 \times 13$

3　$77 = 7 \times 11$; $129 = 3 \times 43$; $221 = 13 \times 17$

4　**a** $900 = 2^2 \times 3^2 \times 5^2$　　　**b** $1800 = 2^3 \times 3^2 \times 5^2$
　c $1350 = 2 \times 3^3 \times 5^2$

5　144　　　　　　$2^2 \times 3^4$
　200　　　　　　$2^4 \times 3^2$
　324　　　　　　$2^3 \times 5^2$
　500　　　　　　$2^2 \times 5^3$

6　**a** $2 \times 3 \times 5 \times 7 = 210$
　b The answer to **a** $\times 11 = 2 \times 3 \times 5 \times 7 \times 11 = 2310$

7　**a** $2^2 \times 3^2 \times 17$　**b** $2 \times 3^2 \times 17$　**c** $2^3 \times 3 \times 17$

8　71, 73 and 79 because they are prime numbers

9　$456\,533 = 7^3 \times 11^3$

1.6　More about HCF and LCM

Exercise 1H

1　**a** $2 \times 3^2 = 18$　　**b** $2^3 \times 3^4 = 648$

2　**a** 35　　**b** 735

3　**a** $2^4 \times 3 \times 5$　**b** $2 \times 3^2 \times 7$　**c** 6　　**d** 5040

4　**a** $2^3 \times 3^2$ and $2^2 \times 3^3$　**b** 36　　**c** 216

5　**a** 16　　**b** 576

6　**a** 33　　**b** 2772

7　**a** $2^2 \times 3 = 12$　　　　**b** $2^3 \times 3^2 \times 5 = 360$

8 a 5   **b** 1575

9 a 15   **b** 23 625

10 a 72   **b** $2^5 \times 3^4$

11 a 1   **b** 12 600

12 a $72 \times 162 = 11664$ **b** $18 \times 648 = 11664$

c You could do a similar calculation for the numbers in questions 2 to 6. You should find that the two products are equal each time.

d There is not a similar result in this case.

1.7   Real numbers

Exercise 1I

1 a yes   **b** no   **c** yes   **d** yes   **e** no

   f yes   **g** no   **h** yes   **i** yes

2 a rational   **b** rational   **c** irrational

   d rational   **e** irrational   **f** rational

   g rational   **h** irrational   **i** irrational

3 a 2.5 and 3.5 is one possible answer

   b 0.4 × 2.5 is one possible answer

   c Not possible

4 $\sqrt{2} \times \sqrt{8}$ is a possible answer

5 π and $4 - \pi$ is a possible answer

6 There are many possible answers. You could just give the same answer as question 5.

Answers to Chapter 2

2.1   Equivalent fractions

Exercise 2A

1 a $\frac{8}{20}$ **b** $\frac{3}{12}$ **c** $\frac{15}{40}$

   d × 6, $\frac{12}{18}$ **e** × 3, $\frac{9}{12}$ **f** × 5, $\frac{25}{40}$

2 a $\frac{2}{3}$ **b** $\frac{4}{5}$ **c** $\frac{5}{7}$ **d** ÷ 6, $\frac{2}{3}$

   e 25 ÷ 5, $\frac{3}{5}$ **f** ÷ 3, $\frac{7}{10}$

3 a $\frac{2}{3}$ **b** $\frac{1}{3}$ **c** $\frac{2}{3}$ **d** $\frac{3}{4}$ **e** $\frac{1}{3}$

   f $\frac{1}{2}$ **g** $\frac{7}{8}$ **h** $\frac{4}{5}$ **i** $\frac{1}{2}$ **j** $\frac{1}{4}$

4 a $\frac{1}{2}, \frac{2}{3}, \frac{5}{6}$ **b** $\frac{1}{2}, \frac{5}{8}, \frac{3}{4}$ **c** $\frac{2}{5}, \frac{1}{2}, \frac{7}{10}$

   d $\frac{7}{12}, \frac{2}{3}, \frac{3}{4}$ **e** $\frac{1}{6}, \frac{1}{4}, \frac{1}{3}$ **f** $\frac{3}{4}, \frac{4}{5}, \frac{9}{10}$

5 a $\frac{1}{3} + \frac{1}{4} = \frac{4}{12} + \frac{3}{12} = \frac{7}{12}$

   Explanations may involve ruling out other combinations

   b $\frac{1}{2}$ as the smallest denominator is the biggest unit fraction

   Diagrams may be used but must be based on equal sized area.

6 a $2\frac{1}{3}$ **b** $2\frac{2}{3}$ **c** $2\frac{1}{4}$

   d $1\frac{3}{7}$ **e** $2\frac{2}{5}$ **f** $1\frac{2}{5}$

7 a $\frac{10}{3}$ **b** $\frac{35}{6}$ **c** $\frac{9}{5}$ **d** $\frac{37}{7}$ **e** $\frac{41}{10}$ **f** $\frac{17}{3}$

   g $\frac{5}{2}$ **h** $\frac{13}{4}$ **i** $\frac{43}{6}$ **j** $\frac{29}{8}$ **k** $\frac{19}{3}$ **l** $\frac{89}{9}$

8 Students check their own answers.

9 $\frac{27}{4} = 6\frac{3}{4}, \frac{31}{5} = 6\frac{1}{5}, \frac{13}{2} = 6\frac{1}{2}$, so $\frac{27}{4}$ is the biggest since $\frac{1}{5}$ is less than $\frac{1}{2}$ and $\frac{3}{4}$ is greater than $\frac{1}{2}$

10 Any mixed number which is between 7.7272 and 7.9. For example $7\frac{4}{5}$

2.2   Fractions and decimals

Exercise 2B

1 a $\frac{7}{10}$ **b** $\frac{2}{5}$ **c** $\frac{1}{2}$ **d** $\frac{3}{100}$ **e** $\frac{3}{50}$

   f $\frac{13}{100}$ **g** $\frac{1}{4}$ **h** $\frac{19}{50}$ **i** $\frac{11}{20}$ **j** $\frac{16}{25}$

2 a 0.5   **b** 0.75   **c** 0.6   **d** 0.9   **e** 0.333

   f 0.625   **g** 0.667   **h** 0.35   **i** 0.636   **j** 0.444

3 a 0.3, $\frac{1}{2}$, 0.6   **b** 0.3, $\frac{2}{5}$, 0.8   **c** 0.15, $\frac{1}{4}$, 0.35

   d $\frac{7}{10}$, 0.71, 0.72   **e** 0.7, $\frac{3}{4}$, 0.8   **f** $\frac{1}{20}$, 0.08, 0.1

   g 0.4, $\frac{1}{2}$, 0.55   **h** 1.2, 1.23, $1\frac{1}{4}$

4 Store A – $\frac{1}{3}$ (0.33) is greater than $\frac{1}{4}$ (0.25)

5 a $\frac{12}{30} = \frac{2}{5}$ **b** 0.4

6 $\frac{7}{8}$ (= 0.875)

7 $\frac{2}{3}$ (= 0.67)

2.3 Recurring decimals

Exercise 2C

1 a 0.333... or $0.\dot{3}$ **b** 0.75   **c** 0.8333... or $0.8\dot{3}$

   d 0.222... or $0.\dot{2}$ **e** 0.65   **f** 0.8181... or $0.8\dot{1}$

   g 0.1875   **h** 0.916 66... or $0.91\dot{6}$

2 a 0.4666... or $0.4\dot{6}$ **b** 0.9333... or $0.9\dot{3}$

3 a 0.1111... **b** 0.1666... **c** 0.2777... **d** 0.0555...

4 $\frac{8}{9}$

5 $\frac{8}{33}$

6 $\frac{11}{30}$

7 $\frac{1}{12}$

8 $2\frac{7}{15}$

9 $0.\dot{2}3076\dot{9}$

10 a $0.\dot{0}\dot{9}$ **b** $0.\dot{1}\dot{8}$ **c** $0.\dot{2}\dot{7}$, $0.\dot{3}\dot{6}$ and $0.\dot{6}\dot{3}$

11 a $0.\dot{2}8571\dot{4}$ **b** $0.\dot{4}2857\dot{1}$ **c** $\frac{4}{7} = 0.\dot{5}7142\dot{8}$, $\frac{5}{7} = 0.\dot{7}1428\dot{5}$ and $\frac{6}{7} = 0.\dot{8}5714\dot{2}$

12 a $\frac{1}{5}$, $\frac{1}{8}$, $\frac{1}{10}$

   b $\frac{1}{N}$ is a terminating decimal if the only prime factors of N are 2 or 5. Otherwise it is a recurring decimal.

2.4 Percentages, fractions and decimals

Exercise 2D

1 a $\frac{2}{25}$ **b** $\frac{1}{2}$ **c** $\frac{1}{4}$ **d** $\frac{7}{20}$ **e** $\frac{9}{10}$ **f** $\frac{3}{4}$

2 a 0.27   **b** 0.85   **c** 0.13   **d** 0.06   **e** 0.8   **f** 0.32

3 a $\frac{3}{25}$ **b** $\frac{2}{5}$ **c** $\frac{9}{20}$ **d** $\frac{17}{25}$ **e** $\frac{1}{4}$ **f** $\frac{5}{8}$

4 a 29%   **b** 55%   **c** 3%   **d** 16%   **e** 60%   **f** 125%

5 a 28%   **b** 30%   **c** 95%   **d** 34%   **e** 27.5%   **f** 87.5%

6 a 0.6   **b** 0.075   **c** 0.76   **d** 0.3125   **e** 0.05   **f** 0.125

7 a 63%, 83%, 39%, 62%, 77%   **b** English

8 34%, 0.34, $\frac{17}{50}$; 85%, 0.85, $\frac{17}{20}$; 7.5%, 0.075, $\frac{3}{40}$; 45%, 0.45, $\frac{9}{20}$; 30%, 0.3, $\frac{3}{10}$; 67%, 0.67, $\frac{2}{3}$; 84%, 0.84, $\frac{21}{25}$; 45%, 0.45, $\frac{9}{20}$; 37.5%, 0.375, $\frac{3}{8}$

2.5 Calculating a percentage

Exercise 2E

1 a 0.88   **b** 0.3   **c** 0.25   **d** 0.08   **e** 1.15

2 a 78%   **b** 40%   **c** 75%   **d** 5%   **e** 110%

3 a $45   **b** $6.30   **c** 128.8 kg   **d** 1.125 kg
   e 1.08 h   **f** 37.8 cm   **g** $0.12   **h** 2.94 m
   i $7.60   **j** 33.88 min   **k** 136 kg   **l** $162

4 $2410

5 a 86%   **b** 215

6 8520

7 287

8 990

9 Mon: 816, Tue: 833, Wed: 850, Thu: 799, Fri: 748

10 a $3.25   **b** 2.21 kg   **c** $562.80
   d $6.51   **e** 42.93 m   **f** $24

11 480 cm^3 nitrogen, 120 cm^3 oxygen

12 13

13 $270

14 More this year as it was 3% of a higher amount than last year.

2.6 Increasing or decreasing quantities by a percentage

Exercise 2F

1 a 1.1   **b** 1.03   **c** 1.2   **d** 1.07   **e** 1.12

2 a $62.40   **b** 12.96 kg   **c** 472.5 g   **d** 599.5 m
   e $38.08   **f** $90   **g** 391 kg   **h** 824.1 cm
   i 253.5 g   **j** $143.50   **k** 736 m   **l** $30.24

3 $29 425

4 1 690 200

5 a Caretaker: $17 325, Driver: $18 165, Supervisor: $20 475 Manager: $26 565
   b 5% of different amounts is not a fixed amount. The more pay to start with, the more the increase (5%) will be.

6 $411.95

7 193 800

8 575 g

9 918

10 60

11 TV: $287.88, microwave: $84.60, CD: $135.13, stereo: $34.66

12 $10

13 c Both the same as $1.05 \times 1.03 = 1.03 \times 1.05$

14 a Shop A, as $1.04 \times 1.04 = 1.0816$, so an 8.16% increase.

15 $540.96

Exercise 2G

1 a 0.92   **b** 0.85   **c** 0.75   **d** 0.91   **e** 0.88

2 a $9.40   **b** 23 kg   **c** 212.4 g   **d** 339.5 m
   e $4.90   **f** 39.6 m   **g** 731 m   **h** 83.52 g
   i 360 cm   **j** 117 min   **k** 81.7 kg   **l** $37.70

3 $5525

4 a 52.8 kg   **b** 66 kg   **c** 45.76 kg

5 Mr Patel $176, Mrs Patel $297.50, Sandeep $341, Priyanka $562.50

6 448

7 705

8 a 66.5 km/h   **b** 73.5 km/h

9 No, as the total is $101. She will save $20.20, which is less than the $25 it would cost to join the club.

10 10% off $50 is $45; 10% off $45 is $40.50; 20% off $50 is $40

11 $765

12 $1.10 \times 0.9 = 0.99$ (99%)

13 Offer A gives 360 grams for $1.40, i.e. 0.388 cents per gram.
Offer B gives 300 grams for $1.12, i.e 0.373 cents per gram, so Offer B is the better offer.
Or Offer A is 360 for 1.40 = 2.6 grams per cent, offer B is 300 for 1.12 = 2.7 grams per cent, so offer B is better.

2.7 Expressing one quantity as a percentage of another

Exercise 2H

1 a 25%   **b** 60.6%   **c** 46.3%   **d** 12.5%
   e 41.7%   **f** 60%   **g** 20.8%   **h** 10%
   i 1.9%   **j** 8.3%   **k** 45.5%   **l** 10.5%

32%

6.5%

33.7%

a 49.2% **b** 64.5% **c** 10.6%

17.9%

4.9%

90.5%

a Brit Com: 20.9%, USA: 26.5%, France: 10.3%, Other 42.3%
b Total 100%, all imports

0 Nadia had the greater percentage increase.
Nadia: (20 − 14) × 100 ÷ 14 = 42.9%.
Imran: (17 − 12) × 100 ÷ 12 = 41.7%

1 Yes, as 38 out of 46 is over 80% (82.6%)

2 Vase 20% loss, radio 25% profit, doll 175% profit, toy train 64% loss

2.8 Simple interest and compound interest

Exercise 2I

7420 dollars

3600 dollars

4 years

a $15 600 **b** $16 224

a $1272 **b** $1348.32 **c** $1429.22

a Amar 3200 $, Mona 3328 $ **b** Mona, 128 $

a $9528.13 **b** £1528.13

£3840

a Simple **b** 6.5%

0 a $13 800 **b** $15 870
c Student's own explanation

1 a 2652.25 and 5304.50 **b** £796.37

2.9 A formula for compound interest

Exercise 2J

$2249.73

$5681.15

3 **a** $5071.50 **b** $5591.33 **c** $6164.44

4 **a** $3589.07 **b** $4458.69

5 $4272.64

6 **a** $3941.57 **b** $441.57

7 8 years

8 The interest in the second five years will be more than the interest in the first five years. The missing number is $5000 \times 1.06^{10} = 8954.24$.

9 **a** $15 000 **b** $16 288.95

10 a $1268.24
b The interest over the year is $268.24.
This is $\frac{268.24}{1000} \times 100\% = 26.824\%$ of $1000.

2.10 Reverse percentage

Exercise 2K

1 **a** 800 g **b** 250 m **c** 60 cm
d $3075 **e** $200 **f** $400

2 80

3 T shirt: $8.40, Tights: $1.20, Shorts: $5.20, Sweater: $10.74, Trainers: $24.80, Boots: $32.40

4 $833.33

5 $300

6 240

7 537.63 dollars

8 4750 blue bottles

9 $2585

10 $1440

11 $2450

12 95 dollars

13 $140

14 $945

15 $1325

16 $1300

17 Lee has assumed that 291.2 is 100% instead of 112%. He rounded his wrong answer to the correct answer of $260.

Answers to Chapter 3

3.1 Order of operations

Exercise 3A

a 11	**b** 6	**c** 10	**d** 12	**e** 11	**f** 13
g 11	**h** 12	**i** 12	**j** 4	**k** 13	**l** 3
a 16	**b** 2	**c** 10	**d** 10	**e** 6	**f** 18
g 6	**h** 15	**i** 9	**j** 12	**k** 3	**l** 8

a (4 + 1) **b** No brackets needed
c (2 + 1) **d** No brackets needed
e (4 + 4) **f** (16 − 4)
g No brackets needed **h** No brackets needed
i (20 − 10) **j** No brackets needed
k (5 + 5) **l** (4 + 2)

m (15 − 5) **n** (7 − 2)
o (3 + 3) **p** No brackets needed
q No brackets needed **r** (8 − 2)

4 No, correct answer is 5 + 42 = 47

5 **a** 2 × 3 + 5 = 11 **b** 2 × (3 + 5) = 16
c 2 + 3 × 5 = 17 **d** 5 − (3 − 2) = 4
e 5 × 3 − 2 = 13 **f** 5 × 3 × 2 = 30

6 4 + 5 × 3 = 19
(4 + 5) × 3 = 27. So 4 + 5 × 3 is smaller

7 (5 − 2) × 6 = 18

8 8 ÷ (5 − 3) = 4

3.2 Choosing the correct operation

Exercise 3B

1 **a** 6000
   b 5 cans cost $1.95, so 6 cans cost $1.95. 32 − 5 × 6 + 2. Cost is $10.53.

2 **a** 288   **b** 16

3 **a** 38
   b Coach price for adults = $8, coach price for juniors = $4, money for coaches raised by tickets = $12 400, cost of coaches = $12 160, profit = $240

4 $34.80
5 (18.81...) Kirsty can buy 18 models.
6 (7.58 ...) Michelle must work for 8 weeks.
7 $8.40 per year, 70 cents per copy
8 $450
9 15
10 Gavin pays 2296.25 − 1840 = $456.25

3.3 Finding a fraction of a quantity

Exercise 3C

1 **a** 18   **b** 10   **c** 18   **d** 28

2 **a** $1800   **b** 128 g   **c** 160 kg
   d $116   **e** 65 litres   **f** 90 min

3 **a** $\frac{5}{8}$ of 40 = 25   **b** $\frac{3}{4}$ of 280 = 210
   c $\frac{4}{5}$ of 70 = 56   **d** $\frac{5}{6}$ of 72 = 60

4 $6080
5 $31 500
6 52 kg
7 **a** 856   **b** 187 675
8 **a** $50   **b** $550
9 **a** $120   **b** $240
10 Lion Autos
11 Offer B

3.4 Adding and subtracting fractions

Exercise 3D

1 **a** $\frac{5}{7}$ **b** $\frac{7}{9}$ **c** $\frac{4}{5}$ **d** $\frac{6}{7}$

2 **a** $\frac{3}{7}$ **b** $\frac{1}{9}$ **c** $\frac{4}{11}$ **d** $\frac{7}{13}$

3 **a** $\frac{6}{8} = \frac{3}{4}$ **b** $\frac{4}{10} = \frac{2}{5}$ **c** $\frac{6}{9} = \frac{2}{3}$ **d** $\frac{2}{4} = \frac{1}{2}$

4 **a** $\frac{4}{8} = \frac{1}{2}$ **b** $\frac{4}{10} = \frac{2}{5}$ **c** $\frac{4}{6} = \frac{2}{3}$ **d** $\frac{8}{10} = \frac{4}{5}$

5 **a** $\frac{12}{10} = \frac{6}{5} = 1\frac{1}{5}$ **b** $\frac{9}{8} = 1\frac{1}{8}$ **c** $\frac{9}{8} = 1\frac{1}{8}$
   d $\frac{13}{8} = 1\frac{5}{8}$ **e** $\frac{11}{8} = 1\frac{3}{8}$ **f** $\frac{7}{6} = 1\frac{1}{6}$
   g $\frac{9}{6} = \frac{3}{2} = 1\frac{1}{2}$ **h** $\frac{5}{4} = 1\frac{1}{4}$

6 **a** $\frac{10}{8} = \frac{5}{4} = 1\frac{1}{4}$ **b** $\frac{6}{4} = \frac{3}{2} = 1\frac{1}{2}$
   c $\frac{5}{5} = 1$ **d** $\frac{16}{10} = \frac{8}{5} = 1\frac{3}{5}$

7 **a** $\frac{5}{8}$ **b** $\frac{5}{10} = \frac{1}{2}$ **c** $\frac{1}{4}$ **d** $\frac{3}{8}$
   e $\frac{1}{4}$ **f** $\frac{3}{8}$ **g** $\frac{4}{10} = \frac{2}{5}$ **h** $\frac{5}{16}$

Exercise 3E

1 **a** $\frac{8}{15}$ **b** $\frac{7}{12}$ **c** $\frac{3}{10}$ **d** $\frac{11}{12}$ **e** $\frac{7}{8}$ **f** $\frac{1}{2}$
   g $\frac{1}{6}$ **h** $\frac{1}{20}$ **i** $\frac{1}{10}$ **j** $\frac{1}{8}$ **k** $\frac{1}{12}$ **l** $\frac{1}{3}$
   m $\frac{1}{6}$ **n** $\frac{7}{9}$ **o** $\frac{5}{8}$ **p** $\frac{3}{8}$ **q** $\frac{1}{15}$ **r** $1\frac{13}{24}$
   s $\frac{59}{80}$ **t** $\frac{22}{63}$ **u** $\frac{37}{54}$

2 **a** $3\frac{5}{14}$ **b** $10\frac{3}{5}$ **c** $2\frac{1}{6}$ **d** $3\frac{31}{45}$
   e $4\frac{47}{60}$ **f** $\frac{41}{72}$ **g** $\frac{29}{48}$ **h** $1\frac{43}{48}$
   i $1\frac{109}{120}$ **j** $1\frac{23}{30}$ **k** $1\frac{31}{84}$

3 $\frac{1}{20}$

4 **a** $\frac{1}{6}$ **b** 30, must be divisible by 2 and 3

3.5 Multiplying and dividing fractions

Exercise 3F

1 **a** $\frac{1}{6}$ **b** $\frac{1}{10}$ **c** $\frac{3}{8}$ **d** $\frac{3}{14}$ **e** $\frac{8}{15}$
   f $\frac{1}{5}$ **g** $\frac{2}{7}$ **h** $\frac{3}{10}$ **i** $\frac{1}{2}$ **j** $\frac{2}{5}$

2 **a** $\frac{3}{32}$ **b** $\frac{3}{8}$ **c** $\frac{7}{20}$
   d $\frac{16}{45}$ **e** $\frac{3}{5}$ **f** $\frac{5}{8}$

3 $\frac{1}{12}$

4 $\frac{3}{8}$

5 **a** $\frac{5}{12}$ **b** $2\frac{1}{12}$ **c** $6\frac{1}{4}$ **d** $2\frac{11}{12}$
   e $3\frac{9}{10}$ **f** $3\frac{1}{3}$ **g** $12\frac{1}{2}$ **h** 30

6 $\frac{2}{5}$ of $6\frac{1}{2} = 2\frac{3}{5}$

Exercise 3G

1 **a** $\frac{3}{4}$ **b** $1\frac{2}{5}$ **c** $1\frac{1}{15}$ **d** $1\frac{1}{14}$ **e** 4
   f 4   **g** 5   **h** $1\frac{5}{7}$ **i** $\frac{4}{9}$ **j** $1\frac{3}{5}$

2 18
3 40
4 15
5 16
6 **a** $2\frac{2}{15}$ **b** 38   **c** $1\frac{7}{8}$ **d** $\frac{9}{32}$ **e** $\frac{1}{16}$ **f** $\frac{256}{625}$

Answers to Chapter 4

4.1 Introduction to directed numbers

Exercise 4A

1 a 0 °C **b** 5 °C **c** –2 °C **d** –5 °C **e** –1 °C
2 a 11 degrees Celsius **b** 9 degrees Celsius
3 8 degrees Celsius

4.2 Everyday use of directed numbers

Exercise 4B

1 –$5
2 –200 m
3 above
4 –5 h
5 –2 °C
6 – 70 km
7 +5 minutes
8 –5 km/h
9 –2
10 a –11 °C **b** 6 degrees Celsius
11 1.54 am

4.3 The number line

Exercise 4C

1 a < **b** > **c** < **d** < **e** > **f** <
 g < **h** > **i** > **j** < **k** < **l** >
2 a < **b** < **c** < **d** > **e** < **f** <
3 a
-5 -4 -3 -2 -1 0 1 2 3 4 5

 c
-25 -20 -15 -10 -5 0 5 10 15 20 25

 d
-10 -8 -6 -4 -2 0 2 4 6 8 10

 e
-50 -40 -30 -20 -10 0 10 20 30 40 50

4 6 °C   –2 °C   –4 °C   2 °C

4.4 Adding and subtracting directed numbers

Exercise 4D

1 a –2° **b** –3° **c** –2° **d** –3° **e** –2° **f** –3°
 g 3 **h** 3 **i** –1 **j** –1 **k** 2 **l** –3
 m –4 **n** –6 **o** –6 **p** –1 **q** –5 **r** –4
 s 4 **t** –1 **u** –5 **v** –4 **w** –5 **x** –5
2 a 7 degrees Celsius **b** –6 °C
3 a 2 – 8

b 2 + 5 – 8 or 2 + 4 – 7 or 8 – 4 – 5 or 8 – 2 – 7 or 5 – 4 – 2
c 2 – 5 – 7 – 8
d 2 + 5 – 4 – 7 – 8
4 250 metres

Exercise 4E

1 a –8 **b** –10 **c** –11 **d** –3 **e** 2 **f** –5
 g 1 **h** 4 **i** 7 **j** –8 **k** –5 **l** –11
 m 11 **n** 6 **o** 8 **p** 8 **q** –2 **r** –1
 s –9 **t** –5
2 a 10 degrees Celsius
 b 7 degrees Celsius
 c 9 degrees Celsius
3 a 2 **b** –3 **c** –5 **d** –7 **e** –10 **f** –20
4 a 2 **b** 4 **c** –1 **d** –5 **e** –11 **f** 8
5 a 13 **b** 2 **c** 5 **d** 4 **e** 11 **f** –2
6 a –10 **b** –5 **c** –2 **d** 4 **e** 7 **f** –4
7 a +6 + +5 = 11 **b** +6 + –9 = –3
 c +6 – –9 = 15 **d** +6 – +5 = 1
8 It may not come on as the thermometer inaccuracy
 might be between 0° and 2° or 2° and 4°
9 –1 and 6

4.5 Multiplying and dividing directed numbers

Exercise 4F

1 a –15 **b** –14 **c** –24 **d** 6 **e** 14 **f** 2
 g –2 **h** –8 **i** –4 **j** 3 **k** –24 **l** –10
 m –18 **n** 16 **o** 36 **p** –4 **q** –12 **r** –4
 s 7 **t** 25 **u** 18
2 a –9 **b** 16 **c** –3 **d** –32 **e** 18 **f** 18
 g 6 **h** –4 **i** 20 **j** 16 **k** 8 **l** –48
 m 13 **n** –13 **o** –8 **p** 0 **q** 16 **r** –42
3 a –2 **b** 30 **c** 15 **d** –27 **e** –7
4 a 4 **b** –9 **c** –3 **d** 6 **e** –4
5 a –9 **b** 3 **c** 1
6 a 16 **b** –2 **c** –12
7 a 24 **b** 6 **c** –4 **d** –2
8 For example: 1 × (–12), –1 × 12, 2 × (–6), 6 × (–2),
 3 × (–4), 4 × (–3)
9 For example: 4 ÷ (–1), 8 ÷ (–2), 12 ÷ (–3), 16 ÷
 (–4), 20 ÷ (–5), 24 ÷ (–6)
10 –5 × 4, 3 × –6, –20 ÷ 2, –16 ÷ –4
11 a 4 **b** 25 **c** 12 **d** 1

Answers to Chapter 5

5.1 Squares and square roots

Exercise 5A

1 **a** 49 **b** 100 **c** 1.44 **d** 6.25 **e** 256 **f** 400
  g 9.61 **h** 20.25 **i** 9 **j** 64 **k** 0.25 **l** 0.25
2 **a** 3 and –3 **b** 10 and –10 **c** 11 and –11
  d 1.2 and –1.2 **e** 20 and –20 **f** 3.5 and –3.5
  g 1 and –1 **h** 100 and –100
3 **a** 5 **b** 6 **c** 10 **d** 7 **e** 8
  f 1.5 **g** 5.5 **h** 1.2 **i** 20 **j** 0.5
4 **a** 81 **b** 40 **c** 100 **d** 14 **e** 36
  f 15 **g** 49 **h** 12 **i** 25 **j** 21
5 **a** 24 **b** 31 **c** 45 **d** 40 **e** 67
  f 101 **g** 3.6 **h** 6.5 **i** 13.9 **j** 22.2
6 $\sqrt{50}$, 3^2, $\sqrt{90}$, 4^2
7 **a** 6^2 is 36 and 7^2 is 49; 40 is between 36 and 49
  b 6.3245553......
8 4 and 5
9 **a** 8 and 9 **b** 9 and 10
  c 12 and 13 **d** 15 and 16
10 $\sqrt{324}$ = 18
11 15

5.2 Cubes and cube roots

Exercise 5B

1 **a** 8 **b** 27 **c** 512 **d** 1000
  e 1.331 **f** 15.625 **g** –27 **h** –125
  i 8000 **j** 68.921 **k** –68.921
2 **a** 2 **b** 5 **c** 9 **d** 1 **e** 3
  f –3 **g** 10 **h** 1.5 **i** 4.5 **j** 0.5
3 **a** 5 and 6 **b** 6 and 7
  c 7 and 8 **d** –8 and –7

4 2^3 because it equals 8, the rest equal 9
5 One possible answer is $8^2 = 4^3$
6 $\sqrt[3]{2000}$, $\sqrt{225}$, 2.5^3, 4^2
7

Number	Square	Cube
10	100	1000
5	25	125
4	16	64
11	121	1331
9	81	729

8 0.8^3, 0.8^2, $\sqrt{0.8}$, $\sqrt[3]{0.8}$.

5.3 Exponential growth and decay

Exercise 5C

1 **a i** 10 million **ii** 20 million **iii** 40 million
  b i 15 million **ii** 45 million **iii** 135 million
2 **a** 6000 **b** 9000 **c** 13 500 **d** 20 250
3 **a** 6000 **b** 1500 **c** 375
4 **a** 4800 **b** 768 **c** 123
5 21.8, 23.8, 25.9
6 18.2, 16.6, 15.1
7 **a** $6312 **b** 6 years
8 **a** $100 000 **b** $195 313
9 The correct value is 150 000 × 1.2^5 = 373 248
10 **a** $20 **b** $20 480
11 **a** 1185 **b** 351
12 272 million
13 **a** $18 **b** $32 **c** $340 **d** $11568

Answers to Chapter 6

6.1 Inequalities

Exercise 6A

1 **a** > **b** < **c** < **d** =
  e = **f** > **g** > **h** <
2 $\frac{1}{3} < \frac{1}{2} < \frac{3}{5}$
3 **a** 4,5,6 **b** 1,2 **c** 6 **d** 1,2,3,4,5
  e 2,3,4 **f** 4,5 **g** 1,2,3 **h** 6
4 **a** underweight **b** overweight
  c normal **d** normal
5 20, 22, 26, 28

6 **a** 49 **b** 45
  c 3,6,9 **d** 16,17,18,19,20
7 **a** true **b** false **c** true
  d true **e** false **f** true
8 **a** 6,7,8 **b** 26, 27, 28 **c** –7, –6, –5, –4
  d –2, –1, 0, 1 **e** there are none **f** 33

6.2 Sets

Exercise 6B

1 **a** days of the week **b** odd numbers **c** planets
  d compass directions **e** numbers on a dice
  (Other descriptions are possible in this question)

2 a {6,7,8,9}   **b** {1,2,3,4,6,12}   **c** {23, 29}   **d** {3,−3}

3 $n(A) = 4$, $n(B) = 6$, $n(C) = 2$, $n(D) = 2$

4 a 12   **b** any three months

5 a {6,7,8}, {6,8,9}, {6,7,9}, {7,8,9}   **b** {6}, {7}, {8}, {9}

6 a ∅, {left}, {right}, {left, right}   **b** {left,right}

7 a 6   **b** 6

8 a i D **ii** E **b** B **c** p

6.3 Venn diagrams

Exercise 6C

1 a i {s,a,n,g}   **ii** {s,p,a,r,e}   **iii** {a,s}
iv {n,g,s,a,p,r,e}   **v** {p,r,e,i,o}   **vi** {i,o}
vii {p,r,e}   **viii** {g,n}
b i 4   **ii** 7   **iii** 4
c Singapore

2 a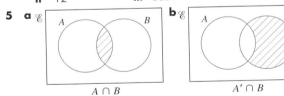

b i {c,n}   **ii** {f,s,c,n}   **iii** {f,s,c,n,i,t,r}
c i 3   **ii** 6

3 a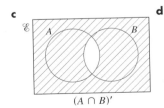

b {multiples of 6} is a possible description

4 a i {1,2,3,4,6,12}   **ii** {1,4,7,10}
iii {1,4}   **iv** {1,3}
v {1,3,4,5,7,9,10,11}   **vi** {2,4,6,8,10,12}
vii {1}   **viii** {1,3,7}
ix {2,6,8,12}   **x** {2,6,12}
b i the missing numbers are 1 and 12
ii 12   **iii** odd

5 a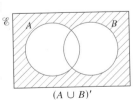

$A \cap B$

b

$A' \cap B$

c

$(A \cap B)'$

d

$(A \cup B)'$

e

$A' \cup B'$

f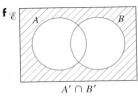

$A' \cap B'$

6 $(A \cap B)'$ is identical to $A' \cup B'$; $(A \cup B)'$ is identical to $A' \cap B'$

7 a   **b**

c   **d**

e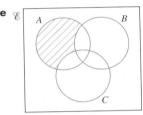

8 a There is only one even prime number. It is 2.
b There are no prime square numbers
c {odd square numbers}
d There are many possible answers. 15 and 33 are two of them

9 a false   **b** true   **c** true   **d** false   **e** true   **f** false

10 63

11 a   **b** ∅   **c** $A \cup B$ **d** $A \cap B$ **e** A

12 a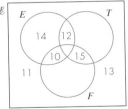

b six
c multiples of 15
d 30 and 60 are two possible values
e {$x : x$ is a multiple of 30} is one possible description

Answers to Chapter 7

7.1 Ratio

Exercise 7A

1 **a** 1 : 3   **b** 1 : 4   **c** 2 : 3   **d** 2 : 1
   e 2 : 5   **f** 2 : 5   **g** 5 : 8   **h** 5 : 1

2 **a** 8 : 1   **b** 12 : 1   **c** 5 : 6   **d** 1 : 24
   e 48 : 1   **f** 5 : 2   **g** 3 : 8   **h** 1 : 5

3 $\frac{7}{10}$

4 $\frac{10}{25} = \frac{2}{5}$

5 **a** $\frac{2}{5}$ **b** $\frac{3}{5}$

6 **a** $\frac{7}{10}$ **b** $\frac{3}{10}$

7 **a** $\frac{1}{2}$ **b** $\frac{7}{20}$ **c** $\frac{3}{20}$

8 3 : 1

9 1 : 4

Exercise 7B

1 **a** 160 g, 240 g   **b** 80 kg, 200 kg
   c 150, 350   **d** 950 m, 50 m
   e 175 min, 125 min   **f** $20, $30, $50
   g $36, $60, $144   **h** 50 g, 250 g, 300 g
   i $1.40, $2, $1.60   **j** 120 kg, 72 kg, 8 kg

2 **a** 175   **b** 30%

3 **a** 28   **b** 42

4 21

5 Joshua $2500, Aicha $3500, Mariam $4000

6 **a** 1 : 400 000   **b** 1 : 125 000   **c** 1 : 250 000
   d 1 : 25 000   **e** 1 : 20 000   **f** 1 : 40 000
   g 1 : 62 500   **h** 1 : 10 000   **i** 1 : 60 000

7 **a** 1 : 1 000 000   **b** 47 km   **c** 8 mm

8 **a** 1 : 250 000   **b** 2 km   **c** 4.8 cm

9 **a** 1 : 20 000   **b** 0.54 km   **c** 40 cm

10 **a** 1 : 1.6   **b** 1 : 3.25   **c** 1 : 1.125
   d 1 : 1.44   **e** 1 : 5.4   **f** 1 : 1.5
   g 1 : 4.8   **h** 1 : 42   **i** 1 : 1.25

Exercise 7C

1 **a** 3 : 2   **b** 32   **c** 80

2 1000 g

3 10 125

4 **a** 14 min   **b** 75 min

5 **a** 11 pages   **b** 32%

6 Ren $2040, Shota $2720

7 **a** lemonade 20 litres, ginger 0.5 litres
   b This one, one-thirteenth is greater than one-fiftieth.

8 100

9 40 cm³

7.2 Increases and decreases using ratios

Exercise 7D

1 **a** 600   **b** 300   **c** 2000   **d** 350   **e** 240   **f** 220

2 **a** 20   **b** 60   **c** 8   **d** 56   **e** 16   **f** 64

3 **a** 160 cm by 120 cm   **b** 60 cm by 45 cm

4 12.5 cm by 15 cm   **b** 3 : 2

5 **a** 9 cm by 15 cm   **b** 6.75 cm by 11.25 cm

6 **a** $7500   **b** 50%   **c** $9375   **d** 25%
   e 87.5%   **f** One way is to cancel 9375:5000

7 **a** One possible answer is to say $\frac{6}{5}$ = 1.2 and this is the multiplier for a 20% increase
   b 11:10   **c** 9:10

7.3 Speed

Exercise 7E

1 18 km/hour

2 440 kilometres

3 52.5 km/hour

4 11.50 am

5 500 s

6 **a** 75 km/hour   **b** 6.5 hours   **c** 175 km   **d** 240 km
   e 64 km/h   **f** 325 km   **g** 4.3 hours (4 h 18 min)

7 **a** 7.75 h   **b** 85.2 km/hour

8 **a** 2.25 h   **b** 157.5 km

9 **a** 1.25 h   **b** 1 h 15 min

10 **a** 48 km/hour   **b** 6 h 40 min

11 **a** 120 km   **b** 48 km/h

12 **a** 30 min   **b** 12 km/h

13 **a** 10 m/s   **b** 3.3 m/s   **c** 16.7 m/s   **d** 41.7 m/s
   e 20.8 m/s

14 **a** 90 km/h   **b** 43.2 km/h   **c** 14.4 km/h   **d** 108 km/h
   e 1.8 km/h

15 **a** 64.8 km/h   **b** 28 s   **c** 8.07 (37 min journey)

16 **a** 6.7 m/s   **b** 66 km   **c** 5 minutes   **d** 133.3 metres

17 6.6 minutes

7.4 Rates

Exercise 7F

1 **a** 3.5 cm   **b** 20 days

2 **a** 15 litres   **b** 20 seconds

3 **a** 125 days   **b** 0.5 litres

4 1600 days (4.38 years)

5 0.5

6 **a** 40   **b** 5 hours   **c** 1.5

7 **a** $50   **b** $90   **c** 4.5   **d** 4.2

8 **a** 31.8 litres   **b** 5.3 litres   **c** 1.06   **d** 0.106
   e 943 km

9 **a** 14.5 kg   **b** 2900 kg   **c** About 8 kg/day

7.5   Direct proportion

Exercise 7G

1 60 g
2 $5.22
3 45
4 $6.72
5 **a** $312.50   **b** 8
6 **a** 56 litres   **b** 350 km
7 **a** 300 kg   **b** 9 weeks
8 40 s
9 **a i** 100 g margarine, 200 g sugar, 250 g flour,
150 g ground rice
ii 150 g margarine, 300 g sugar, 375 g flour,
225 g ground rice
iii 250 g margarine, 500 g sugar, 625 g flour,
375 g ground rice
b 24
10 Pieter's shop as I can buy 24. At Paulo's shop I can only buy 20.

7.6   Inverse proportion

Exercise 7H

1 20 minutes
2 16
3 **a** 36   **b** 48
4 $10.80
5 **a** 16 days   **b** 20
6 The missing times are 5 hours, 1 hour 40 minutes,
1 hour 15 minutes
7 **a** 6   **b** 15

Answers to Chapter 8

8.1   Rounding whole numbers

Exercise 8A

1 **a** 20   **b** 60   **c** 80   **d** 50   **e** 100
f 20   **g** 90   **h** 70   **i** 10   **j** 30
2 **a** 200   **b** 600   **c** 800   **d** 500   **e** 1000
f 100   **g** 600   **h** 400   **i** 1000   **j** 1100
3 **a** 2000   **b** 6000   **c** 8000   **d** 5000
e 10 000   **f** 1000   **g** 6000   **h** 3000
i 9000   **j** 2000
4 **a** True   **b** False   **c** True   **d** True   **e** True   **f** False
5 **a** Highest Germany, lowest Italy
b 36 000, 43 000 , 25 000, 29 000
c 25 499 and 24 500
6 **a** 375   **b** 25 (350 to 374 inclusive)
7 A number between 75 and 84 inclusive added to a number
between 45 and 54 inclusive with a total not equal to 130,
for example 79 + 49 = 128

8.2   Rounding decimals

Exercise 8B

1 **a** 4.8   **b** 3.8   **c** 2.2   **d** 8.3   **e** 3.7
f 46.9   **g** 23.9   **h** 9.5   **i** 11.1   **j** 33.5
2 **a** 5.78   **b** 2.36   **c** 0.98   **d** 33.09   **e** 6.01
f 23.57   **g** 91.79   **h** 8.00   **i** 2.31   **j** 23.92
3 **a** 4.6   **b** 0.08   **c** 45.716   **d** 94.85   **e** 602.1
f 671.76   **g** 7.1   **h** 6.904   **i** 13.78   **j** 0.1
4 **a** 8   **b** 3   **c** 8   **d** 6   **e** 4
f 7   **g** 2   **h** 47   **i** 23   **j** 96

5 3 + 9 + 6 + 4 = 22 dollars
6 3, 3.46, 3.5
7 4.7275 or 4.7282

8.3   Rounding to significant figures

Exercise 8C

1 **a** 50000   **b** 60000   **c** 30000   **d** 90000
e 90000   **f** 0.5   **g** 0.3   **h** 0.006
i 0.05   **j** 0.0009   **k** 10   **l** 90
m 90   **n** 200   **o** 1000
2 **a** 56000   **b** 27000   **c** 80000   **d** 31000
e 14000   **f** 1.7   **g** 4.1   **h** 2.7
i 8.0   **j** 42   **k** 0.80   **l** 0.46
m 0.066   **n** 1.0   **o** 0.0098
3 **a** 60000   **b** 5300   **c** 89.7   **d** 110
e 9   **f** 1.1   **g** 0.3   **h** 0.7
i 0.4   **j** 0.8   **k** 0.2   **l** 0.7
4 **a** 65, 74   **b** 95, 149   **c** 950, 1499
5 Satora 750, 849, Nimral 1150, 1249, Korput 164 500,
165 499
6 One, because there could be 450 then 449.
7 Vashti has rounded to 2 significant figures or nearest 10 000.

8.4   Upper and lower bounds

Exercise 8D

1 **a** 6.5 and 7.5   **b** 115 and 125
c 3350 and 3450   **d** 49.5 and 50.5
e 5.50 and 6.50   **f** 16.75 and 16.85
g 15.5 and 16.5   **h** 14 450 and 14 550
i 54.5 and 55.5   **j** 52.5 and 57.5

2 a $5.5 \leq$ length in cm < 6.5
  b $16.5 \leq$ mass in kg < 17.5
  c $31.5 \leq$ time in minutes < 32.5
  d $237.5 \leq$ distance in km < 238.5
  e $7.25 \leq$ distance in m < 7.35
  f $25.75 \leq$ mass in kg < 25.85
  g $3.35 \leq$ time in hours < 3.45
  h $86.5 \leq$ mass in g < 87.5
  i $4.225 \leq$ distance in mm < 4.235
  j $2.185 \leq$ mass in kg < 2.195
  k $12.665 \leq$ time in minutes < 12.675
  l $24.5 \leq$ distance in metres < 25.5
  m $35 \leq$ length in cm < 45
  n $595 \leq$ mass in g < 605
  o $25 \leq$ time in minutes < 35
  p $995 \leq$ distance in metres < 1050
  q $3.95 \leq$ distance in metres < 4.05
  r $7.035 \leq$ mass in kg < 7.045
  s $11.95 \leq$ time in seconds < 12.05
  t $6.995 \leq$ distance in metres < 7.005

3 a 7.5, 8.5   **b** 25.5, 26.5
  c 24.5, 25.5   **d** 84.5, 85.5
  e 2.395, 2.405   **f** 0.15, 0.25
  g 0.055, 0.065   **h** 250 g, 350 g
  i 0.65, 0.75   **j** 365.5, 366.5
  k 165, 175   **l** 205, 215

4 C: The chain and distance are both any value between 29.5 and 30.5 metres, so there is no way of knowing if the chain is longer or shorter than the distance.

5 2 kg 450 grams

6 a <65.5 g   **b** 64.5 g
  c <2620 g   **d** 2580 g

8.5 Upper and lower bounds for calculations

Exercise 8E

1 65 kg and 75 kg

2 a 12.5 kg   **b** 20

3 9 kg   53.5 – 44.5

4 a 26 cm $\leq$ perimeter < 30 cm
  b 25.6 cm $\leq$ perimeter < 26.0 cm
  c 50.5 cm $\leq$ perimeter < 52.7 cm

5 a 38.25 cm^2 $\leq$ area < 52.25 cm^2
  b 37.1575 cm^2 $\leq$ area < 38.4475 cm^2
  c 135.625 cm^2 $\leq$ area < 145.225 cm^2

6 79.75 m^2 $\leq$ area < 100.75 m^2

7 216.125 cm^3 $\leq$ volume < 354.375 cm^3

8 12.5 metres

9 Yes, because they could be walking at 4.5 km/h and 2.5 km/h meaning that they would cover 4.5 km + 2.5 km = 7 km in 1 hour

10 20.9 m $\leq$ length < 22.9 m (3 sf)

11 a 14.65 s $\leq$ time < 14.75 s
  b 99.5 m $\leq$ length < 100.5 m
  c 6.86 m/s (3 sf)

12 14 s $\leq$ time < 30 s

13 337.75 and 334.21

14 177.3 and 169.4

Answers to Chapter 9

9.1 Standard form

Exercise 9A

1 a 250   **b** 34.5   **c** 0.00467   **d** 34.6
  e 0.020789   **f** 5678   **g** 246   **h** 7600
  i 897 000   **j** 0.008 65   **k** 60 000 000 **l** 0.000567

2 a 2.5×10^2 **b** 3.45×10^{-1} **c** 4.67×10^4
  d 3.4×10^9 **e** 2.078×10^{10} **f** 5.678×10^{-4}
  g 2.46×10^3 **h** 7.6×10^{-2} **i** 7.6×10^{-4}
  j 9.99×10^{-1} **k** 2.3456×10^2 **l** 9.87654×10^1
  m 6×10^{-4} **n** 5.67×10^{-3} **o** 5.60045×10^1

3 2.7797×10^4

4 $3.211\,97 \times 10^5$, 3×10^1, $4.491\,863 \times 10^6$

5 1.298×10^7, 2.997×10^9, 9.3×10^4

6 100

7 61.8 kilometres.

8 7.78×10^8; 5.8×10^7; 5.92×10^9

9.2 Calculating with standard form

Exercise 9B

1 a 5.67×10^3 **b** 6×10^2 **c** 3.46×10^{-1}
  d 7×10^{-4} **e** 5.6×10^2 **f** 6×10^5
  g 7×10^3 **h** 1.6   **i** 2.3×10^7

2 a 1.08×10^8 **b** 4.8×10^6 **c** 1.2×10^9
  d 1.08   **e** 6.4×10^2 **f** 1.2×10^1
  g 2.88   **h** 2.5×10^7 **i** 8×10^{-6}

3 a 2.7×10 **b** 1.6×10^{-2} **c** 2×10^{-1}
  d 4×10^{-8} **e** 2×10^5 **f** 6×10^{-2}

4 2×10^{13}, 1×10^{-10}, mass = 2×10^3 g (2 kg)

5 a (2^{63}) 9.2×10^{18} grains
  b $2^{64} - 1 = 1.8 \times 10^{19}$

6 a 1.0×10^8 sq km
  b 31%

7 3.80×10^7 sq km

8 5×10^4

9 2.3×10^5

10 4.55×10^8 kg or 455 070 tonnes
11 a 100 000 000 (100 million) **b** 1.4%
12 a 2.048×10^6 **b** 4.816×10^6
13 9.41×10^4
14 Any value from 1.00000001×10^8 to 1×10^9 (excluding 1×10^9), i.e. any value of the form $a \times 10^8$ where $1 < a < 10$

15 a India **b** India and Jamaica
  c 2.2×10^7 **d** 21 or 22 **e** 480
16 a Togo **b** Sri Lanka
  c Sri Lanka **d** Russian Federation
  e $\frac{1}{261}$

Answers to Chapter 10

10.1   Units of measurement

Exercise 10A

1 a metres **b** kilometres
  c millimetres **d** kilograms or grams
  e litres **f** tonnes
  g millilitres **h** metres
  i kilograms **j** millimetres
2 Check individual answers.
3 The 5 metre since his height is about 175 cm, the lamp post will be about 525 cm

10.2   Converting between metric units

Exercise 10B

1 a 1.25 m **b** 8.2 cm **c** 0.55 m **d** 4.2 kg
  e 5.75 t **f** 8.5 cl **g** 0.755 kg **h** 0.8 l
  i 2 l **j** 1.035 m³ **k** 0.53 m³ **l** 34 000 m
2 a 3400 mm **b** 135 mm **c** 67 cm **d** 640 m
  e 2400 ml **f** 590 cl **g** 3750 kg **h** 0.000 94 l
  i 2160 cl **j** 15 200 g **k** 14 000 l **l** 0.19 ml
3 He should choose the 2000 mm × 15 mm × 20 mm
4 1 000 000

10.3   Time

Exercise 10C

1 a 1 hour 10 minutes; 2 hours 3 minutes; 2 hours 9 minutes; 1 hour 45 minutes
  b the 0900
2 a 9:45am, 10:36am, 1:33pm, 4:49pm
  b 3 hours and 48 minutes, 6 hours and 13 minutes
3 a 1605 **b** 0715 **c** 6 hours 45minutes
4 a 1050 **b** 1635 **c** 5 hours 45 minutes
5 a 1210 **b** 2hours 50 minutes
6 a 12 minutes **b** 40 minutes **c** 54 minutes
7 1 hour 13 minutes
8 1415
9 0715 the next day

10.4   Currency conversions

Exercise 10D

1 3197.41
2 164
3 The missing values are 4.30, 7.76, 38.78, 193.88, 387.75, 775.50
4 43.01
5 a 224.91 **b** 172.74
6 a i 349.83 **ii** 24692 or 24725 **iii** 432.90 or 433.69
  b 54000 yen, 500 euros, 650 dollars
7 a 2391.38 **b** 3489.75
  c Taiwan dollar **d** 1.4593
8 a 74.7755 **b** 0.14747

10.5   Using a calculator efficiently

Exercise 10E

1 a 144 **b** 108
2 a 12.54 **b** 27.45
3 a 196.48 **b** 1.023 **c** 0.236 **d** 4.219
4 a 3.58 **b** 6
5 a 497.952 **b** 110.978625
6 a 3.12 **b** 0.749 **c** 90.47
  d 184.96 **e** 5.56 **f** 27.52

Answers to Chapter 11

11.1 The language of algebra

Exercise 11A

1 **a** $x + 2$ **b** $x - 6$ **c** $x + k$ **d** $x - t$
  e $x + 3$ **f** $d + m$ **g** $b - y$ **h** $p + t + w$
  i $8x$ **j** hj **k** $x \div 4$ or $\frac{x}{4}$ **l** $2 \div x$ or $\frac{2}{x}$
  m $y \div t$ or $\frac{y}{t}$ **n** wt **o** a^2 **p** g^2

2 **a** $x + 3years$ **b** $x - 4years$

3 $F = 2C + 30$

4 Rule **c**

5 **a** $3n$ **b** $3n + 3$ **c** $n + 1$ **d** $n - 1$

6 Anil: $2n$, Reza: $n + 2$, Dale: $n - 3$, Chen: $2n + 3$

7 **a** $\$4$ **b** $\$(10 - x)$ **c** $\$(y - x)$ **d** $\$2x$

8 **a** $\$75$ **b** $\$15x$ **c** $\$4A$ **d** $\$Ay$

9 $(A - B)$ dollars

10 $\$A \div 5$ or $\$\frac{A}{5}$.

11 **a** Dad: $(72 + x)$ yr, me: $(T + x)$ yr **b** $31yr$

12 **a** $T \div 2$ or $\frac{T}{2}$ **b** $T \div 2 + 4$ or $\frac{T}{2} + 4$ **c** $T - x$

13 **a** $8x$ **b** $12m$ **c** $18t$

14 Andrea: $3n - 3$, Barak: $3n - 1$, Ahmed: $3n - 6$ or $3(n - 2)$, Dina: 0, Emma: $3n - n = 2n$. Hana: $3n - 3m$

15 For example, $2 \times 6m$, $1 \times 12m$, $6m + 6m$, etc.

11.2 Substitution into formulae

Exercise 11B

1 **a** 8 **b** 17 **c** 32

2 **a** 3 **b** 11 **c** 43

3 **a** 9 **b** 15 **c** 29

4 **a** 9 **b** 5 **c** -1

5 **a** 13 **b** 33 **c** 78

6 **a** 10 **b** 13 **c** 58

7 **a** $\$4$ **b** 13 km **c** Yes, the fare is $\$5.00$

8 **a** $2 \times 8 + 6 \times 11 - 3 \times 2 = 76$
  b $5 \times 2 - 2 \times 11 + 3 \times 8 = 12$

9 Any values such that $lw = \frac{1}{2}bh$ or $bh = 2lw$

10 **a** 32 **b** 64 **c** 160

11 **a** 6.5 **b** 0.5 **c** -2.5

12 **a** 2 **b** 8 **c** -10

13 **a** 3 **b** 2.5 **c** -5

14 **a** 6 **b** 3 **c** 2

15 **a** 12 **b** 8 **c** $1\frac{1}{2}$

16 **a** $\frac{1050}{n}$ **b** $\$925$

17 **a i** odd **ii** odd **iii** even **iv** odd
  b Any valid expression such as $xy + z$

18 **a** $\$20$
  b i $-\$40$ **ii** Delivery cost will be zero. **c** 40 kilometres

11.3 Rearranging formulae

Exercise 11C

1 $k = \frac{T}{3}$

2 $y = X + 1$

3 $p = 3Q$

4 $r = \frac{A - 9}{4}$

5 $n = \frac{W + 1}{3}$

6 **a** $m = p - t$
  b $t = p - m$

7 $m = gv$

8 $m = \sqrt{t}$

9 $r = \frac{C}{2\pi}$

10 $b = \frac{A}{h}$

11 $l = \frac{P - 2w}{2}$

12 $p = \sqrt{m - 2}$

13 **a** $-40 - 32 = -72$, $-72 \div 9 = -8$, $5 \times -8 = -40$
  b $68 - 32 = 36$, $36 \div 9 = 4$, $4 \times 5 = 20$
  c student's own demonstration

14 **a** $a = \frac{v - u}{t}$ **b** $t = \frac{v - u}{a}$

15 $d = \sqrt{\frac{4A}{\pi}}$

16 **a** $n = \frac{W - t}{3}$ **b** $t = W - 3n$

17 **a** $y = \frac{x + w}{5}$ **b** $w = 5y - x$

18 $p = \sqrt{\frac{k}{2}}$

19 **a** $t = u^2 - v$ **b** $u = \sqrt{v + t}$

20 **a** $m = k - n^2$ **b** $n = \sqrt{k - m}$

21 $r = \sqrt{\frac{T}{5}}$

22 **a** $w = K - 5n^2$ **b** $n = \sqrt{\frac{K - w}{5}}$

11.4 More complicated formulae

Exercise 11D

1 **a** 2.5 **b** $a = \sqrt{c^2 - b^2}$

2 **a** 60 **b** $a = \frac{2(s - ut)}{t^2}$

3 **a** $b = ac - 2$ **b** $c = \frac{b + 2}{a}$

4 $t = \frac{r}{p} + 3$

5 $e = \left(\frac{12}{d} - 1\right)^2$

6 **a** 5 **b** $u = \sqrt{v^2 - 2as}$ **c** $s = \frac{v^2 - u^2}{2a}$

7 **a** $L = \left(\frac{T}{2\pi}\right)^2 G$ **b** Student's proof

8 **a** $R = \sqrt{\dfrac{D + \pi r^2}{\pi}}$ **b** $r = \sqrt{\dfrac{\pi R^2 - D}{\pi}}$ **c** $\pi = \dfrac{D}{R^2 - r^2}$ **11** $T = \dfrac{b^2 + c^2 - a^2}{2bc}$

9 **a** $x = 5$ or -5 **b** $x = \sqrt{\dfrac{11 + 4y^2}{3}}$ **c** $y = \sqrt{\dfrac{3x^2 - 11}{4}}$ **12 a** 12 **b** $f = \dfrac{uv}{u + v}$

10 **a** $a = \left(\dfrac{T}{2}\right)^2 (c + 3)$ **b** $c = a\left(\dfrac{2}{T}\right)^2 - 3$ **c** $u = \dfrac{fv}{v - f}$ **d** $v = \dfrac{fu}{u - f}$

Answers to Chapter 12

12.1 Simplifying expressions

Exercise 12A

1 **a** $6t$ **b** $15y$ **c** $8w$ **d** $5b^2$ **e** $2w^2$
f $8p^2$ **g** $6t^2$ **h** $15t^2$ **i** $2mt$ **j** $5qt$
k $6mn$ **l** $6qt$ **m** $10hk$ **n** $21pr$

2 **a** All except $2m \times 6m$ **b** 2 and 0

3 $4x$ cm

4 **a** y^3 **b** $3m^3$ **c** $4t^3$ **d** $6n^3$ **e** t^4
f h^5 **g** $12n^5$ **h** $6a^7$ **i** $4k^7$ **j** t^3
k $12d^3$ **l** $15p^6$ **m** $3mp^2$ **n** $6m^2n$ **o** $8m^2p^2$

Exercise 12B

1 **a** $\$t$ **b** $\$(4t + 3)$

2 **a** $10x + 2y$ **b** $7x + y$ **c** $6x + y$

3 **a** $5a$ **b** $6c$ **c** $9e$ **d** $6f$
e $4j$ **f** $3q$ **g** 0 **h** $-w$
i $6x^2$ **j** $5y^2$ **k** 0

4 **a** $7x$ **b** $3t$ **c** $-5x$ **d** $-5k$
e $2m^2$ **f** 0

5 **a** $7x + 5$ **b** $5x + 6$ **c** $5p$ **d** $5x + 6$
e $5p + t + 5$ **f** $8w - 5k$ **g** c **h** $8k - 6y + 10$

6 **a** $2c + 3d$ **b** $5d + 2e$ **c** $f + 3g + 4h$
d $6u - 3v$ **e** $7m - 7n$ **f** $3k + 2m + 5p$
g $2v$ **h** $2w - 3y$ **i** $11x^2 - 5y$
j $-y^2 - 2z$ **k** $x^2 - z^2$

7 **a** $8x + 6$ **b** $3x + 16$ **c** $2x + 2y + 8$

8 Any acceptable answers, e.g. $x + 4x + 2y + 2y$
or $6x - x + 6y - 2y$

9 **a** $2x$ and $2y$ **b** a and $7b$

10 **a** $3x - 1 - x$ **b** $10x$ **c** 25 cm

11 Maria is correct, as the two short horizontal lengths are
equal to the bottom length and the two short vertical
lengths are equal to the side length.

12.2 Expanding brackets

Exercise 12C

1 **a** $6 + 2m$ **b** $10 + 5l$ **c** $12 - 3y$
d $20 + 8k$ **e** $6 - 12f$ **f** $10 - 6w$
g $10k + 15m$ **h** $12d - 8n$ **i** $t^2 + 3t$
j $k^2 - 3k$ **k** $4t^2 - 4t$ **l** $8k - 2k^2$
m $8g^2 + 20g$ **n** $15h^2 - 10h$ **o** $y^3 + 5y$
p $h^4 + 7h$ **q** $k^3 - 5k$ **r** $3t^3 + 12t$
s $15d^3 - 3d^4$ **t** $6w^3 + 3tw$ **u** $15a^3 - 10ab$
v $12p^4 - 15mp$ **w** $12h^3 + 8h^2g$ **x** $8m^3 + 2m^4$

2 **a** $5(t - 1)$ and $5t - 5$
b Yes, as $5(t - 1)$ when $t = 4.50$ is $5 \times 3.50 = \$17.50$

3 He has worked out 3×5 as 8 instead of 15 and he has not
multiplied the second term by 3. Answer should be $15x - 12$.

4 **a** $3(2y + 3)$ **b** $2(6z + 4)$ or $4(3z + 2)$

Exercise 12D

1 **a** $7t$ **b** $9d$ **c** $3e$ **d** $2t$
e $5t^2$ **f** $4y^2$ **g** $5ab$ **h** $3a^2d$

2 **a** $2x$ and $11y$ **b** a and $8b$

3 **a** $2x - 3$ **b** $10x - 16$ or $2(5x - 8)$

4 **a** $22 + 5t$ **b** $21 + 19k$ **c** $22 + 2f$ **d** $14 + 3g$

5 **a** $2 + 2h$ **b** $9g + 5$ **c** $17k + 16$ **d** $6e + 20$

6 **a** $4m + 3p + 2mp$ **b** $3k + 4h + 5hk$
c $12r + 24p + 13pr$ **d** $19km + 20k - 6m$

7 **a** $9t^2 + 13t$ **b** $13y^2 + 5y$ **c** $10e^2 - 6e$ **d** $14k^2 - 3kp$

8 **a** $17ab + 12ac + 6bc$ **b** $18wy + 6ty - 8tw$
c $14mn - 15mp - 6np$ **d** $8r^3 - 6r^2$

9 For x-coefficients, 3 and 1 or 1 and 4; for
y-coefficients, 5 and 1 or 3 and 4 or 1 and 7.

10 $5(3x + 2) - 3(2x - 1) = 9x + 13$

12.3 Factorisation

Exercise 12E

1 **a** $6(m + 2t)$ **b** $3(3t + p)$ **c** $4(2m + 3k)$
d $4(r + 2t)$ **e** $m(n + 3)$ **f** $g(5g + 3)$
g $2(2w - 3t)$ **h** $y(3y + 2)$ **i** $t(4t - 3)$
j $3m(m - p)$ **k** $3p(2p + 3t)$ **l** $2p(4t + 3m)$
m $4b(2a - c)$ **n** $5bc(b - 2)$ **o** $2b(4ac + 3de)$
p $2(2a^2 + 3a + 4)$ **q** $3b(2a + 3c + d)$ **r** $t(5t + 4 + a)$
s $3mt(2t - 1 + 3m)$ **t** $2ab(4b + 1 - 2a)$ **u** $5pt(2t + 3 + p)$

2 **a** Suni has taken out a common factor.
b Because the bracket adds up to $10.
c $30

3 a, d, f and h do not factorise.
b $m(5 + 2p)$ **c** $t(t - 7)$ **e** $2m(2m - 3p)$
g $a(4a - 5b)$ **i** $b(5a - 3bc)$

4 **a** Bernice
b Aidan has not taken out the largest possible common
factor. Craig has taken m out of both terms but there isn't
an m in the second term.

5 There are no common factors.

6 numerator $4x^2 - 12x$, denominator $2x - 6$

12.4 Multiplying two brackets

Exercise 12F

1 $x^2 + 5x + 6$

2 $t^2 + 7t + 12$

3 $w^2 + 4w + 3$

4 $m^2 + 6m + 5$

5 $k^2 + 8k + 15$

6 $a^2 + 5a + 4$

7 $x^2 + 2x - 8$

8 $t^2 + 2t - 15$

9 $w^2 + 2w - 3$

10 $f^2 - f - 6$

11 $g^2 - 3g - 4$

12 $y^2 + y - 12$

13 $x^2 + x - 12$

14 $p^2 - p - 2$

15 $k^2 - 2k - 8$

16 $y^2 + 3y - 10$

17 $a^2 + 2a - 3$

18 $x^2 - 9$

19 $t^2 - 25$

20 $m^2 - 16$

21 $t^2 - 4$

22 $y^2 - 64$

23 $p^2 - 1$

24 $25 - x^2$

25 $49 - g^2$

26 $x^2 - 36$

27 $(x + 2)$ and $(x + 3)$

28 a B: $1 \times (x - 2)$ C: 1×2 D: $2 \times (x - 1)$

   b $(x - 2) + 2 + 2(x - 1) = 3x - 2$

   c Area A $= (x - 1)(x - 2) =$ area of square minus areas
   $$(B + C + D)$$
   $$= x^2 - (3x - 2)$$
   $$= x^2 - 3x + 2$$

29 a $x^2 - 9$

   b i 9991   **ii** 39991

Exercise 12G

1 $6x^2 + 11x + 3$

2 $12y^2 + 17y + 6$

3 $6t^2 + 17t + 5$

4 $8t^2 + 2t - 3$

5 $10m^2 - 11m - 6$

6 $12k^2 - 11k - 15$

7 $6p^2 + 11p - 10$

8 $10w^2 + 19w + 6$

9 $6a^2 - 7a - 3$

10 $8r^2 - 10r + 3$

11 $15g^2 - 16g + 4$

12 $12d^2 + 5d - 2$

13 $8p^2 + 26p + 15$

14 $6t^2 + 7t + 2$

15 $6p^2 + 11p + 4$

16 $-10t^2 - 7t + 6$

17 $-6n^2 + n + 12$

18 $6f^2 - 5f - 6$

19 $-10q^2 + 7q + 12$

20 $-6p^2 - 7p + 3$

21 $-6t^2 + 10t + 4$

22 a $x^2 - 1$ **b** $4x^2 - 1$ **c** $4x^2 - 9$

   d $9x^2 - 25$

23 a $(3x - 2)(2x + 1) = 6x^2 - x - 2$

   $(2x - 1)(2x - 1) = 4x^2 - 4x + 1$

   $(6x - 3)(x + 1) = 6x^2 + 3x - 3$

   $(3x + 2)(2x + 1) = 6x^2 + 7x + 2$

   b Multiply the x terms to match the x^2 term and/or multiply
   the constant terms to get the constant term in the answer.

Exercise 12H

1 $4x^2 - 1$

2 $9t^2 - 4$

3 $25y^2 - 9$

4 $16m^2 - 9$

5 $4k^2 - 9$

6 $16h^2 - 1$

7 $4 - 9x^2$

8 $25 - 4t^2$

9 $36 - 25y^2$

10 $a^2 - b^2$

11 $9t^2 - k^2$

12 $4m^2 - 9p^2$

13 $25k^2 - g^2$

14 $a^2b^2 - c^2d^2$

15 $a^4 - b^4$

16 a $a^2 - b^2$

   b Dimensions: $a + b$ by $a - b$; Area: $a^2 - b^2$

   c Areas are the same, so $a^2 - b^2 = (a + b) \times (a - b)$

17 First shaded area is $(2k)^2 - 1^2 = 4k^2 - 1$
   Second shaded area is $(2k + 1)(2k - 1) = 4k^2 - 1$

Exercise 12I

1 $x^2 + 10x + 25$

2 $m^2 + 8m + 16$

3 $t^2 + 12t + 36$

4 $p^2 + 6p + 9$

5 $m^2 - 6m + 9$

6 $t^2 - 10t + 25$

7 $m^2 - 8m + 16$

8 $k^2 - 14k + 49$

9 $9x^2 + 6x + 1$

10 $16t^2 + 24t + 9$

11 $25y^2 + 20y + 4$

12 $4m^2 + 12m + 9$

13 $16t^2 - 24t + 9$

14 $9x^2 - 12x + 4$

15 $25t^2 - 20t + 4$
16 $25r^2 - 60r + 36$
17 $x^2 + 2xy + y^2$
18 $m^2 - 2mn + n^2$
19 $4t^2 + 4ty + y^2$
20 $m^2 - 6mn + 9n^2$
21 $x^2 + 4x$
22 $x^2 - 10x$
23 $x^2 + 12x$
24 $x^2 - 4x$
25 a Marcela has just squared the first term and the second term. She hasn't written down the brackets twice.
 b Paulo has written down the brackets twice but has worked out $(3x)^2$ as $3x^2$ and not $9x^2$.
 c $9x^2 + 6x + 1$
26 Whole square is $(2x)^2 = 4x^2$.
 Three areas are $2x - 1$, $2x - 1$ and 1.
 $4x^2 - (2x - 1 + 2x - 1 + 1) = 4x^2 - (4x - 1) = 4x^2 - 4x + 1$

12.5 Quadratic factorisation

Exercise 12J
1 $(x + 2)(x + 3)$
2 $(t + 1)(t + 4)$
3 $(m + 2)(m + 5)$
4 $(k + 4)(k + 6)$
5 $(p + 2)(p + 12)$
6 $(r + 3)(r + 6)$
7 $(w + 2)(w + 9)$
8 $(x + 3)(x + 4)$
9 $(a + 2)(a + 6)$
10 $(k + 3)(k + 7)$
11 $(f + 1)(f + 21)$
12 $(b + 8)(b + 12)$
13 $(t - 2)(t - 3)$
14 $(d - 4)(d - 1)$
15 $(g - 2)(g - 5)$
16 $(x - 3)(x - 12)$
17 $(c - 2)(c - 16)$
18 $(t - 4)(t - 9)$
19 $(y - 4)(y - 12)$
20 $(j - 6)(j - 8)$
21 $(p - 3)(p - 5)$
22 $(y + 6)(y - 1)$
23 $(t + 4)(t - 2)$
24 $(x + 5)(x - 2)$
25 $(m + 2)(m - 6)$
26 $(r + 1)(r - 7)$
27 $(n + 3)(n - 6)$
28 $(m + 4)(m - 11)$
29 $(w + 4)(w - 6)$
30 $(t + 9)(t - 10)$
31 $(h + 8)(h - 9)$
32 $(t + 7)(t - 9)$
33 $(d + 1)^2$
34 $(y + 10)^2$
35 $(t - 4)^2$
36 $(m - 9)^2$
37 $(x - 12)^2$
38 $(d + 3)(d - 4)$
39 $(t + 4)(t - 5)$
40 $(q + 7)(q - 8)$
41 $(x + 2)(x + 3)$, giving areas of $2x$ and $3x$, or $(x + 1)(x + 6)$, giving areas of x and $6x$.

Exercise 12K
1 $(x + 3)(x - 3)$
2 $(t + 5)(t - 5)$
3 $(m + 4)(m - 4)$
4 $(3 + x)(3 - x)$
5 $(7 + t)(7 - t)$
6 $(k + 10)(k - 10)$
7 $(2 + y)(2 - y)$
8 $(x + 8)(x - 8)$
9 $(t + 9)(t - 9)$
10 a x^2
  b i $(x - 2)$ **ii** $(x + 2)$ **iii** x^2 **iv** 4
  c $A + B - C = x^2 - 4$, which is the area of D, which is $(x + 2)(x - 2)$.
11 a $x^2 + 4x + 4 - (x^2 + 2x + 1) = 2x + 3$
  b $(a + b)(a - b)$
  c $(x + 2 + x + 1)(x + 2 - x - 1) = (2x + 3)(1) = 2x + 3$
  d The answers are the same.
  e $(x + 1 + x - 1)(x + 1 - x + 1) = (2x)(2) = 4x$
12 $(x + y)(x - y)$
13 $(x + 2y)(x - 2y)$
14 $(x + 3y)(x - 3y)$
15 $(3x + 1)(3x - 1)$
16 $(4x + 3)(4x - 3)$
17 $(5x + 8)(5x - 8)$
18 $(2x + 3y)(2x - 3y)$
19 $(3t + 2w)(3t - 2w)$
20 $(4y + 5x)(4y - 5x)$

Exercise 12L
1 $(2x + 1)(x + 2)$
2 $(7x + 1)(x + 1)$
3 $(4x + 7)(x - 1)$
4 $(3t + 2)(8t + 1)$
5 $(3t + 1)(5t - 1)$
6 $(4x - 1)^2$
7 $3(y + 7)(2y - 3)$
8 $4(y + 6)(y - 4)$
9 $(2x + 3)(4x - 1)$
10 $(2t + 1)(3t + 5)$
11 $(x - 6)(3x + 2)$
12 $(x - 5)(7x - 2)$

13 $4x + 1$ and $3x + 2$

14 a All the terms in the quadratic have a common factor of 6.
b $6(x + 2)(x + 3)$. This has the highest common factor taken out.
c For example, 'A rectangle could be split in many different ways.'

12.6 Algebraic fractions

Exercise 12M

1 a $\frac{5x}{6}$ **b** $\frac{19x}{20}$ **c** $\frac{23x}{20}$ **d** $\frac{3x + 2y}{6}$

e $\frac{x^2y + 8}{4x}$ **f** $\frac{5x + 7}{6}$ **g** $\frac{7x + 3}{4}$ **h** $\frac{13x + 5}{15}$

i $\frac{3x - 7}{4}$ **j** $\frac{5x - 10}{4}$

2 a $\frac{x}{6}$ **b** $\frac{11x}{20}$ **c** $\frac{7x}{20}$ **d** $\frac{3x - 2y}{6}$

e $\frac{xy^2 - 8}{4y}$ **f** $\frac{x - 1}{6}$ **g** $\frac{x + 1}{4}$ **h** $\frac{-7x - 5}{15}$

i $\frac{x - 1}{4}$ **j** $\frac{2 - 3x}{4}$

3 a $\frac{x^2}{6}$ **b** $\frac{3xy}{14}$ **c** $\frac{8}{3}$ **d** $\frac{2xy}{3}$

e $\frac{x^2 - 2x}{10}$ **f** $\frac{1}{6}$ **g** $\frac{6x^2 + 5x + 1}{8}$ **h** $\frac{2x^2 + x}{15}$

i $\frac{2x - 4}{x - 3}$ **j** $\frac{1}{2x}$

4 a x **b** $\frac{x}{2}$ **c** $\frac{3x^2}{16}$ **d** 3

e $\frac{17x + 1}{10}$ **f** $\frac{13x + 9}{10}$ **g** $\frac{3x^2 - 5x - 2}{10}$ **h** $\frac{x + 3}{2}$

i $\frac{2x^2 - 6y^2}{9}$

5 a $\frac{7x + 9}{(x + 1)(x + 2)}$ **b** $\frac{11x - 10}{(x - 2)(x + 1)}$ **c** $\frac{2 - 13x}{(4x + 1)(x + 2)}$

d $\frac{8 - 10x}{(2x - 1)(x + 1)}$ **e** $\frac{x + 1}{(2x - 1)(3x - 1)}$

6 First, he did not factorise and just cancelled the x^2s. Then he cancelled 2 and 6 with the wrong signs. Then he said two minuses make a plus when adding, which is not true.

7 $\frac{2x^2 + x - 3}{4x^2 - 9}$

8 a $\frac{9x + 13}{(x + 1)(x + 2)}$ **b** $\frac{14x + 19}{(4x - 1)(x + 1)}$ **c** $\frac{2x^2 + x - 13}{2(x + 1)}$ **d** $\frac{x + 1}{(2x - 1)(3x - 1)}$

9 a $\frac{x - 1}{2x + 1}$ **b** $\frac{2x + 1}{x + 3}$ **c** $\frac{2x - 1}{3x - 2}$ **d** $\frac{x + 1}{x - 1}$ **e** $\frac{2x + 5}{4x - 1}$

Answers to Chapter 13

13.1 Solving linear equations

Exercise 13A

1 a 56 **b** 2 **c** 6 **d** 3 **e** 4
f $2\frac{1}{2}$ **g** $3\frac{1}{2}$ **h** $2\frac{1}{2}$ **i** 4 **j** 21
k 72 **l** 56 **m** 0 **n** –7 **o** –18
p 36 **q** 36 **r** 60 **s** 7 **t** 11
u 2 **v** 7 **w** 2.8 **x** 1 **y** 11.5
z 0.2

2 a –4 **b** 15

3 a Elif
b Second line: Mustafa subtracts 1 instead of adding 1; fourth line: Mustafa subtracts 2 instead of dividing by 2.

Exercise 13B

1 a 3 **b** 7 **c** 5 **d** 3 **e** 4 **f** 6
g 8 **h** 1 **i** $1\frac{1}{2}$ **j** $2\frac{1}{2}$ **k** $\frac{1}{2}$ **l** $1\frac{1}{5}$
m 2 **n** –2 **o** –1 **p** –2 **q** –2 **r** –1

2 Any values that work, e.g. $a = 2$, $b = 3$ and $c = 30$.

3 55

Exercise 13C

1 a $x = 2$ **b** $y = 1$ **c** $a = 7$ **d** $t = 4$
e $p = 2$ **f** $k = -1$ **g** $m = 3$ **h** $s = -2$

2 $3x - 2 = 2x + 5$, $x = 7$

3 a $d = 6$ **b** $x = 11$ **c** $y = 1$ **d** $h = 4$
e $b = 9$ **f** $c = 6$

4 $6x + 3 = 6x + 10$; $6x - 6x = 10 - 3$; $0 = 7$, which is obviously false. Both sides have $6x$, which cancels out.

5 Check student's example.

13.2 Setting up equations

Exercise 13D

1 90 cents or 0.90 dollars

2 a 1.5 **b** 2

3 a 1.5 cm **b** 6.75 cm^2

4 17

5 8

6 a $8c - 10 = 56$ **b** \$8.25

7 a B: 450 cars, C: 450 cars, D: 300 cars
b 800 **c** 750

8 Length is 5.5 m, width is 2.5 m and area is 13.75 m^2. Tiles cost 123.75 dollars

9 3 years

10 9 years

11 3 cm

12 5

13 a $4x + 40 = 180$ **b** $x = 35°$

14 a $\frac{x + 10}{5} = 9.50$ **b** \$37.50

15 No, as $x + x + 2 + x + 4 + x + 6 = 360$ gives $x = 87°$ so the consecutive numbers (87, 89, 91, 93) are not even but odd

16 $4x + 18 = 3x + 1 + 50$, $x = 33$. Large bottle 1.5 litres, small bottle 1 litre

13.3 Solving quadratic equations by factorisation

Exercise 13E

1 $-2, -5$
2 $-3, -1$
3 $-6, -4$
4 $-3, 2$
5 $-1, 3$
6 $-4, 5$
7 $1, -2$
8 $2, -5$
9 $7, -4$
10 $3, 2$
11 $1, 5$
12 $4, 3$
13 $-4, -1$
14 $-9, -2$
15 $2, 4$
16 $3, 5$
17 $-2, 5$
18 $-3, 5$
19 $-6, 2$
20 $-6, 3$
21 $-1, 2$
22 -2
23 -5
24 4
25 $-2, -6$
26 7
27 a $x(x - 3) = 550$, $x^2 - 3x - 550 = 0$
   b $(x - 25)(x + 22) = 0$, $x = 25$ years
28 $x(x + 40) = 48000$, $x^2 + 40x - 48000 = 0$,
   $(x + 240)(x - 200) = 0$. Fence is $2 \times 200 + 2 \times 240 = 880$ m.
29 $-6, -4$
30 $2, 16$
31 $-6, 4$
32 $-9, 6$
33 $-10, 3$
34 $-4, 11$
35 $-8, 9$
36 $8, 9$
37 1
38 Mario was correct. Sylvan did not make it into a standard quadratic and only factorised the x terms. She also incorrectly solved the equation $x - 3 = 4$.

Exercise 13F

1 a $\frac{1}{3}, -3$ **b** $1\frac{1}{3}, -\frac{1}{2}$ **c** $-\frac{1}{5}, 2$
   d $-2\frac{1}{2}, 3\frac{1}{2}$ **e** $-\frac{1}{6}, -\frac{1}{3}$ **f** $\frac{2}{3}, 4$
   g $\frac{1}{2}, -3$ **h** $\frac{5}{2}, -\frac{7}{6}$ **i** $-1\frac{2}{3}, 1\frac{2}{5}$
   j $1\frac{3}{4}, 1\frac{2}{7}$ **k** $\frac{2}{3}, \frac{1}{8}$ **l** $\pm\frac{1}{4}$
   m $-2\frac{1}{4}, 0$ **n** $\pm 1\frac{2}{5}$ **o** $-\frac{1}{3}, 3$

2 a $-6, 7$ **b** $-\frac{5}{2}, \frac{3}{2}$ **c** $-6, 7$
   d $-1, \frac{11}{13}$ **e** $-2, 3$ **f** $-\frac{2}{5}, \frac{1}{2}$
   g $-\frac{1}{2}, \frac{1}{3}$ **h** $-2, \frac{1}{5}$ **i** 4
   j $-2, \frac{1}{8}$ **k** $-\frac{1}{3}, 0$ **l** $-5, 5$
   m $-\frac{5}{3}$ **n** $-\frac{7}{2}, \frac{7}{2}$ **o** $-\frac{5}{2}, 3$

3 a Both have only one solution: $x = 1$.
   b B is a linear equation, but A and C are quadratic equations.
4 a $(5x - 1)^2 = (2x + 3)^2 + (x + 1)^2$, when expanded and collected into the general quadratics, gives the required equation.
   b $(10x + 3)(2x - 3)$, $x = 1.5$; area = 7.5 cm^2.

13.4 Solving quadratic equations by the quadratic formula

Exercise 13G

1 $1.77, -2.27$
2 $-0.23, -1.43$
3 $3.70, -2.70$
4 $0.29, -0.69$
5 $-0.19, -1.53$
6 $-1.23, -2.43$
7 $-0.41, -1.84$
8 $-1.39, -2.27$
9 $1.37, -4.37$
10 $2.18, 0.15$
11 $-0.39, -5.11$
12 $0.44, -1.69$
13 $1.64, 0.61$
14 $0.36, -0.79$
15 $1.89, 0.11$
16 13
17 $x^2 - 3x - 7 = 0$
18 Hasan gets $x = \frac{4 \pm \sqrt{0}}{8}$ and Mirian gets $(2x - 1)^2 = 0$; each method only gives one solution, $x = \frac{1}{2}$

13.5 Solving quadratic equations by completing the square

Exercise 13H

1 a $(x + 2)^2 - 4$ **b** $(x + 7)^2 - 49$
   c $(x - 3)^2 - 9$ **d** $(x + 3)^2 - 9$
   e $(x - 1.5)^2 - 2.25$ **f** $(x - 4.5)^2 - 20.25$
   g $(x + 6.5)^2 - 42.25$ **h** $(x + 5)^2 - 25$
   i $(x + 4)^2 - 16$ **j** $(x - 1)^2 - 1$
   k $(x + 1)^2 - 1$
2 a $(x + 2)^2 - 5$ **b** $(x + 7)^2 - 54$
   c $(x - 3)^2 - 6$ **d** $(x + 3)^2 - 2$
   e $(x - 1.5)^2 - 3.25$ **f** $(x + 3)^2 - 6$
   g $(x - 4.5)^2 - 10.25$ **h** $(x + 6.5)^2 - 7.25$
   i $(x + 4)^2 - 22$ **j** $(x + 1)^2 - 2$
   k $(x - 1)^2 - 8$ **l** $(x + 1)^2 - 10$

3 **a** $-2 \pm \sqrt{5}$ **b** $-7 \pm 3\sqrt{6}$ **c** $3 \pm \sqrt{6}$
   d $-3 \pm \sqrt{2}$ **e** $1.5 \pm \sqrt{3.25}$ **f** $-3 \pm \sqrt{6}$
   g $4.5 \pm \sqrt{10.25}$ **h** $-6.5 \pm \sqrt{7.25}$ **i** $-4 \pm \sqrt{22}$
   j $-1 \pm \sqrt{2}$ **k** $1 \pm 2\sqrt{2}$ **l** $-1 \pm \sqrt{10}$

4 **a** $1.45, -3.45$ **b** $5.32, -1.32$ **c** $-4.16, 2.16$

5 **a** $x = 1.5 \pm \sqrt{3.75}$ **b** $x = 1 \pm \sqrt{0.75}$
   c $x = -1.25 \pm \sqrt{6.5625}$ **d** $x = 7.5 \pm \sqrt{40.25}$

6 $p = -14, q = -3$

7 **a** 3rd, 1st, 4th and 2nd – in that order

13.6 Simultaneous equations

Exercise 13I

1 **a** $x = 4, y = 1$ **b** $x = 1, y = 4$
   c $x = 3, y = 1$ **d** $x = 5, y = -2$
   e $x = 7, y = 1$ **f** $x = 5, y = \frac{1}{2}$
   g $x = 4\frac{1}{2}, y = 1\frac{1}{2}$ **h** $x = -2, y = 4$
   i $x = 2\frac{1}{2}, y = -1\frac{1}{2}$ **j** $x = 2\frac{1}{4}, y = 6\frac{1}{2}$
   k $x = 4, y = 3$ **l** $x = 5, y = 3$

2 **a** 3 is the first term. The next term is $3 \times a + b$, which equals 14.
   b $14a + b = 47$
   c $a = 3, b = 5$
   d 146, 443

Exercise 13J

1 **a** $x = 2, y = -3$ **b** $x = 7, y = 3$
   c $x = 4, y = 1$ **d** $x = 2, y = 5$
   e $x = 4, y = -3$ **f** $x = 1, y = 7$
   g $x = 2\frac{1}{2}, y = 1\frac{1}{2}$ **h** $x = -1, y = 2\frac{1}{2}$
   i $x = 6, y = 3$ **j** $x = \frac{1}{2}, y = -\frac{3}{4}$
   k $x = -1, y = 5$ **l** $x = 1\frac{1}{2}, y = \frac{3}{4}$

2 **a** They are the same equation. Divide the first by 2 and it is the second, so they have an infinite number of solutions.
   b Double the second equation to get $6x + 2y = 14$ and subtract to get $9 = 14$. The left-hand sides are the same if the second is doubled so they cannot have different values.

Exercise 13K

1 **a** $x = 5, y = 1$ **b** $x = 3, y = 8$
   c $x = 9, y = 1$ **d** $x = 7, y = 3$
   e $x = 4, y = 2$ **f** $x = 6, y = 5$
   g $x = 3, y = -2$ **h** $x = 2, y = \frac{1}{2}$
   i $x = -2, y = -3$ **j** $x = -1, y = 2\frac{1}{2}$
   k $x = 2\frac{1}{2}, y = -\frac{1}{2}$ **l** $x = -1\frac{1}{2}, y = 4\frac{1}{2}$
   m $x = -\frac{1}{2}, y = -6\frac{1}{2}$ **n** $x = 3\frac{1}{2}, y = 1\frac{1}{2}$
   o $x = -2\frac{1}{2}, y = -3\frac{1}{2}$

2 $(1, -2)$ is the solution to equations A and C; $(-1, 3)$ is the solution to equations A and D; $(2, 1)$ is the solution to B and C; $(3, -3)$ is the solution to B and D.

3 Intersection points are $(0, 6)$, $(1, 3)$ and $(2, 4)$. Area is 2 cm^2

4 Intersection points are $(0, 3)$, $(6, 0)$ and $(4, -1)$. Area is 6 cm^2

13.7 Solving inequalities

Exercise 13L

1 **a** $x < 3$ **b** $t > 8$
   c $p \geqslant 10$ **d** $x < 5$
   e $y \leqslant 3$ **f** $t > 5$
   g $x < 6$ **h** $y \leqslant 15$
   i $t \geqslant 18$ **j** $x < 7$
   k $x \leqslant 3$ **l** $t \geqslant 5$

2 **a** 8
   b 6
   c 16
   d 3
   e 7

3 **a** 11
   b 16
   c 16

4 $2x + 3 < 20$, $x < 8.50$, so the most each could cost is $8.49

5 **a** Because $3 + 4 = 7$, which is less than the third side of length 8
   b $x + x + 2 > 10$, $2x + 2 > 10$, $2x > 8$, $x > 4$, so smallest value of x is 5

6 **a** $x = 6$ and $x < 3$ scores -1 (nothing in common), $x < 3$ and $x > 0$ scores 1 (1 in common for example), $x > 0$ and $x = 2$ scores 1 (2 in common), $x = 2$ and $x \geqslant 4$ scores -1 (nothing in common), so we get $-1 + 1 + 1 - 1 = 0$
   b $x > 0$ and $x = 6$ scores $+1$ (6 in common), $x = 6$ and $x \geqslant 4$ scores $+1$ (6 in common), $x \geqslant 4$ and $x = 2$ scores -1 (nothing in common), $x = 2$ and $x < 3$ scores $+1$ (2 in common), $+1 + 1 - 1 + 1 = 2$
   c Any acceptable combination, e.g. $x = 2$, $x < 3$, $x > 0$, $x \geqslant 4$, $x = 6$

7 **a** $x \geqslant -6$ **b** $t \leqslant \frac{8}{3}$
   c $y \leqslant 4$ **d** $x \geqslant -2$
   e $w \leqslant 5.5$ **f** $x \leqslant \frac{14}{5}$

8 **a** $x \leqslant 2$ **b** $x > 38$
   c $x < 6\frac{1}{2}$ **d** $x \geqslant 7$
   e $t > 15$ **f** $y \leqslant \frac{7}{5}$

9 **a** 4   **b** 99   **c** 11   **d** 11   **e** 6

10 **a** $0, 10 - 10$ **b** $x < 16$

11 **a** $x < 9$ **b** $x \geqslant 11$ **c** $x \geqslant 3$

12 **a** $x \geqslant 7.5$ **b** $x \leqslant -2$
   c $x < 6$ **d** $x > 1.5$
   e $x \geqslant -5$ **f** $x < 0.5$

Answers to Chapter 14

14.1 Conversion graphs

Exercise 14A

1 **a i** $8\frac{1}{4}$ kg **ii** $2\frac{1}{4}$ kg **iii** 9 lb **iv** 22 lb
   b 2.2 lb
   c Read off the value for 12 lb (5.4 kg) and multiply this by 4 (21.6 kg)

2 **a i** 10 cm **ii** 23 cm **iii** 2 in **iv** $8\frac{3}{4}$ in
   b $2\frac{1}{2}$ cm
   c Read off the value for 9 in (23 cm) and multiply this by 2 (46 cm)

3 **a i** $320 **ii** $100 **iii** £45 **iv** £78
   b $3.20
   c It would become more steep.

4 **a i** $120 **ii** $82
   b i 32 **ii** 48

5 **a i** $100 **ii** $325
   b i 500 **ii** 250

6 **a i** $70 **ii** $29
   b i $86 **ii** $38

7 **a i** 95 °F **ii** 68 °F **iii** 10 °C **iv** 32 °C
   b 32°F

8 **a** Check student's graph **b** $50

9 **a** Student's own graph **b** about 48 kilometres
   c about 16 miles

10 **a** Student's own graph **b** about 9 centimetres
   c about 4 hours

11 **a** Student's own graph **b** about 23 minutes

14.2 Travel graphs

Exercise 14B

1 **a i** 2 h **ii** 3 h **iii** 5 h
   b i 40 km/h **ii** 120 km/h **iii** 40 km/h
   c 5.30 am

2 **a i** 125 km **ii** 125 km/h
   b i Between 2 pm and 3 pm **ii** About 12 km/h

3 **a** 30 km
   b 40 km
   c 100 km/h

4 **a i** 263 m/min (3 sf)
      ii 15.8 km/h (3 sf)
   b 500 m/min
   c Yuto by 1 minute

5 **a** Patrick ran quickly at first, then had a slow middle section but he won the race with a final sprint. Araf ran steadily all the way and came second. Sean set off the slowest, speeded up towards the end but still came third.
   b i 1.67 m/s **ii** 6 km/h

6 **a**
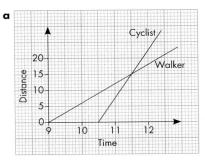

   b At 1130

7 **a i** Because it stopped several times
      ii Ravinder
   b Ravinder at 1558, Sue at 1620, Michael at 1635
   c i 24 km/h
      ii 20.6 km/h
      iii 5

8 **a** 50 metres
   b Student's graph
   c 1 metre/second

9 **a** student's graph
   b 80 km/hour

10 **a** 1300
   b 15 km
   c student's graph
   d For the three stages, 5 km/hour, 4 km/hour and 2 km/hour. For the whole trip 3.75 km/hour

14.3 Speed–time graphs

Exercise 14C

1 **a** 20 m/s **b** 0.5 m/s^2
   c 1 m/s^2 **d** 600 metres
   e 10 m/s

2 **a** 0.6 m/s^2 **b** 750 m

3 **a** 0.2 m/s^2 **b** 0.1 m/s^2
   c 75 metres **d** 2.5 m/s

4 **a** 1 m/s^2 **b** $\frac{2}{3}$ m/s^2
   c 6 kilometres (or 6000 metres)
   d 30 m/s

5 **a** student's graph **b** 0.8 m/s^2 **c** 80 metres

6 **a** 26 m/s **b** student's graph **c** 144 metres

7 **a** $1\frac{1}{3}$ m/s^2
   b They are together. They have both travelled 450 metres

8 **a** 2 m/s^2
   b i after 20 seconds
      ii 100 metres
   c 1150 metres

9 **a** 15 seconds **b** $1\frac{1}{3}$ m/s^2

10 **a** 6 m/s **b** student's own graph
   c 15 metres

Answers to Chapter 15

15.1 Drawing straight line graphs

Exercise 15A

1 Extreme points are (0, 4), (5, 19)
2 Extreme points are (0, –5), (5, 5)
3 Extreme points are (0, –3), (10, 2)
4 Extreme points are (–3, –4), (3, 14)
5 Extreme points are (–6, 2), (6, 6)
6 a Extreme points are (0, –2), (5, 13) and (0, 1), (5, 11)
  b (3, 7)
7 a Extreme points are (0, –5), (5, 15) and (0, 3), (5, 13)
  b (4, 11)
8 a Extreme points are (0, –1), (12, 3) and (0, –2), (12, 4)
  b (6, 1)
9 a Extreme points are (0, 1), (4, 13) and (0, –2), (4, 10)
  b Do not cross because they are parallel
10 a Values of y: 5, 4, 3, 2, 1, 0. Extreme points are (0, 5), (5, 0)
  b Extreme points are (0, 7), (7, 0)
11 a yes **b** no **c** yes **d** no **e** yes **f** no
12 a 6 **b** 3.5 **c** 2
13 a 20 **b** –10

15.2 The equation $y = mx + c$

Exercise 15B

1 a 3 **b** 2 **c** $\frac{1}{2}$
  d 3 **e** $\frac{1}{3}$

2 a $y = 2x - 2$ **b** $y = x + 1$
  c $y = 2x - 3$ **d** $2y = x + 6$
  e $y = x$ **f** $y = 2x$

3 a $y = 2x + 1$, $y = -2x + 1$
  b $5y = 2x - 5$, $5y = -2x - 5$
  c $y = x + 1$, $y = -x + 1$
4 a $y = -2x + 1$
  b $2y = -x$
  c $y = -x + 1$
  d $5y = -2x - 5$
  e $y = -\frac{3}{2}x - 3$ or $2y = -3x - 6$
5 a 3 **b** (0,3)
6 a 4 **b** 4
7 The first and last are parallel because they both have a gradient of 4. The middle one has a gradient of 3.

15.3 Parallel lines

Exercise 15C

1 a $2 \times 3 + 6 = 12$ **b** 2 **c** Student's graph
  d $y = 2x$ **e** $y = 2x + 3$
2 a (0, –1) and (4, 0) **b** $y = \frac{1}{4}x + 3$
3 a –2 **b** (4,0) **c** student's graph
  d $y = -2x$ **e** $y = -2x + 14$
4 a 1 **b** $y = 5x - 9$
5 $y = \frac{2}{3}x - 3$
6 a $y = 2x - 4$ **b** $y = 2x + 8$
7 a $4 + 2 \times 1 = 6$ **b** $y = -2x + 6$
  c –2 **d** $y = -2x$

15.4 Points and lines

Exercise 15D

1 a 3 **b** $\frac{1}{2}$ **c** 4 **d** –1 **e** $-\frac{1}{2}$ **f** $\frac{2}{3}$
2 a $y = 2x - 3$ **b** $y = \frac{1}{2}x + 4$ **c** $y = 4x - 2$ **d** $y = -3x + 8$
3 a (5, 3) **b** (4, 5) **c** (3, 2)
  d (3, 3) **e** (1, 3.5) **f** (–0.5, 0)
4 a student's graph **b** $y = 0.5x + 6.5$ **c** (–1,3)
  d $y = -x + 8$
5 a 5 **b** 13 **c** 10 **d** 17
6 Show that the distance from each point to (2,1) is 5.
7 AB = $\sqrt{32}$, AC = $\sqrt{80}$, BC = $\sqrt{80}$, so two of the sides are the same length.

15.5 Perpendicular lines

Exercise 15E

1 a –2 **b** $-\frac{1}{4}$ **c** $\frac{3}{2}$
2 The gradients are 5 and $-\frac{1}{5}$; $5 \times -\frac{1}{5} = -1$
3 a $-\frac{1}{3}$ **b** –2 **c** 2.5 **d** $-\frac{2}{15}$
4 a 1 **b** 2 **c** $\frac{4}{3}$ **d** –6
5 a $y = -\frac{1}{5}x$ **b** $y = -\frac{1}{5}x + 10$ **c** $y = -\frac{1}{5}x + 2$
6 $y = \frac{3}{5}x + 2$
7 $y = -\frac{1}{4}x + \frac{1}{2}$
8 $y = -5x + 11$
9 $6x + 3y = 7$ is the odd one out
10 $y = 2x + 6$

Answers to Chapter 16

16.1 Quadratic graphs

Exercise 16A

1 **a** x: –3, –2, –1, 0, 1, 2, 3
    y: 11, 6, 3, 2, 3, 6, 11
   b Student's graph

2 **a** x: –3, –2, –1, 0, 1, 2, 3, 4, 5W
    x^2: 9, 4, 1, 0, 1, 4, 9, 16, 25
    –3x: 9, 6, 3, 0, –3, –6, –9, –12, –15
    y: 18, 10, 4, 0, –2, –2, 0, 4, 10
   b 1.8   **c** (1.5, –2.25)
   d x = 1.5   **e** x = 4.2 or –1.2

3 **a** y: 7, 0, –5, –8, –9, –8, –5, 0, 7
   b The graph should give a value of about –8.75
   c The graph should give values of about 4.5 and –2.5

4 **a** y: 10, 4, 0, –2, –2, 0, 4, 10
   b (2.5, –2.25)
   c x = 2.5
   d The graph should give a value of about 6.75
   e The graph should give values of about 0.2 and 4.8

5 **a** x: –4, –3, –2, –1, 0, 1, 2
    y: 7, 2, –1, –2, –1, 2, 7
   b 0.25
   c –2.7, 0.7

6 **a** x: –4, –3, –2, –1, 0, 1, 2, 3, 4
    y: –4, 3, 8, 11, 12, 11, 8, 3, –4
   b 9.75
   c ± 3.5
   d 2.2 and –2.2

7 **a**

x	–5	–4	–3	–2	–1	0	1	2
x^2	25	16	9	4	1	0	1	4
+4x	–20	–16	–12	–8	–4	0	4	8
y	5	0	–3	–4	–3	0	5	12

   b x = –4 and 0
   c –3.8
   d 0.6, –4.6

8 **a** x: –1, 0, 1, 2, 3, 4, 5, 6, 7
    y: 10, 3, –2, –5, –6, –5, –2, 3, 10
   b x = 0.5 or 5.5
   c –5.8
   d –0.3, 6.3

9 **a** y values: –6, 0, 4, 6, 6, 4, 0, –6
   b Student's graph
   c (2.5, 6.25)
   d x = 2.5
   e x = 4.6 and 0.4

16.2 Reciprocal graphs

Exercise 16B

1 **a** y values: 10, 5, 4, 2.5, 2, 1.33, 1, 0.67, 0.5
   b 0.8   **c** 0.3   **d** –1.6

2 **a** y values: 25, 12.5, 10, 5, 2.5, 1, 0.5, 0.33, 0.25
   b Student's graph   **c** 0.5 and –10.5

3 Student's own graph

4 **a** y values 20, 10, 5, 4, 2.5, 2, 1
   b student's graph   **c** student's graph   **d** x = 6.5 or –1.5

16.3 Other graphs

Exercise 16C

1 Student's own graph.

2 **a** y values: –7.81, –4, –1.69, –0.5, –0.06, 0, 0.06, 0.5, 1.69, 4, 7.81
   b 2.3

3 **a** y values: –12.63, –5, –0.38, 2, 2.89, 3, 3.13, 4, 6.38, 11, 18.63
   b –1.4

4 **a** y values: 1, 4.63, 6, 5.88, 5, 4.13, 4, 5.38, 9
   b x = –1.8

5 **a** y values: 20, 5, 2.22, 1.25, 0.8
   b student's graph
   c student's graph, a reflection of the previous one in the y-axis.

6 **a** y values: 4.25, 1.5, –0.22, –1.46, –2.44
   b student's graph
   c About 5.85

7 **a**

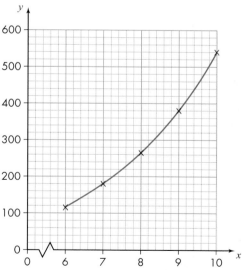

   b 8.3 cm

8 **a** $x^2y = 1000 \Rightarrow y = \dfrac{1000}{x^2}$

b

9 a

b 4.6 cm, 4.6 cm and 9.6 cm

16.4 Exponential graphs

Exercise 16D

1 **a** The missing values are 48, 24, 12 and 6

b

2 **a** 25 000 × 0.8 = 20 000
b The missing values are 16 000, 12 800, 10 240, 8192
c

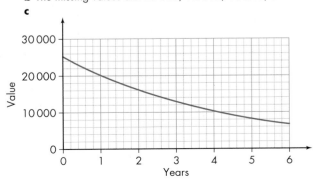

d 3.1 years

3 **a** The missing numbers, rounded to one decimal place, are 79.4, 91.3 and 104.9
b

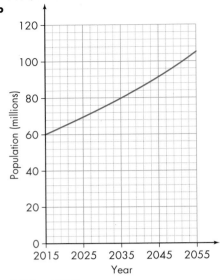

c 2051 or 2052

4 **a** the missing values, to the nearest whole number, are 650, 845, 1099, 1428
b

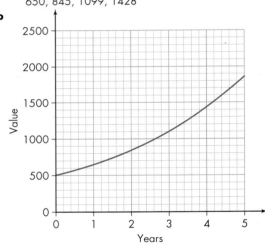

c about 2.6 or 2.7 years

5 **a**

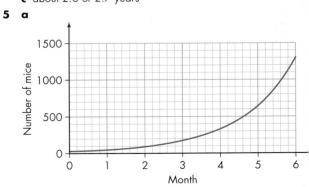

b about 3.3 months

6 a 20% **b**

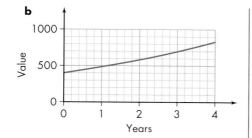

c About $630

7 a The missing values are 4500, 3645 and 3280.5

b

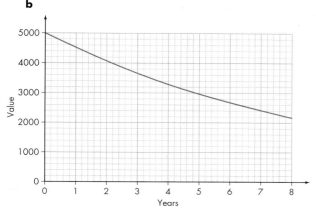

c About 6.6 years

8 It is not exponential growth because the gradient is decreasing. The gradient increases as time passes in exponential growth.

9 a 200 **b** 50% **c** about 2.7 hours

10 a The initial value is $1000. After two years it is $500. After four years it is $250. After six years it is $125. The value halves every two years.

b 29 or 30%

16.5 Estimating gradients

Exercise 16E

Gradients found in this exercise may vary from the answers given due to variations in drawings of the tangents

1 0.67

2 A: 0.5 B: –2

3 a student's drawing
b student's drawing
c about 1.8
d (1, 1.5)

4 a y values: 0, 0.01, 0.1, 0.34, 0.8, 1.56, 2.7
b student's drawing
c student's drawing
d about 1.2

5 a y values: 0.25, 0.5, 1, 2, 4
b student's drawing
c student's drawing
d about 0.7

6 a y values: 2.5, 1.67, 1.25, 1, 0.83
b student's drawing
c about –0.3

Answers to Chapter 17

17.1 Number sequences

Exercise 17A

1 a 9, 11, 13: add 2
b 10, 12, 14: add 2
c 80, 160, 320: double
d 81, 243, 729: multiply by 3
e 28, 34, 40: add 6
f 23, 28, 33: add 5
g 20 000, 200 000, 2 000 000: multiply by 10
h 19, 22, 25: add 3
i 114, 105, 96: subtract 9
j 405, 1215, 3645: multiply by 3
k 25, 12.5, 6.25; halve
l 625, 3125, 15 625: multiply by 5

2 a 16, 22 **b** 26, 37
c 31, 43 **d** 46, 64
e 121, 169 **f** 782, 3907
g 22 223, 222 223 **h** 11, 13
i 33, 65 **j** 78, 108

3 a 48, 96, 192 **b** 33, 39, 45
c 4, 2, 1 **d** 38, 35, 32
e 37, 50, 65 **f** 26, 33, 41
g 14, 16, 17 **h** 25, 22, 19

i 28, 36, 45 **j** 5, 6, 7
k 0.16, 0.032, 0.0064
l 0.0625, 0.031 25, 0.015 625

4 a 21, 34: add previous 2 terms
b 49, 64: next square number
c 47, 76: add previous 2 terms
d 216, 343: cube numbers

5 15, 21, 28, 36

6 61, 91, 127

7 29 and 41

8 No, they both increase by the same number (3).

9 10, 45 and 80

17.2 The nth term of a sequence

Exercise 17B

1 a 3, 5, 7, 9, 11 **b** 1, 4, 7, 10, 13
c 7, 12, 17, 22, 27 **d** 1, 4, 9, 16, 25
e 4, 7, 12, 19, 28 **f** 18, 16, 14, 12, 10

2 a 4, 5, 6, 7, 8 **b** 2, 5, 8, 11, 14
c 3, 8, 13, 18, 23 **d** 0, 3, 8, 15, 24
e 9, 13, 17, 21, 25 **f** 42, 39, 36, 33, 30

3 a 94, 88, 82, 76 **b** the 17th term, –2

4 a 1, 4, 9, 16   **b** 3, 6, 11, 18   **c** 2, 8, 18, 32
   d 3, 15, 35, 63   **e** 199, 196, 191, 184
   f 0.25, 1, 2.25, 4

5 a 1, 8, 27, 64   **b** 2, 9, 28, 65   **c** −1, 6, 25, 62
   d 2, 23, 80, 191   **e** 0.5, 4, 13.5, 32   **f** 108, 101, 82, 45

6 a $305   **b** $600   **c** 3
   d 5 (the amount is $250)

7 $4n - 2 = 3n + 7$ rearranges as $4n - 3n = 7 + 2$, $n = 9$

Exercise 17C

1 a 13, 15, $2n + 1$ **b** 25, 29, $4n + 1$
   c 33, 38, $5n + 3$ **d** 32, 38, $6n - 4$
   e 20, 23, $3n + 2$ **f** 37, 44, $7n - 5$
   g 17, 15; $29 - 2n$ **h** 22, 18; $46 - 4n$
   i 17, 20, $3n - 1$ **j** 42, 52, $10n - 8$
   k 24, 28, $4n + 4$ **l** 29, 34, $5n - 1$

2 a $3n + 1$, 151   **b** $2n + 5$, 105
   c $5n - 2$, 248   **d** $4n - 3$, 197
   e $8n - 6$, 394   **f** $n + 4$, 54
   g $5n + 1$, 251   **h** $8n - 5$, 395
   i $3n - 2$, 148   **j** $3n + 18$, 168
   k $7n + 5$, 355   **l** $8n - 7$, 393

3 a i $4n + 1$ **ii** 401
   b i $2n + 1$ **ii** 201
   c i $3n + 1$ **ii** 301
   d i $2n + 6$ **ii** 206
   e i $4n + 5$ **ii** 405
   f i $5n + 1$ **ii** 501
   g i $3n - 3$ **ii** 297
   h i $6n - 4$ **ii** 596
   i i $8n - 1$ **ii** 799
   j i $2n + 23$ **ii** 223

4 a $8n + 2$ **b** $8n + 1$ **c** $8n$ **d** $8
5 a n^2 **b** $n^2 + 2$ **c** $2n^2$ **d** $n^2 - 1$
6 a n^3 **b** $n^3 + 10$ **c** $0.5n^3$ **d** $10n^3$
7 a $n + 5$ **b** $n^2 + 5$ **c** $n^3 + 5$
   d $5n + 1$ **e** $5n^2$ **f** $5n^3$

17.3 General rules from patterns

Exercise 17D

1 a

   b The missing number is 13
   c $4n - 3$
   d 97
   e 50th diagram

2 a
   b The bottom line is 3, 5, 7, 9, 11
   c $2n + 1$
   d 121
   e 49th set

3 a 18
   b the bottom line is 6, 10, 14, 18
   c $4n + 2$
   d 12

4 a i 24   **ii** $5n - 1$ **iii** 224
   b 25

5 a i 20 cm   **ii** $(3n + 2)$ cm   **iii** 152 cm
   b 332

6 a i 10   **ii** $2n + 2$ **iii** 162
   b 79.8 km

7 a i 14   **ii** $3n + 2$ **iii** 41
   b 66

8 a i 5   **ii** n **iii** 18
   b Formula gives 3 and 6
   c 55

9 a

1	2	3
3	9	27
1	4	13
4	13	40

   b the numbers in column 4 (top to bottom) are: 4, 81, 40, 121. Student's explanation of method.

10 a Student's drawing – one complete recurring sequence should be added to each one.
   b bottom row: 5, 9, 13, 17
   c $4n + 1$
   d $3n + 1$
   e $2n + 1$
   f $9n + 3$

17.4 Further sequences

Exercise 17E

1 a 432   **b** 1053   **c** 1250   **d** 41.472
   e 640 000   **f** 15   **g** 1.6   **h** 32

2 a 6, 1296   **b** 16, 128   **c** 15, 405
   d 20, 2.5   **e** 54, 16

3 a 25×2^n **b** 1.5×2^n **c** 2×3^n **d** 240×0.5^n
   e 50×0.9^n **f** 64×1.25^n

4 a 13 720, 19 208   **b** 5000×1.4^n

5 a $n^2 + 1$ **b** $n^2 + 6$ **c** $n^2 - n$ **d** $3n^2$
   e $3n^2 - 2$ **f** $3n^2 + n$

6 a 15, 21   **b** $n^2 + n$ **c** $0.5(n^2 + n)$ or $0.5n^2 + 0.5n$
   d 210   **e** It is the 50th triangular number because $(50^2 + 50) \div 2 = 1275$

7 a $n^3 - 1$ **b** $n^3 + 50$ **c** $n^3 + n$ **d** $n^3 + 3n$ **e** $4n^3$ **f** $4n^3 - n$

8 a 20   **b** $(4 \times 5 \times 6) \div 6 = 20$ **c** 7 (using 84 oranges)
   d The layers of the tetrahedron are triangular numbers. 20 layers have $(20 \times 21 \times 22) \div 6 = 1540$ oranges.

9 a When $n = 1$ the first term is $a + b$ and this is 6.
   b $2a + 4b = 16$ **c** $a = 4$, $b = 2$; the nth term is $4n + 2n^2$

Answers to Chapter 18

18.1 Using indices

Exercise 18A

1 a 2^4 **b** 3^5 **c** 7^2 **d** 5^3 **e** 10^7
 f 6^4 **g** 4^1 **h** 1^7 **i** 0.5^4 **j** 100^3

2 a $3 \times 3 \times 3 \times 3$
 b $9 \times 9 \times 9$
 c 6×6
 d $10 \times 10 \times 10 \times 10 \times 10$
 e $2 \times 2 \times 2 \times 2 \times 2 \times 2 \times 2 \times 2 \times 2 \times 2$
 f 8
 g $0.1 \times 0.1 \times 0.1$
 h 2.5×2.5
 i $0.7 \times 0.7 \times 0.7$
 j 1000×1000

3 a 16 **b** 243
 c 49 **d** 125
 e 10 000 000 **f** 1296
 g 4 **h** 1
 i 0.0625 **j** 1 000 000

4 a 81 **b** 729
 c 36 **d** 100 000
 e 1024 **f** 8
 g 0.001 **h** 6.25
 i 0.343 **j** 1 000 000

5 125 m^3

6 b 10^2 **c** 2^3 **d** 5^2

7 a 1 **b** 4 **c** 1 **d** 1 **e** 1

8 Any power of 1 is equal to 1.

9 10^6

10 10^6

11 a 1 **b** -1 **c** 1 **d** 1 **e** -1

12 a 1 **b** -1 **c** -1 **d** 1 **e** 1

13 2^{24}, 4^{12}, 8^8, 16^6

18.2 Negative indices

Exercise 18B

1 a $\frac{1}{5^3}$ **b** $\frac{1}{6}$ **c** $\frac{1}{10^5}$ **d** $\frac{1}{3^2}$
 e $\frac{1}{8^2}$ **f** $\frac{1}{9}$ **g** $\frac{1}{w^2}$ **h** $\frac{1}{t}$
 i $\frac{1}{x^m}$ **j** $\frac{4}{m^3}$

2 a 3^{-2} **b** 5^{-1} **c** 10^{-3} **d** m^{-1} **e** t^{-n}

3 a i 2^4 **ii** 2^{-1} **iii** 2^{-4} **iv** -2^3
 b i 10^3 **ii** 10^{-1} **iii** 10^{-2} **iv** 10^6
 c i 5^3 **ii** 5^{-1} **iii** 5^{-2} **iv** 5^{-4}
 d i 3^2 **ii** 3^{-3} **iii** 3^{-4} **iv** -3^5

4 a $\frac{5}{x^3}$ **b** $\frac{6}{t}$ **c** $\frac{7}{m^2}$ **d** $\frac{4}{q^4}$
 e $\frac{10}{y^5}$ **f** $\frac{1}{2x^3}$ **g** $\frac{1}{2m}$ **h** $\frac{3}{4t^4}$
 i $\frac{4}{5y^3}$ **j** $\frac{7}{8x^5}$

5 a $7x^{-3}$ **b** $10p^{-1}$ **c** $5t^{-2}$ **d** $8m^{-5}$ **e** $3y^{-1}$

6 a i 25 **ii** $\frac{1}{125}$ **iii** $\frac{4}{5}$
 b i 64 **ii** $\frac{1}{16}$ **iii** $\frac{5}{256}$
 c i 8 **ii** $\frac{1}{32}$ **iii** $\frac{9}{2}$ or $4\frac{1}{2}$
 d i 1 000 000 **ii** $\frac{1}{1000}$ **iii** $\frac{1}{4}$

7 24 (32 − 8)

8 $x = 8$ and $y = 4$ (or $x = y = 1$)

9 $\frac{1}{2097152}$

10 a x^{-5}, x^0, x^5 **b** x^5, x^0, x^{-5} **c** x^5, x^{-5}, x^0

18.3 Multiplying and dividing with indices

Exercise 18C

1 a 5^4 **b** 5^3 **c** 5^2 **d** 5^3 **e** 5^{-5}

2 a 6^3 **b** 6^0 **c** 6^6 **d** 6^{-7} **e** 6^2

3 a a^3 **b** a^5 **c** a^7 **d** a^4 **e** a^2 **f** a^1

4 a Any two values such that $x + y = 10$
 b Any two values such that $x - y = 10$

5 a 4^6 **b** 4^{15} **c** 4^6
 d 4^{-6} **e** 4^6 **f** 4^0

6 a $6a^5$ **b** $9a^2$ **c** $8a^6$
 d $-6a^4$ **e** $8a^8$ **f** $-10a^{-3}$

7 a $3a$ **b** $4a^3$ **c** $3a^4$
 d $6a^{-1}$ **e** $4a^7$ **f** $5a^{-4}$

8 a $8a^5b^4$ **b** $10a^3b$ **c** $30a^{-2}b^{-2}$
 d $2ab^3$ **e** $8a^{-5}b^7$

9 a $3a^3b^2$ **b** $3a^2c^4$ **c** $8a^2b^2c^3$

10 a Possible answer: $6x^2 \times 2y^5$ and $3xy \times 4xy^4$
 b Possible answer: $24x^2y^7 \div 2y^2$ and $12x^6y^8 \div x^4y^3$

11 12 ($a = 2$, $b = 1$, $c = 3$)

12 $1 = \frac{a^x}{a^x} = a^x \div a^x = a^{x-x} = a^0$

18.4 Fractional indices

Exercise 18D

1 a 5 **b** 10 **c** 8 **d** 9 **e** 25
 f 3 **g** 4 **h** 10 **i** 5 **j** 8
 k 12 **l** 20 **m** 5 **n** 3 **o** 10
 p 3 **q** 2 **r** 2 **s** 6 **t** 6
 u $\frac{1}{4}$ **v** $\frac{1}{2}$ **w** $\frac{1}{3}$ **x** $\frac{1}{5}$ **y** $\frac{1}{10}$

2 a $\frac{5}{6}$ **b** $1\frac{2}{3}$ **c** $\frac{8}{9}$ **d** $1\frac{4}{5}$ **e** $\frac{5}{8}$
 f $\frac{3}{5}$ **g** $\frac{1}{4}$ **h** $2\frac{1}{2}$ **i** $\frac{4}{5}$ **j** $1\frac{1}{7}$

3 $\left(x^{\frac{1}{n}}\right)^n = x^{\frac{1}{n} \times n} = x^1 = x$, but $\left(\sqrt[n]{x}\right)^n = \sqrt[n]{x} \times \sqrt[n]{x} \ldots n$ times $= x$, so $x^{\frac{1}{n}} = \sqrt[n]{x}$

4 $64^{-\frac{1}{2}} = \frac{1}{8}$, others are both $\frac{1}{2}$

5 Possible answer: The negative power gives the reciprocal, so $27^{-\frac{1}{3}} = \frac{1}{27^{\frac{1}{3}}}$. The power one-third means cube root, so you need the cube root of 27 which is 3, so $27^{\frac{1}{3}} = 3$ and $\frac{1}{27^{\frac{1}{3}}} = \frac{1}{3}$

6 Possible answer: $x = 1$ and $y = 1$, $x = 8$ and $y = \frac{1}{64}$.

7 **a** 3   **b** $\frac{1}{3}$ **c** 0   **d** $\frac{1}{2}$

   e $\frac{1}{2}$ **f** $\frac{1}{4}$ **g** $\frac{1}{4}$ **h** $\frac{1}{3}$

   i $\frac{1}{3}$ **j** $\frac{1}{2}$ **k** $\frac{1}{3}$ **l** $\frac{1}{7}$

Exercise 18E

1 **a** 16   **b** 25   **c** 216   **d** 81

2 **a** $t^{\frac{2}{3}}$ **b** $m^{\frac{3}{4}}$ **c** $k^{\frac{2}{5}}$ **d** $x^{\frac{3}{2}}$

3 **a** 4   **b** 9   **c** 64   **d** 3125

4 **a** $\frac{1}{5}$ **b** $\frac{1}{6}$ **c** $\frac{1}{2}$ **d** $\frac{1}{3}$

   e $\frac{1}{4}$ **f** $\frac{1}{2}$ **g** $\frac{1}{2}$ **h** $\frac{1}{3}$

5 **a** $\frac{1}{125}$ **b** $\frac{1}{216}$ **c** $\frac{1}{8}$ **d** $\frac{1}{27}$

   e $\frac{1}{256}$ **f** $\frac{1}{4}$ **g** $\frac{1}{4}$ **h** $\frac{1}{9}$

6 **a** $\frac{1}{100000}$ **b** $\frac{1}{12}$ **c** $\frac{1}{25}$ **d** $\frac{1}{27}$

   e $\frac{1}{32}$ **f** $\frac{1}{32}$ **g** $\frac{1}{81}$ **h** $\frac{1}{13}$

7 $8^{-\frac{2}{3}} = \frac{1}{4}$, others are both $\frac{1}{8}$

8 Possible answer: The negative power gives the reciprocal, so the power one-third means cube root, so we need the cube root of 27 which is 3 and the power 2 means square, so $3^2 = 9$, so $27^{\frac{2}{3}} = 9$ and $\frac{1}{27^{\frac{2}{3}}} = \frac{1}{9}$

9 **a** $\frac{27}{8}$ **b** $\frac{9}{25}$ **c** $\frac{1024}{243}$ **d** $\frac{8}{343}$

   e $\frac{16}{9}$ **f** $\frac{8}{27}$ **g** $\frac{625}{256}$ **h** $\frac{32}{243}$

10 **a** $\frac{25}{9}$ **b** $\frac{27}{64}$ **c** $\frac{125}{729}$ **d** $\frac{243}{32}$

   e $\frac{8}{27}$ **f** $\frac{243}{32}$ **g** $\frac{9}{4}$ **h** $\frac{125}{343}$

   i $\frac{16}{25}$ **j** $\frac{512}{125}$ **k** $\frac{243}{32}$ **l** $\frac{32}{243}$

11 **a** x^4 **b** x^{-1} **c** $4y^2$

   d $10x^2$ **e** $20x^{-1}$ **f** $\frac{1}{3}y$

12 **a** x **b** d^{-1} **c** $t^{\frac{3}{2}}$

   d x^2 **e** $y^{\frac{1}{2}}$ **f** a^4

13 **a** $x^{\frac{1}{2}}$ **b** y^{-1} **c** $a^{\frac{5}{3}}$

   d t^{-2} **e** d^2 **f** 1

Answers to Chapter 19

19.1   Direct variation

Exercise 19A

1 **a** 15   **b** 2
2 **a** 75   **b** 6
3 **a** 150   **b** 6
4 **a** 22.5   **b** 12
5 **a** 175 kilometres   **b** 8 hours
6 **a** 66.50 dollars   **b** 175 kg
7 **a** 44   **b** 84 m^2
8 **a** 33 spaces
   b 66 spaces since new car park has 366 spaces
9 17 minutes 30 seconds

Exercise 19B

1 **a** 100   **b** 10
2 **a** 27   **b** 5
3 **a** 56   **b** 1.69
4 **a** 192   **b** 2.25
5 **a** 25.6   **b** 5
6 **a** 80   **b** 8
7 **a** $50   **b** 225
8 **a** 3.2°C   **b** 10 atm
9 **a** 388.8 g   **b** 3 mm
10 **a** 2 J   **b** 40 m/s

11 **a** 78 dollars   **b** 400 miles
12 4000 cm^3
13 $250
14 **a** B   **b** A   **c** C
15 **a** B   **b** A

19.2   Inverse variation

Exercise 19C

1 $Tm = 12$ **a** 3   **b** 2.5
2 $Wx = 60$ **a** 20   **b** 6
3 $Q(5 - t) = 16$ **a** −3.2   **b** 4
4 $Mt^2 = 36$ **a** 4   **b** 5
5 $W\sqrt{T} = 24$ **a** 4.8   **b** 100
6 $x^3y = 32$ **a** 32   **b** 4
7 $gp = 1800$ **a** $15   **b** 36
8 $td = 24$ **a** 3°C   **b** 12 km
9 $ds^2 = 432$ **a** 1.92 km   **b** 8 m/s
10 $p\sqrt{h} = 7.2$ **a** 2.4 atm   **b** 100 m
11 $W\sqrt{F} = 0.5$ **a** 5 t/h   **b** 0.58 t/h
12 B – This is inverse proportion, as x increases y decreases.

13

x	8	27	64
y	1	$\frac{2}{3}$	$\frac{1}{2}$

Answers to Chapter 20

20.1 Graphical inequalities

Exercise 20A

1 a & b

2 a & b

3 a–c

4 a–c

5

b i Yes   **ii** Yes   **iii** No

6 a & b

7 a & b

8 a & b

9

10 a–d

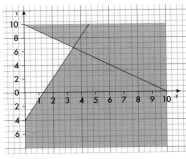

e i No   **ii** Yes   **iii** Yes

20.2 More than one inequality

Exercise 20B

1 a–f

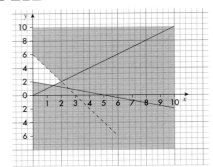

g i No   **ii** No   **iii** Yes

2 a

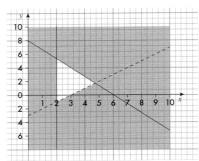

b i No   **ii** Yes   **iii** Yes   **iv** No

3 a & b

4 a & b

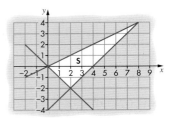

c 4   **d** –2   **e** 12

5 Test a point such as the origin (0, 0), so 0 < 0 + 2, which is true. So the side that includes the origin is the required side.

6 a $x + y \geq 3$, $y \leq \frac{1}{2}x + 3$ and $y \geq 5x - 15$

b 9

c 3 at (3,0)

20.3 Linear programming

Exercise 20C

1 a He buys at least 3 cartons of milk

b $y \geq 2$

c $x + y \leq 8$

d

2 a $x + y \leq 10$

b $2x + 5y \geq 30$

c

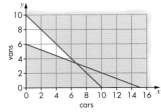

d 6 because cars & vans (x, y) must both be integers

e 4

3 a $x > y$ $x < 2y$ $x + y < 12$

b

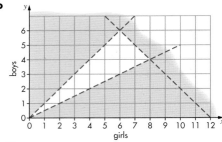

c 7

4 a $x + y \geq 6$ $x + y \leq 12$ $y > x$ $x \geq 2$

b

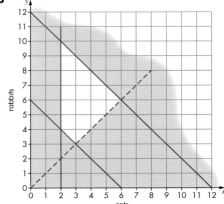

c 4
d 5

5 a Follows from $15x + 10y \geqslant 600$
b $x + y \leqslant 50$
c

d 30

6 a i $x + y \leqslant 15$ **ii** $x + y \geqslant 5$ **iii** $y \geqslant \frac{1}{2}x$
b the number of small coaches is not more than 2 more than the number of large coaches
c

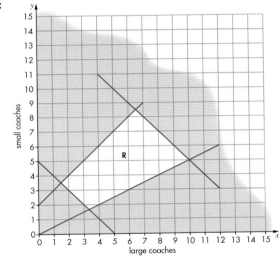

d 3, 4, 5, 6, 7 or 8

7 a $3x + y$
b $3x + y \leqslant 18$ $x > y$ $y \geqslant 2$
c

d 3 wins and 2 draws; 4 wins and 2 draws; 5 wins and 2 draws; 4 wins and 3 draws; 5 wins and 3 draws

Answers to Chapter 21

21.1 Function notation

Exercise 21A

1 a 12   **b** 26   **c** 7
  d −2   **e** 3

2 a 0.5   **b** 5   **c** 50.5
  d 2.5   **e** 0.625 or $\frac{5}{8}$

3 a 5   **b** −3   **c** 999801
  d 1   **e** $\frac{1}{8}$

4 a 4   **b** 32   **c** 1
  d $\frac{1}{2}$ **e** $\frac{1}{8}$

5 a 3   **b** 2   **c** 0
  d −1   **e** 5

6 a 7.5   **b** −2.5   **c** −5

7 a 6   **b** 97   **c** 3.25

8 a 6   **b** at (6, 4)
9 a 3   **b** $-\frac{1}{2}$ **c** $x = 2$
  d $x = 0$

21.2 Inverse functions

Exercise 21B

1 a $x - 7$ **b** $\frac{x}{8}$ **c** $5x$ **d** $x + 3$

2 a 8   **b** 4   **c** 5   **d** −2

3 a $3(x + 2)$ **b** $\frac{x}{4} + 5$ **c** $5x - 4$ **d** $\frac{5x + 6}{3}$
  e $2(\frac{x}{3} - 4)$ **f** $\sqrt[3]{\frac{x}{4}}$

4 a 2   **b** $\frac{1}{2}$ **c** −2.5

5 a $10 - x$ **b** They are identical

6 **a** $\frac{8}{x}$ **b** $\frac{20}{x+1}$ **c** $\frac{2}{x} - 1$

7 **a** $\frac{x+4}{2}$ **b** Student's graph **c** (4,4)

8 5

9 $\frac{x+2}{3}$

21.3 Composite functions

Exercise 21C

1 **a** 6 and 3 **b** 7 and 3.5
  c 10 and 5 **d** $\frac{x+4}{2}$
  e 1 and 5 **f** 1.5 and 5.5
  g −5 and −1 **h** $\frac{x}{2} + 4$

2 **a** 6 and 216 **b** 10 and 1000 **c** $(2x)^3$ or $8x^3$
  d 64 and 128 **e** $2x^3$

3 **a** 1, 9, 25 **b** 1, 3, 5 **c** $\sqrt{2x+1}$

4 **a** 6 and 18 **b** 12 and 36 **c** $9x$

5 **a** $3(x-6)$ **b** $3x - 6$

6 Both are $x - 3$

21.4 More about composite functions

Exercise 21D

1 **a** 3.5 **b** 1 **c** 8 **d** 5.5

2 **a** 20 **b** 9 **c** 8.75 **d** 3

3 **a** 7 **b** 8 **c** 256 **d** 21

4 **a** $6x$ **b** $6x - 5$

5 **a** $9x^2 + 24x + 16$ or $(3x+4)^2$ **b** $6x - 5$
  c $2x + 3$ **d** $4 - 2x$

6 **a** $x - 10$ **b** $x + 10$ **c** x

7 **a** x^4 **b** $\left(\frac{12}{x}\right)^2$ or $\frac{144}{x^2}$
  c $\frac{12}{x^2}$ **d** x

8 **a** 80
  b $2(2x - 1) - 1$ simplified
  c $(2x - 1)^2 + 2(2x - 1)$ simplified

9 **a** $\frac{1}{3x - 3}$ **b** $\frac{x - 1}{x - 4}$

10 **a** $6x - 14$ **b** $\frac{x + 12}{4}$

11 **a** $0.5 (1 + 9) = 5$ **b** 7 **c** 8
  d 8.5, 8.75, 8.875, 8.9375, 8.96875, 8.984375
  e Getting closer and closer to 9, halving the difference from 9 each time.

12 Student's own description of the convergence towards 9.

Answers to Chapter 22

22.1 Angle facts

Exercise 22A

1 **a** 48° **b** 307° **c** 108° **d** 52°
  e 59° **f** 81° **g** 139° **h** 58°

2 **a** 82° **b** 105° **c** 75°

3 45° + 125° = 170° and for a straight line it should be 180°.

4 **a** $x = 100°$ **b** $x = 110°$ **c** $x = 30°$

5 **a** $x = 55°$ **b** $x = 45°$ **c** $x = 12.5°$

6 **a** $x = 34°, y = 98°$ **b** $x = 70°, y = 120°$ **c** $x = 20°, y = 80°$

7 $6 \times 60° = 360°$; imagine six of the triangles meeting at a point.

8 $x = 35°, y = 75°$; $2x = 70°$ (opposite angles), so $x = 35°$ and $x + y = 110°$ (angles on a line), so $y = 75°$

22.2 Parallel lines

Exercise 22B

1 **a** 40° **b** $b = c = 70°$
  c $d = 75, e = f = 105°$ **d** $g = 50°, h = i = 130°$
  e $j = k = l = 70°$ **f** $n = m = 80°$

2 **a** $a = 50°, b = 130°$ **b** $c = d = 65°, e = f = 115°$
  c $g = i = 65°, h = 115°$ **d** $j = k = 72°, l = 108°$
  e $m = n = o = p = 105°$ **f** $q = r = s = 125°$

3 **a** $a = 95°$ **b** $b = 66°, c = 114°$

4 **a** $x = 30°, y = 120°$ **b** $x = 25°, y = 105°$
  c $x = 30°, y = 100°$

5 **a** $x = 50°, y = 110°$ **b** $x = 25°, y = 55°$
  c $x = 20, y = 140°$

6 290°; x is double the angle allied to 35°, so is $2 \times 145°$

7 $a = 66$

8 Angle PQD = 64° (alternate angles), so angle DQY = 116° (angles on a line = 180°)

9 Use alternate angles to see b, a and c are all angles on a straight line, and so total 180°.

10 Third angle in triangle equals q (alternate angle), angle sum of triangle is 180°.

22.3   Angles in a triangle

Exercise 22C

1 **a** 70°   **b** 50°   **c** 80°   **d** 60°
   e 75°   **f** 109°   **g** 38°   **h** 63°

2 **a** No, total is 190° **b** Yes, total is 180° **c** No, total is 170°
   d Yes, total is 180° **e** Yes, total is 180° **f** No, total is 170°

3 **a** 60°   **b** Equilateral triangle **c** Same length

4 **a** 70° each   **b** Isosceles triangle **c** Same length

5 **a** 109°   **b** 130°   **c** 135°

6 Isosceles triangle; angle DFE $\angle 30°$ (opposite angles), angle
   DEF $\angle 75°$ (angles on a line), angle FDE $\angle 75°$ (angles in a
   triangle), so there are two equal angles in the triangle and
   hence it is an isosceles triangle

7 a is $\angle 80°$ (opposite angles), b is 65° (angles on a line), c is
   35° (angles in a triangle)

8 Missing angle = y, $x + y = 180°$ and $a + b + y = 180°$ so
   $x = a + b$

22.4   Angles in a quadrilateral

Exercise 22D

1 **a** $a = 110°$, $b = 55°$ **b** $c = e = 105°$, $d = 75°$
   c $f = 135°$, $g = 25°$ **d** $h = i = 94°$
   e $j = l = 105°$, $k = 75°$ **f** $m = o = 49°$, $n = 131°$

2 **a** $x = 25°$, $y = 15°$ **b** $x = 7°$, $y = 31°$ **c** $x = 60°$, $y = 30°$

3 **a** $x = 50°$: 60°, 70°, 120°, 110° – possibly trapezium
   b $x = 60°$: 50°, 130°, 50°, 130° – parallelogram or
                                          isosceles trapezium
   c $x = 30°$: 20°, 60°, 140°, 140° – possibly kite
   d $x = 20°$: 90°, 90°, 90°, 90° – square or rectangle

4 52°

5 Both 129°

6 $y = 360° – 4x$

22.5   Regular polygons

Exercise 22E

1 **a i** 45°   **ii** 8   **iii** 1080°
   b i 20°   **ii** 18   **iii** 2880°
   c i 15°   **ii** 24   **iii** 3960°
   d i 36°   **ii** 10   **iii** 1440°

2 **a i** 172°   **ii** 45   **iii** 7740°
   b i 174°   **ii** 60   **iii** 10 440°
   c i 156°   **ii** 15   **iii** 2340°
   d i 177°   **ii** 120   **iii** 21 240°

3 **a** Exterior angle is 7°, which does not divide exactly into
        360°
   b Exterior angle is 19°, which does not divide exactly into
        360°
   c Exterior angle is 11°, which does divide exactly into 360°
   d Exterior angle is 70°, which does not divide exactly into
        360°

4 **a** 7° does not divide exactly into 360°
   b 26° does not divide exactly into 360°
   c 44° does not divide exactly into 360°
   d 13° does not divide exactly into 360°

5 $x = 45°$, they are the same, true for all polygons

6 **a** 36°   **b** 10

22.6   Irregular polygons

Exercise 22F

1 **a** 1440°   **b** 2340°   **c** 17 640°   **d** 7740°

2 **a** 9   **b** 15   **c** 102   **d** 50

3 **a** 130°   **b** 95°   **c** 130°

4 **a** 50°   **b** 40°   **c** 59°

5 Hexagon

6 **a** Octagon   **b** 89°

7 **a i** 71°   **ii** 109°   **iii** Equal
   b If S = sum of the two opposite interior angles, then $S + I = 180°$
        (angles in a triangle), and we know $E + I = 180°$ (angles on a
        straight line), so $S + I = E + I$, therefore $S = E$

8 144°; $360 – (2 \times 108)°$

9 Three angles are 135° and two angles are 67.5°

22.7   Tangents and diameters

Exercise 22G

1 **a** 38°   **b** 110°   **c** 15°   **d** 45°

2 **a** $x = 12°$, $y = 156°$
   b $x = 100°$, $y = 50°$
   c $x = 62°$, $y = 28°$
   d $x = 30°$, $y = 60°$

3 $\angle OCD = 58°$ (triangle OCD is isosceles), $\angle OCB = 90°$
   (tangent/radius theorem), so $\angle DCB = 32°$, hence triangle
   BCD is isosceles (2 equal angles)

22.8   Angles in a circle

Exercise 22H

1 **a** 56°   **b** 62°   **c** 105°   **d** 55°
   e 45°   **f** 30°   **g** 60°   **h** 145°

2 **a** 55°   **b** 52°   **c** 50°   **d** 24°
   e 39°   **f** 80°   **g** 34°   **h** 30°

3 **a** 41°   **b** 49°   **c** 41°

4 **a** 72°   **b** 37°   **c** 72°

5 $\angle AZY = 35°$ (angles in a triangle), $a = 55°$ (angle in a
   semicircle = 90°)

6 **a** $x = y = 40°$ **b** $x = 131°$, $y = 111°$
   c $x = 134°$, $y = 23°$ **d** $x = 32°$, $y = 19°$
   e $x = 59°$, $y = 121°$ **f** $x = 155°$, $y = 12.5°$

7 68°

8 $\angle ABC = 180° – x$ (angles on a line), $\angle AOC = 360° – 2x$
   (angle at centre is twice angle at circumference), reflex
   $\angle AOC = 360° – (360° – 2x) = 2x$ (angles at a point)

9 **a** x
   b $2x$
   c From part **b**, angle AOD = $2x$
        Similarly, angle COD = $2y$
        So angle AOC = AOD + COD = $2x + 2y = 2(x + y)$
        = 2 × angle ABC

22.9   Cyclic quadrilaterals

Exercise 22I

1 **a** $a = 50°$, $b = 95°$ **b** $c = 92°$, $x = 90°$
c $d = 110°$, $e = 110°$, $f = 70°$ **d** $g = 105°$, $h = 99°$
e $j = 89°$, $k = 89°$, $l = 91°$ **f** $m = 120°$, $n = 40°$
g $p = 44°$, $q = 68°$ **h** $x = 40°$, $y = 34°$

2 **a** $x = 26°$, $y = 128°$ **b** $x = 48°$, $y = 78°$
c $x = 133°$, $y = 47°$ **d** $x = 36°$, $y = 72°$
e $x = 55°$, $y = 125°$ **f** $x = 35°$
g $x = 48°$, $y = 45°$ **h** $x = 66°$, $y = 52°$

3 **a** $x = 49°$, $y = 49°$ **b** $x = 70°$, $y = 20°$
c $x = 80°$, $y = 100°$ **d** $x = 100°$, $y = 75°$

4 **a** $x = 50°$, $y = 62°$ **b** $x = 92°$, $y = 88°$
c $x = 93°$, $y = 42°$ **d** $x = 55°$, $y = 75°$

5 **a** $x = 95°$, $y = 138°$ **b** $x = 14°$, $y = 62°$
c $x = 32°$, $y = 48°$ **d** $52°$

6 **a** $71°$ **b** $125.5°$ **c** $54.5°$

7 **a** $x + 2x - 30° = 180°$ (opposite angles in a cyclic quadrilateral), so $3x - 30° = 180°$
b $x = 70°$, so $2x - 30° = 110°$ ∠DOB = 140° (angle at centre equals twice angle at circumference), $y = 80°$ (angles in a quadrilateral)

8 **a** x **b** $360° - 2x$
c ∠ADC = $\frac{1}{2}$reflex ∠AOC = $180° - x$, so ∠ADC + ∠ABC = 180°

9 Let ∠AED = x, then ∠ABC = x (opposite angles are equal in a parallelogram), ∠ADC = $180° - x$ (opposite angles in a cyclic quadrilateral), so ∠ADE = x (angles on a line)

10 Let ∠ABC = x and ∠EFG = y.
Then ∠ADC = $180 - x°$ (opposite angles in a cyclic quadrilateral) and ∠EDG = $180 - y°$.
But ∠ADC = ∠EDG (opposite angles).
$180 - x° = 180 - y°$ and therefore $x = y$.

Answers to Chapter 23

23.1   Measuring and drawing angles

Exercise 23A

1 **a** $40°$ **b** $125°$ **c** $340°$ **d** $225°$
2 Student's drawings of angles.
3 AC and BE; AD and CE; AE and CF.
4 Yes, the angle is $75°$.
5 Any angle between $90°$ and $100°$.
6 **a** $80°$ **b** $50°$ **c** $25°$

23.2   Bearings

Exercise 23B

1 **a** $110°$ **b** $250°$ **c** $091°$ **d** $270°$ **e** $130°$ **f** $180°$
2 Student's sketches.
3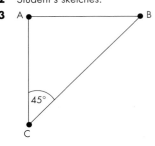

4 **a** $090°, 180°, 270°$ **b** $000°, 270°, 180°$
5

Leg	Actual distance	Bearing
1	50 km	060°
2	70 km	355°
3	65 km	260°
4	46 km	204°
5	60 km	130°

6 **a** $045°$ **b** $286°$
7 **a** $250°$ **b** $325°$ **c** $144°$
8 **a** 900 m **b** $280°$
c ∠NHS = 150° and HS = 3 cm
9 $108°$
10 $255°$

23.3   Nets

Exercise 23C

1 **a** **b** **c**

2 Yes.

3

4 a **b**

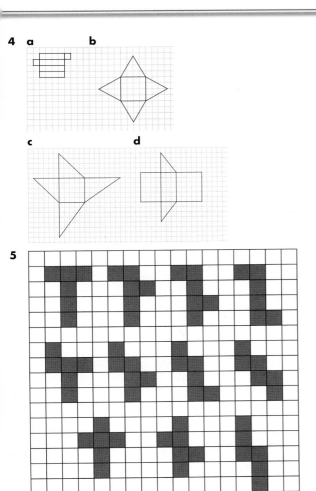

5

6 a and b

23.4 Congruent shapes

Exercise 23D

1 **a** yes **b** yes **c** no **d** yes **e** no **f** yes
2 **a** triangle ii **b** triangle iii **c** sector i
3

PQR to QRS to RSP to SPQ;
SXP to PXQ to QXR to RXS

4

EGF to FHE to GEH to HFG;
EFX to HGX; EXH to FXG

5

ABC to CDA; BDC to DBA;
BXA to DXC; BXC to DXA

6

AXB to AXC

23.5 Similar shapes

Exercise 23E

1 **a** 2 **b** 3
2 **a** Yes, 4
  b No, corresponding sides have different ratios.
3 **a** PQR is an enlargement of ABC
  b 1 : 3 **c** Angle R **d** BA
4 **a** Sides in same ratio
  b Angle P **c** PR
5 **a** Same angles **b** Angle Q **c** AR
6 **a** 8 cm **b** x = 45cm, y = 9cm
  c x = 19.5 cm, y = 10.8 cm
  d 4.2 cm
7 **a** The angles are all 90 degrees. The sides of a square
    are all equal so the ratio between sides of two different
    squares will be the same, whatever two sides are chosen.
  b No. They will only be similar if they have the same ratio of
    length to width.
8 5.2 m

23.6 Areas of similar triangles

Exercise 23F

1 **a** 2.5 **b** 125 cm^2
2 **a** All equilateral triangles are similar **b** 3.8 cm^2 (to 2 sf)
3 40.32 cm^2
4 75 cm^2
5 **a** 144 cm^2 **b** 69.4 cm^2
6 **a** All angles are the same
  b 247.7 cm^2
7 **a** 2 **b** 10 cm **c** 7.1 cm
8 354.9 cm^2
9 It will double the area
10 28.3 cm^2

23.7 Areas and volumes of similar shapes

Exercise 23G

1 **a** 4 : 25 **b** 8 : 125
2 **a** 16 : 49 **b** 64 : 343

3

Linear scale factor	Linear ratio	Linear fraction	Area scale factor	Volume scale factor
2	1 : 2	$\frac{2}{1}$	4	8
3	1 : 3	$\frac{3}{1}$	9	27
$\frac{1}{4}$	4 : 1	$\frac{1}{4}$	$\frac{1}{16}$	$\frac{1}{64}$
5	1 : 5	$\frac{5}{1}$	25	125
$\frac{1}{10}$	10 : 1	$\frac{1}{10}$	$\frac{1}{100}$	$\frac{1}{1000}$

4 135 cm^2

5 a 56 cm^2 **b** 126 cm^2

6 a 48 m^2 **b** 3 m^2

7 a 2400 cm^3 **b** 8100 cm^3

8 4 litres

9 1.38 m^3

10 $6

11 4 cm

12 8 × 0.60 = $4.80 which is greater than $4.00 so the large tub is better value

13 a 3 : 4 **b** 9 : 16 **c** 27 : 64

14 720 ÷ 8 = 90 cm^3

Exercise 23H

1 6.2 cm, 10.1 cm

2 4.26 cm, 6.74 cm

3 9.56 cm

4 3.38 m

5 8.4 cm

6 26.5 cm

7 16.9 cm

8 a 4.33 cm, 7.81 cm **b** 143 g, 839 g

9 53.8 kg

10 1.73 kg

11 8.8 cm

12 7.9 cm and 12.6 cm

13 b

Answers to Chapter 24

24.1 Constructing shapes

Exercise 24A

1 a BC = 2.9 cm, ∠B = 53°, ∠C = 92°
  b EF = 7.4 cm, ED = 6.8 cm, ∠E = 50°
  c ∠G = 105°, ∠H = 29°, ∠I = 46°
  d ∠J = 48°, ∠L = 32°, JK = 4.3 cm
  e ∠N = 55°, ON = OM = 7 cm
  f ∠P = 51°, ∠R = 39°, QP = 5.7 cm

2 a Students can check one another's triangles.
  b ∠ABC = 44°, ∠BCA = 79°, ∠CAB = 57°

3 5.9 cm

4 Student drawing.

5 Student drawing.

6 Student drawing.

7 BC can be 2.6 cm or 7.8 cm.

8 4.3 cm

9 4.3 cm

10 a Right-angled triangle constructed with sides 3, 4, 5 and 4.5, 6, 7.5, and scale marked 1 cm : 1 m
  b Right-angled triangle constructed with 12 equally spaced dots.

11 An equilateral triangle of side 4 cm.

12 Even with all three angles, you need to know at least one length.

24.2 Bisectors

Exercise 24B

1–9 Practical work; check students' constructions

10 The centre of the circle

24.3 Scale drawings

Exercise 24C

1 a pond: 40 m × 10 m, fruit: 50 m × 10 m, trees: 20 m × 20 m, lawn: 30 m × 20 m, vegetables: 50 m × 20 m
  b pond: 400 m^2, fruit: 500 m^2, trees: 400 m^2, lawn: 600 m^2, vegetables: 1000 m^2

2 a 33 cm **b** 9 cm

3 a 30 cm × 30 cm **b** 40 cm × 10 cm **c** 20 cm × 15 cm
  d 30 cm × 20 cm **e** 30 cm × 20 cm **f** 10 cm × 5 cm

4 a Student's scale drawing. **b** 39 plants

5 a 8.4 km **b** 4.6 km **c** 6.2 km
  d 6.4 km **e** 7.6 km **f** 2.4 km

6 a student's drawing **b** 12.9 metres

7 a 900 km **b** 1100 km **c** 860 km

8 c – 7 cm represents 210 m, so 1 cm represents 30 m

24.4 Drawing a locus

Exercise 24D

1 Circle with radius: **a** 2 cm **b** 4 cm **c** 5 cm

2 a ... **b** ...

c ... **d** ...

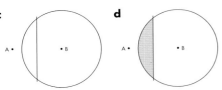

3 a Circle with radius 4 cm

b

4 a ... **b** ... **c** ...

d ... **e** ... **f** ...

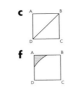

5 Construct the bisector of angle BAC and the perpendicular bisector of the line AC.

6

7

8

A and B are the two points

9

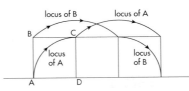

Answers to Chapter 25

Exercise 25A

1 10.3 cm
2 5.9 cm
3 8.5 cm
4 20.6 cm
5 18.6 cm
6 17.5 cm
7 5 cm
8 13 cm
9 10 cm
10 The smaller square in the first diagram and the two smaller squares in the second have the same area.

Exercise 25B

1 a 15 cm **b** 14.7 cm **c** 6.3 cm **d** 18.3 cm
2 a 20.8 m **b** 15.5 cm **c** 15.5 m **d** 12.4 cm
3 a 5 m **b** 6 m **c** 3 m **d** 50 cm
4 There are infinite possibilities, e.g. any multiple of 3, 4, 5 such as 6, 8, 10; 9, 12, 15; 12, 16, 20; multiples of 5, 12, 13 and of 8, 15, 17.
5 42.6 cm

Exercise 25C

1 a 0.682 **b** 0.829 **c** 0.922 **d** 1

2 a 0.731 **b** 0.559 **c** 0.388 **d** 0
3 a i 0.574 **ii** 0.574
b i 0.208 **ii** 0.208
c i 0.391 **ii** 0.391
d They are the same
e i sin 15° is the same as cos 75°
ii cos 82° is the same as sin 8°
iii sin x is the same as cos $(90° - x)$
4 a 0.933 **b** 1.48 **c** 2.38 **d** Infinite
e 1 **f** 0.364 **g** 0.404 **h** 0
5 Tan has values > 1
6 a 3.56 **b** 8.96 **c** 28.4 **d** 8.91
7 a 5.61 **b** 7.08 **c** 1.46 **d** 7.77
8 a $\frac{4}{5}, \frac{3}{5}, \frac{4}{3}$

b $\frac{5}{13}, \frac{12}{13}, \frac{5}{12}$

c $\frac{7}{25}, \frac{24}{25}, \frac{7}{24}$

Exercise 25D

1 a 30° **b** 51.7° **c** 39.8°
d 61.3° **e** 87.4° **f** 45.0°
2 a 60° **b** 50.2° **c** 2.6°
d 45.0 **e** 78.5° **f** 45.6°

3 a 31.0°   **b** 20.8°   **c** 41.8°
   d 46.4°   **e** 69.5°   **f** 77.1°
4 Error message, largest value 1, smallest value −1
5 a i   17.5°   **ii** 72.5°   **iii** 90°
   b Yes

Exercise 25E
1 a 17.5°   **b** 22.0°   **c** 32.2°
2 a 5.29 cm   **b** 5.75 cm   **c** 13.2 cm
3 a 4.57 cm   **b** 6.86 cm   **c** 100 cm
4 a 5.12 cm   **b** 9.77 cm   **c** 11.7 cm   **d** 15.5 cm

Exercise 25F
1 a 51.3°   **b** 75.5°   **c** 51.3°
2 a 5.35 cm   **b** 14.8 cm   **c** 12.0 cm   **d** 8.62 cm
3 a 5.59 cm   **b** 46.6°   **c** 9.91 cm   **d** 40.1°

Exercise 25G
1 a 33.7°   **b** 36.9°   **c** 52.1°
2 a 9.02 cm   **b** 7.51 cm   **c** 7.14 cm   **d** 8.90 cm
3 a 13.7 cm   **b** 48.4°   **c** 7.03 cm   **d** 41.2°

Exercise 25H
1 a 12.6   **b** 59.6   **c** 74.7   **d** 16.0
   e 67.9   **f** 20.1
2 a 44.4°   **b** 39.8°   **c** 44.4°   **d** 49.5°
   e 58.7°   **f** 38.7°
3 a 67.4°   **b** 11.3   **c** 134   **d** 28.1°
   e 39.7   **f** 263   **g** 50.2°   **h** 51.3°
   i 138   **j** 22.8
4 a Sides of right-hand triangle are sine θ and cosine θ
   b Pythagoras' theorem
   c Students should check the formulae

Exercise 25I
1   10.1 km
2   22°
3   429 m
4 a 156 m
   b No. the new angle of depression is $\tan^{-1}\left(\frac{200}{312}\right) = 33°$ and half of 52° is 26°
5 a 222 m   **b** 42°
6 a 21.5 m   **b** 17.8 m
7   13.4 m
8   19°
9   The angle is 16° so Cara is not quite correct.

Exercise 25J
1   25.1°
2 a 25cm   **b** 58.6°   **c** 20.5 cm
3 a 3.46 m   **b** 75.5°   **c** 73.2°
4 a 24.0°   **b** 48.0°   **c** 13.5 cm   **d** 16.6°
5   It is 44.6°; use triangle XDM where M is the midpoint of BD; triangle DXB is isosceles, as X is over the point where the diagonals of the base cross; the length of DB is $\sqrt{656}$. and the cosine of the required angle is $0.5\sqrt{656} \div 18$

Exercise 25K
1 a The bottom row of the table is 0.174, 0.5, 0.766, 0.996, 1, 0.996, 0.766, 0.5, 0.174.
   b

   c It has reflection symmetry. The line of symmetry is $x = 90$.
   d You could choose 10° and 170°, 30° and 150°, 50° and 130° or 85° and 95°
2   30° and 150°.
3   46° and 134°.
4   122.9°
5 a The bottom row of the table is 0.966, 0.819, 0.5, 0.174, 0, −0.174, −0.5, −0.819, −0.966.
   b

   c It has rotational symmetry of order 2 about the point (90, 0)
6 a 31.8°   **b** 148.2°   **c** 120°
   d 90°   **e** 82.8°   **f** 97.2°
7 a 53°   **b** 104°   **c** 49°, 131°   **d** 90°
   e 90°   **f** 72°, 108°   **g** no solution   **h** 45°

Exercise 25L
1 a 3.64 m   **b** 8.05 cm   **c** 19.4 cm
2 a 46.6°   **b** 112.0°   **c** 36.2°
3 a i 30°   **ii** 40°
   b 19.4 m
4   36.5 m
5   22.2 m
6   3.47 m
7   64.6 km
8   134°

Exercise 25M
1 a 7.71 m   **b** 29.1 cm   **c** 27.4 cm
2 a 76.2°   **b** 125.1°   **c** 90°
   d Right-angled triangle

3 5.16 cm

4 65.5 cm

5 **a** 10.7 cm **b** 41.7° **c** 38.3° **d** 6.69 cm

6 58.4 km at 092.5°

7 21.8°

8 42.5 km

9 111°; the largest angle is opposite the longest side

Exercise 25N

1 **a** 8.60 m **b** 90° **c** 27.2 cm
     d 26.9° **e** 27.5° **f** 62.4 cm
     g 90.0° **h** 866 cm **i** 86.6 cm

2 7 cm

3 11.1 km

4 **a** A = 90°; this is Pythagoras' theorem
     b A is acute
     c A is obtuse

5 142 m

Exercise 25O

1 **a** 24.0 cm^2 **b** 26.7 cm^2 **c** 243 cm^2
     d 21 097 cm^2 **e** 1224 cm^2

2 4.26 cm

3 **a** 42.3° **b** 49.6°

4 2033 cm^2

5 21.0 cm^2

6 726 cm^2

7 149 km^2

8 **a** 66.4 m **b** 118.9° **c** 1470 m^2

Answers to Chapter 26

26.1 Perimeter and area of a rectangle

Exercise 26A

1 **a** 35 cm^2, 24 cm **b** 33 cm^2, 28 cm
     c 45 cm^2, 36 cm **d** 70 cm^2, 34 cm
     e 56 cm^2, 30 cm **f** 10 cm^2, 14 cm

2 **a** 53.3 cm^2, 29.4 cm **b** 84.96 cm^2, 38 cm

3 39

4 **a** 4 **b** 1 h 52 min

5 40 cm

6 Area B, 44 cm^2; perimeter B, 30 cm

7 Never (the area becomes four times greater).

8 **a** 28 cm, 30 cm^2 **b** 28 cm, 40 cm^2
     c 40 cm, 51 cm^2 **d** 30 cm, 35 cm^2
     e 32 cm, 43 cm^2 **f** 34 cm, 51 cm^2
     g cannot tell what the perimeter is; 48 cm^2
     h 34 cm, 33 cm^2

9 72 cm^2

10 48 cm

26.2 Area of a triangle

Exercise 26B

1 **a** 21 cm^2 **b** 12 cm^2 **c** 14 cm^2
     d 55 cm^2 **e** 90 cm^2 **f** 140 cm^2

2 **a** 28 cm^2 **b** 8 cm **c** 4 cm
     d 3 cm **e** 7 cm **f** 44 cm^2

3 64 cm^2

4 **a** 40 cm^2 **b** 65 m^2 **c** 80 cm^2

5 **a** 65 cm^2 **b** 50 m^2

6 For example: height 10 cm, base 10 cm; height 5 cm, base 20 cm; height 25 cm, base 4 cm; height 50 cm, base 2 cm

7 Triangle c; a and b each have an area of 15 cm^2 but c has an area of 16 cm^2

26.3 Area of a parallelogram

Exercise 26C

1 **a** 96 cm^2 **b** 70 cm^2 **c** 20 m^2
     d 125 cm^2 **e** 10 cm^2 **f** 112 m^2

2 No, it is 24 cm^2, she used the slanting side instead of the perpendicular height.

3 16 cm

4 **a** 500 cm^2 **b** 15

26.4 Area of a trapezium

Exercise 26D

1 **a** 30 cm^2 **b** 77 cm^2 **c** 24 cm^2 **d** 42 cm^2
     e 40 cm^2 **f** 6 cm **g** 3 cm

2 **a** 27.5 cm, 36.25 cm^2
     b 33.4 cm, 61.2 cm^2
     c 38.6 m, 88.2 m^2

3 Any pair of lengths that add up to 10 cm. For example: 1 cm, 9 cm; 2 cm, 8 cm; 3 cm, 7 cm; 4 cm, 6 cm; 4.5 cm, 5.5 cm

4 Shape c. Its area is 25.5 cm^2

5 Shape a. Its area is 28 cm^2

6 a

7 2 cm

8 1.4 m^2

26.5 Circumference and area of a circle

Exercise 26E

1 **a** 31.4 cm, 78.5 cm^2 **b** 18.8 cm, 28.3 cm^2
     c 9.4 cm, 7.1 cm^2 **d** 25.1 cm, 50.3 cm^2

2 **a** 25.1 cm and 50.3 cm^2
     b 15.7 cm and 19.6 cm^2
     c 28.9 cm and 66.5 cm^2
     d 14.8 cm and 17.3 cm^2

3 **a** **i** 56.5 cm   **ii** 254.5 cm^2
    b **i** 69.1 cm   **ii** 380.1 cm^2
    c **i** 40.8 cm   **ii** 132.7 cm^2
    d **i** 88.0 cm   **ii** 615.8 cm^2

4 **a** 19.1 cm   **b** 9.5 cm
    c 286.5 cm^2 (or 283.5 cm^2)

5 962.9 cm^2 (or 962.1 cm^2)

6 **a** 30.8 cm and 56.5 cm^2 **b** 17.9 cm and 19.6 cm^2

7 **a** 50.3 m^2 **b** 44.0 cm^2 **c** 28.3 cm^2

8 141.4 cm^2; A = $\pi \times 9^2 - \pi \times 6^2 = 141.4$ cm^2

9 $a^2 = \pi r^2$, so $r^2 = \frac{a^2}{\pi}$ therefore $r = \frac{a}{\sqrt{\pi}}$

10 21.5 cm^2

26.6 Surface area and volume of a cuboid

Exercise 26F

1 **a** **i** 198 cm^3 **ii** 234 cm^2
    b **i** 90 cm^3 **ii** 146 cm^2
    c **i** 1440 cm^3 **ii** 792 cm^2
    d **i** 525 cm^3 **ii** 470 cm^2

2 24 litres

3 **a** 160 cm^3 **b** 480 cm^3 **c** 150 cm^3

4 **a** **i** 64 cm^3 **ii** 96 cm^2
    b **i** 343 cm^3 **ii** 294 cm^2
    c **i** 1000 mm^3 **ii** 600 mm^2
    d **i** 125 m^3 **ii** 150 m^2
    e **i** 1728 m^3 **ii** 864 m^2

5 86

6 **a** 180 cm^3 **b** 5 cm   **c** 6 cm
    d 10 cm   **e** 81 cm^3

7 1.6 m

8 48 m^2

9 **a** 3 cm   **b** 5 m   **c** 2 mm   **d** 1.2 m

10 **a** 148 cm^3 **b** 468 cm^3

11 If this was a cube, the side length would be 5 cm, so total surface area would be 5 × 5 × 6 = 150 cm^2; no this particular cuboid is not a cube.

12 600 cm^2

26.7 Volume of a prism

Exercise 26G

1 **a** **i** 21 cm^2 **ii** 63 cm^3
    b **i** 48 cm^2 **ii** 432 cm^3
    c **i** 36 m^2 **ii** 324 m^3

2 **a** 432 m^3 **b** 225 m^3 **c** 1332 m^3

3 **a** A cross-section parallel to the side of the pool always has the same shape.
    b About $3\frac{1}{2}$ hours

4 7.65 m^3

5 **a** 21 cm^3, 210 cm^3
    b 54 cm^2, 270 cm^2

6 146 cm^3

7 78 m^3 (78.3 m^3)

8 327 litres

9 10.2 tonnes

10 672 cm^3

26.8 Volume and surface area of a cylinder

Exercise 26H

1 **a** **i** 226 cm^3 **ii** 207 cm^2
    b **i** 14.9 cm^3 **ii** 61.3 cm^2
    c **i** 346 cm^3 **ii** 275 cm^2
    d **i** 1060 cm^3 **ii** 636 cm^2

2 **a** **i** 72π cm^3 **ii** 48π cm^2
    b **i** 112π cm^3 **ii** 56π cm^2
    c **i** 180π cm^3 **ii** 60π cm^2
    d **i** 600π m^3 **ii** 120π m^2

3 665 cm^3

4 Label should be less than 10.5 cm wide so that it fits the can and does not overlap the rim and more than 23.3 cm long to allow an overlap.

5 332 litres

6 There is no right answer. Students could start with the dimensions of a real can. Often drinks cans are not exactly cylindrical. One possible answer is height of 6.6 cm and diameter of 8 cm.

7 About 127 cm

8 A diameter of 10 cm and a length of 5 cm give a volume close to 400 cm^3 (0.4 litres).

26.9 Arcs and sectors

Exercise 26I

1 **a** **i** 5.59 cm   **ii** 22.3 cm^2
    b **i** 8.29 cm   **ii** 20.7 cm^2
    c **i** 16.3 cm   **ii** 98.0 cm^2
    d **i** 15.9 cm   **ii** 55.6 cm^2

2 2π cm, 6π cm^2

3 **a** 73.8 cm
    b 20.3 cm

4 **a** 107 cm^2
    b 173 cm^2

5 43.6 cm

6 (36π − 72) cm^2

7 36.5 cm^2

8 **a** 13.9 cm
    b 7.07 cm^2

26.10 Volume of a pyramid

Exercise 26J

1 **a** 56 cm^3 **b** 168 cm^3 **c** 1040 cm^3
    d 84 cm^3 **e** 160 cm^3

2 270 cm^3

3 **a** Put the apexes of the pyramids together. The 6 square bases will then form a cube.

b If the side of the base is a then the height will be $\frac{1}{2}a$.
Total volume of the 6 pyramids is a^3.
Volume of one pyramid is
$\frac{1}{6}a^3 = \frac{1}{3} \times \frac{1}{2} \times a \times a^2 = \frac{1}{3} \times$ height × base area

4 6.9 m (height of cuboid)

5 **a** 73.3 m^3 **b** 45 m^3 **c** 3250 cm^3

6 1.5 g

7 5.95 cm

8 14.4 cm

9 260 cm^3

26.11 Volume and surface area of a cone

Exercise 26K

1 **a i** 3560 cm^3 **ii** 1430 cm^2
   b i 314 cm^3 **ii** 283 cm^2
   c i 1020 cm^3 **ii** 679 cm^2

2 24π cm^2

3 **a** 816π cm^3 **b** 720π mm^3

4 **a** 4 cm   **b** 6 cm
   c Various answers, e.g. 60° gives 2 cm, 240° gives 8 cm

5 24π cm^2

6 If radius of base is r, slant height is $2r$.
Area of curved surface = $\pi r \times 2r = 2\pi r^2$, area of base = πr^2

7 140 g

8 2.81 cm

26.12 Volume and surface of a sphere

Exercise 26L

1 **a** 36π cm^3 and 36 π cm^2
   b 288π cm^3 and 144 π cm^2
   c 1330π cm^3 and 400 π cm^2

2 65 400 cm^3, 7850 cm^2

3 **a** 1960 cm^2
   b 8180 cm^3

4 125 cm

5 6231

6 7.8 cm

7 48%

Answers to Chapter 27

27.1   Lines of symmetry

Exercise 27A

1 **a**   **b**   **c**

d   **e**   **f**

g

2 **a i** 5   **ii** 6   **iii** 8
   b 10

3 2, 1, 1, 2, 0

4

5 **a** 1   **b** 5   **c** 1   **d** 6

6

27.2   Rotational symmetry

Exercise 27B

1 **a** 4   **b** 2   **c** 2   **d** 3   **e** 6
2 **a** 4   **b** 5   **c** 6   **d** 4   **e** 6
3 **a** 2   **b** 2   **c** 2   **d** 2   **e** 2

4 **a** 6
   b 9 (the small red circle surrounded by nine 'petals') and 12 (the centre pattern)

5 For example:

27.3 Symmetry of special two-dimensional shapes

Exercise 27C

1

2 **a** kite
  b rectangle 2, square 4, equilateral triangle 3, rhombus 2
3 **a** isosceles   **b** no
4 **a** parallelogram   **b** square
5 **a** rectangle and rhombus   **b** no
6 **a** B and D   **b** AB and AD; CB and CD   **c** kite
7 **a** diameter   **b** infinite   **c** infinite
8 **a** A and C; B and D
  b AD and BC; AB and DC
  c Parallelogram
9 It will have two pairs of equal angles

27.4 Symmetry of three-dimensional shapes

Exercise 27D

1

2 **a** Diagrams to show axes going through the centres of all three pairs of opposite faces
  b 2
3 Two are similar to the one shown, dividing the end triangles in two. The other goes through the centre of each of the long edges, parallel to the end triangles.
4 **a** 3 about AB; 2 about CD
  b they are similar to CD, each passing through the centre of a rectangular face.

5 **a** 4   **b** no
6 There are four. All pass through the vertex. Two pass through opposite corners of the square. Two pass through the mid points of opposite sides of the square
7 **a** Any plane dividing each circle in half or the circular plane exactly half way up the cylinder
  b any line at right angles to the one shown, passing through the centre of the cylinder
8 **a** one   **b** infinite
9 **a** six through the centre of each hexagon; one parallel to the hexagons passing through the centre of the prism
  b 4

27.5 Symmetry in circles

Exercise 27E

1 **a–d** Student's own drawing
  e because the perpendicular bisector of any chord passes through the centre of the circle
  f Here is one method: draw two chords; construct the perpendicular bisectors; they meet at the centre
2 **a** Isosceles because OA and OB are radii
  b OA = OC; OB = OD; AB = CD so corresponding sides are equal
  c 50°
3 **a** EM = FM (given); OE = OF (radii); OM is common to both. Corresponding sides are equal
  b EMO and FMO are equal and add up to 180° (because EMF is a straight line) so they must both be 90°
  c 18°
4 **a** Angle between a radius and a tangent
  b XP = YP (tangents from a point are equal); OX = OY (radii); OP is common. So corresponding sides are equal
  c 146°
5 5 cm (use Pythagoras' theorem)
6 40°

Answers to Chapter 28

28.1 Introduction to vectors

Exercise 28A

1 a i $\begin{pmatrix} 4 \\ -1 \end{pmatrix}$ **ii** $\begin{pmatrix} 3 \\ 1 \end{pmatrix}$ **iii** $\begin{pmatrix} 2 \\ 3 \end{pmatrix}$

 iv $\begin{pmatrix} -2 \\ 4 \end{pmatrix}$ **v** $\begin{pmatrix} 2 \\ -4 \end{pmatrix}$ **vi** $\begin{pmatrix} -1 \\ 2 \end{pmatrix}$

b Both are $\begin{pmatrix} 1 \\ -2 \end{pmatrix}$. D is the midpoint of AC.

2 a

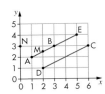

b i $\begin{pmatrix} -2 \\ 4 \end{pmatrix}$ **ii** $\begin{pmatrix} 2 \\ -4 \end{pmatrix}$ **iii** $\begin{pmatrix} 3 \\ 4 \end{pmatrix}$

 iv $\begin{pmatrix} -1 \\ 2 \end{pmatrix}$ **v** $\begin{pmatrix} -2 \\ -6 \end{pmatrix}$

3

4 The diagrams should show the following vectors:

 a $\begin{pmatrix} 3 \\ -1 \end{pmatrix}$ **b** $\begin{pmatrix} -1 \\ 4 \end{pmatrix}$ **c** $\begin{pmatrix} 1 \\ 7 \end{pmatrix}$

 d $\begin{pmatrix} 3 \\ -1 \end{pmatrix}$ **e** $\begin{pmatrix} -1 \\ -7 \end{pmatrix}$ **f** $\begin{pmatrix} 2 \\ -8 \end{pmatrix}$

5 a $\begin{pmatrix} -1 \\ 2 \end{pmatrix}$ **b** $\begin{pmatrix} 6 \\ 12 \end{pmatrix}$ **c** $\begin{pmatrix} -5 \\ -6 \end{pmatrix}$

 d $\begin{pmatrix} 5 \\ 6 \end{pmatrix}$ **e** $\begin{pmatrix} -12 \\ -8 \end{pmatrix}$ **f** $\begin{pmatrix} 1 \\ 6 \end{pmatrix}$

6 a–e

 f k is 4.

28.2 Using vectors

Exercise 28B

1 a Any three, of: $\overrightarrow{AC}$, $\overrightarrow{CF}$, $\overrightarrow{BD}$, $\overrightarrow{DG}$, $\overrightarrow{GI}$, $\overrightarrow{EH}$, $\overrightarrow{HJ}$, $\overrightarrow{JK}$
 b Any three of: $\overrightarrow{BE}$, $\overrightarrow{AD}$, $\overrightarrow{DH}$, $\overrightarrow{CG}$, $\overrightarrow{GJ}$, $\overrightarrow{FI}$, $\overrightarrow{IK}$

 c Any three of: $\overrightarrow{AO}$, $\overrightarrow{CA}$, $\overrightarrow{FC}$, $\overrightarrow{IG}$, $\overrightarrow{GD}$, $\overrightarrow{DB}$, $\overrightarrow{KJ}$, $\overrightarrow{JH}$, $\overrightarrow{HE}$
 d Any three of: $\overrightarrow{BO}$, $\overrightarrow{EB}$, $\overrightarrow{HD}$, $\overrightarrow{DA}$, $\overrightarrow{JG}$, $\overrightarrow{GC}$, $\overrightarrow{KI}$, $\overrightarrow{IF}$

2 a 2a **b** 2b **c** 3a + 2b **d** a + 2b
 e a + b **f** 2a + 2b **g** 3a + b

3 $\overrightarrow{AI}$, $\overrightarrow{BJ}$, $\overrightarrow{DK}$

4 $\overrightarrow{OF}$, $\overrightarrow{BI}$, $\overrightarrow{EK}$

5

6 a a – b **b** 3a – b **c** 2a – b **d** a – b
 e a + b **f** –a – b **g** 2a – b **h** –a – 2b
 i a + 2b **j** –a + b **k** 2a – 2b **l** a – 2b

7 a $\overrightarrow{BJ}$, $\overrightarrow{CK}$
 b $\overrightarrow{EB}$, $\overrightarrow{GO}$, $\overrightarrow{KH}$

8

9 a i 3a + 2b **ii** 3a + b
 iii 2a – b **iv** 2b – 2a
 b $\overrightarrow{DG}$ and $\overrightarrow{BC}$

10 a 2a + b **b** 2b – a **c** a + 1.5b
 d 0.5a + 0.5b **e** 1.5a + 1.5b **f** 1.5a – 0.5b

11 a i –a + b **ii** $\frac{1}{2}$(–a + b)
 iii **iv** $\frac{1}{2}$a + $\frac{1}{2}$b

 b

 c M is midpoint of parallelogram of which OA and OB are two sides.

12 a i –a + b **ii** $\frac{1}{3}$(–a + b) **iii** $\frac{2}{3}$a + $\frac{1}{3}$b
 b $\frac{3}{4}$a + $\frac{1}{4}$b

13 a i $\frac{2}{3}$b **ii** $\frac{1}{2}$a + $\frac{1}{2}$b **iii** –$\frac{2}{3}$b
 b $\frac{1}{2}$a – $\frac{1}{6}$b
 c $\overrightarrow{DE}$ = $\overrightarrow{DO}$ + $\overrightarrow{OE}$ = $\frac{3}{2}$a – $\frac{1}{2}$b
 d $\overrightarrow{DE}$ parallel to $\overrightarrow{CD}$ = (multiple of $\overrightarrow{CD}$) and D is a common point

28.3 The magnitude of a vector

Exercise 28C

1 a

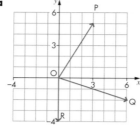

b $\sqrt{34}$; $\sqrt{40}$; 4
c $\sqrt{58}$
d $\sqrt{40}$

2 a 10 and 13   b $\begin{pmatrix} 11 \\ -4 \end{pmatrix}$ c $\sqrt{137}$

   d No. 10 + 13 does not equal $\sqrt{137}$
   e $\sqrt{401}$ f $\sqrt{401}$
   g Yes. They are vectors in opposite directions but the same length.

3 a 10, 10, 10
   b Because they are all the same distance from A. The radius is 10.

4 a $\sqrt{17}$ b $\sqrt{261}$ c 13   d 10

Answers to Chapter 29

29.1 Translations

Exercise 29A

1 a i $\begin{pmatrix} 1 \\ 3 \end{pmatrix}$ ii $\begin{pmatrix} 4 \\ 2 \end{pmatrix}$ iii $\begin{pmatrix} 2 \\ -1 \end{pmatrix}$

  b i $\begin{pmatrix} 4 \\ -2 \end{pmatrix}$ ii $\begin{pmatrix} -2 \\ 3 \end{pmatrix}$ iii $\begin{pmatrix} 3 \\ 3 \end{pmatrix}$

  c i $\begin{pmatrix} -4 \\ -2 \end{pmatrix}$ ii $\begin{pmatrix} 1 \\ -1 \end{pmatrix}$ iii $\begin{pmatrix} 0 \\ 4 \end{pmatrix}$

  d i $\begin{pmatrix} -2 \\ -7 \end{pmatrix}$ ii $\begin{pmatrix} 5 \\ 0 \end{pmatrix}$ iii $\begin{pmatrix} 1 \\ -5 \end{pmatrix}$

2

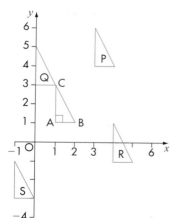

3 a $\begin{pmatrix} -3 \\ -1 \end{pmatrix}$ b $\begin{pmatrix} 4 \\ -4 \end{pmatrix}$ c $\begin{pmatrix} -5 \\ -2 \end{pmatrix}$

  d $\begin{pmatrix} 4 \\ 7 \end{pmatrix}$ e $\begin{pmatrix} -1 \\ 5 \end{pmatrix}$ f $\begin{pmatrix} 1 \\ 6 \end{pmatrix}$

  g $\begin{pmatrix} -4 \\ 4 \end{pmatrix}$ h $\begin{pmatrix} -4 \\ -7 \end{pmatrix}$

4 $\begin{pmatrix} -x \\ -y \end{pmatrix}$

5 $\begin{pmatrix} -1 \\ 4 \end{pmatrix}$

29.2 Reflections

Exercise 29B

1

2 a-e

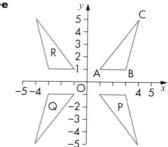

f Reflection in the y-axis

3 a–b

c *y*-value changes sign

d (*a*, −*b*)

4 a–b

c *x*-value changes sign

d (−*a*, *b*)

5 Possible answer: Take the centre square as ABCD then reflect this square each time in the line, AB, then BC, then CD and finally AD.

6 *x* = −1

29.3 Further reflections

Exercise 29C

1 Possible answer:

2

3 a–i

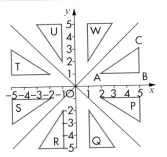

i A reflection in *y* = *x*

4

5 a–c

d Coordinates are reversed: *x* becomes *y* and *y* becomes *x*

e (*b*, *a*)

6 a–c

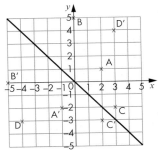

d Coordinates are reversed and the sign changes: *x* becomes −*y* and *y* becomes −*x*

e (−*b*, −*a*)

29.4 Rotations

Exercise 29D

1 a

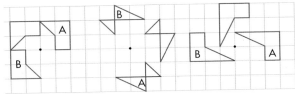

b i Rotation 90° anticlockwise

ii Rotation 180°

2

3 Possible answer: If ABCD is the centre square, rotate about A 90° anticlockwise, rotate about new B 180°, now rotate about new C 180°, and finally rotate about new D 180°.

4

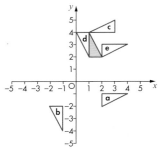

5 **a** (4,5) 180°
   b (5,5) 90° anticlockwise
   c (3,3) 180°
   b (3,5) 90° clockwise

6 **a** E   **b** H

7 **i**

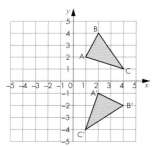

  ii A' (2, −1), B' (4, −2), C' (1, −4)
  iii Original coordinates (x, y) become (y, −x)
  iv Yes

8 **i**

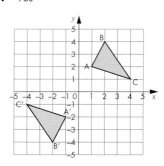

  ii A' (−1, −2), B' (−2, −4), C' (−4, −1)
  iii Original coordinates (x, y) become (−x, −y)
  iv Yes

9 Show by drawing a shape or use the fact that (a, b) becomes (a, −b) after reflection in the x-axis, and (a, −b) becomes (−a, −b) after reflection in the y-axis, which is equivalent to a single rotation of 180°.

29.5 Further rotations

Exercise 29E

1 **a-c**

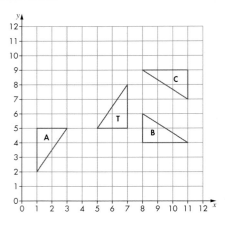

  d Rotation of 180° about (9.5, 6.5).

2 **a** (3,0)   **b** (0,0)   **c** (6,0)

3 **a** (0, −1.5) 180°   **b** (−0.5, −1.5) 90° clockwise
  c (−3,5,2.5) 90° anti clockwise
  d (0.5, 2) 180°

4 Show by drawing a shape or use the fact that (a, b) becomes (b, a) after reflection in the line y = x, and (b, a) becomes (−a, −b) after reflection in the line y = −x, which is equivalent to a single rotation of 180°.

5 **a**

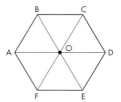

  b **i** Rotation 60° clockwise about O
     ii Rotation 120° clockwise about O
     iii Rotation 180° about O
     iv Rotation 240° clockwise about O
  c **i** Rotation 60° clockwise about O
     ii Rotation 180° about O

6 Rotation 90° anticlockwise about (3, −2).

29.6 Enlargements

Exercise 29F

1

2 a   **b**

c

3

4 a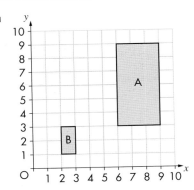

b 3 : 1   **c** 3 : 1   **d** 9 : 1

29.7 Further Enlargements

Exercise 29G

1 a   **b**

2

3 a-c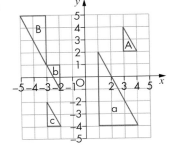

d Scale factor $-\frac{1}{2}$, centre (1, 3)

e Scale factor −2, centre (1, 3)

f Scale factor −1, centre (−2.5, −1.5)

g Scale factor −1, centre (−2.5, −1.5)

h Same centres, and the scale factors are reciprocals of each other

4 Enlargement, scale factor −2, about (1, 3)

29.8 Combined transformations

Exercise 29H

1 a (−4, −3)

2 a (−5, 2)   **b** Reflection in *y*-axis

3 A: translation $\begin{pmatrix} 1 \\ -2 \end{pmatrix}$,

B: reflection in *y*-axis,

C: rotation 90° clockwise about (0, 0),

D: reflection in *x* = 3,

E: reflection in *y* = 4,

F: enlargement by scale factor 2, centre (0, 1)

4 a T_1 to T_2: rotation 90° clockwise about (0, 0)

b T_1 to T_6: rotation 90° anticlockwise about (0, 0)

c T_2 to T_3: translation $\begin{pmatrix} 2 \\ 2 \end{pmatrix}$

d T_6 to T_2: rotation 180° about (0, 0)

e T_6 to T_5: reflection in *y*-axis

f T_5 to T_4: translation $\begin{pmatrix} 4 \\ 0 \end{pmatrix}$

5 a-c

d Reflection in the line $y = -x$

6 Reflection in x-axis, translation $\begin{pmatrix} 0 \\ -5 \end{pmatrix}$, rotation 90° clockwise clockwise about (0, 0)

7

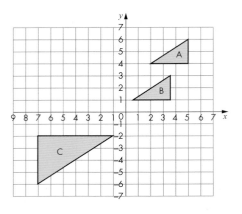

b Enlargement, scale factor $-\frac{1}{2}$, centre (1, 2)

Answers to Chapter 30

30.1 Introduction to matrices

Exercise 30A

1 $\begin{pmatrix} 7 & 11 \\ 14 & 11 \\ 17 & 12 \\ 38 & 34 \end{pmatrix}$ is one possibility. The matrix must have 4 rows and 2 columns.

2 a 2×5 **b** 4×3 **c** 1×4 **d** 3×1

3 a $\begin{pmatrix} 5 & 4 \\ -1 & 1 \end{pmatrix}$ **b** $\begin{pmatrix} 43 \\ 2 \\ 4 \end{pmatrix}$ **c** $\begin{pmatrix} 2 & 4.5 & 6.5 \\ -5 & -5.5 & -2 \end{pmatrix}$

d $\begin{pmatrix} 2 & 4.5 & 6.5 \\ 6 & -5.5 & -2 \end{pmatrix}$ **e** $(-5 \quad 4 \quad 7 \quad -1)$

4 a $\begin{pmatrix} 5 & 7 \\ 12 & 5 \\ 9 & -1 \end{pmatrix}$ **b** Not possible **c** $\begin{pmatrix} 10 & 11 & -2 \\ -1 & 3 & 2 \end{pmatrix}$

d $\begin{pmatrix} 3 & -3 \\ 8 & 1 \\ -7 & -1 \end{pmatrix}$ **e** $\begin{pmatrix} 6 & -9 & 0 \\ 7 & -1 & 2 \end{pmatrix}$ **f** $\begin{pmatrix} 3 & 15 \\ 6 & 6 \\ 24 & 0 \end{pmatrix}$

g $\begin{pmatrix} 1 & 5 & -0.5 \\ -2 & 1 & 0 \end{pmatrix}$ **h** $\begin{pmatrix} 11 & 1 \\ 28 & 7 \\ -5 & -3 \end{pmatrix}$ **i** $\begin{pmatrix} 28 & 23 & -5 \\ 1 & 7 & 6 \end{pmatrix}$

j $\begin{pmatrix} 30 & 33 & -6 \\ -3 & 9 & 6 \end{pmatrix}$

5 $u = 7, v = -4, w = -1, x = -2, y = 4$

6 a $\begin{pmatrix} 3 & 3 \\ 6 & -2 \end{pmatrix}$ **b** $\begin{pmatrix} 2 & 2.5 \\ 4.5 & 1.5 \end{pmatrix}$

30.2 Multiplying matrices

Exercise 30B

1 a $\begin{pmatrix} 9 & 12 \\ 14 & 19 \end{pmatrix}$ **b** $\begin{pmatrix} 15 \\ 27 \end{pmatrix}$ **c** $\begin{pmatrix} 5 & -4 \\ 8 & 11 \end{pmatrix}$

d $\begin{pmatrix} 5 & 8 & 10 & 5 \\ 8 & -4 & 16 & 8 \end{pmatrix}$

2 a $\begin{pmatrix} 9 \\ 5 \end{pmatrix}$ **b** $(7 \quad 14 \quad -1)$ **c** $\begin{pmatrix} 2 & 4 \\ 13 & 5 \end{pmatrix}$

d $\begin{pmatrix} 4 & 6 \\ 9 & 10 \end{pmatrix}$ **e** $\begin{pmatrix} 7 & 6 & 1 & 2 \\ 5 & 5 & 1 & 1 \end{pmatrix}$

3 a i $\begin{pmatrix} 7 & 4 & 7 \\ 2 & 4 & 6 \end{pmatrix}$ **ii** $\begin{pmatrix} 7 & 3 \\ 6 & 2 \end{pmatrix}$

iii $\begin{pmatrix} 7 & 4 & 7 \\ 10 & 5 & 9 \end{pmatrix}$ **iv** $\begin{pmatrix} 4 & 4 \\ 6 & 5 \end{pmatrix}$

b The columns of B and the rows of C are different lengths.

c No

4 a $\begin{pmatrix} 10 & 5 \\ 5 & 5 \end{pmatrix}$ **b** $\begin{pmatrix} 2 & 10 \\ 5 & 27 \end{pmatrix}$

c $\mathbf{C}^2 = \begin{pmatrix} 5 & 4 \\ 4 & 5 \end{pmatrix}$; $\mathbf{C}^3 = \begin{pmatrix} 13 & 14 \\ 14 & 13 \end{pmatrix}$

5 a 2×2 **b** 5×5
c No. You cannot multiply a row with 5 elements by a column with 2 elements.

6 $x = -2$, and $y = 2$.

7 a $\mathbf{AB} = (12)$ $\mathbf{BA} = \begin{pmatrix} 4 & 8 & 16 \\ 2 & 4 & 8 \\ 1 & 2 & 4 \end{pmatrix}$

b $\mathbf{AB} = \begin{pmatrix} 2 & 2 & 6 \\ 11 & 21 & 8 \end{pmatrix}$ $\mathbf{BA}$ is impossible

c $\mathbf{AB} = \begin{pmatrix} 1 & 0 & -3 \\ -2 & 0 & 6 \\ 3 & 0 & -9 \end{pmatrix}$ $\mathbf{BA} = \begin{pmatrix} -7 & 7 \\ 1 & -1 \end{pmatrix}$

d $\mathbf{AB}$ is impossible, $\mathbf{BA}$ is impossible

e $\mathbf{AB}$ is impossible, $\mathbf{BA} = (-1 \quad -1 \quad 7)$

f $\mathbf{AB} = \begin{pmatrix} 1 & 0 \\ 0 & 1 \end{pmatrix}$ $\mathbf{BA} = \begin{pmatrix} 1 & 0 \\ 0 & 1 \end{pmatrix}$

8 Sometimes true. **7f** is an example where it is true. **3c** is an example where it is not true.

30.3 2 × 2 matrices

1 a $\begin{pmatrix} 2 & 3 \\ 1 & 2 \end{pmatrix}$ **b** $\begin{pmatrix} 2 & 3 \\ 1 & 2 \end{pmatrix}$ **c** $\begin{pmatrix} -5 & 9 \\ 7 & -3 \end{pmatrix}$

d $\begin{pmatrix} -12 & 0 \\ 10 & 4 \end{pmatrix}$ **e** $\begin{pmatrix} 14 & 5 \\ -10 & -1 \end{pmatrix}$ **f** $\begin{pmatrix} 10 & -2 \\ -3 & 7 \end{pmatrix}$

2 a P   **b** Q   **c** Z   **d** Q   **e** Q   **f** Z

3 a Z   **b** I

4 a $\begin{pmatrix} 8 & 2 \\ 2 & 4 \end{pmatrix}$ **b** $\begin{pmatrix} 17 & 6 \\ 6 & 5 \end{pmatrix}$ **c** $\begin{pmatrix} 5 & 1 \\ 1 & 3 \end{pmatrix}$

d $\begin{pmatrix} 26 & 8 \\ 8 & 10 \end{pmatrix}$ **e** $\begin{pmatrix} 25 & 8 \\ 8 & 9 \end{pmatrix}$ **f** $\begin{pmatrix} 26 & 8 \\ 8 & 10 \end{pmatrix}$

5 a $\begin{pmatrix} 1 & 1 \\ 8 & 0 \end{pmatrix}$ **b** $\begin{pmatrix} 9 & 1 \\ 8 & 8 \end{pmatrix}$

6 Both equal $\begin{pmatrix} -9 & 16 \\ -12 & -9 \end{pmatrix}$

7 a $\begin{pmatrix} 1 & 0 \\ 0 & 1 \end{pmatrix}$ **b** $\begin{pmatrix} 1 & 0 \\ 0 & 1 \end{pmatrix}$ **c** Both equal I

8 a $\begin{pmatrix} 0 & 0 \\ 0 & 0 \end{pmatrix}$ **b** $\begin{pmatrix} 0 & 0 \\ 0 & 0 \end{pmatrix}$ **c** Both equal Z

30.4 The inverse of a 2 × 2 matrix

1 a 2, 4, 1
b $\frac{1}{2}\begin{pmatrix} 1 & -1 \\ -4 & 6 \end{pmatrix}$, $\frac{1}{4}\begin{pmatrix} 2 & -1 \\ -4 & 4 \end{pmatrix}$, $\begin{pmatrix} 1 & 1 \\ 2 & 3 \end{pmatrix}$

2 Student demonstration

3 The determinant is zero

4 a and **c** are singular; **b** and **d** are non-singular with inverses
$\frac{1}{30}\begin{pmatrix} 3 & -5 \\ 3 & 5 \end{pmatrix}$ and $\frac{-1}{10}\begin{pmatrix} -5 & -5 \\ -1 & 1 \end{pmatrix}$

5 a $M + N = \begin{pmatrix} 6 & -1 \\ 2 & 5 \end{pmatrix}$; $|M + N| = 32$
b No
c T

6 a $\begin{pmatrix} 2 & -3 \\ -5 & 8 \end{pmatrix}$ **b** $\begin{pmatrix} -5 & 2 \\ -3 & 1 \end{pmatrix}$ **c** $\frac{1}{2}\begin{pmatrix} 1 & -3 \\ -2 & 8 \end{pmatrix}$

d $\frac{1}{5}\begin{pmatrix} -5 & -10 \\ 2 & 3 \end{pmatrix}$ **e** $\frac{1}{10}\begin{pmatrix} 8 & 10 \\ 3 & 5 \end{pmatrix}$ **f** No inverse

7 $x = -5$

8 a Both equal $\begin{pmatrix} 1 & 0 \\ 0 & 1 \end{pmatrix}$

b $A^{-1} = B$ and $B^{-1} = A$

30.5 Matrices and transformations

1 a and **d**

b $\begin{pmatrix} -2 & -5 & -5 & -2 \\ 0 & 3 & 5 & 6 \end{pmatrix}$

c $\begin{pmatrix} 0 & 3 & 5 & 6 \\ 2 & 5 & 5 & 2 \end{pmatrix}$

2 a $\begin{pmatrix} 0 & 1 \\ 1 & 0 \end{pmatrix}\begin{pmatrix} 2 & 2 & 5 \\ 3 & 5 & 5 \end{pmatrix} = \begin{pmatrix} 3 & 5 & 5 \\ 2 & 2 & 5 \end{pmatrix}$

b

c Same, reflection in the line $y = x$.

3 a and **b**
$\begin{pmatrix} 0 & 1 \\ -1 & 0 \end{pmatrix}\begin{pmatrix} 1 & 1 & 3 \\ 0 & 4 & 4 \end{pmatrix} = \begin{pmatrix} 0 & 4 & 4 \\ -1 & -1 & -3 \end{pmatrix}$

c Rotation 90° clockwise about the origin

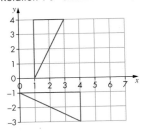

4 a Reflection in y-axis
b Rotation 180° about the origin
c Reflection in x-axis
d Enlargement, centre the origin, factor 2
e Enlargement, centre the origin, factor −2
f Rotation 90° anticlockwise about the origin
g Reflection in $y = x$
h Reflection in the line $y = -x$

30.6 Finding a matrix

1 Student's drawing.

2 a $\begin{pmatrix} 1 & 0 \\ 0 & -1 \end{pmatrix}$ **b** $\begin{pmatrix} -1 & 0 \\ 0 & 1 \end{pmatrix}$ **c** $\begin{pmatrix} 4 & 0 \\ 0 & 4 \end{pmatrix}$

d $\begin{pmatrix} 0 & 1 \\ -1 & 0 \end{pmatrix}$ **e** $\begin{pmatrix} -1 & 0 \\ 0 & -1 \end{pmatrix}$

3 a $\begin{pmatrix} 0 & 1 \\ -1 & 0 \end{pmatrix}$ **b** $\begin{pmatrix} -1 & 0 \\ 0 & 1 \end{pmatrix}$ **c** $\begin{pmatrix} 0 & 1 \\ 1 & 0 \end{pmatrix}$

d $\begin{pmatrix} 1 & 0 \\ 0 & -1 \end{pmatrix}$ **e** $\begin{pmatrix} 0 & 1 \\ 1 & 0 \end{pmatrix}$ **f** $\begin{pmatrix} 1 & 0 \\ 0 & -1 \end{pmatrix}$

4 a i $C = \begin{pmatrix} 0 & 1 \\ -1 & 0 \end{pmatrix}$ **ii** $A = \begin{pmatrix} 0 & -1 \\ 1 & 0 \end{pmatrix}$

b $C^2 = \begin{pmatrix} -1 & 0 \\ 0 & -1 \end{pmatrix}$ 180° rotation about the origin.

c Student demonstration

d $\begin{pmatrix} 1 & 0 \\ 0 & 1 \end{pmatrix}$ The identity matrix.

e **AC** = **CA** = **I** which means that the inverse of **C** is **A**.

Answers to Chapter 31

31.1 Frequency tables

Exercise 31A

1 a

Goals	0	1	2	3
Frequency	6	8	4	2

b 1 goal
c 22

2 a

Temperature (°C)	14–16	17–19	20–22	23–25	26–28
Frequency	5	10	8	5	2

b 17–19° C
c Getting warmer in the first half and then getting cooler towards the end.

3 a

Score	1	2	3	4	5	6
Frequency	5	6	6	6	3	4

b 30
c Yes, frequencies are similar.

4 a

Height (cm)	151–155	156–160	161–165	166–170
Frequency	2	5	5	7
Height (cm)	171–175	176–180	181–185	186–190
Frequency	5	4	3	1

b 166–170 cm
c Student's survey results.

5 Various answers such as 1–10, 11–20, etc. or 1–20, 21–40, 41–60

6 The ages 20 and 25 are in two different groups.

31.2 Pictograms

Exercise 31B

1

2

3 a May 10 h, Jun 11 h, Jul 12 h, Aug 11 h, Sep 10 h
b Visual impact, easy to understand.
4 a Simon   **b** $165
c Difficult to show fractions of a symbol.
5 a i 12   **ii** 6   **iii** 13
b Check students' pictograms.
c 63

31.3 Bar charts

Exercise 31C

1 a Swimming **b** 74

2 a

b $\frac{40}{100} = \frac{2}{5}$

c Easier to read the exact frequency.

3 a

b Amir got more points overall, but Hasrul was more consistent.

4 a

Time (min)	1–10	11–20	21–30	31–40	41–50	51–60
Frequency	4	7	5	5	7	2

b

c For example: Some live close to the school. Some live a good distance away and probably travel to school by bus.

5 a

b

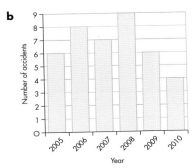

c Use the pictogram because an appropriate symbol makes more impact.

6 Yes. If you double the minimum temperature each time, it is very close to the maximum temperature.

31.4 Pie charts

Exercise 31D

1 a

b

c

2 Pie charts with following angles:
  a 36°, 90°, 126°, 81°, 27°
  b 168°, 52°, 100°, 40°
3 Pie chart with these angles: 60°, 165°, 45°, 15°, 75°
4 **a** 36
  b Pie chart with these angles: 50°, 50°, 80°, 60°, 60°, 40°, 20°
  c Student's bar chart.
  d Bar chart, because easier to make comparisons.
5 **a** Pie charts with these angles: 124°, 132°, 76°, 28°
  b Split of total data seen at a glance.
6 **a** 55°   **b** 22
7 **a** Pie charts with these angles: Strings: 36°, 118°, 126°, 72°, 8°
      Brass: 82°, 118°, 98°, 39°, 23°
  b Overall, the strings candidates did better, as a smaller proportion failed. A higher proportion of Brass candidates scored very good or excellent.
8 $\frac{1}{9}$

31.5 Scatter diagrams

Exercise 31E

1 **a** No correlation
  b Positive correlation
2 **a** No relationship between temperature and speed of cars.
  b As people get older, they have more money in the bank.
3 **a** and **b** Student's scatter diagram and line of best fit.
  c about 20 cm/s
  d about 35 cm
4 **a** Student's scatter diagram.
  b Yes, usually (good correlation).
5 **a** and **b** Student's scatter diagram and line of best fit.
  c Greta
  d about 70
  e about 72
6 **a** Student's scatter diagram.
  b no, because there is no correlation.
7 **a** and **b** Student's scatter diagram and line of best fit.
  c about 2.4 km
  d 8 minutes
8 23 kilometres/hour
9 Points showing a line of best fit sloping down from top left to bottom right.

31.6 Histograms

Exercise 31F

1

2 a

b Monday

3 a The frequencies are 6, 4, 5, 3, 2

b

4 a

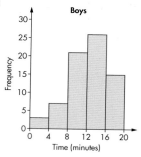

b 72 boys and 72 girls
c Nobody took longer than 20 minutes
d Student's own comments

5 a and **b**

c More girls were underweight

6 a 18
b 36
c The shortest time was at least 10 minutes. The longest time was at most 70 minutes.
d

e Yes. Student's own explanation.

31.7 Histograms wirh bars of unequal width

1 The respective frequency densities on which each histogram should be based are:

a 2.5, 6.5, 6, 2, 1, 1.5

b 4, 27, 15, 3

c 17, 18, 12, 6.67

d 0.4, 1.2, 2.8, 1

e 9, 21, 13.5, 9

2 a

3

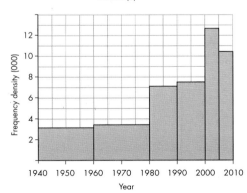

4 a 775   **b** 400

5 a

Age, y (years)	9 < y ≤ 10	10 < y ≤ 12	12 < y ≤ 14
Frequency	4	12	8
Age, y (years)	14 < y ≤ 17	17 < y ≤ 19	19 < y ≤ 20
Frequency	9	5	1

b

Temperature, t (°C)	10 < t ≤ 11	11 < t ≤ 12	12 < t ≤ 14
Frequency	15	15	50
Temperature, t (°C)	14 < t ≤ 16	16 < t ≤ 19	19 < t ≤ 21
Frequency	40	45	15

c

Mass, m (kg)	50 < m ≤ 70	70 < m ≤ 90	90 < m ≤ 100
Frequency	160	200	120
Mass, m (kg)	100 < m ≤ 120	120 < m ≤ 170	
Frequency	120	200	

6 a

b 45

7 a

b 33 plants

8 a

Speed, v (mph)	0 < v ≤ 40	40 < v ≤ 50	50 < v ≤ 60
Frequency	80	10	40
Weight, w (kg)	60 < v ≤ 70	70 < v ≤ 80	80 < v ≤ 100
Frequency	110	60	60

b 360

9 a 80

b 31.25%

10 a

b Student's own description.

Answers to Chapter 32

32.1 The mode

Exercise 32A

1 a 4   **b** 48   **c** −1
  d $\frac{1}{4}$ **e** no mode   **f** 3.21

2 a red   **b** Sun   **c** β   **d** ★

3 a 32   **b** 6   **c** no
  d no; boys generally take larger shoe sizes

4 a 5
  b no; more than half the form got a higher mark

5 The mode will be the most popular item or brand sold
  in a shop.

6 a 28
  b i brown   **ii** blue   **iii** brown
  c Both students had blue eyes.

7 a May lose count.
  b Put in a table, or arrange in order
  c 4

32.2 The median

Exercise 32B

1 a 5   **b** 33   **c** $7\frac{1}{2}$ **d** 24
  e $8\frac{1}{2}$ **f** 0   **g** 5.25

2 a $2.20   **b** $2.25   **c** median, because
  it is the central value

3 a 5
  b i 15   **ii** 215   **iii** 10   **iv** 10

4 a 13, Ella   **b** 162 cm, Pat   **c** 40 kg, Elisa
  d Ella, because she is closest to the 3 medians

5 a 12   **b** 13

6 Answers will vary

7 12, 14, 14, 16, 20, 22, 24

8 a Possible answer: 11, 15, 21, 21 (one below or equal to
  12 and three above or equal)
  b Any four numbers higher than or equal to 12, and any two
  lower or equal
  c Eight, all 4 or under

9 A median of $8 does not take into account the huge value of
  the $3000 so is in no way representative.

32.3 The mean

Exercise 32C

1 a 6   **b** 24   **c** 45   **d** 1.57   **e** 2
2 a 55.1   **b** 324.7   **c** 58.5   **d** 44.9   **e** 2.3
3 a 61   **b** 60   **c** 59   **d** Badru   **e** 2
4 42 min
5 a $200   **b** $260   **c** $278
  d Median, because the extreme value of $480 is not taken
  into account
6 a 35   **b** 36
7 a 6
  b 16; all the numbers and the mean are 10 more than those
  in part a
  c i 56   **ii** 106   **iii** 7
8 Possible answers: Speed – Kath, James, John, Joseph; Roberts
  – Frank, James, Helen, Evie. Other answers are possible.
9 36
10 24

32.4 The range

Exercise 32D

1 a 7   **b** 26   **c** 5   **d** 2.4   **e** 7
2 a 5°, 3°, 2°, 7°, 3°
  b Variable weather over England
3 a $31, $28, $33
  b $8, $14, $4
  c Not particularly consistent
4 a 82 and 83
  b 20 and 12
  c Fay, because her scores are more consistent
5 a 5 min and 4 min
  b 9 min and 13 min
  c Number 50, because times are more consistent
6 a Isaac, Oliver, Evrim, Chloe, Lilla, Badru and Isambard
  b 70 cm to 92 cm
7 a Teachers because they have a high mean and students
  could not have a range of 20.
  b Year 11 students as the mean is 15–16 and the range is 1.

32.5   Which average to use

1 a i 29   **ii** 28   **iii** 27.1
  b 14
2 a i Mode 3, median 4, mean 5
    ii 6, 7, $7\frac{1}{2}$
    iii 4, 6, 8
  b i Mean: balanced data
    ii Mode: 6 appears five times
    iii Median: 28 is an extreme value
3 a Mode 73, median 76, mean 80
  b The mean, because it is the highest average
4 a 150   **b** 20
5 a Mean   **b** Median
  c Mode   **d** Median
  e Mode   **f** Mean
6 No. Mode is 31, median is 31, and mean is 31½.
7 a Median   **b** Mode   **c** Mean
8 Tom mean, David median, Mohamed mode
9 Possible answers: **a** 1, 6, 6, 6, 6   **b** 2, 5, 5, 6, 7
10 Boss chose the mean while worker chose the mode.
11 11.6
12 52.7 kg

32.6   Using frequency tables

1 a i 7   **ii** 6   **iii** 6.4
  b i 4   **ii** 4   **iii** 3.7
  c i 8   **ii** 8.5   **iii** 8.2
  d i 0   **ii** 0   **iii** 0.3
2 a 668   **b** 1.9   **c** 0   **d** 328
3 a 2.2, 1.7, 1.3   **b** Better dental care
4 a 0   **b** 0.96
5 a 7   **b** 6.5   **c** 6.5
6 a 1   **b** 1   **c** 0.98
7 a Roger 5, Brian 4
  b Roger 3, Brian 8
  c Roger 5, Brian 4
  d Roger 5.4, Brian 4.5
  e Roger, because he has the smaller range
  f Brian, because he has the better mean
8 Possible answers: 3, 4, 15, 3 or 3, 4, 3, 15 …
9 Add up the weeks to see she travelled in 52 weeks of the year, the median is in the 26th and 27th week. Looking at the weeks in order, the 23rd entry is the end of 2 days in a week so the median must be in the 3 days in a week.

32.7   Grouped data

1 a i $30 < x \leq 40$ **ii** 29.5
  b i $0 < y \leq 100$ **ii** 158.3

c i $5 < z \leq 10$ **ii** 9.43
d i 7–9   **ii** 8.4 weeks
2 a $100 < m \leq 120$ g   **b** 10 860 g   **c** 108.6 g
3 a 207   **b** 19–22 cm   **c** 20.3 cm
4 a 160   **b** 52.6 min   **c** modal group
  d 65%
5 a $175 < h \leq 200$ **b** 31%   **c** 193.25
  d No: mode, mean and median are all less than 200 hours
6 Average price increases: Soundbuy 17.6p, Springfields 18.7p, Setco 18.2p
7 Yes average distance is 11.7 miles per day.
8 The first 5 and the 10 are the wrong way round.
9 $740
10 As we do not know what numbers are in each group, we cannot say what the median is.

32.8   Cumulative frequency diagrams

1 a Cumulative frequency 1, 4, 10, 22, 25, 28, 30
  b

  c 54 secs, 16 secs
2 a Cumulative frequency 1, 3, 5, 14, 31, 44, 47, 49, 50
  b

  c 56 secs, 17 secs
  d Pensioners, median closer to 60 secs
3 a Cumulative frequency 12, 30, 63, 113, 176, 250, 314, 349, 360
  b

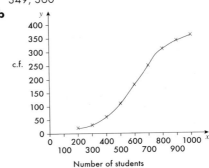

  c 605 students, 280 students

d 46–47 schools
e about 830
f about 550

4 **a** Cumulative frequency 2, 5, 10, 16, 22, 31, 39, 45, 50
b

Temperature °C

c 20.5°C, 10°C
d 10.5°C

5 **a**

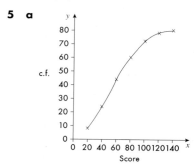

Score

b 56, 43
c about 17.5%

6 **a** Cumulative frequency 6, 16, 36, 64, 82, 93, 98, 100
b

Pocket money (c)

c 225c, 90c
d about 120 cents and about 340 cents

7 **a** Paper A 68, Paper B 57
b Paper A 28, Paper B 18
c Paper B is the harder paper, it has a lower median and a lower upper quartile.
d **i** Paper A 43, Paper B 45 **ii** Paper A 78, Paper B 67

8 **a** about 40%   **b** about 6 minutes

9 Find the top 10% on the cumulative frequency scale, read along to the graph and read down to the marks. The mark seen will be the minimum mark needed for this top grade.

Answers to Chapter 33

33.1 Probability scale

Exercise 33A

1 **a** unlikely   **b** unlikely   **c** impossible
d very likely   **e** even chance

2

Impossible   certain
e Student's own estimate

3 Student's own estimate
4 Student to provide own answers.
5 No. What happens today does not depend on what happened yesterday.

33.2 Calculating probabilities

Exercise 33B

1 **a** $\frac{1}{10}$ **b** $\frac{4}{10}$ or $\frac{2}{5}$ **c** $\frac{7}{10}$
d $\frac{1}{2}$ **e** 0

2 **a** $\frac{1}{8}$ **b** $\frac{5}{8}$ **c** $\frac{1}{2}$

3 **a** 0   **b** 1

4 **a** $\frac{1}{10}$ **b** $\frac{1}{2}$ **c** $\frac{2}{5}$ **d** $\frac{1}{5}$ **e** $\frac{2}{5}$

5 **a** $\frac{6}{11}$ **b** $\frac{5}{11}$ **c** $\frac{6}{11}$

6 **a** $\frac{1}{5}$ **b** $\frac{1}{2}$ **c** $\frac{1}{2}$ **d** $\frac{7}{10}$

7 $\frac{1}{25}$

8 **a** AB, AC, AD, AE, BC, BD, BE, CD, CE, DE
b 1   **c** $\frac{1}{10}$ **d** 6   **e** $\frac{3}{5}$ **f** $\frac{3}{10}$

9 **a** **i** $\frac{12}{25}$ **ii** $\frac{7}{25}$ **iii** $\frac{6}{25}$
b They add up to 1.
c All possible outcomes are mentioned.

10 35%

11 0.5

12 Class U

13 There might not be the same number of boys as girls in the class.

33.3 Probability that an event will not happen

Exercise 33C

1 **a** $\frac{3}{4}$ **b** 0.55   **c** 0.2

2 **a** $\frac{3}{4}$ **b** $\frac{17}{20}$ **c** $\frac{19}{20}$

3 a i $\frac{1}{4}$ **ii** $\frac{3}{4}$

b i $\frac{3}{11}$ **ii** $\frac{8}{11}$

4 Because it might be possible for the game to end in a draw.

33.4 Probability in practice

Exercise 33D

1 a 0.2, 0.08, 0.1, 0.105, 0.148, 0.163, 0.1645
   b 6   **c** 1   **d** $\frac{1}{6}$ **e** 1000

2 a 0.095, 0.135, 0.16, 0.265, 0.345
   b 40   **c** No; all numbers should be close to 40.

3 a 0.2, 0.25, 0.38, 0.42, 0.385, 0.3974
   b 8

4 a 6   **b** and **c** Student to provide own answers.

5 a Caryl, threw the greatest number of times.
   b 0.39, 0.31, 0.17, 0.14
   c Yes; all answers should be close to 0.25.

6 The missing top numbers are 4 and 5, the bottom two numbers are both likely to be close to 20.

7 Thursday

8 Although he might expect the probability to be close to $\frac{1}{2}$ giving 500 heads, the actual number of heads is unlikely to be exactly 500, but should be close to it.

33.5 Combined events

Exercise 33E

1 a 7   **b** 2 and 12
   c $\frac{1}{36}, \frac{1}{18}, \frac{1}{12}, \frac{1}{9}, \frac{5}{36}, \frac{1}{6}, \frac{5}{36}, \frac{1}{9}, \frac{1}{12}, \frac{1}{18}, \frac{1}{36}$
   d i $\frac{1}{12}$ **ii** $\frac{1}{3}$ **iii** $\frac{1}{2}$ **iv** $\frac{7}{36}$
    v $\frac{5}{12}$ **vi** $\frac{5}{18}$

2 a $\frac{1}{12}$ **b** $\frac{11}{36}$ **c** $\frac{1}{6}$ **d** $\frac{5}{9}$

3 a $\frac{1}{36}$ **b** $\frac{11}{36}$ **c** $\frac{5}{18}$

4

a $\frac{5}{18}$ **b** $\frac{1}{6}$ **c** $\frac{1}{9}$
d 0   **e** $\frac{1}{2}$

5 a $\frac{1}{4}$ **b** $\frac{1}{2}$
   c $\frac{3}{4}$ **d** $\frac{1}{4}$

6

a 6
b i $\frac{4}{25}$ **ii** $\frac{13}{25}$ **iii** $\frac{1}{5}$ **iv** $\frac{3}{5}$

7 a

b $\frac{8}{64} = \frac{1}{8}$

8 $\frac{7}{36}$: a diagram will help him to see all possible outcomes

9 a

Type of rose	Colour of rose					
	white	red	orange	yellow	pink	copper
dwarf	✗	✓	✗	✗	✓	✗
small	✗	✓	✗	✗	✓	✗
medium	✗	✓	✗	✗	✓	✗
large climbing	✗	✗	✗	✗	✗	✗
rambling	✗	✗	✗	✗	✗	✗

b i $\frac{6}{30} = \frac{1}{5}$ **ii** $\frac{4}{5}$

33.6 Tree diagrames

Exercise 33F

1 a $\frac{1}{4}$
   b $\frac{1}{2}$
   c $\frac{3}{4}$

2 a $\frac{1}{4}$
   b $\frac{1}{4}$
   c i $\frac{1}{16}$
     ii $\frac{7}{16}$

3 **a** $\frac{2}{3}$ **b** $\frac{1}{2}$

c

d **i** $\frac{1}{6}$ **ii** $\frac{1}{2}$ **iii** $\frac{5}{6}$

e 15 days

4 **a** $\frac{2}{5}$

    b **i** $\frac{4}{25}$ **ii** $\frac{12}{25}$

5 **a** $\frac{1}{8}$ **b** $\frac{3}{8}$ **c** $\frac{7}{8}$

6 **a**

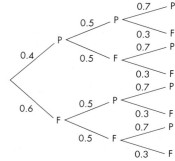

    b 0.14
    c 0.41
    d 0.09

7 **a** $\frac{3}{5}$

    b Student's explanation (there are 5 Smudge fans left and 9 girls to choose from)

    c   1st girl   2nd girl

S = Smudge
M = Mirage

    d **i** $\frac{1}{3}$ **ii** $\frac{7}{15}$ **iii** $\frac{8}{15}$

8 **a** $\frac{1}{6}$ **b** 0

    c Student's tree diagram
    d **i** $\frac{2}{3}$ **ii** $\frac{1}{3}$ **iii** 0

9 **a** **i** $\frac{3}{8}$ **ii** $\frac{5}{8}$

    b **i** $\frac{5}{12}$ **ii** $\frac{7}{12}$

10 **a** **i** $\frac{5}{13}$ **ii** $\frac{8}{13}$

    b **i** $\frac{15}{91}$ **ii** $\frac{4}{13}$

11 **a** $\frac{1}{120}$ **b** $\frac{7}{40}$ **c** $\frac{21}{40}$ **d** $\frac{7}{24}$

12 **a** $\frac{1}{9}$ **b** $\frac{2}{9}$ **c** $\frac{2}{3}$ **d** $\frac{7}{9}$

13 **a** 0.54   **b** 0.38   **c** 0.08   **d** 1

INDEX

William Collins' dream of knowledge for all began with the publication of his first book in 1819. A self-educated mill worker, he not only enriched millions of lives, but also founded a flourishing publishing house. Today, staying true to this spirit, Collins books are packed with inspiration, innovation and practical expertise. They place you at the centre of a world of possibility and give you exactly what you need to explore it.

Collins. Freedom to teach.

Published by Collins
An imprint of HarperCollins*Publishers*
The News Building
1 London Bridge Street
London
SE1 9GF

Browse the complete Collins catalogue at
www.collins.co.uk

ISBN-13 978-0-00-815037-2

Chris Pearce asserts his moral rights to be identified as the author of this work.

British Library Cataloguing in Publication Data
A Catalogue record for this publication is available from the British Library.

Commissioned by Katie Sergeant
Edited and proofread by Joan Miller
Answer check by Steven Matchett
Illustrations by Ann Paganuzzi
Cover design by Angela English
Cover concept design by Angela English
Design concept by
Jordan Publishing Design Ltd
Design and typesetting by
Jouve India Private Ltd
Production by Niccolò de Bianchi
Printed and bound by Grafica Veneta S.p.A

The syllabus content is reproduced by permission of Cambridge International Examinations.

® IGCSE is the registered trademark of Cambridge International Examinations.

Acknowledgements

The publishers wish to thank Cambridge International Examinations for permission to reproduce questions from past IGCSE ® Mathematics papers. Cambridge International Examinations bears no responsibility for the example answers to questions taken from its past paper questions. These have been written by the author.

The publishers wish to thank the following for permission to reproduce photographs. Every effort has been made to trace copyright holders and to obtain their permission for the use of copyright material. The publishers will gladly receive any information enabling them to rectify any error or omission at the first opportunity.

Cover: corbis_fancy/Fotolia, p6t Jitalia17/iStock, p6ct Rawpixel Ltd/iStock, p6cb Izabela Habur/iStock, p6b Elena Talberg/iStock, p20t weareadventurers/iStock, p20ct eugeph/iStock, p20cb digitalskillet/iStock, p20b Lisa F. Young/iStock, p48tl andy_lim/iStock, p48tr Irtati Hasan Wibisono/iStock, p48c tejerophotography/iStock, p48t bojan fatur/iStock, p48b Sean Locke/iStock, p48bl Viktor Kitaykin/iStock, p48r nullplus/iStock, p48br Joas Kotzsch/iStock, p62tl Uyen Le/iStock, p62tr Peter van Wagner/iStock, p62l Andrew Howe/iStock, p62bl craftvision/iStock, p62tc Martin McCarthy/iStock, p62br Tan Kian Khoon/iStock, p74t Vitali Khamitsevich/iStock, p74c rzdeb/iStock, p74b Billcasselman/WIKI, p80t Pal Teravagimov/Shutterstock, p80b michaeljung/Shutterstock, p81 Mikadun/Shutterstock, p82 Andy Dean Photography/Shutterstock, p83t Artush/Shutterstock, p83b Per Bengtsson/Shutterstock, p84t Darren Hendley/iStock, p84b gerenme/iStock, p94t Pete Saloutos/iStock, p94c Darren Pearson/iStock, p94br Graeme Purdy/iStock, p94bl Andrew Howe/iStock, p112tl Lya_Cattel/iStock, p115l Rido/Shutterstock, p115r Hans Kim/Shutterstock, p124 Christian Anthony/iStock, p132cl Patricia Burch/iStock, p132bl Terry Wilson/iStock, p132br Just Keep Drawing/Shutterstock, p162t Panagiotis Risvas/Dreamstime, p162c Konstantin Sutyagin/Dreamstime, p188t P J Morley/iStock, p188b Dane Wirtzfeld/iStock, p190t tab62/Shutterstock, p190c kurhan/Shutterstock, p190b S_L/Shutterstock, p192l kurhan/Shutterstock, p192c sireonio/Shutterstock, p192r Alexander Raths/Shutterstock, p195l Leah-Anne Thompson/Shutterstock, p195r michaeljung/Shutterstock, p250tl Andreas Weber/iStock, p250br Philip Beasley/iStock, p268t Rick Szczechowski/iStock, p268b Derek Dammann/iStock, p286t Manfred Konrad/iStock, p286b Rubén Hidalgo/iStock, p300t Ivan Kmit/iStock, p300ctl Scott Leigh/iStock, p300ctc zxvisual/iStock, p300ctr Dan Barnes/iStock, p300cbl SilentWolf/iStock, p300cbr Soundsnaps/iStock, p300bl Gary Martin/iStock, p300br Ivan Stevanovic/iStock, p300bc porcorex/iStock, p310t narvikk/iStock, p310b Goktug Gurellier/iStock, p320tr Tyler Olson/iStock, p320cl Eric Hood/iStock, p320cr Eric Hood/iStock, p320b Ugurhan Betin/iStock, p338bl ranplett/iStock, p338br LilKar/Shutterstock, p366t Karim Hesham/iStock, p366c Luis Carlos Torres/iStock, p366b magaliB/iStock, p388t Alfonso de Tomas/Shutterstock, p388cl Branislav/iStock, p388cr BranislaviStock, p388b Gene Chutka/iStock, p404t fotoVoyager/iStock, p404c Matthias Weinrich/iStock, p436t Robert Churchill/iStock, p436b Wendell Franks/iStock, p466tl Ludmila Yilmaz/iStock, p466tc Carmen Martínez Banús/iStock, p466tr agoxa/iStock, p466cl Jill Britton/iStock, p466cr Dirk Freder/iStock, p466b Giorgio Fochesato/iStock, p476tr Tungphoto/Shutterstock, p476cl Mark Evans/iStock, p476cr David Joyner/iStock, p476bl Stefan Weichelt/iStock, p488l Roberto Gennaro/iStock, p488r Andrey Prokhorov/iStock, p510t bubaone/iStock, p510bl Robert Churchill/iStock, p510br Kenishirotie/iStock, p534tr Kevin Russ/iStock, p562t JenDen2005/iStock, p562b Domenico Pellegriti/iStock, p590l George Clerk/iStock, p590r Owen Price/iStock.